INTERNATIONAL

REVIEW OF CYTOLOGY

SUPPLEMENT 14

Intracellular Symbiosis

INTERNATIONAL

Review of Cytology

EDITED BY

G. H. BOURNE
St. George's University School of Medicine
St. George's, Grenada
West Indies

J. F. DANIELLI
Worcester Polytechnic Institute
Worcester, Massachusetts

ASSISTANT EDITOR

K. W. JEON
Department of Zoology
University of Tennessee
Knoxville, Tennessee

SUPPLEMENT 14

Intracellular Symbiosis

EDITED BY

KWANG W. JEON
Department of Zoology
University of Tennessee
Knoxville, Tennessee

ACADEMIC PRESS 1983
A Subsidiary of Harcourt Brace Jovanovich, Publishers
New York London
Paris San Diego San Francisco São Paulo Sydney Tokyo Toronto

ACADEMIC PRESS, INC.
111 Fifth Avenue, New York, New York 10003

United Kingdom Edition published by
ACADEMIC PRESS, INC. (LONDON) LTD.
24/28 Oval Road, London NW1 7DX

LIBRARY OF CONGRESS CATALOG CARD NUMBER: 74-17773

ISBN 0-12-364375-9

PRINTED IN THE UNITED STATES OF AMERICA

83 84 85 86 9 8 7 6 5 4 3 2 1

Contents

Some Eco-evolutionary Aspects of Intracellular Symbiosis

F. J. R. TAYLOR

Integration of Bacterial Endosymbionts in Amoebae

KWANG W. JEON

Perspective on Algal Endosymbionts in Larger *Foraminifera*

JOHN J. LEE

The Biology of the Xenosome, an Intracellular Symbiont

A. T. SOLDO

Endosymbionts of *Euplotes*

KLAUS HECKMANN

Endonuclear Symbionts in Ciliates

HANS-DIETER GÖRTZ

Metabolic Interchange in Algae–Invertebrate Symbiosis

CLAYTON B. COOK

The Molecular Biology of *Rhizobium*–Legume Symbiosis

DESH PAL S. VERMA AND SHARON LONG

Analysis of Possible Gene Transfer between an Insect Host and Its Bacteria-like Endocytobionts

W. SCHWEMMLER

Cellular and Molecular Mechanisms of Intracellular Symbiosis in Leishmaniasis

K.-P. Chang

Symbiotic Interaction between *Legionella pneumophila* and Human Leukocytes

Marcus A. Horwitz

Plastids—Past, Present, and Future

Jean M. Whatley

Contributors

Numbers in parentheses indicate the pages on which the authors' contributions begin.

K.-P. Chang (267), *The Rockefeller University, New York, New York 10021*

Clayton B. Cook (177), *Marine Science Institute, Northwestern University, Nahant, Massachusetts 01905*

Hans-Dieter Görtz (145), *Zoological Institute, University of Münster, D-4400 Münster, Federal Republic of Germany*

Klaus Heckmann (111), *Zoological Institute, University of Münster, D-4400 Münster, Federal Republic of Germany*

Marcus A. Horwitz (307), *Laboratory of Cellular Physiology and Immunology, The Rockefeller University, New York, New York 10021*

Kwang W. Jeon (29), *Department of Zoology, University of Tennessee, Knoxville, Tennessee 37996*

John J. Lee (49), *Department of Biology, City College, City University of New York, New York, New York 10031, and Department of Invertebrates, American Museum of Natural History, New York, New York 10024*

Sharon Long (211), *Department of Biological Sciences, Stanford University, Stanford, California 94305*

W. Schwemmler (247), *Institut für Pflanzenphysiologie und Zellbiologie, Freie Universität Berlin, Berlin, Federal Republic of Germany*

A. T. Soldo (79), *Research Laboratories of the Veterans Administration Medical Center and the Department of Biochemistry, University of Miami School of Medicine, Miami, Florida 33125*

F. J. R. Taylor (1), *Department of Botany and Oceanography, University of British Columbia, Vancouver, British Columbia V6T 1W5, Canada*

Desh Pal S. Verma (211), *Department of Biology, McGill University, Montreal, Quebec H3A 1B1, Canada*

Jean M. Whatley (329), *Botany School, University of Oxford, Oxford OX1 3RA, England*

Preface

Recently there has been a heightened interest on studies of cellular symbiosis, as exemplified by the four international conferences held between 1978 and 1980, the proceedings of which were published later. Since then, more information has been gathered on the physiological and biochemical interactions between intracellular symbiotic partners, as a result of applying new techniques in cell and molecular biology. The purpose of the present volume is to bring together new results on intracellular symbiosis and to review symbiotic systems not well covered in recent publications.

In the first article, Taylor presents a general overview of the significance of intracellular symbiosis from the point of view of evolution and ecology. In the next group of articles, Jeon, Lee, Soldo, Heckmann, and Görtz present detailed descriptions of known examples of host–symbiont interactions in free-living cells such as amoebae and ciliates. The topics covered include newly discovered instances of symbiosis, mechanisms for establishment of symbiosis, specificity, and mutual influence between the host and symbionts. In the article by Cook, the flow of metabolites between hosts and symbionts is described, as observed in the algae–invertebrate symbiosis.

Recent findings on *Rhizobium*–legume symbiosis are presented by Verma and Long, and a possible transfer of genes between bacterial endosymbionts and cicada cells is discussed by Schwemmler. Chang and Horwitz cover pathogenic protozoan and bacterial symbionts infecting mammalian cells. One remarkable aspect of these symbioses is that the symbionts live inside macrophages and leukocytes that are specialized to phagocytose and destroy invading foreign organisms. In the last article, the current views on chloroplast evolution from symbiotic algae are critically examined by Whatley, together with a discussion on possible further changes of plastids.

The endosymbionts described range from bacteria (all Gram-negative) to eukaryotic algae and protozoa, while their hosts are as diverse as ciliate nuclei, plant cells, or mammalian cells. In the past, cellular responses to intruding foreign genes, whether they be in the form of naked DNAs, viruses, or cells, were usually studied as pathological phenomena. Perhaps, this has been because most of the microbes associated with cells of higher organisms are pathogenic. Recently, however, it has been recognized that the more significant aspect of cellular symbiosis is in its contribution to cellular divergence and variation. For example, genetic changes arising from intracellular symbiosis are far greater in magnitude than those arising from other intrinsic causes such as mutation or hybridization. Thus, it is inevitable for any discussion of intracellular symbiosis to involve the origin of new cell components as is done in many of the articles in this volume. At the same time, however, cellular symbiotic systems are good models to study other aspects of cell biology, including physiological and biochemical interactions between organelles, cellular adaptation, cellular defense mechanisms, and membrane differentiation. It is hoped that this Supplement will serve to stimulate studies on this important phenomenon in cell biology further and to enhance our understanding of various unknown aspects of intracellular symbiosis.

I thank Professor J. F. Danielli for his advice during the initial planning of this volume and all the authors for their patience and kind cooperation. I am also grateful to the staff at Academic Press for their expert and willing assistance which made my job easy and pleasant.

KWANG W. JEON

Foreword

Over the last fifty years, it has slowly been recognized that genes move between species by the actions of gene vectors, and that symbionts may coevolve with host cells until the symbiont is better described as an organelle of the host cell. These two mechanisms are of profound importance in evolution, and must be a major concern of all cell biologists who wish to understand the cells as a unit of life and evolution. This volume will be of interest to all working on cell biology, on evolution, on pathogenesis, and on the cellular aspects of synthetic biology and biotechnology.

JAMES F. DANIELLI

INTERNATIONAL REVIEW OF CYTOLOGY

SUPPLEMENT 1

Z. Hruban and M. Recheigl, Jr., Microbodies and Related Particles: Morphology, Biochemistry, and Physiology, 1969

SUPPLEMENT 2

Peter Luykx, Cellular Mechanisms of Chromosome Distribution, 1970

SUPPLEMENT 3

Andrew S. Bajer and J. Molè-Bajer, Spindle Dynamics and Chromosome Movements, 1972

SUPPLEMENT 4

G. H. Bourne, J. F. Danielli, and K. W. Jeon, eds., Aspects of Nuclear Structure and Function, 1974

SUPPLEMENT 5

G. H. Bourne, J. F. Danielli, and K. W. Jeon, eds., Aspects of Cell Control Mechanisms, 1977

SUPPLEMENT 6

G. H. Bourne, J. F. Danielli, and K. W. Jeon, eds., Studies in Ultrastructure, 1977

SUPPLEMENT 7

G. H. Bourne, J. F. Danielli, and K. W. Jeon, eds., Neuronal Cells and Hormones, 1978

SUPPLEMENT 8

G. H. Bourne, J. F. Danielli, and K. W. Jeon, eds., Aspects of Genetic Action and Evolution, 1978

SUPPLEMENT 9

J. F. Danielli and M. A. DiBerardino, eds., Nuclear Transplantation, 1979

SUPPLEMENT 10

Warren W. Nichols and Donald G. Murphy, eds., Differentiated Cells in Aging Research, 1979

SUPPLEMENT 11A

Indra K. Vasil, ed., Perspectives in Plant Cell and Tissue Culture, 1980

SUPPLEMENT 11B

Indra K. Vasil, ed., Perspectives in Plant Cell and Tissue Culture, 1980

SUPPLEMENT 12

A. L. Muggleton-Harris, ed., Membrane Research: Classic Origins and Current Concepts, 1981

SUPPLEMENT 13

Kenneth L. Giles and Alan G. Atherly, eds., Biology of the Rhizobiaceae, 1981

INTERNATIONAL REVIEW OF CYTOLOGY, SUPPLEMENT 14

Some Eco-evolutionary Aspects of Intracellular Symbiosis

F. J. R. TAYLOR

Departments of Botany and Oceanography, University of British Columbia, Vancouver, British Columbia, Canada

We may soon be confronted by the information that our cells are little communities, ecosystems, made up of all kinds of little symbiotic animals at work together, some of them commuters (Lewis Thomas, 1970).

I. Introduction

Most papers on intracellular symbiosis deal, appropriately enough, with the physiological interactions of the partners. Evolutionary and ecological aspects have been less frequently considered, or only in passing, although they can provide a useful framework through which to view the phenomenon. Recently evolutionary topics have been included in several symposia on symbiosis (Taylor, 1979, 1980; D. L. Taylor, 1981). The possible or probable role of symbiosis in the evolution of the eukaryotic condition was the subject of an entire symposium (Fredrick, 1981) and books by Li (1979, in Chinese) and Margulis (1981). A very thorough review of the biochemical evidence relating to the origin of the eukaryotes has been provided recently by Gray and Doolittle (1982).

Although this volume is directed principally at a cell biology readership, it is hoped that this article may encourage evolutionists and ecologists to apply some basic principles of their fields to this sharply defined microenvironment.

ISBN 0-12-364375-9

II. The Partners

Cytobionts (intracellular symbionts: Taylor, 1979; endocytobionts, Schwemmler, 1979, 1980) and their hosts can be categorized according to their basic organization, i.e., prokaryotic or eukaryotic, or according to function, although the latter may not be readily determined and can vary with time or circumstances.

Stanier (1970) and others have noted that one of the characteristics of prokaryotes is their inability to harbour cellular cytobionts. This has been attributed to several factors. The obligate presence of a wall and absence of cytosis (e.g., phagocytosis, pinocytosis) act as a barrier, with no host participation in the entry of foreign particulates into the cell as there is in eukaryotes. Not that the prokaryotic cell is inviolate: bdellovibrios penetrate the wall, but they are parasites, destroying the cells they have entered (Stolp, 1979) and, of course, viruses of many types enter prokaryotic cells.

All contemporary examples of cytobiosis involve either prokaryotic or eukaryotic cytobionts living within eukaryotic hosts.

Although a fairly wide spectrum of cytobionts have been identified in various hosts there are a few species which seem to be particularly successful in inhabiting the cells of many species of animals, both unicellular and multicellular and are of extremely wide, circumglobal distribution. A few photosynthetic genera, the chlorococcalean green alga *Chlorella* and the dinoflagellates *Symbiodinium* and *Zooxanthella,* have formed stable symbioses with an extraordinary variety of hosts, particularly among protists and coelenterates. Why have these been so successful? In what ways are they "preadapted" to this environment?

They are photosynthetic, thus offering the optimum reciprocal recycling of photosynthate and animal wastes, but there are a multitude of other photosynthe-

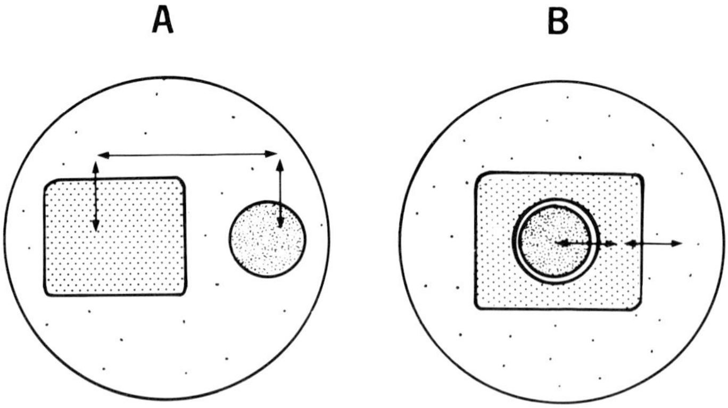

FIG. 1. Diagrammatic representation of the contrasting interactional relationships between two organisms and the environment in a conventional ecosystem (A) and when one is cytobiotic within the other (B).

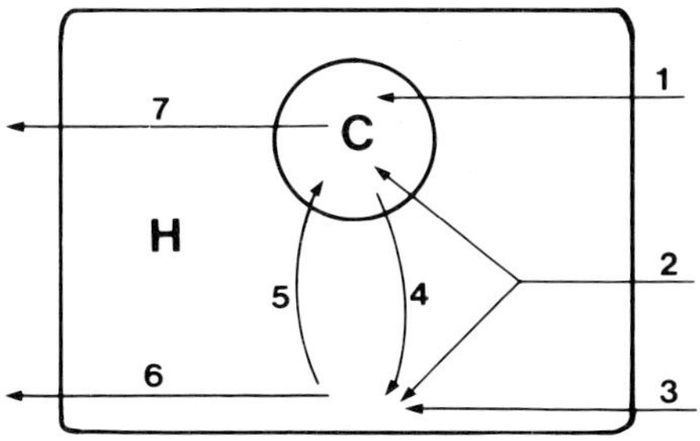

FIG. 2. Pathways of materials (and energy in some cases) between the external environment, the host cell and the cytobiont. (1) External materials and/or energy for the cytobiont only. (2) Materials competed for by the host and cytobiont. (3) Materials and/or energy for the host only. (4) Cytobiont-derived nutrients for the host. (5) Host-derived nutrients for the cytobiont. (6) Host wastes not cycled to the cytobiont. (7) Cytobiont wastes not of use to the host.

tic protists which could play this role. Many of the latter require one or more of the vitamins thiamin, biotin, or B_{12} (Droop, 1963), which would be a nutrient requirement competitive with their hosts (pathway 2 of Fig. 2). Perhaps significantly, neither *Chlorella* nor *Symbiodinium* has been shown to have absolute vitamin requirements. This is particularly striking since most other marine phytosynthetic dinoflagellates which have been tested require vitamin B_{12}. McLaughlin and Zahl (1966) did find a reduction in the proportion of small cells in *S. microadriaticum* with B_{12} depletion and one strain of a dinoflagellate cytobiotic with the foraminiferan *Sorites* also required B_{12} (review by Lee, 1980).

These cytobionts also have another unusual feature which may help them avoid digestion in the initial infection stage. Although *Symbiodinium* and *Zooxanthella* have a naked, flagellated phase they also have a cyst-like phase in which the flagella are lost and the cell is surrounded by a continuous wall. Many other coastal dinoflagellates are capable of forming cysts but usually as a dormant, resistant phase. In these two cytobionts the cells remain photosynthetically active and, instead of dividing in the motile phase like most dinoflagellates, they divide in the coccoid cyst state (a comparable but nonsymbiotic state is found in the tropical genus *Pyrocystis*). *Chlorella* spends its entire life cycle as coccoid cells. The walls of both the zooxanthellae and the zoochlorellae contain cellulose and it is thought that this might be protective in carnivorous hosts lacking cellulase (Droop, 1963). However, perhaps significantly, in both the walls also have sporopollenin, an extremely resistant material also found in pollen grains. Not all *Chlorella* or other green algae have this layer (Pickett-Heaps, 1975), but all zoochlorellae do.

In some zooxanthellae the wall becomes reduced, but this can be viewed as secondary once the association is established, probably as an integrative development that facilitates exchange. Other factors are also clearly involved (see next section).

There is little or no probability of a nonphotosynthetic eukaryotic cytobiont existing in nonparasitic cytobiosis with a nonphotosynthetic host because of too great a similarity, and therefore competition, in nutritional requirements. Virtually all nonphotosynthetic cytobionts are therefore prokaryotic.

In order for stable cytobiosis to exist the relationship between the partners must either become selectively neutral or favorable to one or both, although the initial stages may be hazardous to one or both. The relationship is cemented by progressive interdependence, the partners becoming more and more integrated and dependent with time. This reaches an irreversible state, with a loss of genetic information by one of the partners. Typically it is the cytobiont genome which becomes most reduced with time, but in the early stages this may not be the case. For example, in the *x*D bacterium/*Amoeba proteus* association it is the host which becomes dependent first (Jeon, 1980).

The loss of information from the cytobiont and the gain by the host is in accordance with the "Mark Principle" (New Testament, Mark *iv,* 25) which predicts that, in a system where there is interaction across a boundary (membranes in this case), the side with the higher variety will dominate and grow at the expense of the other (Canny, 1981). Interestingly, this applies also to protoplast fusion experiments. Ohiwa (1981) found that he could not get a stable 1:1 relationship between the organelles (nuclei and chloroplasts) of artificially fused cells of *Zygnema, Spirogyra,* or *Mougeotia* in pairwise combinations. Only when there are more of one type of parent organelle than its counterpart from the other do stable fusion products survive.

The vast majority of host eukaryotes are phagocytic, which is presumably a function of their exposure to potential cytobionts: their feeding mechanism continually exposes them to the risk of infection (see below). Furthermore, there appear to be many more examples of cytobiosis in unicellular protists than in multicellular hosts, which is probably a reflection of the greater likelihood of transmission of the cytobionts to the offspring in the former (Taylor, 1979). Transfer through the gametes is very rare in multicellular organisms and maternal, if present (Smith, 1980).

In a sense, gamete production in a multicellular organism is a "filter" which protects the next generation from infections acquired by the parents but, in symbioses in which the host benefits, it is the host which has evolved a mechanism for getting around this normally useful bottleneck. The commonest means is by concentrating symbionts on or under the protective covering of the egg, so that the newly hatched larva will be infected almost immediately by the most appropriate strain of symbiont. This can be seen in both plant–animal symbioses and those in which both partners are nonphotosynthetic (e.g., the bacterial cytobionts of the leafhopper *Euscelis incisus:* Schwemmler, 1981). In an interesting variation on this, only the swimming ephyra larva and adult medusa of *Cassiopaea* contain zooxanthellae. The gametes, and the planula and benthic scyphistoma stage, do not contain *Symbiodinium* (the dinoflagellate zooxanthella). Strobilation occurs only when the scyphistoma becomes infected during feeding (Trench *et al.*, 1981).

Even within unicellular hosts the gametes may lack the cytobionts. In radiolaria and foraminifera, for example, each new sexual generation must be reinfected. However, in those protists in which the sexual cells differ little from normal individuals, e.g., ciliates or dinoflagellates, the cytobionts can mingle during cytoplasmic fusion.

III. Initial Infection Phase

All intracellular symbioses essentially can be thought of as long-term, nonpathogenic infections. It is tempting to assume that it is benefit to the host which leads to retention of the cytobiont, the resulting consortium being more successful than the partners if they had remained alone. Furthermore, there is an almost traditional view that mutualists are more sophisticated in their relationships with the host than parasites because they do not destroy their hosts, i.e., the mutualist is a "perfect parasite" (Caullery, 1952). A close examination of modern cytobioses suggests quite a different picture.

The great majority of hosts of nonparasitic symbionts are phagotrophic feeders, whether they are multicellular or unicellular. In the former it is the digestive tissues which are first infected, with varying spread to other tissues. The cytobionts are acquired from the external environment during normal feeding activities.

Unlike most parasites (see Bannister, 1979, for a review of intracellular parasite adaptions), nonparasitic cytobionts do not seem to have evolved any special mode of entry into their hosts (Smith, 1980) although the cyanobacterium *Richelia,* found in some diatoms, may be an exception (Taylor, 1982). The usual fate of ingested material is to be digested, following fusion of a lysosome or similar digestive enzyme container with the phagocytic vacuole. It is reasonable to assume that natural selection has acted to make this as efficient as possible. Not only does this provide nutrition, but it also constitutes the principal defensive mechanism against accidentally ingested foreign organisms. The undigested residue is egested.

By one means or another parasitic and nonparasitic cytobionts avoid both these fates. It is tempting to suggest various clever adaptions by mutualistic cytobionts to overcome these processes. In some hosts, such as the ciliate *Paramecium bursaria,* it has been shown that, although some *Chlorella* may be digested, there is a disruption in the introduction of digestive enzymes into the perialgal vacuoles (Karakashian and Karakashian, 1973) and the vacuole membranes are altered (Meier *et al.*, 1980). This is also the case for the parasite *Toxoplasma gondii* (Jones and Hirsch, 1972). There are indications that surface proteins (antigens) may be involved in the recognition of strains which have previously inhabited a host, as well as altered physiological features, for example in the comparison of uptake and retention of strains of *Chlorella* by green hydra (Pool, 1979). Reisser (1981) and his co-workers have treated chlorellae, isolated (''ex-symbiotic'') from *P. bursaria,* with a cellulase–pectinase mixture, or have coated their surfaces with antibodies. In both cases, most of the treated cells were digested when offered to aposymbiotic *P. bursaria* (hosts that have been rendered symbionts free), thus suffering the same fate as wild, nonsymbiotic chlorellae. Weis (1981) has shown that algal maltose release is also involved in the process.

Protozoans and invertebrates lack an immune system of the vertebrate type (Bang, 1973) but this does not mean that they are incapable of distinguishing like from non-like. They may simply lack the ability to do much about it.

Many of the incorporated eukaryotic cytobionts are, or were, flagellates, but switch to a coccoid phase, if they are not already in it, on ingestion. A few exceptions are found in the zooxanthellae of some foraminiferans and some prasinomonad flagellates in microplankton (examples in Taylor, 1982). These appear to be only slightly integrated with their hosts and some, at least, divide in the motile state in culture whereas the others divide in the coccoid stage. Although the latter is clearly advantageous in a cytobiotic situation, it also occurs in some free-living dinoflagellates and many coccoid green algae. Thus, these cannot readily be termed adaptions to an intracellular existence although some may see them as ''preadapted'' to this life style.

Organisms which play complimentary roles in the cycling of materials in ecosystems are also more likely to be involved in a nonparasitic relationship, with a cycling of materials between the two. This is most obvious in plant–animal symbioses (Buchner, 1965) in which nitrogenous wastes and CO_2 are used by the plant and the animal obtains ''leaked'' photosynthate in return. Other benefits may accrue, such as enhanced calcification by the host resulting from a reduction in phosphorous levels in the host by the photosynthetic cytobiont, or reduced predation pressure, such as that in *Paramecium* (Berger, 1980) or green hydra (Stiven, 1965).

Two autotrophic partners (a rare combination) may not be competitors for all nutrients. The filamentous cyanobacterium (blue-green alga) *Richelia* which is cytobiotic in planktonic marine diatoms, can fix nitrogen although its host cannot (Mague *et al.*, 1974). Are they competitors in all other respects? The infected diatoms may not be very active photosynthetically (Weare *et al.*, 1974). The nutritional interactions of many bacterial cytobionts with their hosts are poorly understood although in a few cases a nutritional benefit to the host has been demonstrated, the classic example being the satisfaction of the requirement for hemin, lysine, and folic acid by bacteria in the trypanosome *Crithidia oncopelti* (see Chang, this volume).

Essentially, then, these cases of beneficial cytobiosis can be regarded as fortuitous indigestion, the cytobionts being imprisoned within the cell, some to experience periodic freedom when their host liberates them or dies (multicellular hosts particularly), others to become unfit for the outside world.

I prefer this mechanistic view of the events leading to cytobiosis, rather than one which sees them as steps to a "fitter" host:cytobiont consortium. Once established, however, selection no doubt favors retention and integration. It is the parasites which show obvious, strong adaptions in nearly all respects. The casual photosynthetic mutualists, which form the most widely encountered consortia, show little by comparison with their free-living relatives, not having arrived at that state via the amelioration of parasitism. Where both partners are heterotrophic or autotrophic we may observe something closer to the classic progression. The bacterial cytobiosis of the *x*D strain of *Amoeba proteus* studied by Jeon and his colleagues (Jeon, 1980, and this volume) began as a severely pathogenic bacterial infection which killed most of the strains it entered. However, in those cells that survived, the association came to be tolerated, and then required by the host, all in an astonishing short period of time (as little as 200 amoeba generations), a transition which has been repeated with other noninfected strains.

This very high rate of change in organisms which are usually much more constant fits the "punctuated equilibrium" model of evolution (Eldredge and Gould, 1972) very nicely. It should be noted, however, that rapid dependence does not always follow in cytobiotic infections. Gigantism in benthic foraminifera, and skeletal banding in corals have been used as rather weak arguments for the presence of symbionts at least as far back as the Triassic (Lee *et al.*, 1979; D. L. Taylor, 1981) and yet in neither of these associations is there a continuous obligate symbiotic requirement. There is evidently a selective advantage to these highly successful organisms in maintaining supplementary external feeding.

IV. Maintenance Phase

Once they have been "accepted" by the host cell the cytobionts may occupy several different locations within it. The least integrated are those occurring in groups in vacuoles scattered throughout the cytoplasm (e.g., *Chlorella* in *Paramecium*). Many bacteria and some eukaryotic cytobionts such as those occupying the dinoflagellates *Peridinium balticum* or *Kryptoperidinium* (*Glenodinium*) *foliaceum* seem to become embedded directly in the cytoplasm, not in obvious vacuoles. In the latter cases there may only be a single membrane between the host and cytobiont cytoplasms.

In some the location within the host is very specific. Photosynthetic cytobionts are usually found near the periphery of the host cells. The remarkably precise alignment of the ectosymbiotic bacteria on the flagellate *Myxotricha paradoxa* and its relatives is well known (see Margulis, 1981), but intracellular bacteria may also have precise cortical arrangements. Those in *Trichonympha turkestanica,* also a termite-gut inhabitant, lie in a position very similar to the cortical mitochondria of ciliates, and seem to have contact with the flagellar basal bodies. Gromov *et al.* (1977) have suggested that they may play a role in flagellar movement. Bacteria may also occur within cell organelles, having been seen within the "closed" nuclei of euglenoids, dinoflagellates, and ciliates (Preer and Preer, 1977; Görtz, 1980, and this volume). Although a little is known of the mechanism by which species of the bacterium *Holospora* enter ciliate macro- and micronuclei (Ossipov and Podlipaer, 1977), the nature of their interaction(s) with their hosts is unknown. The premitotic, perinuclear association of bacteria in the giant amoeba *Pelomyxa* is an excellent way to ensure propagation during cell division (Chapman-Andresen, 1971; Whatley, 1976).

The means by which cytobiont numbers are regulated by the host have been of considerable interest to those studying intracellular symbioses ("endocytobiologists"?). Cytobiont numbers may be regulated by mitotic synchronization, expulsion, digestion, or growth inhibition (Muscatine and Pool, 1979), but the detailed mechanisms are little understood. They were the chief topic of the recent meetings at Göttingen (Weissner and Lorenzen, 1981) and are also detailed elsewhere in this volume. Control of a key cytobiont nutrient seems to be the mechanism in at least some cases, e.g., in green *Hydra* (Muscatine and Neckelmann, 1981).

V. The Ecology of the "Cytocosm"

A host cell constitutes the ultimate microhabitat, the "cytocosm" (my epithet), a minute subset of larger, more conventionally recognized ecosystems. Although its limits are more readily defined than those of many ecological boundaries it is more difficult to deal with in a conventional ecological way for several reasons and here I will only try to identify a few primary features of this unique habitat.

Homeostasis is a well-known feature that distinguishes living organisms from nonliving systems, this being accomplished through the expenditure of energy and elaborate, interconnected feedback mechanisms. "Stability" is a rather difficult-to-define property of conventional systems which refers to their resistance to perturbation and is thought to be a function of the complexity of the trophic links in the system (May, 1975; McNaughton, 1977). In this sense a cytobiont comes to live within a system in which stability is a major factor, the system attempting to compensate for perturbations caused by the activities of the cytobiont. Succession, except on an extremely slow time scale, will destroy the system. Thus the best possibility for acceptance of the newcomer will be if its activities either resemble those of a part of the internal system already present (such as an organelle) or if it provides an internal material relationship similar to one formerly existing only externally (cycling).

When the cytobiont is located within a vacuole the host has the opportunity to control the exchange of materials between its cytoplasm and the vacuole by selective permeability, the use of pumps, etc. However, once the cytobiont becomes embedded in the cytoplasm this possibility is lost.

Partly because an ecological approach to this internal habitat is not usually taken, and partly because of the difficulties of measurement, there is very little information which can be obtained about the details of the environment in which a formerly free-living organism may find itself.

Perhaps the cytobiont-to-be is a phytoplankter, adapted to the uptake of inorganic nutrient ions in low concentrations from the surrounding water, keeping its osmotic balance, while trying to remain in the sunlit euphotic zone. Abruptly it becomes imprisoned in an animal host. Its movements, if any, are now involuntary. The environment is close to isosmotic. The pH is radically different, although now similar to its own internal environment. Organic ions are now the rule. How do its pumps now operate? How quickly can it adjust to the new nutrient levels and ionic forms? There are no satisfactory answers to any of these questions at present. For years most ecological physiology has focused on the interactions between whole organisms or cells and the environment and only recently has more attention begun to be paid to the intracellular location of nutrient pools.

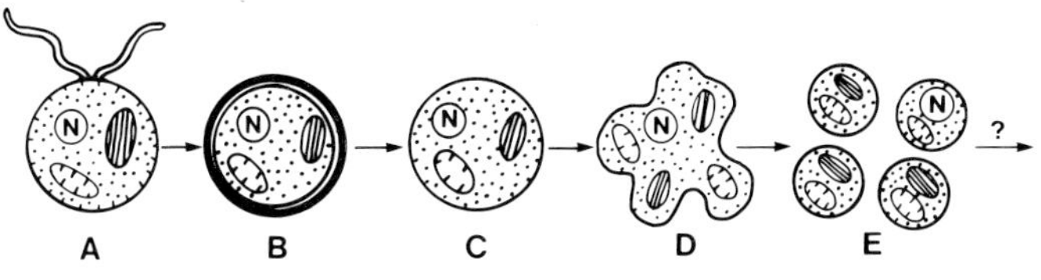

FIG. 3. Degrees of integrative modifications observed in photosynthetic, eukaryotic cytobionts (the arrows indicating the progression inferred from examples of each state). (A) An unmodified flagellate, possibly in some foraminiferan hosts (Taylor, 1982). (B) Coccoid/encysted stage with continuous wall, reduction in other peripheral structures, and division in this state, e.g., *Symbiodinium, Chlorella*. (C) Reduction of the wall, e.g., some *Symbiodinium*. (D) Ramification and organellar amplification, e.g., in *Peridinium balticum*, and in some *Mesodinium rubrum*. (E) Fragmentation of the cytobiont, e.g., *Mesodinium rubrum* (Oakley and Taylor, 1978).

Liebigean principles should apply equally to the government of growth in cells inside cells as to those outside (Droop, 1963), but which are the limiting nutrients most likely to act on cytobionts?

A comparison of aposymbiotic with normal, bacterial-cytobiont-containing cicadas by Schwemmler (1974, and this volume) showed that the pH and osmotic pressure in the hemolymph and egg cytoplasm were higher in the latter and he has drawn parallelisms between the relationships between mitochondria and the nucleocytoplasm of the cicadas, and cytobionts and the host nucleocytoplasm (Schwemmler, 1981).

The colonization of cells by foreign organisms clearly has many unique features by comparison with that of other habitats and yet there are certain parallels to other systems which may reflect some key features of the ecology of cytobiosis.

Moulder (1974, 1979) has been foremost in encouraging such comparisons although his interest has been principally with parasitic forms. Because of the limited number of species which can survive such an environment and the adaptions which may suit organisms to this environment but which may hamper their survival in other habitats, he suggested that the intracellular habitat can be compared to an extreme environment such as a hot spring or desert, salt lake, or extreme cold habitat. On the other hand it could also be looked on as the most benign.

To Schoenberg (1980) there seemed to be a parallel with the colonization and community development of an intertidal rock surface by larvae, a major difference being that, instead of the settling larvae selecting a suitable substrate, it is the cell which directly selects the cytobionts which inhabit it.

After a talk I was once asked if a useful analogy could be drawn between cytobioses and the development of young cuckoos in the nests of other birds. Although at first sight this may seem an inappropriate level at which to make comparisons it may be as apt as the analogies above. When the cytobionts become established they come to be "surrogate organelles," resembling chloroplasts or mitochondria (depending on whether they are photosynthetic or not). If there are already organelles with similar trophic properties, the cytobionts are essentially in competition for nutrients with those organelles, i.e., both are "fed" by the host, the cytobiont mimicking the host's organelles, its survival dependent on its ability to do this. However, if it outcompetes the host organelles it will destroy the system which nurtures it unless it can provide the host with the same or better service than its organelles. In this way a cytobiont may replace a set of host organelles. In the dinoflagellates *Peridinium balticum* and *Kryptoperidinium foliaceum* a photosynthetic, eukaryotic symbiont appears to have replaced the host chloroplasts with its own, the former being reduced to a vestige containing the eyespot (Taylor, 1979). In trichomonad flagellates living in low-oxygen, gut environments, mitochondria appear to have been replaced by "hydrogenosomes" (Muller, 1975), DNA-containing bodies which seem to be former cytobionts whose metabolism was more appropriate under anaerobic conditions than mitochondria, producing ATP by substrate phosphorylation while oxidizing pyruvate to acetate, with hydrogen gas as an anaerobic by-product. They resemble clostridial bacteria in their metabolism.

Of course, it is not known if the chloroplasts or mitochondria above became reduced due to some random mutational events prior to the acquisition of the cytobionts, but competitive suppression seems to be more likely.

Competition in cytobioses may involve not only cytobionts with host organelles, but also with other cytobionts. Several authors (e.g., Preer and Preer, 1977; Görtz, 1980, and this volume; Heckmann, this volume) have noted that it is rare for more than one type of bacterium to be present in a ciliate host such as *Paramecium aurelia* or *Euplotes,* even though many types have been identified. This seems most probably to be due to an established cytobiont outcompeting others attempting to gain entry, and it has been suggested that their relatively nonpathogenic cytobionts may offer the host some protection against other less benign microbes. Is this analagous to the conflict of similar types, but genetically heterogeneous, of organelles brought into "conflict" by syngamy (Eberhard, 1981)?

There has been a recent debate as to whether ecosystems can be regarded as cybernetic (feedback control systems) or not. Engleberg and Boyarsky (1979) concluded that they are not, in that they lack a communication network linking all components and they are not normally sufficiently closed for an impact to feedback onto all parts, leading to compensatory stability. The latter involves goals, i.e., the function of design. They accepted that an organism generally fits the criteria, small control signals, such as hormones or nerve impulses, causing major effects. Patten and Odum (1981) argued that ecosystems are basically cybernetic although no specific controllers are involved. Without going into the details of the latter's arguments (which eventually lead to a view rather like Lovelock and Margulis' "Gaia Hypothesis" in which there is a global homeostatic force: Lovelock, 1979), it is evident that the cytocosm, and the interaction between it and any cytobionts present, fits the narrower view of a cybernetic system as outlined by Engleberg and Boyarsky. What is not clear is whether this contributes any practical suggestions for the analysis of cytobioses. It does emphasize the identification of the control system or systems involved: whether the regulation of cytobiont number or biomass is regulated simply by its trophic relationships with the host or not.

MacArthur and Wilson (1967) introduced the concept of "*r*-" and "*K*-selection" to the eco-evolutionary field (see also Pianka, 1976; Cavalier-Smith, 1980). A species can be viewed as manifesting its selection in a spectrum that ranges from *r*-selection, in which reproduction rate (*r*) is maximized, to *K*-selection, in which energy and materials are channeled into other methods of increasing species survival (care of young, large size, defense, etc.). The spectrum is relative to the ecosystem being analyzed. *r*-Selection is looked on as the most basic form of natural selection. It is interesting to examine intracellular symbioses from this point of view.

It is immediately apparent that reproductive rate has to be very closely coupled between host and cytobiont. Thus, although zoochlorellae or zooxanthellae often have high photosynthetic efficiencies, even at low light intensities, the increased productivity does not produce greater cytobiont reproduction or growth but is channeled to the host as exudate instead. Similarly, being in a relatively constant environment the cytobionts tend to remain in a single life-cycle phase (most commonly equivalent to a cyst although they are metabolically active in this coccoid state and divide in it). Cytobiont sexuality has not been observed within a host, probably because of this, although it may occur during the free-swimming liberated state, e.g., in *Symbiodinium microadriaticum* (Taylor, 1973). Thus in several, albeit unconventional ways, cytobionts can be considered to be strongly *K*-selected, although they "leak" photosynthate or other materials rather than growing to a large size.

VI. Lamarckianism and Cytobiosis

Once a cytobiont becomes established in a host, it will be transmitted to daughter cells together with the surrounding cytoplasm cleaved during cytokinesis. In a multicellular host this can cause a spread in the infected tissue, but only rarely do the cytobionts enter the gametes to be transmitted sexually. However in unicellular eukaryotes (protists), cytobionts are readily transmitted to offspring during asexual division and may also be transmitted sexually when the host cell fuses with another. Integrative processes usually lead to changes in the genomes of both the host and cytobionts with the passage of time.

This truly is "the inheritance of an acquired characteristic" by both partners and, as such, represents the clearest example of a semi-Lamarckian evolutionary mechanism (Wallin, 1927; Taylor, 1979). It is not Lamarckian in the strictest sense, since "needs" are not essential to the process and Lamarck did not use the expression above (Cannon, 1959; Waddington, 1975). However, interestingly enough, "use and disuse," which were central to Lamarck's thinking, are a major feature of the integrative changes which follow the initial infection, unrequired structures, such as flagella or a wall, being rapidly lost or reduced by the cytobiont.

This form of Lamarckian mechanism might be dismissed by the macroevolutionist as being of little consequence in the grand scheme of things. However, if the SET is even partially true, then *one of the greatest developments of life on earth, the origin of the eukaryotic condition, arose by a Lamarckian-like mechanism!*

VII. From Cytobiont to Organelle

Newly colonizing cytobionts are potentially of two types: those which have been in hosts previously and have been expelled for one reason or another, e.g., as part of a regular rhythm in the anemone *Aiptasia*, (Steele, 1976) or in response to an unusual stress (Goreau, 1964) such as low salinity or high temperature (Fankboner and Reid, 1981); and those which have no history of an intracellular existence. There is a possibility of some small adaptive changes occurring in the first type which may enhance their probability of reinfecting a host, but they cannot change too far, otherwise they will not be able to survive their exposure to a free existence during the interim (Smith, 1979).

Continued isolation from their former gene pool is likely to lead to divergent genetic change in the small cytobiont populations by processes similar to those thought to operate on other small, isolated populations. However, there is another major characteristic of cytobioses which leads to a more unusual result. The host and the cytobionts may eventually lose their ability to exist independently. Contemporary examples suggest that it is usually the host which first becomes dependent (e.g., the *x*D amoeba/bacterial association), but mutual dependence follows, even while the cytobionts still morphologically resemble their free-living counterparts (e.g., *omikron* bacteria in *Euplotes*).

Once a cytobiont becomes essential for the survival of its host cells its distinction from an organelle becomes academic (not referable to known organelle-type?).

With the further passage of time it is likely that there is a loss of information in the cytobiont, the host genome coding for an increasing percentage of the proteins required by the cytobiont until a situation similar to the coding for chloroplast and mitochondrial proteins in the nucleus arises, assuming that the Serial Endosymbiosis Theory (SET) is at least partly correct (see below). Interestingly, the likelihood of the host becoming absolutely dependant on the cytobiont seems to be greater if the cytobiont is nonphotosynthetic than photosynthetic. Even after what has been assumed to be a considerable period of association a *Euglena* can still survive loss of its chloroplasts, reverting to organic nutrition. Many contemporary symbioses involving photosynthetic cytobionts are facultative (Buchner, 1965), the hosts surviving their loss providing adequate alternative food is available. However, many of these also occur in very oligotrophic, tropical waters, the hosts being able to thrive there due to the habitual presence of the photosynthetic cytobionts (''photobionts'') and so to exploit a region that they might otherwise not to be able to.

Why does the informational loss occur and how does the nucleus come to code for cytobiont/organelle features? It seems to be a universal property of systems that those with high information increase at the expense of those with lower information. Redundant genes are likely to be lost from the cytobiont. However, the cytobiont has many features, such as flagella or a wall, which became nonfunctional when it changed from an external to an intracellular habitat and the information for these is probably also lost.

One of the principal criticisms of the SET has been the necessity of a transfer of genes from the cytobiont to the host genome (Raff and Mahler, 1975). While several possible mechanisms have been proposed (see Weeden, 1981) some involving vectors such as viruses,[1] as in crown gall formation (Schell *et al.*, 1979), there may be a much simpler mechanism. Thornley and Harrington (1981) have noted that some individuals of the cytobiont may well be digested during the initial infective phase, or even later perhaps. Is all the DNA and RNA broken down into meaningless nucleotides or do some pieces remain intact enough to contain information and could this become incorporated in the host genome? If foreign nucleoproteins are injected into the cytoplasm of frog oocytes or amoebae some migrate to and accumulate in the nucleus of the injected cells. DeRobertis *et al.* (1982) have shown this to occur with small foreign nuclear RNAs and 5 S RNA. 7 S RNA and tRNAs appear to remain in the cytoplasm. Blascheck (1979) and Leber and Hemleben (1979) have found that plant nuclei can incorporate foreign DNA. Information by ingestion/injection?

Another interesting feature which may be involved has been noted by Soldo and Godoy (1973). Free-living prokaryotes do not usually have multicopy genomes although these are typical of organelles. However, the cytobiotic bacteria *lambda* and *pi* in ciliates do possess multiple copies, perhaps 10–20 in the former, 5–6 in the latter. Is this amplification also adaptive in cytobionts becoming surrogate organelles?

VIII. The Origin of Eukaryotes

This has been the single most important evolutionary event attributed to symbiosis, leading to a new hierarchical unit (Taylor, 1979). The theory invoking symbiosis in this context has been termed the Serial Endosymbiosis Theory (Taylor, 1974).

This topic will not be considered in much detail here for it has received extensive treatment recently (see particularly the reviews by Dodson, 1979; Taylor, 1980; and Gray and Doolittle, 1982; Fredrick, 1981; Margulis, 1981). Nevertheless, a "status report" on the theories seems to be appropriate.

[1]In this connection it is interesting to note that observations of viruses in cytobionts are becoming commonplace. The best known case is the presence of phage in "kappa" bacteria in *Paramecium aurelia*, where the toxic, ribbonlike "R body" may be a viral–host product (reviewed by Preer and Preer, 1977). Van Etten and his colleagues (Van Etten *et al.*, 1982) have found phage-like viruses to be present in all the zoochlorellae they have isolated from both *Hydra* and *P. bursaria* and suggested that the viruses may play a role in determining the acceptability of the *Chlorella* to the host.

It is accepted by all that a prokaryote-like stock gave rise to a wall-less protoeukaryote that increased in size, with concomitant elaboration of its membrane systems and an ability to perform cytosis (endocytosis and exocytosis, dependent on membrane translocations, ATP, and an actomyosin system). Most accept that the nucleus represents an amplification of the original host genome, the nuclear envelope, endoplasmic reticulum, Golgi apparatus, lysosomes, and vacuoles being elaborations of the internal membranes: the "Endomembrane System." A few have suggested that the nucleus had a symbiotic origin (e.g., Kunicki-Goldfinger, 1980) but there appears to be little compelling reason(s) for this, the nucleoplasm clearly being continuous with the cytoplasm during mitosis in most eukaryotes (not in a few flagellates such as dinoflagellates or euglenoids, which nevertheless have nuclear pores).

Several lines of evidence have led to the wide-spread acceptance of the view that the chloroplast arose by an intracellular symbiotic association between one or more cyanobacteria-like stocks and the protoeukaryote, entry presumably having been by phagocytic engulfment. Biochemical evidence, reviewed *in extenso* by Gray and Doolittle (1982), has been particularly influential. The presence of organelle-specific DNA and RNA, ribosomes of similar size to prokaryotes (70 S) rather than the surrounding cytoplasm (80 S), sensitivity of protein synthesis to similar inhibitors (e.g., chloramphenicol, streptomycin, erythromycin) and insensitivity to cycloheximide, and the existence of contemporary analogs to the postulated cytobionts ("cyanelles") have all been suggestive of the role of endosymbiosis. However, the most striking evidence has come from DNA–RNA hybridization comparisons between chloroplast, nucleocytoplasm, and cyanobacteria (Pigott and Carr, 1972), partial sequencing of small ribosomal subunit (16/18 S) rRNA (Bonen and Doolittle, 1975), complete sequencing of large subunit (5 S) rRNA (Phillips and Carr, 1981), and the amino acid sequences of various proteins (Dayhoff and Schwartz, 1980; Schwartz and Dayhoff, 1981). In essence, all the latter studies show a considerable similarity between the organelle and free-living cyanobacteria, in some cases as much as between genera of living cyanobacteria, and little or none with the surrounding nucleocytoplasm.

Mitochondria are thought to have arisen by a symbiotic liason most probably involving a cytobiont resembling the Rhodospirillaceae (purple nonsulfur bacteria), at an earlier date than the chloroplast (near universal occurrence in eukaryotes, greater integration with the host), showing a similar spatial relationship to the surrounding cell (permanently closed double-membrane envelope, occasional connections with ER), smaller ribosomes, and some similar antibiotic sensitivities and insensitivities. The case has been more difficult to make because there is less DNA in mitochondria, the ribosomes are variable in size, and 5 S rRNA is lacking in fungal or animal mitochondria (but present in plant mitochondria). Indeed there are many features of mitochondria which are unique, not the least being a change in the codon UGA, usually a stop codon in both prokaryotes and eukaryotes but a secondary codon for tryptophane in mitochondria (review by Mahler, 1980, 1981). Although mitochondria use fMet tRNA as an initiator like prokaryotes (and unlike eukaryotic cytoplasmic initiator tRNA) there are some structural differences between the two. Their messenger RNA contains neither "caps," "tails" (eukaryotic), nor "Shine-Dalgarno" sequences (prokaryote-like). Parthier (1980) found that a comparison of the respective tRNA sequences produced equivocal results, either no more similar to prokaryotes or eukaryotes, or only very slightly more like the former.

An unexpected discovery, which provided the only evidence for a greater similarity between mitochondria and their surrounding nucleocytoplasm than between them and prokaryotes, was the demonstration of intervening sequences ("introns") in their mRNA genes, with posttranscriptional processing (excision of the introns and splicing of the exons). This was thought to be a uniquely eukaryotic, nuclear phenomenon whose significance has been debated at length. Introns have been found in yeast mitochondria (Locker *et al.*, 1981) but apparently are absent from animal mitochondria (Brown, 1981). The strength of this feature as evidence for direct differentiation of mitochondria from the cytoplasm of the protoeukaryote has been weakened, however, by the interpretations of Doolittle (1978), Darnell (1978), Desnizky and Pinevich (1980), Gray and Doolittle (1982), and others who have suggested that introns were present in the stock ancestral to all these components and have been lost in prokaryotes and the mitochondria of higher organism. An insertion sequence has recently been reported by Luehrsen *et al.* (1981) from a member of the Archaebacteria, thought to have had a very early divergence.

The strongest evidence for a symbiotic origin for the mitochondrion again comes from protein and nucleic acid sequencing. Dayhoff and Schwartz (1981) found that, in a phyletic "tree" constructed from information on c-type cytochromes, ferredoxin, and 5 S rRNA sequences mitochondria occurred on a branch distinct from the eukaryotic nucleocytoplasm, closest to *Rhodopseudomonas*-like photosynthetic bacteria. Bonen *et al.* (1977) compared the oligonucleotide catalogues of 18 S rRNA from wheat cytosol, mitochondria, and 16 S rRNA from prokaryotes and obtained a similar picture to that for chloroplasts, i.e., a much greater resemblance between the prokaryotes, both photosynthetic and nonphotosynthetic, and mitochondria despite the S value difference, than between mitochondria and cytosol 18 S rRNA. This was confirmed from a comparison of complete sequences for these rRNAs from the mitochondria of *Aspergillus nidulans*, yeast, man, and mouse; the nuclei of yeast and animals, the chloroplast of maize, and *E. coli*. Küntzel and Köchel (1981) found several regions of conserved primary stucture in all, but a tree constructed from the data clearly supported a symbiotic origin of the mitochondria from prokaryotes.

Küntzel *et al.* (1981) also maintained that the sequencing of the 5 S rRNAs showed a similar relationship for mitochondria but there are problems in the alignment which weaken their case (Gray and Spencer, 1981) and the 5 S rRNA of wheat mitochondria is probably best viewed as unusual, not particularly favoring any hypothesis based on present evidence.

Early versions of the SET postulated some aerobic, nonphotosynthetic bacterium as the protomitochondrion. Later John and Whatley (1975) drew attention to the very close resemblances, particularly in the electron transport systems, between the denitrifying bacterium *Paracoccus* and mitochondria. There are also many resemblances between purple nonsulfur (PNS) bacteria (Rhodospirallaceae) and mitochondria (Dayhoff and Schwarz, 1981; Dickerson, 1980) and cytochrome c sequencing and 16 S rRNA oligonucleotide cataloguing now indicates that *Paracoccus* should be assigned to the Rhodospirallaceae (this and some incongruities reviewed by Gray and Doolittle, 1982).

The view of Dickerson and Dayhoff and Schwarz that mitochondria arose from PNS protomitochondrian with loss of photosynthesis is an attractive one. Although not discussed by them it raises the interesting possiblity that the original selective advantage to the host (protoeukaryote) derived from the photosynthetic abilities of its cytobiont, rather than its more efficient respiration potential. Furthermore, PNS bacteria are facultatively anaerobic/aerobic, being photosynthetic under the former conditions but their photosynthesis is suppressed by oxygen, with fewer membrane invaginations developed. Essentially, then, a scenario emerges in which, as long as the host remained in an anaerobic environment it had a photosynthetic cytobiont/organelle but when it migrated into the aerobic environment created by the cyanobacteria the photosynthetic abilities of its cytobiont were suppressed. Instead it got more efficient respiration which allowed it to be more active but also required it to be a more active phagotroph in order to satisfy this demand. This is quite a different picture from that proposed earlier in which the inefficient, facultatively aerobic host acquired a more efficiently respiring, nonphotosynthetic cytobiont. In another variation Pinevich and Desnizky (1979) proposed that both mitochondria and chloroplasts originated from the same cyanobacterial stock.

At present the weight of evidence favors a symbiotic origin for the mitochondrion when the numerous unique features of mitochondria and other peculiarities which could be thought of as the products of a long divergence from the parent stock(s) are subtracted from the picture.

There are ways to account for the resemblances using an autogenous model similar to those of Bogorad (1975), Cavalier-Smith (1975), or Taylor (1976), but these are not the simplest explanations for the observed data. The problem of resemblances which has emerged, particularly from protein and nucleic acid sequencing, is essentially just as one would expect from the SET although this does not "prove" the latter. In view of the long established metabolic similarities between chloroplast and cyanobacteria on the one hand, and mitochondria and Krebs Cycle-containing eubacteria on the other, the discovery that their genomes are similar is hardly surprising. The conservation of the pathways and electron transport chains should be reflected in their genomes. Coding for many organellar proteins by the nuclear genome is still more readily explained by an autogenous model although fairly plausible ways to account for it have been advanced (e.g., Weeden, 1981). What is striking in the sequence data, at least at first sight, is how *little* resemblance there is between organellar constituents and their surrounding constituents. If they were once undifferentiated (subsequent to the divergence of protoeukaryote from other prokaryotes), can this lack be reasonably explained by changing, specialized function, selection, and loss of unused genes?

Essentially it is the consistency of the whole pattern, involving evidence of many types, which has made the SET the most probable mechanism by which the eukaryotes arose even though absolute proof will never be possible.

Margulis (1981) and her co-workers have come up with numerous provocative ectosymbiotic associations between contemporary spirochetes and flagellates to bolster her proposal (Sagan, 1967) that eukaryotic flagella (''undulipodia'' in her terminology) and homologous structures such as centrioles and other microtubule centers (MCs or MTOCs) arose by the integration of similar ectosymbionts and the protoeukaryote. Microtubule-like structures are present in spirochetes, especially in those of the large *Pillotinas,* although it has not yet been confirmed that these are homologous with tubulin rather than the flagellin of prokaryote flagella. Unfortunately, although RNA is present in basal bodies there is no DNA in them. Eukaryotic MCs are fully integrated spatially into cytosol and can be formed *de novo,* unlike chloroplasts and mitochondria. At present the symbiotic origin of eukaryotic flagella and MCs remains an intriguing possibility but lacking in a number of the features possessed by mitochondria and chloroplasts which make their cases more compelling (e.g., organelle-specific DNA, permanent isolation from the host cytoplasm by membranes, never arising *de novo* and therefore permanently lost by antibiotic treatment, greater similarity to homologous molecules in free-living prokaryotes than in host cytoplasm, etc.). There seems to be little reason to think that the host was unable to produce microtubules. Furthermore, there is another possibility of a foreign origin. Taylor (1979) has drawn attention to some interesting similarities between centriole synthesis and that of RNA viruses.

The nature of the stock that gave rise to the protoeukaryote is still on open question. Gray and Doolittle (1982) have extensively reviewed the available data and hypotheses. The divergence of the Archaebacterial line from the Eubacterial line at an early stage in prokaryotic evolution seems to be reasonably well established (Woese 1981; Fox, 1981) although how early is not clear. *Ther-*

moplasma acidophilum, the peculiar mycoplasma-like bacterium possessing many features that resemble the postulated protoeukaryote (but being minute and living under extremely high temperatures and very low pH) is clearly a member of the Archaebacteria, but Woese and Fox do not think there is much in favor of deriving the protoeukaryote from an archaebacterial stock (although they do not list all the features that the two have in common, emphasizing the differences instead) preferring a "tripartite" split between Eubacteria, Archaebacteria, and the eukaryotic lineage at a very early stage. Halophilic bacteria, also members of the Archaebacteria, show some of the properties one might expect in a group related to the ancestral protoeukaryote. Fox (1981) concluded "if the archaebacterial genomes are found to contain intervening sequences and to lack operons, the case for an archaebacterial protoeukaryote would be greatly strengthened." Shortly after this an intervening sequence was found in the 5 S rRNA gene member of the Archaebacteria, *Halococcus morrhuae* (Luehrsen *et al.*, 1981).

IX. Multiple Origins for Chloroplasts and Mitochondria?

Having concluded that a symbiotic origin is most probable for these organelles it is reasonable to ask if it occurred more than once for each. Taylor (1974) distinguished between primary organellar sources (the original symbiotic events involving prokaryotic protoorganelles and the host) and secondary sources involving the acquisition of the organelles of a *eukaryotic* cytobiont by the host.

The possibility of multiple primary origins was raised by Mereschkowsky (1910). It was later revived for chloroplasts by Raven (1970) who suggested that the three basic pigment-chloroplast types: red (chlorophyll $a + d$), green ($a + b$), and brown ($a + c$), arose by symbioses involving at least three different prokaryotic stocks. At the time it was known that red algal chloroplasts are very similar to cyanobacteria (blue-green algae) in many respects, including the possession of phycobilin pigments in "phycobilisomes" located on the outer surface of the thylakoids. However there were no free-living counterparts to the other two chloroplast types. It was, therefore, a dramatic event when *Prochloron,* a photosynthetic prokaryote living symbiotically with didemnid ascidians and possessing chlorophylls *a* and *b,* lacking phycobilins, and exhibiting thylakoidal stacking, was discovered (Lewin and Withers, 1975; Newcomb and Pugh, 1975; Whatley, 1977). Here was an ideal analogue for the $a + b$ type chloroplast. Indeed, the occurrence of chlorophyll *b* in a prokaryote requires a multiple symbiotic origin unless the origin of chlorophyll *b* occurred twice. Furthermore, Bonen *et al.* (1979), on the basis of the partial sequencing (oligonucleotide cataloguing) of 16 S rRNA, found that *Porphyridium* (red alga) chloroplasts were more similar to some free-living cyanobacteria than to those of *Euglena.*

Similarly, in the trees of Schwartz and Dayhoff (1981) various chloroplasts grouped on different lines, together with assorted cyanobacteria, based on 5 S rRNA, 2Fe-2S ferredoxins and cytochrome c_6.

Some phytoflagellate groups, most notably cryptomonads and to a less extent euglenoids, seem to have chloroplasts which are incongruous with their other cell features when compared with other groups in a phyletic scheme (Taylor, 1976, 1978). Gibbs (1978, 1981) and Gillot and Gibbs (1980) have produced ultrastructural evidence to support the view that in both these groups the chloroplasts may have been secondarily derived from the acquisition and degeneration of eukaryotic symbionts somewhat like those occurring in the dinoflagellates *Peridinium balticum* or *Kryptoperidinium foliaceum* (summarized by Taylor, 1979). The presence of the "nucleomorph," a small body resembling a vestigial nucleus (Gillot and Gibbs, 1980; Gibbs, 1981) in the perichloroplastic space in cryptomonads, is particularly provocative in this regard, although it has not yet been shown to contain DNA. The extraordinary, cryptomonad-derived cytobiont in the ciliate *Mesodinium rubrum* (summarized by Taylor, 1979) is capable of fragmenting into numerous discrete islands of cytoplasm within its host (Fig. 3), each containing a chloroplast with nucleomorph and mitochondrion, one of the islands containing a large cryptomonad nucleus (Oakley and Taylor, 1978). Perhaps this state is only possible because of the presence of nucleomorphs in the fragments.

Dodge (1979, 1980) and Whatley and Whatley (1980) have produced phyletic schemes by incorporating primary and secondary multiple chloroplast origins. Taylor (1978), on the other hand, found that in his arrangement, which did not invoke symbiosis, chloroplast types were predictably consistent with mitochondrial morphology (cristal type) and flagellar features, suggesting that they had co-evolved, although cryptomonads were a striking anomaly, as noted above. Furthermore, if one looks at carotenoid composition (Liaaen-Jensen, 1977) the picture is not compatible with the "three-chloroplast" origin schemes. For example, although euglenoid chlorophylls resemble green algae ($a + b$) their carotenoids are completely different (early records of the presence of zeaxanthin being erroneous), being more similar to xanthophytes, and the cryptomonad carotenoid spectrum is quite unlike red algae even though they possess phycobilin pigments (not in phycobilisomes).

More recent studies on *Prochloron* reveal a different picture from that suggested by the earlier data. Fatty acid composition lipid content (Perry *et al.*, 1978) and 16 S rRNA oligonucleotide cataloguing (Seewaldt and Stackebrandt, 1982) suggest that *Prochloron* is simply an anomalously pigmented cyanobacterium (Antia, 1977) with few similarities to green algal chloroplasts, other than chloropyll *b*, "stacking" and negative features. In cyanobacteria, stacking of the thylakoids is not possible due to the presence of the phycobilisomes on the outside of the thylakoids, but the loss of the latter in *Prochloron* seems to permit the close proximity of the thylakoids, mimicking that in green algae.

At first sight the origin of chlorophyll *b* twice seems to be highly improbable. However, this is less so when it is realized that the sole difference between chlorophylls *a* and *b* lies in the substitution of formyl (CHO) for methyl (CH_3) at position 3 on ring II in the porphyrin moiety (Meeks, 1974). How this occurs is not known but it might involve an intermediate hydroxy-methyl (CH_2OH) step or, if the folic acid cycle is involved, only a single step. Thus *Prochloron* seems to have had little to do directly with the ancestry of green chloroplasts, serving instead as a model to show how their transition from a cyanobacterium might have occurred.

Several enzymes are the cooperative product of nuclear and organellar genes. For example ribulose biphosphate carboxylase (RuBPase) consists of a small and large subunit originating in the above fashion. It would be interesting to know if this is the case for all chloroplast types, current information being available only for "*a* + *b*" chloroplasts (Gillham, 1978). Coding in similar locations for all chloroplasts types might be improbable in a multiple-origin scheme although there may be a mechanistic reason why the large subunit cannot be coded for in the nucleus (impermeability of the chloroplast envelope).

In mitochondria there *is* a difference in this regard. The membrane factor of F1 ATPase is coded for by the mitochondrial genome in yeast but by the nucleus in another fungus, *Neurospora* (Gillham and Boynton, 1981). There is considerable variability in other mitochondrial features, including ribosome size and the presence (in yeast) or absence (in mammals) of "introns" (Mahler, 1981) although they are morphologically and metabolically conservative. Cytochrome c and ferredoxin sequences suggest multiple mitochondrial origins (Dayhoff and Schwartz, 1981) but it is possible to interpret the "trees" as indicating only a single origin. Recently Küntzel and Köchel (1981) found that the small subunit rRNA sequencing also suggested a multiple origin.

It seems to be too early to draw reasonable conclusions from the evidence relating to multiple origins. The cases for the primary symbiotic origins of chloroplasts and mitochondrial are based on many features, including gross differences in RNA and amino acid sequences. However, is sequence information reliable and conservative enough to recognize finer relationships? Uzzell and Spolsky (1981) found that *Euglena* and *Porphyridium* chloroplasts grouped closer together than to the cyanobacteria *Aphanocapsa* and *Anacystis*. There are some problems in protein sequence tree constructions (Demoulin, 1979; Ambler *et al.*, 1979a,b). Can alignment technique differences undermine confidence in the sensitivity of the data, such as that in recent comparisons of mitochondrial 5 S rRNA sequences (Küntzel *et al.*, 1981; Gray and Spencer, 1981)? It is one thing to argue from gross distinctions between sequence or other biochemical data, and quite another to say that fine distinctions, which nevertheless place organisms or organelles on separate branches, have similar validity. The best arguments seem to be those involving evidence from several independent sources but these leave the question open at present.

X. Hierarchical Consequences of the SET

> In a very speculative mood, we shall make bold to suggest here the possible relevance of these considerations to the cell theory. . . . it might be useful to think of the cell, not as a genetic unit but rather as an assembly of several independent genetic organelles which have become thoroughly integrated (Dubos and Kessler, 1963, p. 10).

This "bold suggestion" by Dubos and Kessler, two decades ago, has ceased to be all that speculative. As a result of the studies on organellar genetics and molecular biology referred to in previous sections it is clear that *eukaryotic "cells" are not hierarchically equivalent to prokaryotic "cells."* A nonphotosynthetic eukaryotic cell consists of *two* genome-containing, membrane bound compartments: the nucleocytoplasmic and the mitochondrial compartments (the host and Krebs cycle-containing symbionts, respectively, if the SET is correct), *each* of which is hierarchically equivalent to a prokaryotic cell (Fig. 4). A plastid-containing cell is equivalent to three distinct, prokaryotic cells, each of which is still permanently isolated by a membrane.

Thus, the term *cell,* considered to be such a fundamental unit of life, is ambiguous and the Cell Theory requires modification. Although it is true that all organisms capable of metabolism consist of cells, and cells are defined as genome-containing, membrane-bounded entities (the only definition that includes all the organisms we think of as cellular), eukaryotic cells are multiples of prokaryotic cells. Note that this is regardless of the mode of origin of the former, although the SET makes the nonequivalence more evident and readily understandable. This conceptual revision was explicitly stated and developed by Taylor (1974). An analytical terminology, monad, dyad, triad, polymonad, polydyad, etc., was used to refer to the hierarchical state of an organism (Fig. 4) based on the number of distinct, different genome-containing compartments, and developed further by Margulis (1976) and Schwemmler (1979). Contrary to the impression given by Margulis (1981) it was not expected that these terms would replace "cell" in everyday use. However, there is a relatively simple modification to the term which conveys the distinction between the two basic types of cell and which is more readily introduceable into general terminology: *procell* and *eucell,* or procyte and eucyte. The former pair has been used, although not with hierarchical connotations, by Brown and Bertke (1974), the latter by Schwemmler (1979).

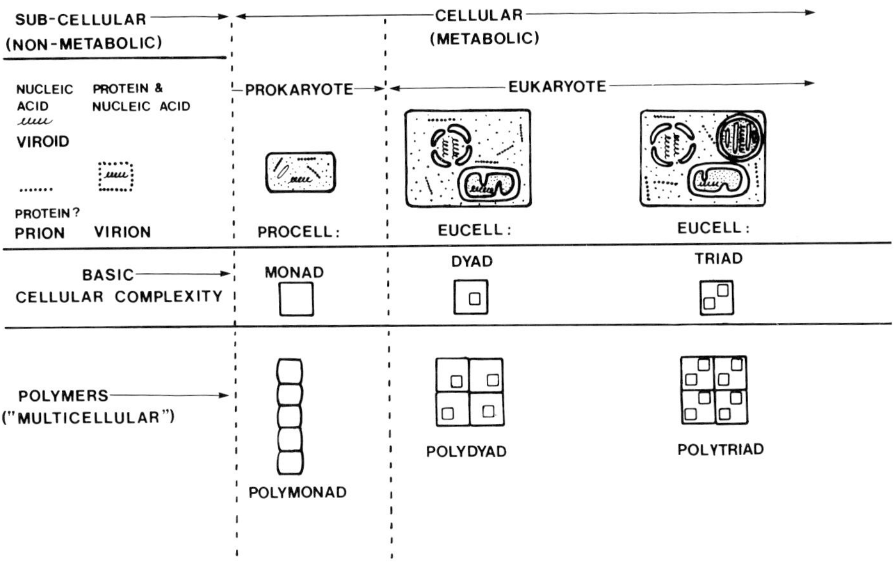

FIG. 4. Hierarchy of the lower organisms. Modified from Taylor (1974). Subcellular "organisms" consist of those existing as a single molecular type (*viroids*, plant viruses consisting of "naked" nucleic acid; *prions*, newly postulated infectious agents causing scrapie and other long-incubation diseases, consisting of protein with no detectable nucleic acid; Prusiner, 1982) and viruses. *Procell:* a basically monogenomic membrane-bound, metabolizing unit (additional plasmids not in separate compartments). *Eucell:* a polygenomic, multicompartment, metabolizing unit. Animal cells are dyadic, plant cells triadic. "Multicellular" prokaryotes are polymeric monads, whereas multicellular eukaryotes are polydyads or polytriads. Cytobioses add additional, membrane-bound genomic units to dyads or triads. A cyanelle in a nonphotosynthetic eukaryotic host would form a triadic organization (e.g., *Cyanophora paradoxa*). See also Margulis (1981).

Although not recognizing the dyad/triad distinction, these terms could be used to refer to monogenomic and polygenomic membrane-bound entities, respectively, so that new students of biology would not get the false impression of equivalence between prokaryotic and eukaryotic cells which current statements of the Cell Theory create.

XI. Conclusion

Intracellular symbioses are the most intimate interspecies associations known. They therefore have great ecological and evolutionary interest. The intracellular environment, the *cytocosm*, represents a sharply bounded microhabitat with extreme stability (over the life-span of the host), and extreme selection against colonization. Its constancy and closeness to the internal composition of the cytobiont render it benign to those who can overcome its severe initial barriers, providing that the materials that are cycled do not reduce the host's ability to survive. Within the host the cytobionts play a role equivalent to the host organelles, a resemblance which increases with time. Once they are obligately required, they *are* organelles, albeit of an unconventional type. The cytobiont presence may competitively suppress the host organelles most similar to their dominant metabolic mode, essentially replacing them.

In dy- or triadic hosts (protists) the cytobionts are distributed directly to the daughter cells during cytokinesis. As such they display an indisputable example of the inheritance of acquired characters. In multicellular hosts the cytobionts rarely penetrate the gametes although mechanisms to ensure the rapid reinfection of the offspring are usually present. Because of the direct cytoplasmic continuity from generation in protists cytobionts can readily evolve to become permanently dependent on an intracellular existence. However, in multicellular hosts in which gametic infection is not achieved the cytobionts must retain a free-living capability in order to survive host death.

The potential for dramatic, quantum evolutionary change in cytobioses is evident. However, many might assume that it is too unusual a phenomenon to have a broad impact on macroevolution. The probable origin of the eukaryotic cellular condition by symbiotic events (the Serial Endosymbiosis Theory) indicates that this crucial event (to us) was of such a type. Furthermore, it means that the Cell Theory, as currently taught, requires modification in order to recognize that "cells" involve at least two hierarchical levels. The distinction between cells consisting of a single genome-containing compartment ("procells" or "procytes") from those consisting or two or more ("eucells" or "eucytes") does not require acceptance of the SET although the latter offers the simplest explanation for it.

Not only have some organelles had a primary symbiotic origin, but there are indications (considered convincing by some but not by this author) that chloroplasts and even mitochondria may have originated more than once. Additionally, symbiosis may have been involved in the secondary acquisition of one or more organellar types through the degeneration of a eukaryotic cytobiont in a host. The most plausible case of the latter can be found in cryptomonads. Although a few metazoans and protists seem to be able to maintain chloroplasts from ingested food ("organellar maintenance"), the foreign organelles cannot survive indefinitely in the foreign host cytoplasm and so this does not appear to be a source of permanent organelles at present, although it might have been in the past.

REFERENCES

Ambler, R. P., Daniel, M., Hermoso, J. H., Meyer, T. E., Bartsch, R. G., and Kamen, R. D. (1979a). *Nature (London)* **278,** 659–660.

Ambler, R. D., Meyer, T. E., and Kamen, R. D. (1979b). *Nature (London)* **287,** 661–662.
Antia, N. J. (1977). *Br. Phycol. J.* **12,** 271–276.
Bang, F. B. (1973). *Bioscience* **23**(10), 584–589.
Bannister, L. H. (1979). *Proc. R. Soc. London Ser. B* **204,** 141, 163.
Berger, J. (1980). *J. Gen. Microbiol.* **118,** 397–404.
Blascheck, W. (1979). *Plant Sci. Lett.* **15,** 139–149.
Bogorad, L. (1975). *Science* **188,** 891–898.
Bonen, L., and Doolittle, W. F. (1975). *Proc. Natl. Acad. Sci. U.S.A.* **72,** 2310–2314.
Bonen, L., Cunningham, R. S., Gray, M. W. and Doolittle, W. F. (1977). *Nucleic Acids Res.* **4,** 663–671.
Bonen, L., Doolittle, W. F., and Fox, G. E. (1979). *Can. J. Biochem.* **57,** 879–888.
Brown, G. G. and Costra, J. F. (1981). *Ann. N.Y. Acad. Sci.* **361,** 135–153.
Brown, W. M. (1981) *Ann. N.Y. Acad. Sci.* **361,** 119–134.
Brown, W. V., and Bertke, E. M. (1974). "Textbook of Cytology" 2nd ed., pp. 1–528. Mosby, St. Louis, Missouri.
Buchner, P. (1965). "Endosymbiosis of Animals with Plant-like Microorganisms," pp. 1–909. Wiley (Interscience), New York.
Cannon, H. G. (1959). "Lamarck and Modern Genetics," pp. 1–152. Manchester Univ. Press, Manchester.
Canny, M. J. (1981). *Proc. Ecol. Soc. Aust.* **11,** 1–11.
Caullery, M. J. G. C. (1952). "Parasitism and Symbiosis," pp. 1–340. Sidgewick & Jackson, London.
Cavalier-Smith, T. (1975). *Nature (London)* **256,** 463–468.
Cavalier-Smith, T. (1980). *BioSystems* **12,** 43–59.
Chapman-Andresen, C. (1971). *Annu. Rev. Microbiol.* **25,** 27–68.
Darnell, J. E. (1978). *Science* **202,** 1257–1260.
Dayhoff, M. O., and Schwartz, R. M. (1980). *Proc. Int. Colloq. Endosymb. Cell Res., Tübingen* pp. 63–83.
Dayhoff, M. O., and Schwartz, R. M. (1981). *Ann. N.Y. Acad. Sci.* **361,** 92–104.
Demoulin, V. (1979). *Science* **205,** 1036–1039.
DeRobertis, E. M., Lienhard, S., and Parisot R. (1982). *Nature (London)* **295,** 572–577.
Desnizky, A. G., and Pinevich, A. V. (1980). *Vestnik Leningr. Univ.* (21), 80–85.
Dickerson, R. E. (1980). *Sci. Am.* **242,** 136–153.
Dodge, J. D. (1979). *In* "Biochemistry and Physiology of Protozoa" (S. Hunter and M. Levandowsky, eds.), 2nd Ed., Vol. 1, pp. 1–57. Academic Press, New York.
Dodge, J. D. (1980). *Proc. Int. Colloq. Endosymb. Cell Res., Tübingen* pp. 33–50.
Dodson, E. O. (1979). *Can. J. Microbiol.* **25,** 652–674.
Doolittle, W. F. (1978). *Nature (London)* **272,** 581–582.
Droop, M. R. (1963). *Symp. Soc. Gen. Microbiol., 13th* pp. 1–356.
Dubos, R., and Kessler, A. (1963). *Symp. Soc. Gen. Microbiol., 13th* pp. 1–356.
Eberhard, W. G. (1981). *Ann. N.Y. Acad. Sci.* **361,** 44–52.
Eldredge, N., and Gould S. J. (1972). *In* "Models in Paleobiology" (T. J. M. Schopf, ed.) pp. 82–115. Freeman, San Francisco, California.
Engleberg, J. and Boyarsky, L. L. (1979). *Am. Nat.* **114,** 317–324.
Fankboner, P. V., and Reid, R. G. (1981). *Experientia* **37,** 251–252.
Fox, G. E. (1981). *In* "Evolution Today", (G. E. Scudder and J. L. Reveal, eds.), pp. 235–244. Hunt Institute, Pittsburgh, Pennsylvania.
Fredrick, J. F. (1981). *Ann. N.Y. Acad. Sci.* **361,** pp. 1–512.
Gibbs, S. P. (1978). *Can. J. Bot.* **56,** 2883–2889.
Gibbs, S. P. (1981). *Ann. N.Y. Acad. Sci.* **361,** 193–208.
Gillham, N. W. (1978). "Organelle Heredity," pp. 1–602. Raven, New York.

Gillham, N. W., and Boynton, J. E. (1981). *Ann. N.Y. Acad. Sci.* **361,** 20–43.
Gillot, M. A., and Gibbs, S. P. (1980). *J. Phycol.* **16,** 558–568.
Goreau, T. (1964). *Science* **145,** 383–386.
Görtz, H.-D. (1980). *Proc. Int. Colloq. Endosymb. Cell Res., Tübingen* pp. 381–392.
Gray, M. W., and Doolittle, W. F. (1982). *Microbiol. Rev.* **46** (1), 1–42.
Gray, M. W., and Spencer D. F. (1981). *Nucleic Acids Res.* **9** (14), 3523–3528.
Gromov, B. V., Seravin, L. N., and Gerasimova, Z. P. (1977). *Mikrobiologia* **46,** 971–972.
Jeon, K. W. (1980). *In* "Cellular Interaction in Symbiotic and Parasitic Relationships" (C. B. Cook, P. Pappas, and E. Rudolph, eds.), pp. 245–262. Ohio State Univ. Press, Columbus.
John, P., and Whatley, F. R. (1975). *Nature (London)* **254,** 495–498.
Jones, T. C., and Hirsch, J. G. (1972). *J. Exp. Med.* **136,** 1173–1194.
Karakashian, M., and Karakashian, S. J. (1973). *Exp. Cell Res.* **81,** 111–119.
Kunicki-Goldfinger, W. J. H. (1980). *Proc. Int. Colloq. Endosymb. Cell Res., Tübingen* pp. 969–985.
Küntzel, H., and Köchel, H. G. (1981). *Nature (London)* **293,** 751–755.
Küntzel, H., Heidrich, M., and Piechulls, B. (1981). *Nucleic Acids Res.* **9,** 1451–1461.
Leber, B., and Hemleben, V. (1979). *Z. Pflanzenphys.* **91,** 309–316.
Lee, J. J. (1980). *In* "Biochemistry and Physiology of the Protozoa" (M. Levandowsky and S. Hubner, eds.), Vol. 3, Ch. 2, pp. 43–66. Academic Press, New York.
Lee, J. J., McEnerny, M. E., and Kahn, E. G. (1979). *Micropalentology* **25,** 118–140.
Lewin, R. A. (1981). *Ann. N.Y. Acad. Sci.* **361,** 325–329.
Lewin, R. A., and Withers, N. W. (1975). *Nature (London)* **256,** 735–737.
Li, J.-Y. (1979). "The Generation of Cells in the Evolutionary History of Life: The Origin of Eukaryotic Cells in Evolution," pp. 1–237. Science Press, Beijing. (In Chinese).
Liaaen-Jensen, S. (1977). *In* "Marine Natural Products Biochemistry" (D. J. Faulkner and W. J. Fenical, eds.), pp. 239–259. Plenum, New York.
Locker, J., Synenki, R. M., Merten, S., and Rabinowitz, M. (1981). *Ann. N.Y. Acad. Sci.* **361,** 105–118.
Lovelock, J. E. (1979). "Gaia—A New Look at Life on Earth," pp. 1–157. Oxford Univ. Press, London and New York.
Luehrsen, K. R., Nicholson, D. E., Eubanks, D. C., and Fox, G. E. (1981). *Nature (London)* **293,** 755–756.
MacArthur, R. H., and Wilson, E. O. (1967). "The Theory of Island Biogeography," pp. 1–362. Princeton Univ. Press, Princeton, New Jersey.
McLaughlin, J. J. A., and Zahl, P. A. (1966). *In* "Association of Microorganisms, Plants and Marine Organisms" (S. M. Henry, ed.), Vol. 1, pp. 257–298. Academic Press, New York.
McNaughton, S. J. (1977). *Am. Nat.* **111,** 515–525.
Mague, T. H., Weare, N. M., and Holm-Hansen, O. (1974). *Mar. Biol.* **24,** 109–119.
Mahler, H. R. (1980). *Proc. Int. Colloq. Endosymb. Cell Res., Tübingen* pp. 869–893.
Mahler, H. R. (1981). *Ann. N.Y. Acad. Sci.* **361,** 53–75.
Margulis, L. (1975). *Biosystems* **7,** 266–292.
Margulis, L. (1976). *Exp. Parasitol.* **34,** 277–349.
Margulis, L. (1981). "Symbiosis in Cell Evolution," pp. 1–419. Freeman, San Francisco, California.
May, R. M. (1975). "Stability and complexity in Model Ecosystems," 2nd ed., pp. 1–235. Princeton Univ. Press, Princeton, New Jersey.
Meeks, J. C. (1974). *In* "Algal Physiology and Biochemistry" (W. D. P. Stewart, ed.), Ch. 5, pp. 161–175. Blackwell, Oxford.
Meier, R., Reisser, W., and Weissner, W. (1980). *Z. Naturforsch.* **35**C, 1107–1110.
Mereschkowsky, C. (1910). *Biol. Centralbl.* **30,** 321–325, 363–364.

Moulder, J. W. (1974). *J. Infect. Dis.* **130,** 300–306.
Moulder, J. W. (1979). *Proc. R. Soc. London Ser. B* **204,** 199–210.
Muller, M. (1975). *Annu. Rev. Microbiol.* **29,** 467–483.
Muscatine, L., and Neckelmann, N. (1981). *Ber. Dtsch. Bot. Ges.* **94**(3), 571–583.
Muscatine, L., and Pool, R. R. (1979). *Proc. R. Soc. London Ser. B* **204,** 131–139.
Muscatine, L., Poole, R. R., and Trench, R. K. (1975). *Trans. Am. Microsc. Soc.* **94**(4), 450–469.
Newcomb, E. H., and Pugh, T. D. (1975). *Nature* (London) **253,** 533–543.
Oakley, B. R., and Taylor, F. J. R. (1978). *BioSystems* **10,** 361–369.
Ohiwa, T. (1981). *Bot. Mag. Tokyo* **94,** 261–271.
Ossipov, D. V., and Podlipaer, S. A. (1977). *Acta Protozool.* **16,** 289–308.
Parthier, B. (1980). *In* "Cell Compartmentation and Metabolic Channeling" (L. Nover, F. Lynen, and K. Mothes, eds.), pp. 107–121. Elsevier, Amsterdam.
Patten, B. C., and Odum, E. P. (1981). *Am. Nat.* **118,** 886–895.
Perry, G. J., Gillian, F. T., and Johns, R. B. (1978). *J. Phycol.* **14,** 369–371.
Phillips, D. O., and Carr, N. G. (1981). *Ann. N.Y. Acad. Sci.* **361,** 298–311.
Pianka, E. R. (1976). *Am. Zool.* **16,** 775–784.
Pickett-Heaps, J. D. (1975). "Green Algae: Structure, Reproduction and Evolution in Selected Genera," pp. 1–606. Sinnauer, Sunderland, Massachusetts.
Pigott, G. H., and Carr, N. G. (1972). *Science* **175,** 1259–1261.
Pinevich, A. V., and Desnizky, A. G. (1979). *Fitologia* **21,** 755–766.
Pool, R. R., Jr. (1979). *J. Cell Sci.* **35,** 367–379.
Preer, L. B., and Preer, J. R., Jr. (1977). *In* "Cell Biology: A Comprehensive Treatise" (L. Goldstein and D. M. Prescott, eds.), Ch. 8, pp. 329–373. Academic Press, New York.
Prusiner, S. B. (1982). *Science* **216,** 136–144.
Raff, R. A., and Mahler, H. R. (1975). *Symp. Soc. Exp. Biol.* **29,** 41–92.
Raven, P. H. (1970). *Science* **169,** 641–646.
Reanney, D. C. (1974). *Int. Rev. Cytol.* **37,** 21–52.
Reisser, A. (1981). *Ber. Dtsch. Bot. Ges.* **94**(3), 557–565.
Richmond, M., and Smith D. C. (1979). *Proc. R. Soc. London Ser. B* **204,** 113–286).
Sagan, L. (1967). *J. Theor. Biol.* **14,** 225–274.
Schell, J., Van Montagu, M., DeBeuckeleer, M., DeBlock, M., Depicker, A., DeWilde, M., Engler, G., Genetello, C., Hernalsteens, J. P., Holsters, M., Seurinck, J., Silva, B., Van Vliet, F., and Villarroel, R. (1979). *Proc. R. Soc. London Ser. B* **204,** 251–266.
Schiff, J. A., ed. (1982). "On the Origin of Chloroplasts," in press.
Schoenberg, D. A. (1980). *Proc. Int. Colloq. Endosymb. Cell Res., Tübingen* pp. 145–154, 155–162.
Schwartz, R. M., and Dayhoff, M. O. (1981). *Ann. N.Y. Acad. Sci.* **361,** 260–273.
Schwemmler, W. (1974). *J. Insect Physiol.* **20,** 1467–1474.
Schwemmler, W. (1979). "Mechanismen der Zellevolution", pp. 1–276. deGruyter, Berlin.
Schwemmler, W. (1980). *BioSystems* **12,** 111–122.
Schwemmler, W. (1981). *Ber. Dtsch. Bot. Ges.* **94**(3), 591–599.
Schwemmler, W. and Schenk, H. E. A. (1980). *Proc. Int. Colloq. Endosymb. Cell Res., Tübingen* pp. 1–1060.
Seewaldt, E., and Stackebrandt, E. (1982). *Nature (London)* **295,** 618–620.
Smith, D. C. (1979). *Proc. R. Soc. London Ser. B* **204,** 115–130.
Smith, D. C. (1980). *In* "Cellular Interactions in Symbiotic and Parasitic Associations" (C. B. Cook, E. D. Rudolph, and I. Pappas eds.), pp. 1–305. Ohio State Univ. Press, Columbus.
Soldo, A. T., and Godoy, G. P. (1973). *J. Mol. Biol.* **73,** 93–108.
Spencer, D. F., Bonen, L., and Gray, M. W. (1981). *Biochemistry* **20,** 4022–4029.
Stanier, R. Y. (1970). *Microb.* **20,** 1–38.

Steele, R. D. (1976). *J. Zool. London* **179,** 387–405.
Stiven, A. E. (1965). *J. Invert. Pathol.* **7,** 356–367.
Stolp, H. (1979). *Proc. R. Soc. London Ser. B* **204,** 211–217.
Taylor, D. L. (1973). *Adv. Mar. Biol.* **11,** 1–56.
Taylor, D. L. (1981). *Ber. Dtsch. Bot. Ges* **94**(3), 583–591.
Taylor, F. J. R. (1974). *Taxon* **23,** 229–258.
Taylor, F. J. R. (1976). *Taxon* **25,** 377–390.
Taylor, F. J. R. (1978). *BioSystems* **10,** 1–23.
Taylor, F. J. R. (1979). *Proc. R. Soc. London Ser. B* **204,** 267–286.
Taylor, F. J. R. (1980). *BioSystems* **13,** 65–108.
Taylor, F. J. R. (1982). *Ann. Inst. Oceanogr. Paris* **57,** 37–66.
Thomas, L. (1970). *Keynote Addr. Symp. 3rd, Int. Inflammation Club* pp. 1–11.
Thornley, A. L. and Harrington, A. (1981). *J. Theor. Biol.* **91,** 515–523.
Trench, R. K. (1979). *Annu. Rev. Plant Physiol.* **30,** 485–531.
Trench, R. K., Colley, N. J., and Fitt, W. K. (1981). *Ber. Dtsch. Bot. Ges.* **94**(3), 529–547.
Uzzell, T., and Spolsky C. (1981). *Ann. N.Y. Acad. Sci.* **361,** 481–499.
Van Etten, J. L., Meints, R. H., Kuczarski, D., Burbank, D. E., and Lee, K. (1982). *Proc. Natl. Acad. Sci. U.S.A.,* in press.
Waddington, C. H. (1975). "The Evolution of an Evolutionist," pp. 1–328. Cornell Univ. Press, Ithaca, New York.
Wallin, I. E. (1927). "Symbionticism and the Origin of Species," pp. 1–171. Bailliere, London.
Weare, N. M., Azam, F., Mague, T. H., and Holm-Hansen, O. (1974). *J. Phycol.* **10,** 369–371.
Weeden, N. F. (1981). *J. Mol. Evol.* **17,** 133–139.
Weis, D. L. (1981). *Ber. Dtsch. Bot. Ges.* **94**(3), 547–557.
Weissner, W., and Lorenzen, H. (1981). *Ber. Dtsch. Bot. Ges.* **94**(3), 325–333.
Whatley, F. R., and Whatley, J. M. (1980). *Proc. Int. Colloq. Endosymb. Cell Res., Tübingen* pp. 849–858.
Whatley, J. M. (1976). *New Phytol.* **76,** 111–118.
Whatley, J. M. (1977). *New Phytol.* **79,** 309–313.
Whatley, J. M., and Whatley, F. R. (1981). *New Phytol.* **87,** 233–247.
Withers, N. W., Alberte, R. S., Lewin, R. A., Thornber, J. P., Britton, G., and Goodwin, T. W. (1978). *Proc. Natl. Acad. Sci. U.S.A.* **75,** 2301–2305.
Woese, C. R. (1981). *Sci. Am.* **244,** 98–122.

Integration of Bacterial Endosymbionts in Amoebae

KWANG W. JEON

Department of Zoology, University of Tennessee, Knoxville, Tennessee

I. Introduction

It has long been recognized that cells can be suitable habitats for smaller microbes, and intracellular symbionts are found in a wide variety of host cells (e.g., Jennings and Lee, 1975; Richmond and Smith, 1979; Cook *et al.*, 1980). Many of them are parasitic, while others are innocuous without known functions (Beale and Jurand, 1969; Preer, 1971) or even beneficial (Karakashian, 1963; Muscatine and Lenhoff, 1965; Van Wagtendonk, 1963; Chang *et al.*, 1975). In any case, hose cells provide a ''shelter'' and supply important material needs for the symbionts. Following a long period of symbiosis, the partners may become mutually dependent and interact as a single functioning unit. Such functional integration is regarded as an important mechanism whereby cells can increase their degree of complexity and become diversified (see Taylor, 1974, 1979; Margulis, 1976, 1981; Whatley *et al.*, 1979).

The establishment of stable endosymbiosis is always preceded by a series of complex interactions between the prospective hosts and symbionts. These interactions include surface recognition, entry of one partner into the other, avoidance of destruction by the host partner, and sustained multiplication of both partners (cf. Bannister, 1979). The whole process of establishing stable symbiosis usually requires myriads of cell generations, and elucidation of the under-

ISBN 0-12-364375-9

lying mechanisms is rarely possible. In a few cases, however, the process of symbiont integration has been studied in some detail, e.g., amoebae (Ahn and Jeon, 1979), hydra (Muscatine *et al.*, 1975; Cook, 1980), legumes (Beringer *et al.*, 1979), and paramecia (Karakashian, 1975; Weis and Ayala, 1979).

The purpose of this article is to review known facts about the process and results of symbiont integration using the amoeba–bacteria symbiosis as a model, in which the two partners developed a stable symbiosis within a short period of time under laboratory conditions. In 1966, a strain of *Amoeba proteus* which had been under continuous laboratory culture for over 20 years became spontaneously infected with a large number of rod-shaped bacteria (Jeon and Lorch, 1967). The infecting bacteria were initially virulent to the host cells. However, some amoebae survived and within a few years the infecting bacteria appeared to have become less virulent in that the initially observed adverse effects on their host diminished. Symbiotic bacteria became integrated in xD amoebae as required cell components by 1971 (Jeon, 1972), that is, the host amoebae were unable to survive without their bacterial endosymbionts. Currently, symbiotic xD amoebae harbor an average of 42,000 symbionts each and grow well (Ahn and Jeon, 1979). In addition, it has been demonstrated that a different strain of amoebae (nonsymbiotic D) can be induced to establish obligatory symbiosis with X-bacteria isolated from the original xD amoebae (Jeon and Ahn, 1978). The new host becomes dependent on symbionts in about 200 cell generations. Other physiological changes are brought about much sooner, e.g., within a few cell divisions, following the introduction of X-bacteria into nonsymbiotic amoebae (Lorch and Jeon, 1981, 1982). This amoeba–bacteria symbiosis offers extraordinary opportunities for studying the establishment of new symbiotic associations and host–symbiont interactions.

Throughout the article, the following symbols and terms will be used: D amoebae, nonsymbiotic D strain of *A. proteus;* xD amoebae, symbiotic amoebae, originally derived from D amoebae; X-bacteria, symbiotic bacteria present in xD amoebae; SVs, symbiont-containing vesicles in the cytoplasm of xD amoebae; suffixes n and c, nucleus and cytoplasm, respectively, used in micromanipulation experiments.

II. Symbiotic Bacteria

The symbiotic bacteria have not been characterized taxonomically because it has not been possoble to grow them *in vitro* (Section VII). The X-bacteria are gram-negative rods with an average size of 0.5×2 μm. Their ultrastructure is like that of normal *E. coli* (Morgan *et al.*, 1967) or symbiotic bacteria present in other cells (e.g., Chang, 1974; Preer *et al.*, 1974), with dense cytoplasmic matrix and clear nucleoid zones. These bacteria are enclosed in vesicles of

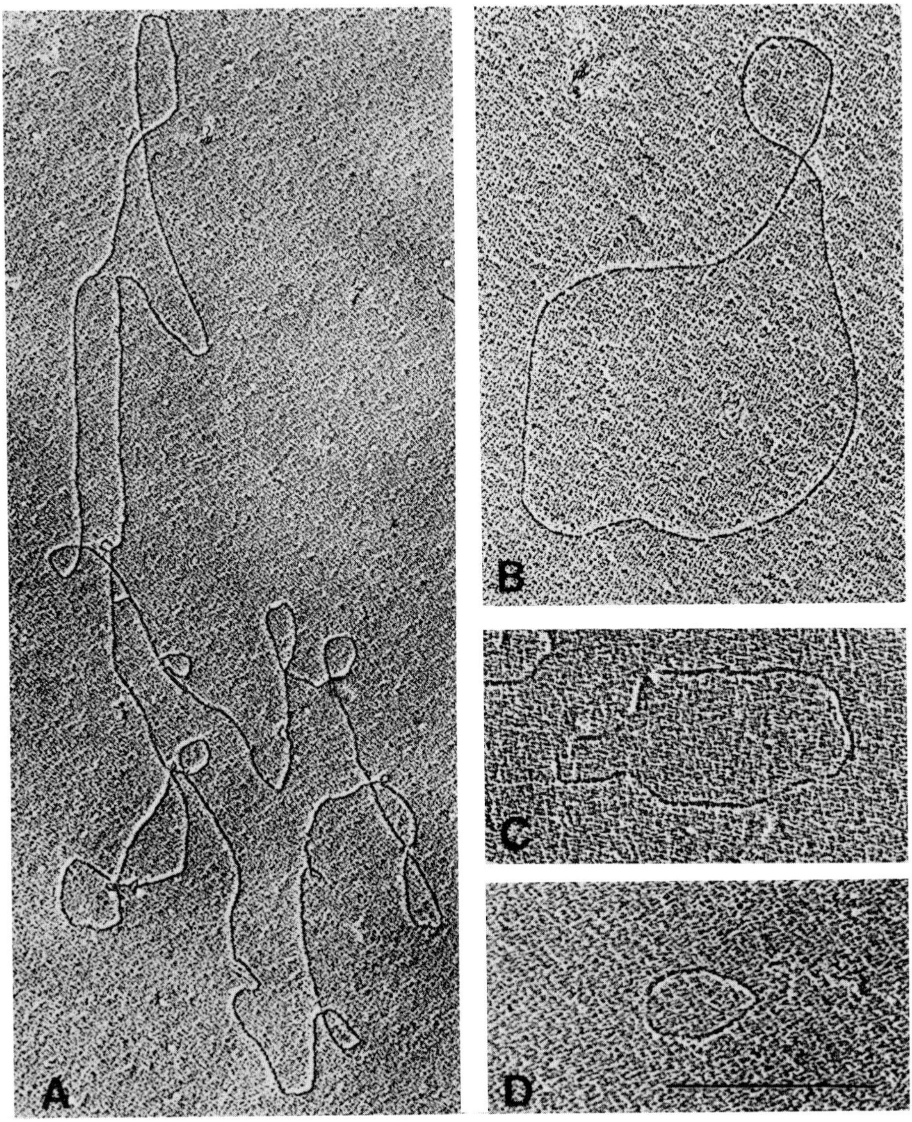

FIG. 1. Electron micrographs of plasmid DNAs from X-bacteria and *E. coli*. (A) Large plasmid DNA (pHJ11) of X-bacteria (contour length, 19.08 μm). (B) Small plasmid DNA (pHJ12) of X-bacteria (contour length, 6.84 μm). (C) RSF2124 plasmid of *E. coli* (contour length, 3.66 μm). (D) pBR322 plasmid of *E. coli* (contour length, 1.33 μm). Bar represents 1 μm. Reproduced from Han and Jeon (1980) by permission.

varying sizes. In large vesicles, bacteria are embedded among irregular membrane-bounded vesicles containing fibrous material of unknown nature (Jeon and Hah, 1977).

The symbiotic bacteria contain two kinds of plasmid DNAs, designated as pHJ11 and pHJ12, with apparent molecular weights of 39 and 14 million (Fig. 1; Han and Jeon, 1980). Symbionts are sensitive to antibacterial agents such as chloramphenicol (Jeon and Hah, 1977) or trimethoprim (Jeon, 1977). The adverse effect of trimethoprim, an inhibitor of dihydrofolate reductase, underscores the finding that X-bacteria do not utilize exogenous thymine during DNA synthesis (J. Freesh and Jeon, unpublished work).

Individual X-bacteria can be isolated from xD amoebae by combining the procedures of freezing and thawing, and ultrasonication with centrifugation on a Ficoll gradient (Han and Jeon, 1980). Up to 95% of the total X-bacteria present can be released and collected by this procedure. When isolated X-bacteria are injected into rabbits, antibodies are produced that specifically react with X-bacteria as tested by the indirect fluorescent antibody test using fluorescein-conjugated sheep anti-rabbit IgG (J. Tullock and Jeon, unpublished work). However, X-bacteria appear to be less antigenic than other free-living bacteria since there are no agglutinating antibodies in the immune sera. The anti-X-bacteria serum was tested for cross-reactivity for the purpose of identifying a possible precursor strain of bacteria that changed to X-bacteria, but none of the 15 strains of free-living gram-negative bacteria in the amoeba culture dishes reacted positively. Thus, it has not been possible to identify a potential precursor strain of X-bacteria which may be still living in the amoeba culture media.

III. Entry of Endosymbionts into the Host

Various aspects of the entry of symbionts into the host cells have been discussed by several authors (Bannister, 1979; Cook, 1980; Smith, 1979, 1980). The common modes of entry are phagocytic or nonphagocytic engulfment and active penetration by the symbiont. It has been pointed out that the initial surface interaction between the symbiont and host precedes the ensuing entry of symbionts in many cases, e.g., chlorella and hydra (Muscatine *et al.*, 1975; Pool and Muscatine, 1980), chlorella and paramecia (Weis, 1980), legumes and *Rhizobium* (Beringer *et al.*, 1979), protozoa (Bannister, 1979), and spirochaetes and protozoa (Margulis *et al.*, 1979).

In amoebae, no specific surface interaction appears to be used for recognizing potential symbionts. Any exogenous material present is ingested during phagocytosis, although certain stimuli are needed for the induction of phagocytosis (Ahn and Jeon, 1979). It is common to find free-living bacteria ingested together with food organisms even in species such as *A. proteus* that do not use bacteria as their major nutritional source. There has been no report of any prospective symbionts actively penetrating the amoeba membrane to gain entry.

The only alternative mode of entry possible in amoebae is experimental injection of symbionts into the cytoplasm. Various strains of bacteria have been introduced into amoebae for different purposes (Casely-Smith and Savanat, 1966; Jeon and Lorch, 1967; Jeon and Jeon, 1976; Ahn and Jeon, 1979). When free-living bacteria are injected into the amoeba cytoplasm, they become enclosed in vaculoe membranes and are eventually digested. However, symbiotic bacteria from the xD strain of amoeba survive in the new host cell even when they become initially enclosed in vesicles following injection. When injecting X-bacteria isolated from xD amoebae into other strains of amoebae, it is important to wash the bacteria free of xD amoeba cytoplasm so that the injected hosts will not be affected by the strain-specific lethal factor present in each strain of amoebae (Jeon and Lorch, 1969). Otherwise, the injected host amoebae lose viability before establishing symbiosis with the injected X-bacteria.

IV. Establishment of Symbiosis

It has been well recognized that one of the prerequisites for endosymbionts to establish themselves in a new host is to evade intracellular destruction by the host cell. Then the symbionts must multiply, keeping a numerical balance with their host.

A. Avoidance of Digestion

Several different mechanisms have been proposed by which symbionts may evade digestion by their hosts. They include resistance to lysosomal enzymes, inhibition of lysosomal fusion, and escape from the vaculoe system (Bannister, 1979). Among these possible mechanisms, endosymbionts in amoebae appear to use the first two methods: resistance to lysosomal enzymes and inhibition of lysosomal fusion.

1. *Resistance to Lysosomal Enzymes*

Following entry into an amoeba by induced phagocytosis, bacterial symbionts are found in phagosomes together with other ingested material (Ahn and Jeon, 1979; Fig. 2). Digestible material is hydrolyzed within the secondary lysosomes and the residue is defecated within a few days. In contrast, symbiotic bacteria become segregated in separate vesicles in which they multiply. Thus, it is clear that at least some of the bacteria are resistant to lysosomal enzymes. In addition, the symbiotic bacteria have been found to resist the actions of common lysing agents such as lysozyme and weak detergents *in vitro* (Han and Jeon, 1980),

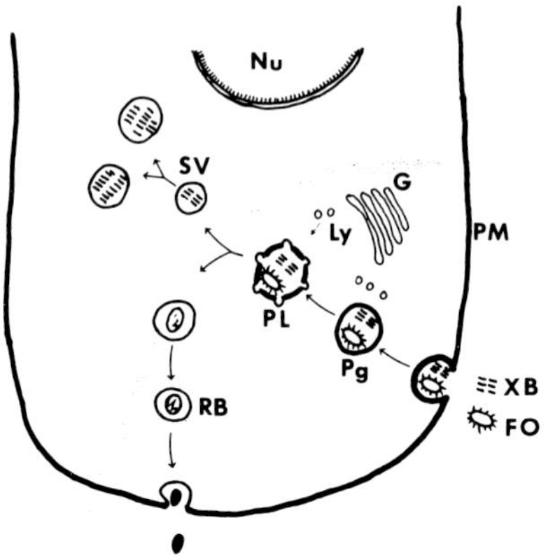

FIG. 2. Schematic diagram to show the apparent membrane differentiation and flow in xD amoebae. FO, Food organism; G, Golgi complex; Ly, lysosomes; Nu, nucleus; Pg, phagosome; PL, phagolysosome; PM, plasmalemma; RB, residual bodies; SV, symbiont-containing vesicles.

suggesting that the walls of symbiotic bacteria are nonspecifically resistant to lysing actions of various agents.

It is not known what contributes to the observed resistance of X-bacteria to digestive enzymes of amoebae *in vivo* and to lysing agents *in vitro*. Meanwhile, the suggestion has been made that the plasmid DNAs found in X-bacteria (Han and Jeon, 1980) encode some protein components of the bacterial outer membranes, as has been found to be the case in other organisms (Costerton, 1977; Olsen *et al.*, 1977). Further work is needed to identify such components and clarify their roles. Another useful approach would be to introduce the plasmid DNAs from X-bacteria into other bacteria such as *E. coli* which are known to be digested by amoebae and see if they could be transformed to resist digestion by lysosomal enzymes in amoebae.

A few other kinds of protozoan parasites such as *Leishmania* (Alexander and Vickerman, 1975; Chang and Dwyer, 1976; Chang, 1981), *Trypanosoma* (Nogueira and Cohn, 1976; Kress *et al.*, 1977), and *Mycobacterium* (Brown *et al.*, 1969; Armstrong and Hart, 1975) have been shown to resist digestion by lysosomal enzymes. These organisms are considered to avoid being digested either by producing an enzyme inhibitor or by having resistant membrane component(s) in their surfaces.

2. *Inhibition of Lysosomal Fusion*

Once segregated into separate vesicles, the symbiotic bacteria of xD amoebae do not come into contact with lysosomal enzymes because lysosomes do not fuse with vesicles containing these bacteria (Jeon and Jeon, 1976; Ahn and Jeon,

1982; Fig. 2). Thus, these bacteria somehow prevent the fusion between the SV membranes and lysosomes. It is only when xD amoebae are kept at or above 26°C that symbionts within vesicles are digested by amoebae (Lorch and Jeon, 1980). It is interesting to note that the symbionts from xD amoebae cultured at elevated temperatures lack plasmids (Han and Jeon, unpublished work), suggesting that the plasmids play a role in the prevention of lysosomal fusion with SV membranes.

There are other examples in which endosymbionts are known to prevent lysosomal fusion, e.g., *Toxoplasma* (Jones and Hirsch, 1972), *Encephalitozoon* (Weidner, 1976), and *Paramecium* (Karakashian and Karakashian, 1973; Karakashian and Rudzinska, 1981).

The lack of lysosomal fusion with SVs might be due to either the presence of a component in the vesicle membranes which actively inhibits lysosomal recognition and fusion or the absence of a component needed in the fusion process. In an attempt to elucidate the mechanism for the prevention of lysosomal fusion, Ahn and Jeon (1982) compared the protein composition of the SV membranes with that of the plasmalemma and of induced phagolysosomal membranes by means of SDS–polyacrylamide gel electrophoresis and autoradiography. They isolated the plasmalemma, SVs, and phagolysosomes containing polystyrene spheres, and subjected solubilized membranes to electrophoresis. They found that SV membranes contained a high-molecular-weight polypeptide which was not present in either the plasmalemma or phagolysosomal membranes (Fig. 3). This polypeptide was located on the cytoplasmic side of SV membranes as determined by labeling isolated SVs with ^{125}I before solubilizing the membrane proteins for electrophoresis. However, the origin of and specific roles played by this unique polypeptide have not been determined.

The vacuole–membrane system in xD amoebae represents a useful model with which to study membrane differentiation. The plasmalemma, which apparently does not fuse with lysosomes, invaginates to form phagocytic vaculoes that normally fuse with lysosomes. A portion of the phagolysosomal membranes gives rise to SV membranes that contain unique polypeptide components and avoid lysosomal fusion (Fig. 2). It should be possible to determine at what stage of SV formation the SV-specific polypeptide components appear, for example, by examining the membrane polypeptides at intervals following experimental induction of SVs and to clarify the roles played by these components in the prevention of lysosomal fusion with SVs.

B. Symbiont Multiplication

The multiplication of symbionts within established xD amoebae is difficult to study due to the lack of a suitable method, but newly infected D amoebae can be used effectively to examine the kinetics of symbiont multiplication. Once segre-

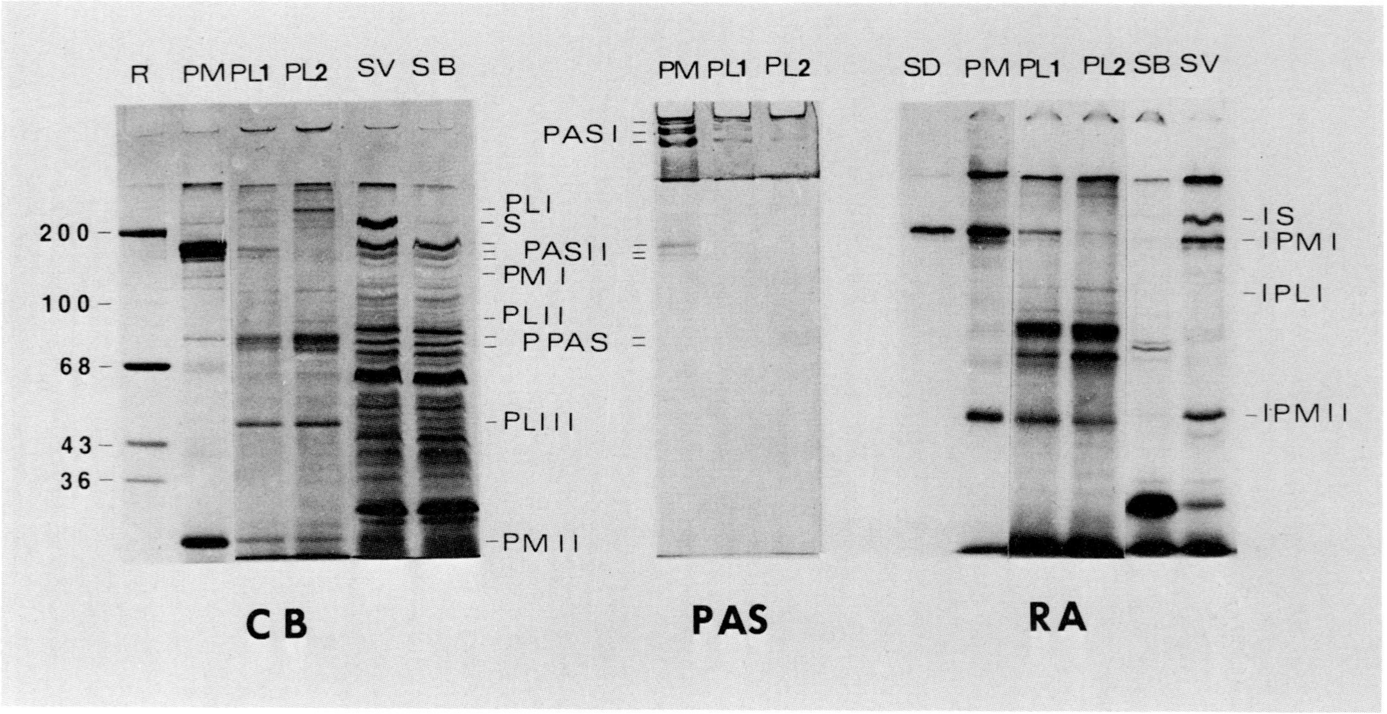

FIG. 3. Electrophoretic pattern of isolated subcellular fractions of amoebae, after staining with Coomassie brilliant blue (CB) or periodic acid–Schiff reagent (PAS), and after autoradiography (RA). R, Molecular weight standards (1×10^{-3}); PM, isolated plasmalemma; PL1, phagolysosomes isolated 1 hr after ingestion; PL2, phagolysosomes isolated 2 hr after ingestion; SV, isolated symbiont-containing vesicles; SB, isolated symbiotic X-bacteria; SD, sediment of amoebae (centrifugation for 10 minutes at 1400 *g*) that were iodinated before lysis, and thus only the polypeptides on the external surface would have been labeled. Reproduced from Ahn and Jeon (1982) by permission.

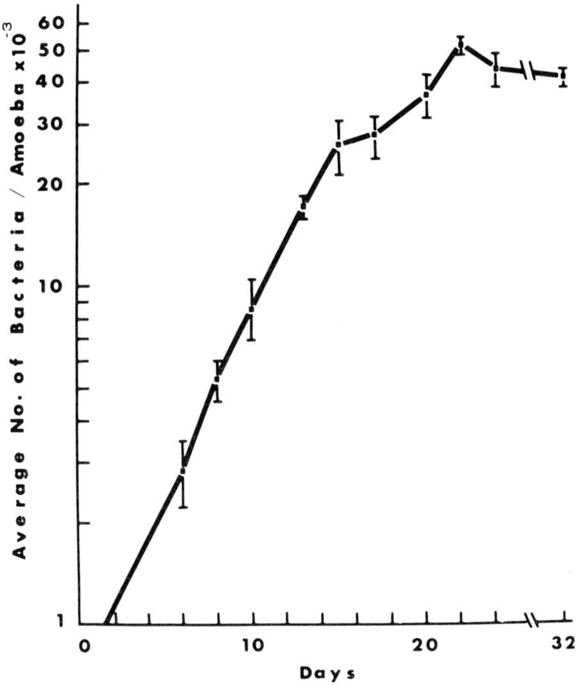

FIG. 4. Graph to show the increase in the average number of X-bacteria per newly infected D amoeba. The mean doubling time of the average number of X-bacteria per amoeba was 1.22 days for the first 15 days, during which the newly infected D amoebae multiplied with a mean generation time of 2.55 days. Reproduced from Ahn and Jeon (1979) by permission.

gated in separate SVs from phagolysosomes, symbiotic bacteria divide by the usual binary fission, multiplying at a rate faster than that of the host until the maximum carrying number is reached (Ahn and Jeon, 1979). When the initial number of bacteria introduced into each D amoeba is about 300, they divide twice as fast as the host amoebae and reach the maximum number of 42,000 per cell within 3 weeks (Fig. 4). It is not known if all the introduced bacteria survive and multiply. The minimum number of bacteria required to infect an amoeba is also unknown. A maximum carrying number of about 42,000 symbionts per amoeba is maintained both in established xD amoebae and in newly infected amoebae. Thus, it is obvious that the host exerts a control over the symbiont population, but the mechanism for such regulation is not known.

A host's control over the symbiont population is expected to be operative in any stable symbiotic association, and at least three mechanisms have been proposed to account for the observed control: expulsion of symbionts (Taylor, 1969; Steele, 1976), digestion (Fankboner, 1971; Weis, 1976), and regulation of symbiont growth (Pardy and Heacox, 1976; Muscatine and Pool, 1979; Reisser,

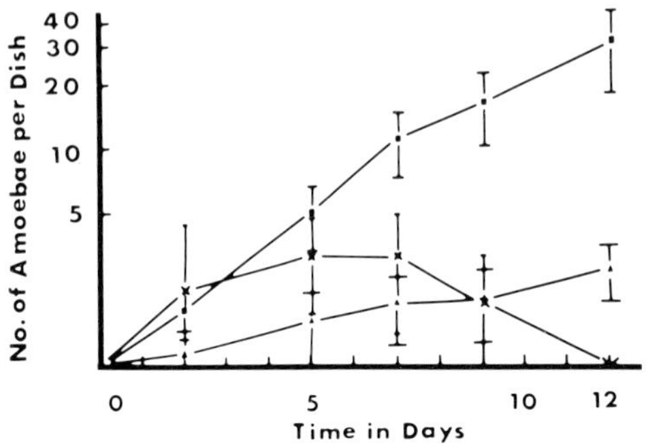

FIG. 5. Growth rates of xD amoebae at different temperatures: 13°C (▲), 20°C (■), and 26°C (×). During the 12-day culture period the mean generation time of cells grown at 13 and 20°C was 8.1 and 2.3 days, respectively. Cells grown at 26°C died out by the twelfth day. Reproduced from Jeon and Ahn (1978) by permission.

1980; McAuley, 1981). It has been difficult to decide which one or combination of these mechanisms might operate in the amoeba–bacteria association, but there is indirect evidence that some symbionts are released into the medium and that the symbionts' growth is regulated. For example, D amoebae become infected with X-bacteria after being kept with xD amoebae in the same dish for a few days (Jeon, unpublished work). When xD amoebae are cultured at a lower temperature, e.g., 15°C, the number of symbionts per amoeba increases (Fig. 5; Jeon and Ahn, 1978), indicating that amoebae do not have the same control over symbiont population when culture conditions are different.

V. Influence of Symbionts on the Host

The presence of endosymbionts in a host cell is bound to affect the latter. Thus, for example, kappa particles confer specific resistance to paramecia bearing them against the kappa toxin (Preer *et al.*, 1974) and symbiotic bacteria of *Crithidia* cause a change in the surface membrane carbohydrates of the hosts (Dwyer and Chang, 1976). In the case of amoebae, several morphological, physiological, and biochemical effects have been found.

A. Effect on Cell Morphology

The gross morphology of symbiotic amoebae seen under a dissecting microscope is not noticeably different from that of nonsymbiotic amoebae. However, an xD amoeba with the full complement of symbiotic bacteria can easily be distinguished from a nonsymbiotic amoeba by the presence of numerous vesicles visible under dark-field illumination or a phase-contrast microscope (at ×450 or higher magnifications) (Jeon and Lorch, 1967; Ahn and Jeon, 1979). Small SVs with only a few X-bacteria can be detected a week after experimental infection of nonsymbiotic amoebae. Under the electron microscope, however, such small vesicles can be detected as early as 2 days after infection. Apart from the presence of SVs, there are no detectable structural changes that can be attributed to the symbionts.

B. Physiological Effects

1. *Growth Rates*

During the initial phase of experimental infection for up to 12 months, infected amoebae have been found to grow faster than either established symbiotic amoebae or nonsymbiotic amoebae (Jeon and Ahn, 1978). In one series of experiments, the mean generation time of newly infected amoebae was 1.8 days as compared to 2.1 days for nonsymbiotic amoebae and 2.5 days for established xD amoebae. The same effect has been consistently observed in newly infected D amoebae (Lorch and Jeon, unpublished work), but reasons for the enhanced growth are not known.

2. *Sensitivity to Starvation*

Sensitivity to starvation was one of the first adverse effects noticed during the initial bacterial infection (Jeon and Lorch, 1967), and is still exhibited by symbiotic amoebae although to a lesser degree. In 1966, 50% of xD amoebae died after a weeks starvation, as compared to 5% of noninfected D amoebae. In one series of experiments conducted in 1982, 13% of xD amoebae cytolyzed after a week without feeding, while 5% of control amoebae did so (Jeon, unpublished work). The reason for the increased sensitivity to starvation shown by symbiotic amoebae is not known, but it appears that nutritional requirement of the endosymbionts is an added burden to the host cells.

3. *Sensitivity to Overfeeding*

Continued overfeeding with *Tetrahymena* has been known to be harmful to amoebae (Griffin, 1960, 1973; Prescott and Carrier, 1964). However, overfed amoebae can be rescued by a thorough washing, if they have not been damaged beyond recovery. The symbiotic xD amoebae are extremely sensitive to overfeeding such that many of them become irreversibly damaged after being with excess *Tetrahymena* for only 2 consecutive days. They become detached from the substratum and round up with dark condensed cytoplasm. Then, most of them cytolyze within a few days without returning to normal forms. It appears that this increased sensitivity is related to the overall fragility of the plasmalemma of symbiotic amoebae, which is especially noticeable during micrurgical operations (Lorch and Jeon, unpublished work).

4. *Effect of Elevated Culture Temperature*

Several strains of amoebae have been known to grow well at temperatures up to 27°C (Hawkins and Danielli, 1963). It has been found that xD amoebae cannot grow at temperatures above 26°C (Fig. 6) and all die within 2 weeks (Jeon and Ahn, 1978). The death of amoebae is preceded by the disappearance of symbiotic bacteria, apparently by digestion (Lorch and Jeon, 1980). Since these amoebae are dependent on their symbionts for survival (Section VII), it follows that they die after losing their symbionts. The symbiotic bacteria appear to be digested by amoebae in turn after the symbionts have lost the plasmids they normally harbor (cf. Fig. 1; Han and Jeon, 1980). Thus, it appears that the temperature sensitivity of symbiotic amoebae can be traced back to the psychrophilic properties of the plasmid DNAs present in the symbiotic bacteria.

5. *Sensitivity to Crowding*

Symbiotic amoebae are much more sensitive to crowding such as under the mass-culturing conditions (Griffin, 1973; Goldstein and Ko, 1976) than are normal amoebae. Thus, while D amoebae may keep growing at a cell density of over 3000 amoebae/cm^2 for several days, the maximum density at which xD amoebae may remain healthy for more than a few days is 1000 amoebae/cm^2; when the density is higher than this, many cells detach from the substratum, round up, and cytolyze within a few days. The reasons for the increased sensitivity are unknown.

6. *Nucleocytoplasmic Compatibility*

A small percentage (7%) of the nuclei of established xD amoebae was found to be compatible with the cytoplasm of D amoebae in 1971 (Jeon, 1972). At present, few of the nuclei of established xD amoebae are compatible with D cytoplasm (Lorch and Jeon, 1980,1981,1982).

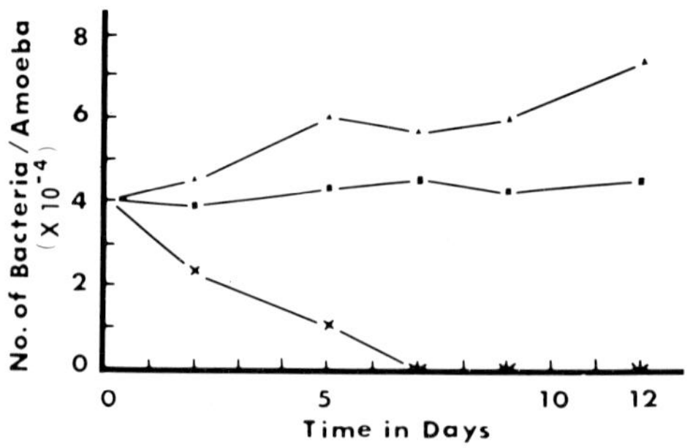

FIG. 6. The number of X-bacteria in xD amoebae grown at different temperatures: 13°C (▲), 20°C (■), and 26°C (×). The average number of endosymbionts per amoeba grown at 20°C was about 42,000. Reproduced from Jeon and Ahn (1978) by permission.

In order to determine how long it takes for the nuclei of D amoebae to become incompatible with D cytoplasm under the influence of X-bacteria, Lorch and Jeon (1981) backtransferred nuclei of D amoebae that had been in symbiosis for 4 to 15 weeks into D cytoplasm. They found that fewer than 5% of transplanted nuclei were viable. When D nuclei were subjected to the influence of a full complement of X-bacteria by transplanting the former into xD cytoplasm, most of the nuclei (over 95%) became incompatible with the D cytoplasm within a few days (Lorch and Jeon, 1982). In contrast, many of the affected nuclei were compatible with xD cytoplasm, indicating that these nuclei had not lost viability. Thus, the influence of X-bacteria in causing the changes in nucleocytoplasmic compatibility appears to be almost instant, but the mechanism remains unknown.

7. *Nuclear Lethal Effect*

One of the more dramatic effects of the presence of symbionts on amoeba nuclei is the acquisition by the nuclei of the property of exerting a lethal effect on D amoebae, the original strain. When the nucleus of an xD amoeba is transplanted into a D amoeba with its own nucleus, the latter loses viability (Lorch and Jeon, 1981). If D amoebae are infected with X-bacteria by intracellular injection, within 4 to 9 weeks, their nuclei change to have a strong lethal effect on D amoebae. In contrast, D nuclei transplanted into the cytoplasm of xD amoebae acquire the same lethal effect much sooner; more than 70% of transplanted D nuclei become lethal to D amoebae after only a few mitotic cycles (Lorch and Jeon, 1982).

While the precise mechanism is unknown, the interstrain lethal effect has been attributed to strain-specific high-molecular-weight proteins, called lethal factors (Jeon and Lorch, 1969), that are present in amoeba nuclei as well as in the cytoplasm (Jeon and Lorch, 1970). If the observed lethal effect of xD amoeba nuclei on D amoebae were caused by a similar protein, it would mean that xD amoebae synthesize a new protein. Indeed, the cytoplasm of newly infected D amoebae as well as that of established xD amoebae has been found to contain a unique polypeptide with a molecular weight of about 29,000, which is detectable by two-dimensional electrophoresis in polyacrylamide (Jeon *et al.*, 1981). Preliminary results suggest that this polypeptide is produced by xD amoebae and not by symbiotic bacteria. It appears that X-bacteria cause a change in the protein synthesis by their new hosts. There are other examples where exogenous components such as viruses (Croce, 1981), infective microorganisms (Schell *et al.*, 1979; Trench, 1980), and transferred cell constituents (Ruddle, 1977; Celis *et al.*, 1980), alter cellular synthetic activities.

VI. Specificity of Symbiotic Relationship

The specificity of the symbiotic relationships between symbionts and their hosts varies widely depending on organisms involved. Some protozoan parasites have a rather wide range of host species and host cells (Bannister, 1979), while others have a more strict specificity (Pearch and Lowrie, 1972; Muscatine *et al.*, 1975). An example of an intermediate degree of specificity is found in *Paramecium bursaria* (Karakashian, 1975; Weis, 1980).

Specificity of symbiotic relationship could operate at various steps during the establishment of symbiosis, viz., initial recognition, uptake, retention, and subsequent regulation. For example, hydra phagocytes distinguish their native symbiotic algae from others first by ingesting more of them and then by preferred retention afterward (Muscatine *et al.*, 1975; Pool and Muscatine, 1980).

In the case of amoebae, since potential symbionts are indiscriminately ingested, whether they establish a stable symbiosis or not would depend on their ability to resist digestion or expulsion by the host and on their subsequent multiplication. The X-bacteria have been introduced to several different types of cells ranging from giant amoebae to amphibian eggs. The symbionts of xD amoebae can establish stable symbiosis with other strains of *A. proteus* (Jeon and Jeon, 1976; Jeon, unpublished work) and *Chaos carolinensis* (Jeon and Jeon, 1982). However, they do not infect ciliates such as *Tetrahymena, Colpidium, Blepharisma,* and *Paramecium* or *Xenopus* eggs (Jeon, unpublished work). It appears that X-bacteria can multiply only in amoeboid protozoans.

VII. Mutual Dependence between the Host and Symbionts

Symbiotic bacteria in amoebae have been dependent on their host for survival since the early stage of their association with amoebae, as evidenced by their failure to grow in various media under all conditions tested (Jeon and Lorch, 1967). The tested media include the charcoal–yeast extract agar used to grow *Legionella* bacteria (Jeon, unpublished work).

The first indication that amoebae were dependent on their symbionts came from nuclear transplantation experiments (Jeon, 1972). When nuclei of xD amoebae were transferred into the cytoplasm of D amoebae, over 90% of the resulting xD_nD_c cells were nonviable whereas the reciprocal heterotransplants (D_nxD_c) formed viable clones. The nonviable xD_nD_c cells could be resuscitated only by transferring live X-bacteria into them, e.g., by microinjection (Jeon and Jeon, 1976). Later, other methods were adopted to selectively remove X-bacteria from symbiotic amoebae, such as treatment with chloramphenicol (Jeon and Hah, 1977), trimethoprim (Jeon, 1977), or culturing amoebae at or above 26°C (Jeon and Ahn, 1978).

At elevated temperatures (Section V,B,4), symbionts completely disappear after 7 days, the death of xD amoebae following within a week. Thus, when cultured at 26°C, xD amoebae yield a growth pattern (Fig. 7) very similar to that of aging cultured mammalian cell lines, i.e., Phase III phenomenon (Hayflick,

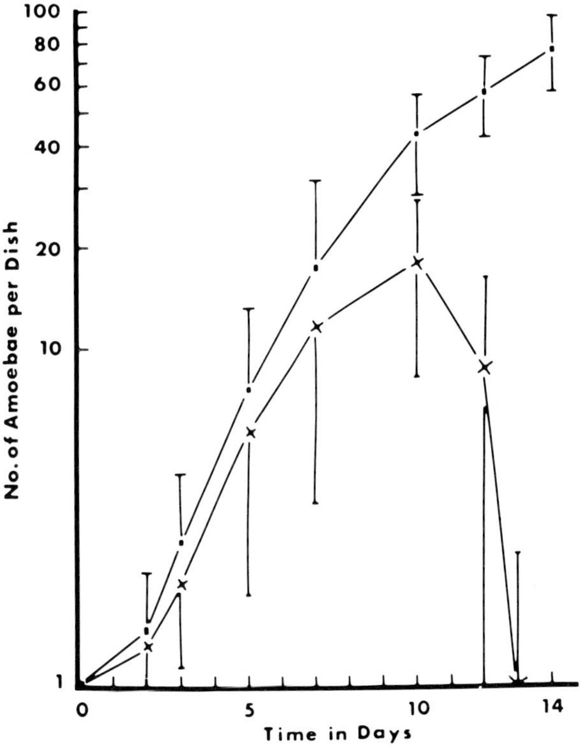

FIG. 7. Growth rates of D (■) and xD (×) amoebae at 26°C. All xD amoebae cytolyzed by the thirteenth day. Reproduced from Jeon and Ahn (1978) by permission.

1965). When survival at an elevated culture temperature is used as the criterion for dependence, it is found that any D amoebae that have been in symbiosis for 18 months or longer are dependent on their symbionts. Amoebae that have been in symbiosis for shorter periods of time are able to go on multiplying after losing their symbionts. Other methods of removing X-bacteria, e.g., treatment with chloramphenicol, have not been tried on the newly infected amoebae to check their viability after the removal of symbionts. However, it is assumed that the viability of aposymbiotic amoebae thus obtained would be the same as that of heat-treated amoebae. While the mode of the development of the host's dependence on its symbionts is not known, it is interesting to note that the compatibility of the cell nucleus of an infected amoeba with the cytoplasm of the original strain changes much sooner, viz., within a few weeks (Section IV,B,6). In other words, the nucleus of a newly infected D amoeba is viable in its own cytoplasm deprived of endosymbionts but is not in the cytoplasm of another D amoeba. This suggests that the cytoplasm of an amoeba that has once harbored X-bacteria is no longer the same as that prior to infection with X-bacteria.

VIII. Changes in Endosymbionts

Since their initial association with amoebae in 1966, X-bacteria have shown detectable changes as follows:

1. The number of X-bacteria per xD amoeba has decreased from 60,000–150,000 to 42,000. This suggests that the symbionts have become more controllable by their hosts. It must be a change in X-bacteria since the maximum number per amoeba is the same in newly infected D amoebae as in established ones (Ahn and Jeon, 1979).
2. The adverse effects of X-bacteria on amoebae have diminished both in established xD amoebae and in newly infected D amoebae. This decrease in virulence may be related in part to the reduction in the number of symbionts per host cell as described above.
3. The proportion of symbionts with dense matrix as seen under the electron microscope has decreased. Initially, more than half of symbionts present in SVs appeared dense (Jeon and Lorch, 1967). At present, only a small proportion of symbionts are electron dense (Jeon and Hah, 1977). The significance of the electron density of some X-bacteria is not clear.
4. The vesicular membranes enclosing X-bacteria seem to have become more resistant to osmotic changes of the surrounding medium. Initially, most of the SVs ruptured when the plasmalemma of xD amoebae was broken (Jeon and Lorch, 1967). At present, many of the SVs remain intact after xD amoebae have been homogenized and they can be collected free of other cell organelles (Ahn and Jeon, 1982).

IX. Concluding Remarks

The symbiosis between X-bacteria and amoebae represents a unique system for studying host–symbiont interactions at the cellular and molecular levels. The X-bacteria have been integrated as hereditary cell components and are different from other nonobligate or transitory symbionts. In addition, X-bacteria can establish a stable, obligate symbiosis with new hosts under experimental conditions. Their hosts, amoebae, are also unusual in that micrurgical manipulations such as nuclear transplantation and transfer of cytoplasm can be carried out with relative ease. These advantages could be utilized to study various aspects of symbiont integration that have not been clarified yet.

Concerning the symbionts' avoidance of digestion and subsequent establishment of a stable symbiosis, the following assumptions could be made: (1) X-Bacteria are resistant to lysosomal enzymes of amoebae or to other lysing agents because of resistant components such as glycoproteins that might be present in their cell walls. The protein moieties of these membrane components may be coded by the plasmid DNAs found in X-bacteria. (2) The failure for amoeba lysosomes to fuse with SVs may be due to an altered chemical composition or structure of the vesicular membranes. Both X-bacteria and their host amoebae may contribute to such alterations. (3) Both the host and symbionts continue to change structurally and physiologically, further stabilizing their relationship. The symbionts' plasmid DNAs may play a role in the change of symbionts.

Further studies are needed to substantiate these assumptions. Meanwhile, one of the puzzling questions is why amoebae become dependent on X-bacteria in spite of the fact that amoebae do not seem to derive any benefit from their symbionts. On the contrary, many of the detectable effects of X-bacteria on amoebae are detrimental (Section V). It remains to be seen whether the host–symbiont relationship will develop into a mutually beneficial one. A continued observation of such developments will give us further insight to cell variation and the origin of cellular organelles.

Acknowledgments

The author thanks Drs. C. A. Shivers and N. Greenberg for their critical reading of the manuscript. The research work of the author and his associates described in this article was supported by grants from the National Institutes of General Medical Sciences and the National Science Foundation.

References

Ahn, T. I., and Jeon, K. W. (1979). *J. Cell. Physiol.* **98,** 49–58.

Ahn, T. I., and Jeon, K. W. (1982). *Exp. Cell Res.* **137,** 253–268.

Alexander, J., and Vickerman, K. (1975). *J. Protozool.* **25,** 502–508.

Armstrong, J. A., and Hart, P. D. (1975). *J. Exp. Med.* **142.** 1–16.

Bannister, L. H. (1979). *Proc. R. Soc. London Ser. B* **204,** 141–163.

Beale, G. H., and Jurand, A. (1969). *J. Cell Sci.* **5,** 65–91.

Beringer, J. E., Brewin, N., Johnston, A. W. B., Schuman, H. M., and Hopwood, D. A. (1979). *Proc. R. Soc. London Ser. B* **204,** 219–233.

Brown, C. A., Draper, P., and Hart, P. D. (1969). *Nature (London)* **221,** 658–660.

Casley-Smith, J. R., and Savanat, T. (1969). *Aust. J. Exp. Biol. Med. Sci.* **44,** 111–122.

Celis, J. E., Graessmann, A., and Loyter, A., eds. (1980). "Transfer of Cell Constitutuents into Eukaryotic Cells." Plenum, New York.

Chang, K. P. (1974). *J. Protozool.* **21,** 699–709.

Chang, K. P. (1981). *Am. J. Trop. Med. Hyg.* **30,** 334–339.

Chang, K. P., and Dwyer, D. M. (1976). *Science* **193,** 678–680.

Chang, K. P., Chang, C. S., and Sassa, S. (1975). *Proc. Natl. Acad. Sci. U.S.A.* **72,** 2979–2983.
Cook, C. B. (1980). *In* "Cellular Interactions in Symbiosis and Parasitism" (C. B. Cook, P. W. Pappas, and E. D. Rudolph, eds.), pp. 47–73. Ohio State Univ. Press, Columbus.
Cook, C. B., Pappas, P. W., and Rudolph, E. D., eds. (1980). *In* "Cellular Interactions in Symbiosis and Parasitism" (C. B. Cook, P. W. Pappas, and E. D. Rudolph, eds.), pp. 47–73. Ohio State Univ. Press, Columbus.
Costerton, J. W. (1977). *In* "Microbiology-1977" (D. Schlessinger, ed.), pp. 151–157. Amer. Soc. Microbiol., Washington, D.C.
Croce, C. M. (1981). *Int. Rev. Cytol.* **71,** 1–17.
Dwyer, D. M., and Chang, K. P. (1976). *Proc. Natl. Acad. Sci. U.S.A.* **73,** 852–856.
Fankboner, P. V. (1971). *Biol. Bull.* **141,** 222–234.
Goldstein, L., and Ko, C. (1976). *Methods Cell Biol.* **13,** 239–246.
Griffin, J. L. (1960). *Exp. Cell Res.* **21,** 170–178.
Griffin, J. L. (1973). *In* "The Biology of Amoeba" (K. W. Jeon, ed.), pp. 83–98. Academic Press, New York.
Han, J. H., and Jeon, K. W. (1980). *J. Bacteriol.* **141,** 1466–1469.
Hawkins, S. E., and Danielli, J. F. (1963). *Q. J. Microsc. Sci.* **104,** 45–50.
Hayflick, L. (1965). *Exp. Cell Res.* **37,** 614–636.
Jennings, D. H., and Lee, D. L., eds. (1975). *Symb. Symp. Soc. Exp. Biol.* **29.**
Jeon, K. W. (1972). *Science* **176,** 1122–1123.
Jeon, K. W. (1977). *Int. Congr. Protozool., 5th* p. 445.
Jeon, K. W., and Ahn, T. I. (1978). *Science* **202,** 635–637.
Jeon, K. W., and Danielli, J. F. (1971). *Int. Rev. Cytol.* **30,** 49–89.
Jeon, K. W., and Hah, J. C. (1977). *J. Protozool.* **24,** 289–293.
Jeon, K. W., and Jeon, M. S. (1976). *J. Cell. Physiol.* **89,** 337–344.
Jeon, K. W., and Jeon, M. S. (1982). *J. Protozool.* **29,** 84.
Jeon, K. W., and Lorch, I. J. (1967). *Exp. Cell Res.* **48,** 236–240.
Jeon, K. W., and Lorch, I. J. (1969). *Exp. Cell Res.* **56,** 233–238.
Jeon, K. W., and Lorch, I. J. (1970). *J. Cell. Physiol.* **75,** 193–198.
Jeon, K. W., Ahn, T. I., and Lorch, I. J. (1981). *J. Protozool.* **27,** 50A.
Jones, T. C., and Hirsch, J. C. (1972). *J. Exp. Med.* **136,** 1173–1294.
Karakashian, M. W. (1975). *Symp. Soc. Exp. Biol.* **29,** 145–173.
Karakashian, M. W., and Karakashian, S. J. (1973). *Exp. Cell Res.* **81,** 111–119.
Karakashian, S. J. (1963). *Physiol. Zool.* **36,** 52–68.
Karakashian, S. J., and Rudzinska, M. A. (1981). *Exp. Cell Res.* **131,** 387–393.
Kress, Y., Tanowitz, H., Bloom, B., and Wittner, M. (1977). *Exp. Parasitol.* **51,** 385–396.
Lorch, I. J., and Jeon, K. W. (1980). *J. Protozool.* **27,** 423–426.
Lorch, I. J., and Jeon, K. W. (1981). *Science* **211,** 949–951.
Lorch, I. S., and Jeon, K. W. (1982). *J. Protozool.* **29,** 468–470.
McAuley, P. J. (1981). *J. Cell Sci.* **47,** 197–206.
Margulis, L. (1976). *Exp. Parasitol.* **39,** 277–349.
Margulis, L. (1981). "Symbiosis in Cell Evolution". Freeman, San Francisco, California.
Margulis, L., Chase, D., and To, L. P. (1979). *Proc. R. Soc. London Ser. B* **204,** 189–198.
Morgan, C., Rosenkranz, H. S., Carr, H. S., and Rose, H. M. (1967). *J. Bacteriol.* **93,** 1987–2002.
Muscatine, L., and Lenhoff, H. M. (1965). *Biol. Bull.* **129,** 316–328.
Muscatine, L., and Pool, R. R. (1979). Proc. R. Soc. London *Ser. B* **204,** 131–139.
Muscatine, L., Cook, C. B., Pardy, R. L., and Pool, R. R. (1975). *Symp. Soc. Exp. Biol.* **29,** 175–203.
Nogueira, N., and Cohn, Z. (1976). *J. Exp. Med.* **143,** 1402–1420.
Olsen, R. H., Siak, J. S., and Shipley, P. L. (1977). *In* "Microbiology—1977" (D. Schlessinger, ed.), pp. 138–144. Amer. Soc. Microbiol., Washington, D.C.

Pardy, R. L., and Heacox, A. E. (1976). *Nature (London)* **260,** 809–810.
Pearce, J. H., and Lowrie, D. B. (1972). *Symp. Soc. Gen. Microbiol.* **22,** 193–316.
Pool, R. R., and Muscatine, L. (1980). *In* "Endocytobiology: Endosymbiosis and Cell Biology" (W. Schwemmler and H. E. A. Schenk, eds.), pp. 223–238. De Gruyter, Berlin.
Preer, J. R. (1971). *Annu. Rev. Genet.* **5,** 361–406.
Preer, J. R., Preer, L. B., and Jurand, A. (1974). *Bacteriol. Rev.* **38,** 113–163.
Prescott, D. M., and Carrier, R. F. (1964). *Methods Cell Biol.* **1,** 85–95.
Reisser, W. (1980). *In* "Endocytobiology: Endosymbiosis and Cell Biology" (W. Schwemmler and H. E. A. Schenk, eds.), pp. 97–104. De Gruyter, Berlin.
Richmond, M. H., and Smith, D. C., eds. (1979). "The Cell as a Habitat." Royal Soc., London.
Ruddle, F. E. (1977). *In* "International Cell Biology" (B. R. Brinkley and K. R. Potter, eds.), pp. 449–452. Academic Press, New York.
Schell, J., Van Montagu, M., De Bruckeleer, M., De Block, M., Depicker, A., De Wilde, M., Engler, G., Genetello, C., Hernalsteens, J. P., Holsters, M., Seurinck, J., Silva, B., Van Vliet, F., and Villarroel, R. (1979). *Proc. R. Soc. London Ser. B* **204,** 251–266.
Smith, D. C. (1979). *Proc. R. Soc. London Ser. B* **204,** 115–130.
Smith, D. C. (1980). *In* "Endocytobiology: Endosymbiosis and Cell Biology" (W. Schwemmler and H. E. A. Schenk, eds), pp. 317–322. De Gruyter, Berlin.
Steele, R. D. (1976). *J. Zool (London)* **179,** 387–405.
Taylor, D. L. (1969). *J. Mar. Biol. Assoc. UK* **49,** 1057–1065.
Taylor, F. J. R. (1974). *Taxon* **23,** 229–258.
Taylor, F. J. R. (1979). *Proc. R. Soc. London Ser. B* **204,** 267–286.
Trench, R. K. (1980). *In* "Cellular Interactions in Symbiosis and Parasitism" (C. B. Cook, P. W. Pappas, and E. D. Rudolph, eds.), pp. 275–297. Ohio State Univ. Press, Columbus.
Van Wagtendonk, W. J., Clark, J. A. D., and Goddoy, G. A. (1963). *Proc. Natl. Acad. Sci. U.S.A.* **50,** 835–838.
Weis, D. (1976). *J. Protozool.* **23,** 527–529.
Weis, D. (1980). *In* "Endocytobiology: Endosymbiosis and Cell Biology" (W. Schwemmler and H. E. A. Schenk, eds.), pp. 105–112. De Gruyter, Berlin.
Weis, D. S., and Ayala, A. (1979). *J. Protozool.* **26,** 245–248.
Whatley, J. M., John, P., and Whatley, F. R. (1979). *Proc. R. Soc. London Ser. B* **204,** 165–187.
Weidner, E. (1976). *In* "Comparative Pathobiology" (L. A. Bulla and T. C. Cheng, eds), Vol. 1, pp. 111–126. Plenum, New York.

INTERNATIONAL REVIEW OF CYTOLOGY, SUPPLEMENT 14

Perspective on Algal Endosymbionts in Larger *Foraminifera*

JOHN J. LEE

Department of Biology, City College, City University of New York, and Department of Invertebrates, American Museum of Natural History, New York, New York

I. Overview

Although Hyman's (1940) view of the acellularity of protozoa has long been rejected, those who have spent some time studying protozoan cytology or fine structure still appreciate her perspective on the complexity of the group. The fact that five of the chapters in this volume describe particular cases of intracellular symbiosis in protozoa (and many other interesting cases have not been covered) is testimony enough to the physiological and cytological complexity which has been attained within their cells, which, after all, have been evolving as long, or perhaps longer, than other animal groups.

Judging by their present abundance in the world's oceans and by their rich fossil record, foraminifera have been among the most successful animal groups (~35,000 described species). In many ways foraminafera, and in particular the larger foraminifera, are atypical protozoa. They attain large cellular size (in the millimeter to centimeter range) and have very long life spans (months to years) linked to lunar or annual cycles. Cellular giantism would seem to bring many

ISBN 0-12-364375-9

problems related to cell cycle, metabolic integration and control, trophic dynamics, and nutrient exchange, yet since the mesozoic there has been continuous selection among various foraminiferal groups in tropical seas for this character (Newell, 1959). One giant nummulite from the Eocene, *Nummulites gizehensis,* reached more than 12 cm in diameter! The modern larger foraminifera of tropical and semitropical seas are so diverse that they are placed in 10 distinct families (Ross, 1974); those in the fossil record have been placed in an additional 12 families. In recent years it has become generally appreciated that algal endosymbiosis is common to all of the modern larger foraminiferal species and that adaptation for algal symbiosis in larger foraminifera was probably a key driving force in their evolution (Lee, 1974, 1980; Lee *et al.,* 1979; Ross, 1974; Haynes, 1965).

In the sense that they are the hosts for a wide variety of different algal types, larger foraminifera are more broadly adapted for symbiosis than any other group of marine invertebrates (recently reviewed by Lee, 1980). Specific members of the porcellaneous family Soritidae are hosts for dinoflagellates (e.g., *Marginopora vertebralis, Amphisorus hemprichii, Sorites marginalis, S. orbitolites,* and *S. orbiculus*), unicellular chlorophytes (*Archaias angulatus* and *Cyclorbiculina compressa*) and unicellular rhodophytes (*Peneroplis planatus* and *Spirulina arietina*). Members of the perforate families Amphisteginidae (e.g., *Amphistegina lessonii, A. lobifera,* and *A. papillosa*) and Nummulitidae (e.g., *Heterocyclina tuberculata, Heterostegina depressa,* and *Operculina ammonoides*) are the hosts for a variety of small pennate diatoms (Leutenegger, 1977). Larger foraminifera are not the only foraminifera with endosymbiotic algae. Symbiotic dinoflagellates are found in near-surface-dwelling planktonic foraminifera and several small perforate benthic species (*Elphidium williamsoni, E. excavatum, Nonion germanicum,* and *Metarotaliella parva*) are known to husband chloroplasts derived from diatoms (Lopez, 1979, and unpublished; Lanners, unpublished).

II. The Contributions of Morphological, Cytological, and Fine Structural Analyses to Our Understanding of Functional Relationships

A. Shape of the Animal, Architecture of the Shell, and Experiments Aimed at Demonstrating the Functional Aspects of Shell Architecture in Symbiosis

The shells of larger foraminifera are morphologically well adapted to serve as greenhouses for endosymbiosis (Haynes, 1965; Chapronier, 1975; Lee, 1974; Hullock, 1979; Ross, 1974, 1977). In overall appearance the animals tend to be coin-like discs, lenticulate, oblate spheroids, or cylindrical. There has been a tendency in all of the groups of larger foraminifera toward thinning of the side walls of the animal and increasing the surface area to volume ratio. The thinning of the walls is interpreted as an adaptation to decrease light attenuation and thereby optimize photosynthetic activities of the endosymbionts. Quite obviously other morphological adaptations have evolved which maintain and strengthen the integrity of the shell while permitting the window-like thinning. In some groups the chambers are subdivided into chamberlets which have pillars and septa which form an internal framework which supports the thin lateral chamber walls. Members of the suborder Miliolina tend to have very large ratios of surface area to volume. Soritids, a family in the suborder, have an overall appearance which suggests an analogy to a circular waffle wrapped in thin plastic (Fig. 1). The thin plate-like crystals of calcite form a tile-like roof on the upper and lower surfaces (Fig. 2) of the animal while the major thickness of the shell is formed of thin elongate calcite needles. The random orientation of the crystals in the surface gives dry shells a porcellaneous appearance in reflected light and a translucent appearance in sea water. Symbionts are clustered under the thin windows in the shell. The pseudopods of the animal emerge from the periphery of the animal (the edge of the coin) through slit-like apertures (Fig. 3) and form an extensive pseudopodial feeding web.

The calcite of aragonite crystals in the walls of animals in the suborder Rotaliina tend to be oriented perpendicular to the surface (radial walls) a feature which tends to attenuate light. The surfaces of shells are perforated by fine pores (Fig. 4), however, and each of the families with large genera has modifications which subdivide the chambers of the shell into smaller units with corresponding thinning of exterior surfaces. The particular mode of subdivision of chamber cavities and the passages between them and to the exterior of the shell are often quite complex (see review by Hottinger, 1978). In many symbiont-bearing genera (e.g., *Amphistegina, Heterostegina*) the undersides (pore rims) of the pores are expanded into cup-like vaults (Fig. 5). Individual endosymbiotic diatoms fit inside the cups and are crowded in a thin layer just below the illuminated surface of the animal (Hansen and Buchardt, 1977; Hottinger and Dreher, 1974; Leutenegger, 1977; McEnery and Lee, 1981). Most pseudopods are extended from the aperture(s) located on the umbilical side of these planoconvex or biconvex animals. Some small thin pseudopodia also emerge through the sieve plates which cover the pores (Hansen, 1972) of at least the one organism which has been examined in this respect.

Fine structural observations and radionuclide tracer studies suggest that the pores of perforate foraminifera and the ribs and pits in the shells of imperforate species may be physiologically important for endosymbionts (Leutenegger and Hansen, 1979). Clusters of mitochondria have been observed close to the inner end of pores of a number of non-symbiont-bearing species (*Buliminella tenuata, Bolivina argentea, Loxostomum pseudobeyrichi,* and *Cassidulinoids cornuta*) collected in deep (750–898 m) box-cores from the inner continental shelf area of

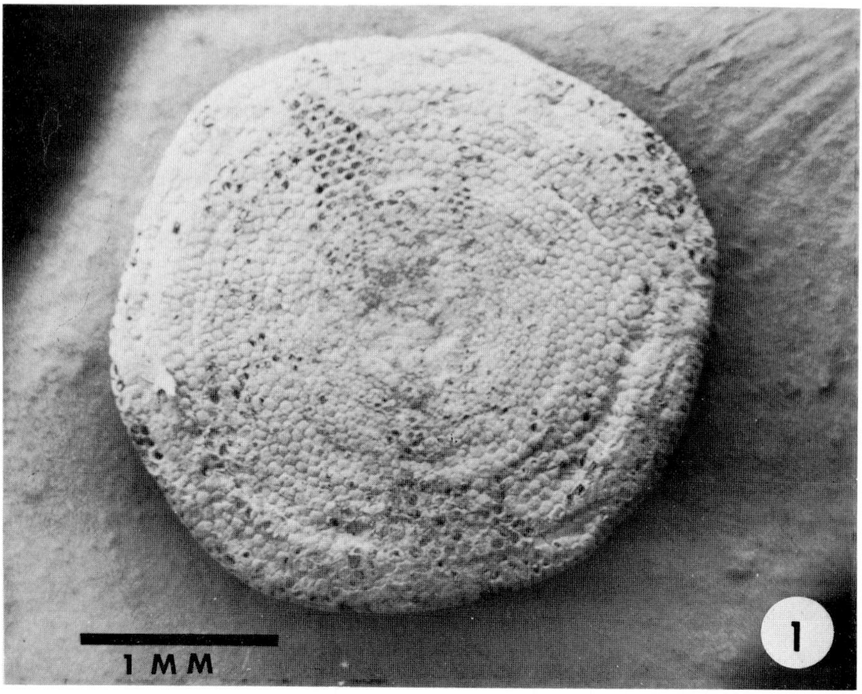

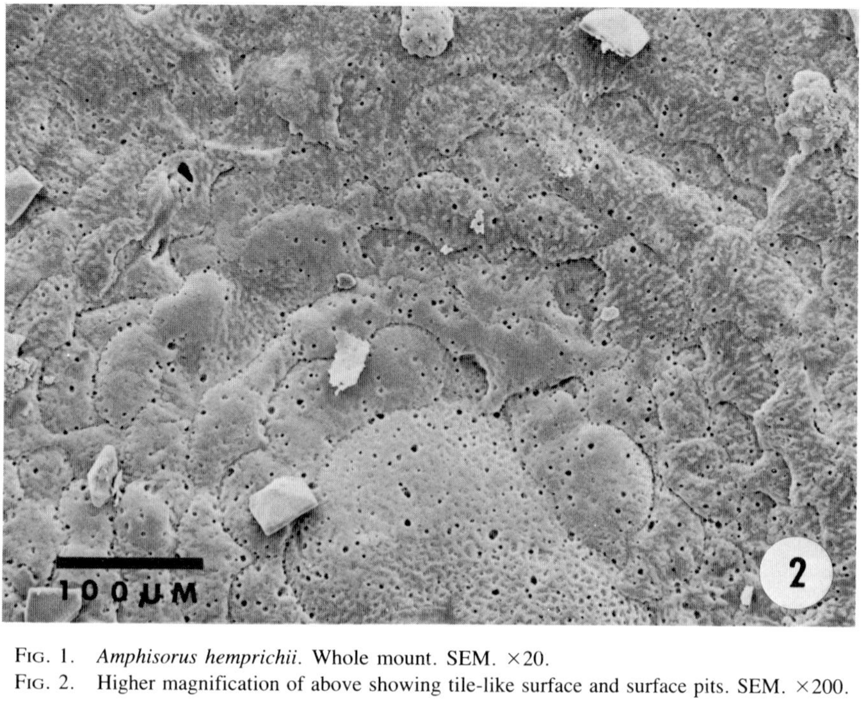

FIG. 1. *Amphisorus hemprichii*. Whole mount. SEM. ×20.
FIG. 2. Higher magnification of above showing tile-like surface and surface pits. SEM. ×200.

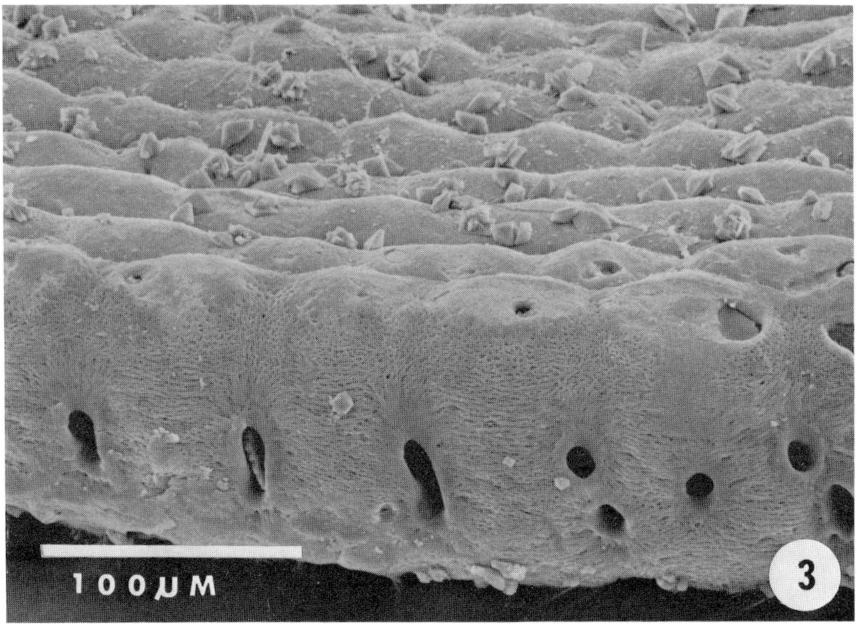

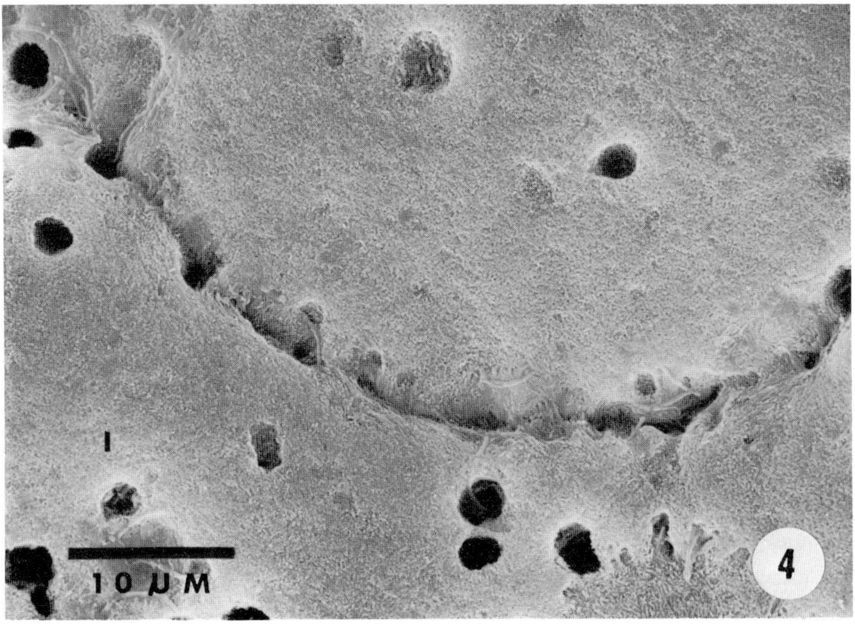

FIG. 3. Lateral edge of *Amphisorus hemprichii* showing apertures. SEM. ×350.
FIG. 4. Upper surface showing junction between plates and surface pits. SEM. ×2000.

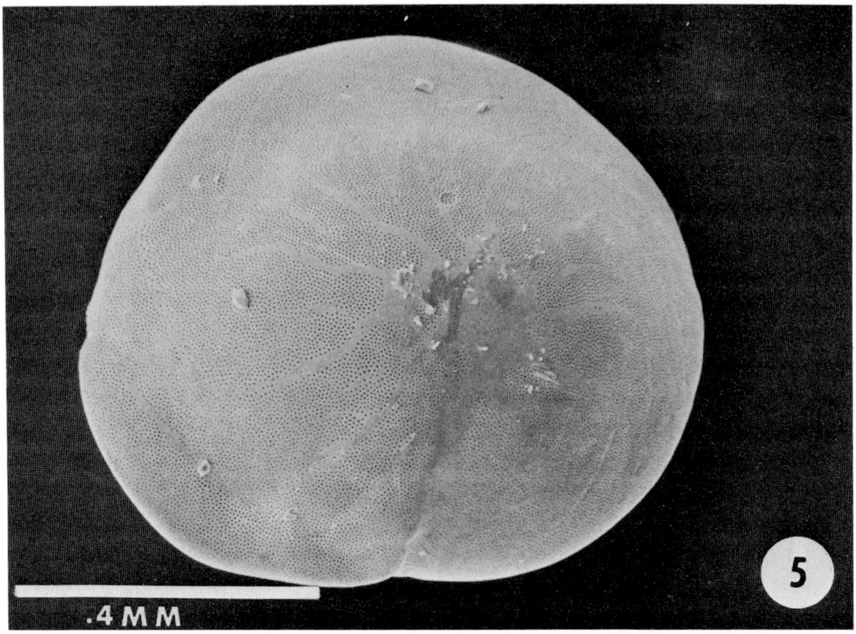

FIG. 5. *Amphistegina lessonii.* Whole mount. Showing surface pores. SEM. ×10.

FIG. 6. Higher magnification of above. Shell has been broken to show outer surface (bottom); innter surface with cuplike pore rims (top right); edge of shell (top left). SEM. ×100.

San Pedro Basin California, an environment with low oxygen (~0.3 ml O_2/liter). While the ultrastructure of the pores and the inner organic lining of the shell varies among these species and several others studied (e.g., *Nonionella stella* and *Bolivina* cf. *subexcavata*) the distributional relationship between mitochondria and pores suggests that the pores may function actively in oxygen uptake.

As mentioned above the algal endosymbionts are clustered just below the pores in larger foraminifera. Although there seems to be some disagreement between the fine structural observations of Hansen (1972) and Leutenegger (1977) about the completeness of the pore plates and organic membranes which may or may not always form a total barrier between internal cytoplasm and the external sea water in *Amphistegina* (a symbiont-bearing species) their collaborative radionuclide tracer experiments suggested that $^{14}CO_2$ was readily taken up through the pores during a short incubation in the light. Curiously radionuclide-labeled glucose was not taken up (Leutenegger and Hansen, 1979). They used a technique devised by Berthold (1976) to hold the animals rigid and to separate the apertural region (where the main pseudopods emerge) from the rest of the shell. Berthold demonstrated that neutral red was picked up through the pores of *Patellina corrugata,* a non-symbiont-bearing form. Leutenegger and Hansen (1979) were also able to show some uptake of neutral red through the pores of *Amphistegina lobifera*. Another piece of evidence for pore function comes from the fine structural study of calcarinid foraminifera which live in particularly oxygen-rich environments (Hottinger and Leutenegger, 1980). The pore pits in *Calcarina calcar* are completely filled by an organic liner and its endosymbiotic diatoms are randomly distributed in the cytoplasm. The causal negative inference made by Hottinger and Leutenegger (1979) is that since the pores are strongly sealed off in *C. calcar* and the influx of possible algal favoring metabolites is unlikely, the endosymbionts in the animal are uniformly distributed.

Scanning electron micrographic studies of the shell surfaces of many porcellaneous (imperforate) symbiont-bearing species have shown that although the animals lack pores, their surfaces are pitted and furrowed, a structural adaptation which should substantially reduce the pathway for diffusion through the lateral test walls to the symbionts (Hansen and Dalberg, 1979). Hansen and Dalberg (1979) were able to experimentally support their ideas by demonstrating the uptake of $H^{14}CO_3^-$ through the walls of *Amphisorus hemprichii* using Berthold's (1976) technique for isolating apertural pseudopods, and holding and mounting the animals. Work in our own laboratory (in progress) supports their observations. We have examined a great many pits in the surface of *Amphisorus hemprichi*. Most pits are just that . . . pits. They end blindly (Fig. 7). However we have found some on every animal examined which are completely open to the cytoplasm below (Fig. 8). Furthermore, we have found evidence in glutaraldehyde-fixed and critical point-dried specimens, of short, very slender pseudopods

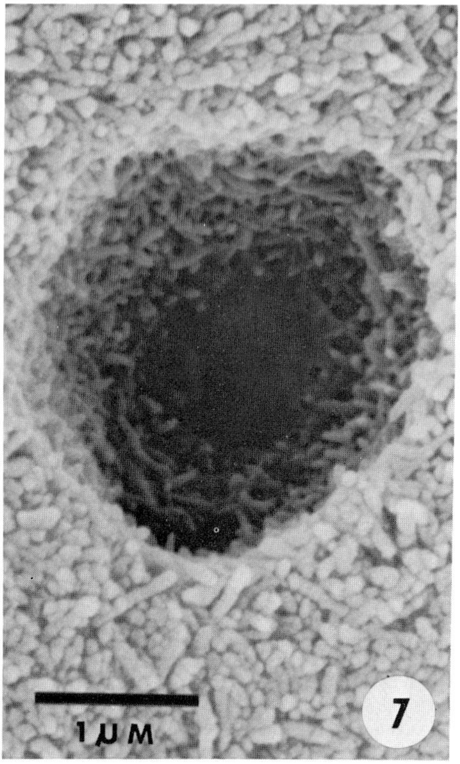
1 µM
7

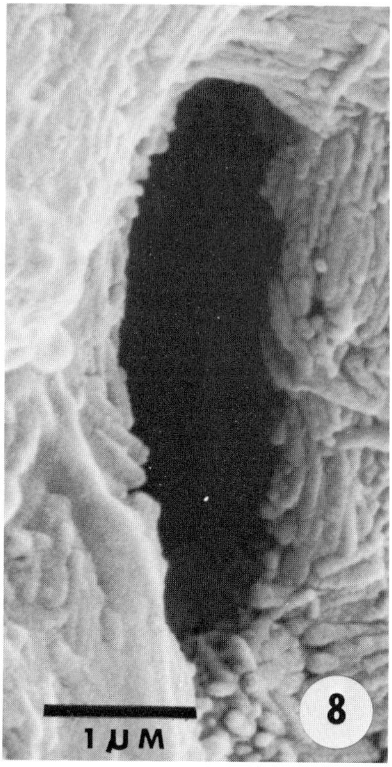
1 µM
8

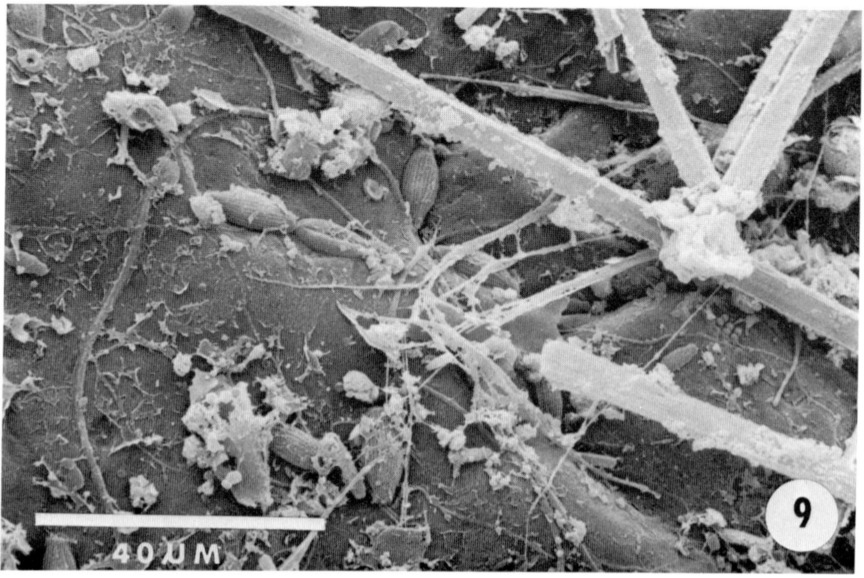
40 µM
9

emerging from some pores and crevices at the sutures between plates on the surface of the animal (Fig. 9). The pseudopods seem to be involved with feeding on the surface of the animal. The relative importance of open pits and pseudopods emerging from crevices has to be evaluated critically. Since they are not widely broadcast over the entire surface they may be less important in the overall nutrient/metabolite economy of this species than are the pits.

Several workers have observed changes in the thickness and shape of symbiont-bearing larger foraminifera which are correlated with depth (Hottinger and Dreher, 1974; Hottinger, 1977; Larsen, 1976; Larsen and Drooger, 1977; Hallock, 1979; Hallock and Hansen, 1979). In general the oblate species tend to have increased surface area (become flatter) with increasing depth. Very rotund species are abundant at depths < 10 m (e.g., *Baculogypsina spaerulata, Calcarina spengleri*) which are subject to some turbulence. At intermediate depths (5–20 m) *Amphistegina lessonii,* a less oblate species, is dominant (Hansen and Buchardt, 1977; Hallock, 1979). Even flatter species (e.g., *Heterostegina depressa*) tend to be found in waters > 15 m or in shaded areas (Rottger, 1976). In addition there is a tendency among species with laminated shells (e.g., *Amphistegina*) toward reduction in thickness of secondary laminae with depth (Hallock and Larsen, 1979). Since these changes in shape and wall thickness occur in specimens collected in the Gulf of Elat, an environment which has fairly uniform hydrographic conditions (e.g., temperature, salinity, pO_2) throughout the photic zone (Klinker *et al.*, 1977), most workers suggest that the changes observed in shells are causally linked to gradients in light quality and diminished intensity with depth. The organisms seem to be balancing their surface area to the amount of incoming light (Hallock, 1979). Working with laboratory clones of *A. lessonii* and *A. lobifera,* Hallock (1979) was able to show that these animals produced significantly thicker tests when grown at high light levels (2600 $\mu W/cm^2$) than when grown under reduced light (300 $\mu W/cm^2$). Her finding was particularly important in light of Haynes' (1965) well-reasoned arguments that the shape of larger foraminifera is a compromise between hydrodynamic factors and metabolic requirements. Hallock's (1979) data also seem to support the findings of Lee *et al.* (1980) and M. J. Lee *et al.* (1982) which suggest that the endosymbiotic algae are photoinhibited at high light levels. Thicker calcareous shells would attenuate more light than thinner ones.

FIG. 7. A deep blind pit on the surface of *Amphisorus hemprichii.* ×22,000.

FIG. 8. An open crevice at the juncture of two plates on *Amphisorus hemprichii.* SEM. ×19,000.

FIG. 9. Pseudopods on the surface of *Amphisorus hemprichii* which have emerged from a crevice. SEM. ×950.

B. Regionalization within the Cell

Our knowledge of the cytology and fine structure of larger foraminifera is quite fragmentary. A few of the most abundant species, two dinoflagellate-bearing ones, *Sorites marginalis* and *Amphisorus hemprichi,* and two diatom-bearing ones, *Amphistegina lobifera* and *Heterostegina depressa,* are better known than are the others (Leutenegger, 1977; Hansen and Dalberg, 1979; Leutenegger and Hansen, 1979; Hottinger and Dreher, 1974; McEnery and Lee, 1980; Spindler, 1978a,b; Hottinger and Leutenegger, 1980).

Sorites marginalis and *Amphisorus hemprichi* are morphologically quite similar to each other; both are coin-like discs ~ 3–5 mm in diameter. Their shells are subdivided into chamberlets, many of which are packed so full of dinoflagellates that more than 80% of their volume is taken by endosymbiotic algae (Muller-Merz and Lee, 1976; McEnery and Lee, 1981). Many of the algae are pressed against the thin lateral walls of their hosts (Hansen and Dalberg, 1979). The dinoflagellates in both hosts are roughly spherical, measuring 6–11 μm in diameter, and have a single lobate peripheral chloroplast. The pyrenoid of the alga has a single stalk and is surrounded by a broad starch cap (Muller-Merz and Lee, 1976; Leutenegger, 1977; McEnery and Lee, 1981). The algae within their host seem identical to *Symbiodinium microadriaticum* (Freudenthal, 1962). Our own observations (McEnery and Lee, 1980) of dividing zooxanthellae in *A. hemprichii,* and those of Leutenegger (1977) in *Sorites orbiculus,* suggest that the endosymbionts are well established in their hosts and not just temporarily farmed. Both animals are multinucleate and heterokaryotic.

Of the two animals *S. marginalis* is more highly regionalized (Muller-Merz and Lee, 1976). In it the hundreds of small generative nuclei (micronuclei) are restricted in their distribution to innermost (embryonic apparatus and inner) chambers. Most of the several score of somatic nuclei (macronuclei) are located in the intermediate chambers (Fig. 20) along with the majority of symbiotic dinoflagellates. Some symbionts and somatic nuclei are also found in the inner and outer chambers. The outer chambers of *S. marginalis* are filled with food vacuoles and residua from food vacuoles. Diatom frustrules are recognizable among the residua.

Amphisorus hemprichii differs cytologically from *S. marginalis* only in the degree of regional differention. Zooxanthellae were found in every chamber of the animal. Fewer were found in the embryonic, inner, and outermost chambers (~ 10% of the total volume) than in the intermediate chambers, where they were densely packed (~ 79% of the total volume). Food vacuoles were most dense in the outer chambers but they were also found in the intermediate chambers. A small fraction of the zooxanthellae, particularly those in the outer chambers, were surrounded by larger vacuoles. The fine structure of the dinoflagellates in the vacuoles suggested that they were either being digested or undergoing autolysis. In most of the other phagolysosomes diatoms with frustrules were easily recognized and identifiable. The diversity of diatom species found in the vacuoles and as residua suggests that the animals have a varied diet. During the course of a study in progress on endosymbiont recruitment potential, we had occasion to examine in the SEM specimens of *Amphisorus hemprichii* fixed and critically dried in place on the surfaces of *Halophila* leaves. The animals surround themselves with "midden heaps," of empty diatom shells and debris (Fig. 10). Pseudopodia are in evidence throughout the heaps (Fig. 11) and beyond them on the surfaces of the leaves. Some of the organisms in the pseudopodial nets are disrupted (Fig. 13). This suggests either fixation artifact or extracellular digestion. Pseudopodia also seem to be actively feeding on the bacteria and algae on the surface of the shell (Fig. 12). In addition to the short slender pseudopodia emerging from pits and crevices between plates (mentioned in the previous section) the margins and peripheral upper surfaces of the shell are coated by pseudopodia emerging from the apertures. No one has yet carefully examined the extrathalamous pseudopodal activities of *Sorites marginalis* so that comparison with *A. hemprichii* must be cautiously qualified. It is possible, however, that the differences between the two species in regionalization are attributable to relative differences in extrathalmous digestive activities. Another, perhaps insignificant, cytological difference noted between *A. hemprichii* and *S. marginalis* is that the former seems to have fewer nuclei (~100–200 generative nuclei per cell) than the latter. There also seem to be more gradations in the sizes of the somatic nuclei in *A. hemprichii* (McEnery and Lee, 1981).

In contrast to the dinoflagellate zooxanthellae of *S. marginalis* and *A. hemprichii* the spherical frustrule-less diatom zooxanthellae of *Amphistegina lessonii* and *A. lobifera* are densely packed on the dorsal and dorsal lateral surfaces of the shells in a thin layer closely opposed to the cup-like pore rims mentioned in the previous section (Leutenegger, 1977; McEnery and Lee, 1981) Fig. 6). The cytoplasm in the ventral and middle region of the organisms is filled with digestive vacuoles and residua (McEnery and Lee, 1981). Diatoms with frustrules are often observed in food vacuoles and cytoplasmic residua, piquing curiosity as to the mechanisms by which the endosymbiotic diatoms avoid the same fate. Except for a fine structural histochemical dissertation project just begun in our laboratory (R. Koestler) this important aspect of endosymbiosis has not been examined.

Heterostegina depressa is the least cytologically specialized of the larger foraminifera thus far examined. In it the endosymbiotic diatoms tend to be oval to fusiform (Hottinger and Dreher, 1974; Dietz-Elbrachter, 1971; Leutenegger, 1977; McEnery and Lee, 1981). Diatom endosymbionts are fairly uniformly distributed throughout the animals. The symbionts tend to be slightly more elongate in the outer chambers. Food vacuoles containing algae have not been observed in this animal. The cytoplasm in the outer chambers and exterior

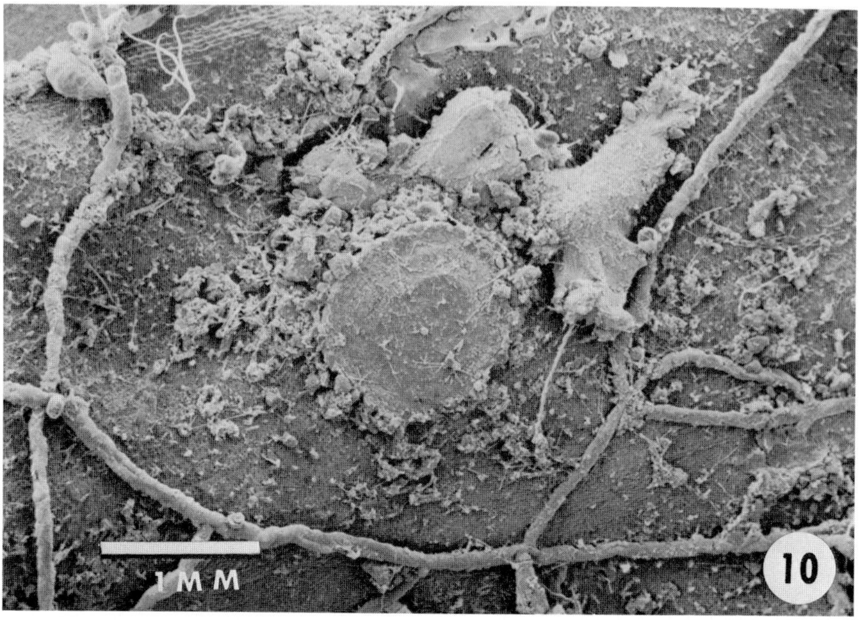

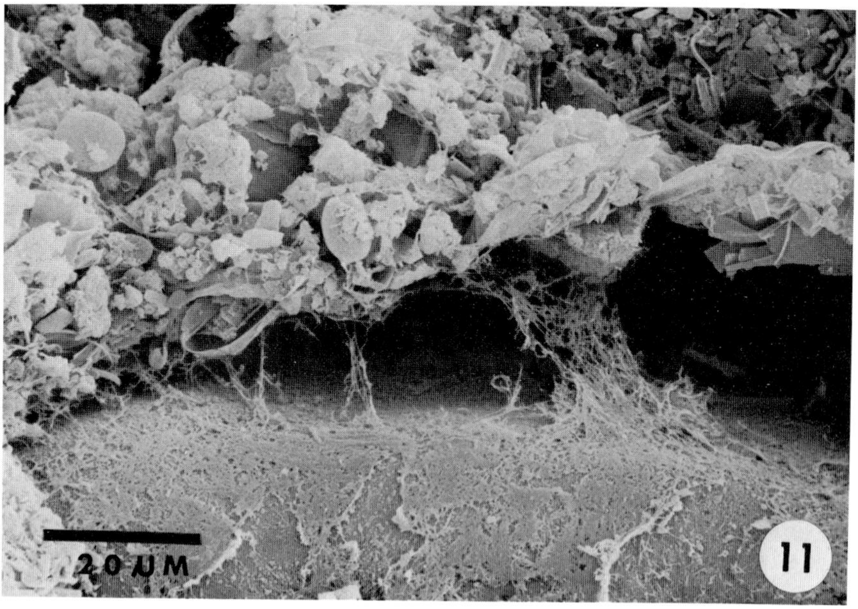

FIG. 10. *Amphisorus hemprichii* surrounded by a midden-heap on the surface of a *Halophila* leaf from the Red Sea. SEM.

FIG. 11. Higher magnification of the specimen above showing the pseudopods between the midden-heap (above) and animal (below). The separation between the two may be a fixation artifact. SEM. ×1050.

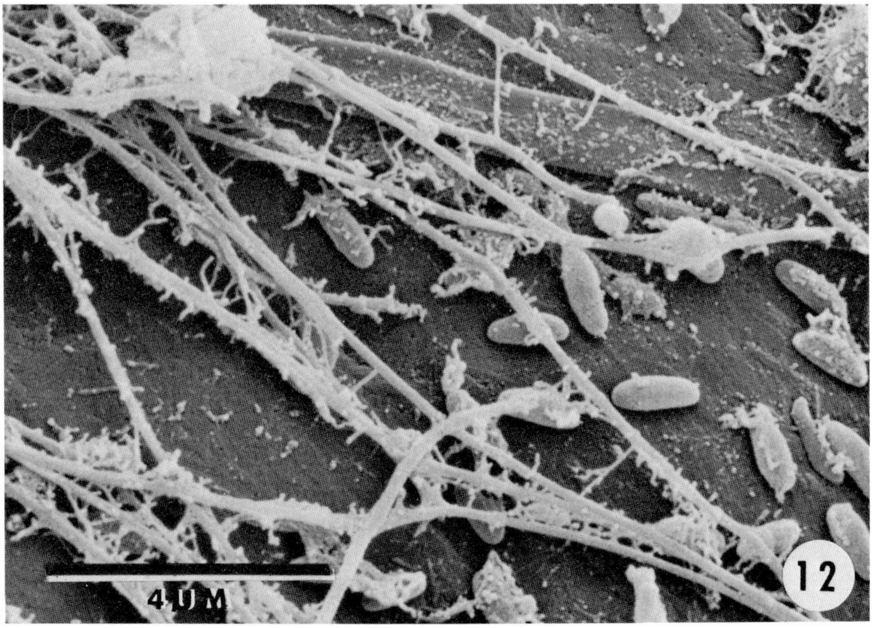

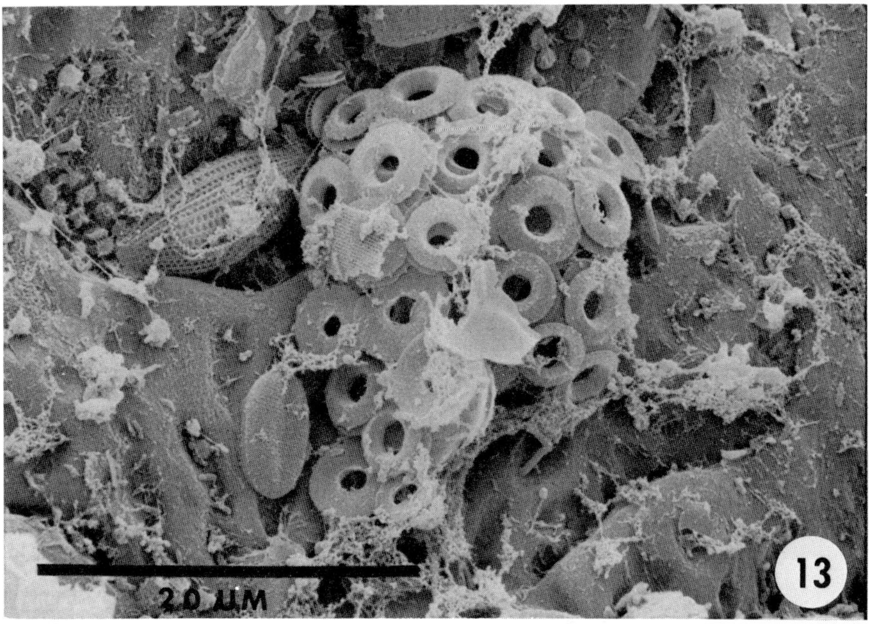

FIG. 12. Pseudopodia on the surface of *Halophila* capturing bacteria. SEM. ×9250.

FIG. 13. Diatoms and a coccolithophorid being digested on the surface of an *A. hemphrichii*. SEM.

surfaces of the shell is more coarsely granular than in other parts of the animal and vacuoles, possibly containing bacteria being digested, have been reported (Hottinger and Dreher, 1974; Leutenegger, 1977). It is quite possible that these deeper (Fermont, 1977; Hottinger, 1977) and/or shade dwelling (Röttger, 1976) animals feed on bacteria and digest them extracellularly. This begs systematic investigation.

The sequestration and persistence of algal chloroplasts in several other foraminifera, *Elphidium williamsoni, E. excavatum, Nonion germanicum, Metarotalliella parva,* and *M. simplex,* are also noteworthy (Lopez, 1979, and unpublished; Lanners, unpublished). In all of these species fine structural observations suggest that the chloroplasts are found throughout the protoplasm of the animals but tend to be concentrated in the outermost parts. Individual chloroplasts are surrounded by a pair of double membranes (Lopez, 1979; Lanners, unpublished, Fig. 17). Presumably the innermost membrane is the chloroplast envelope and the outer one is a vacuole derived from the host. The experimental work of Lopez (in progress) suggests that *E. williamsoni* is relatively nonselective since chloroplast-depleted foraminiferans became reinfected when fed with natural populations of benthic microphytes or with any of three different species of algae, *Amphora coffeaeformis, Amphora* sp., and *Melosira varians,* out of 15 isolates (clones?) tested. The identity of the donor of chloroplasts to *Metarotaliella parva* has just recently been uncovered in materials brought from Tübingen (Lanners and Lee, in progress). The diatom donor is clearly an unusually small species of *Entomoneis* (Fig. 15). For reasons that are not clear, all the animals in some cultured stocks of the same collection from Villefranche-Sur-Mer sequestered chloroplasts while the animals in other stocks did not.

Fine structural studies have shown that the endosymbiotic algae within their hosts have reduced or lack their typical cell envelopes (e.g., theca frustrules, walls; Hottinger and Dreher, 1974; Hansen and Dalberg, 1979; Muller-Merz and Lee, 1976; Leutenegger, 1977; Berghold, 1978; Schmaljohann and Rottger, 1978; McEnery and Lee, 1981). Most individual algae lie in vacuoles surrounded by two unit membranes with little or no perialgal space lying between them. The outer membrane is presumed to be a host-derived phagosome membrane; the inner one is presumed to be the cell membrane of the endosymbiont. Since chloroplasts and mitochondria occupy a large part of the periphery of many symbionts stacks of three closely applied membranes are characteristically visualized in many electron micrographs. Several exceptions are noted. The dinoflagellate zooxanthellae in *Amphisorus hemprichii* and *Sorites orbiculus* seem to

FIG. 14. *Metarotaliella parva.* Whole mount. ×150. Figure courtesy of Ellen Lanners, Institute of Zoology, University of Tübingen, Tübingen, Federal Republic of Germany.

FIG. 15. *Entomoneis* sp. the donor of chloroplasts to *M. parva.* SEM.

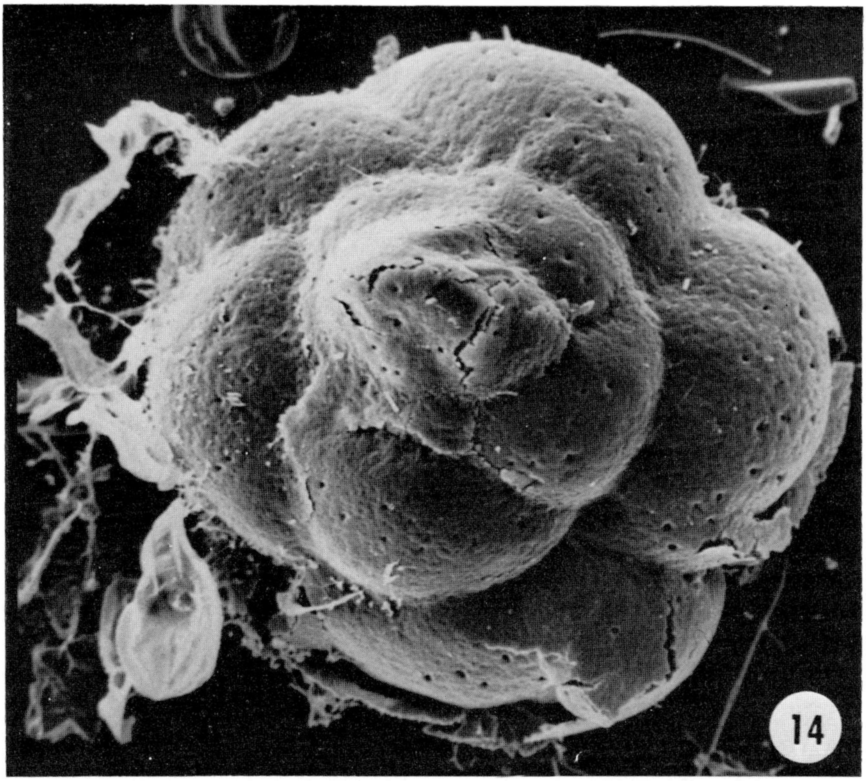
14

20 μm
15

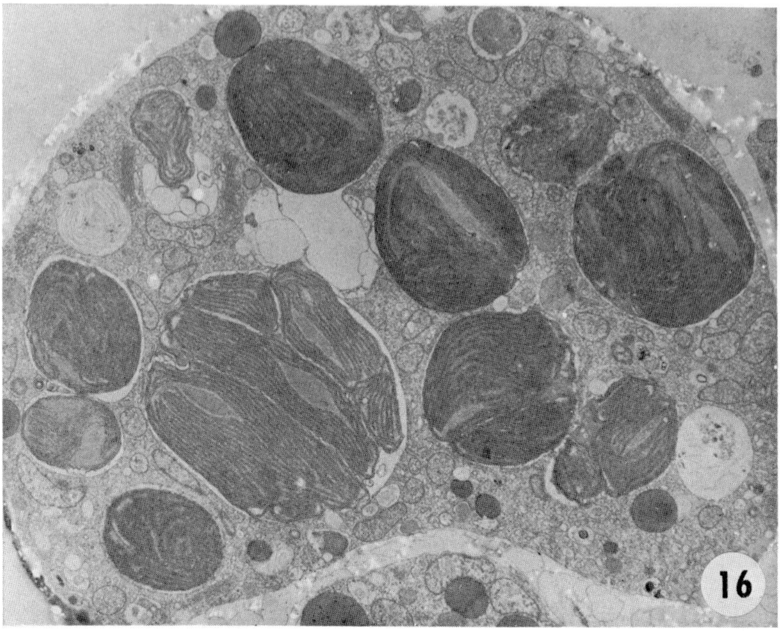
16

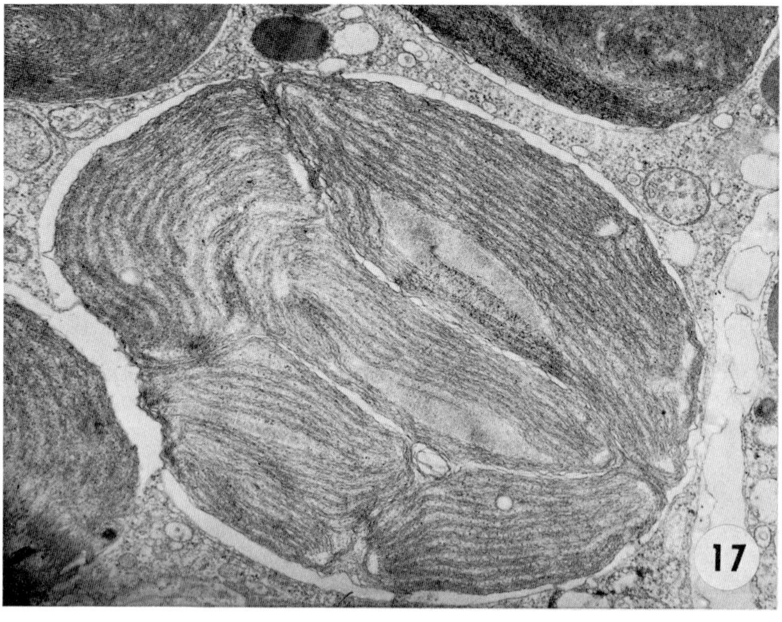
17

have rudimentary amphaesmal vessicles. Several of the TEM micrographs of the endosymbionts in *Calcarina calcar* show some cell wall material also present in these algae as well (Hottinger and Leutenegger, 1980; Figs. 9 and 10). Dinoflagellate zooxanthellae and Chlamydomonad zoochlorellae do not form flagella within their hosts (Leutenegger, 1977; Muller-Merz and Lee, 1976). Flagellar basal bodies, however, have been observed in the zooxanthellae inside *A. hemprichii*. Since the endosymbiotic algae form typical cell envelopes, and appropriate species form flagella, once they are released and cultivated (e.g., Lee *et al.*, 1974, 1979a,b, 1980a,b,c) we must presume that these traits are generally suppressed within foraminiferal hosts. This aspect of foraminiferal endosymbiosis needs study.

In many respects the cytological picture of the algal endosymbionts in larger foraminifera is similar to that found in symbiosis in *Paramecium bursaria* and *Hydra viridis*, both much more extensively studied systems. After endosymbiotic *Chlorella* sp. are engulfed by *Hydra* digestive cells the algae surrounded by phagosome membranes migrate to the base of the cells or region of the cell not normally associated with digestive activity (Pardy and Muscatine, 1973; Muscatine *et al.*, 1975; Muscatine and Pool, 1979) where they are sequested whereas noninfective *Chlorella vulgaris* collect into large apical phagosomes and eventually are exocytosed. The endosymbiotic *Chlorella* sp. in *Paramecium bursaria* are also surrounded by individual perialgal vacuoles which are widely separated by host cytoplasm (Karakashian, 1975). Experimental studies led Weis (1976) to suggest that algae sequestered in the peripheral cytoplasm are not digested whereas those in the central cytoplasm are more vulnerable in *P. bursaria*. Several lines of experimental evidence led to the hypothesis that natural symbiont strains of *Chlorella* interfere with the normal course of host digestion (reviewed in Karakashian, 1975). A recent fine structual study with thorotrast labeling showed that this was indeed the case (Karakashian and Rudzinska, 1981). Secondary lysosomes fused with food vacuoles containing bacteria and not with vacuoles containing algae, leading the investigators to conclude that the symbiotic algae alter the membrane of the vacuole which surrounds them thereby inhibiting fusion with secondary lysosomes. On the basis of the *Paramecium bursaria* and *Hydra viridis* experimental work it is reasonable to speculate that the sequestration of the algal endosymbionts from the regions of active digestion is probably one mechanism for symbiont survival in larger foraminifera. As in

---

FIG. 16. Section of a chamber of *Metarotaliella parva* showing endosymbiotic chloroplasts. ×8250. Figure courtesy of Ellen Lanners, Institute of Zoology, University of Tübingen, Tübingen, Federal Republic of Germany.

FIG. 17. Higher magnification of above showing individual chloroplast. ×19,500. Figure courtesy of Ellen Lanners, Institute of Zoology, University of Tübingen, Tübingen, Federal Republic of Germany.

Paramecium bursaria it is also possible that some strains of algae can inhibit fusion with secondary lysosomes. If it can be found that a limited but varied group of small pennate diatoms can do it, then it would help explain the diversity of diatom endosymbiont species found in some of the larger foraminifera.

The presence of free starch grains in the cytoplasm and disintegrating algal symbionts has been noted in some regions of a number of larger foraminifera: *Sorites marginalis, Archaias angulatus, Amphisorus hemprichii, Amphistegina lessonii, Amphistegina lobifera,* and *Heterostegina depressa* (Muller-Merz and Lee, 1976; McEnery and Lee, 1981; Lee and McEnery, 1982; Dietz-Elbrachter, 1971). These observations can be interpreted in a number of ways as mechanisms for (1) pruning or culling senescent algal cells; (2) harvesting some crop from a symbiont garden; (3) regulating endosymbiont numbers in cells; (4) or any combination of these. As a phenomenon it seems to be fairly common since similar observations have been made in a wide variety of algal–invertebrate symbiotic associations (recently reviewed by Muscatine and Pool, 1979). The question as to whether one observes autolysis of senescent algae or host digestion of some fraction of the endosymbionts seems open (Muscatine and Pool, 1979).

Although mechanisms for the regulation of algal cell numbers per cell have been investigated in a number of algal–invertebrate systems this has not yet been examined in larger foraminifera. In *Paramecium bursaria* the numbers of algae per animal are influenced by light intensity and food supply (reviewed by Karakashian, 1975; Reisser, 1980). One new model for regulation of algae in this ciliate is based on space limitations, amount of organic and inorganic nutrients, and light intensity for the algae (Reisser, 1980). The division rate of the ciliate is influenced by the amount of food and maltose excreted by its symbiotic algae. The balance in relative numbers of algae per animal cell is thus achieved by a mechanism which does not require the host to digest or expel excess symbionts or limit the CO_2 supply to them.

The *Hydra* system may be dissimilar. When *Hydra vividis* are grown in constant light with daily feeding the numbers of algae per digestive cell are maintained in a steady state (reviewed by Muscatine *et al.*, 1975). Changes in illumination and nutrition alter the steady-state ratio. One model proposed for regulation of cell numbers suggests that host and algal cell mitosis are closely linked (McAuley, 1980). Division of the zoochlorellae occurs when the host cells are stimulated to divide but is otherwise inhibited.

III. Nutritional, Radionuclide Tracer, and Stable Isotope Experiments on Functional Aspects of Symbiosis

Several different approaches have given some insight into the functional contributions of the algal endosymbionts to larger foraminifera (recently reviewed by Lee and McEnery, 1982). Radionuclide tracer feeding experiments and respirometry have been used to estimate the carbon budgets of *Archaias angulatus* and *Sorites marginalis* from Key Large Sound, Florida (Lee and Bock, 1976) and *Amphistegina lessonii, A. lobifera,* and *Heterostegina depressa* from the Red Sea at Elat (Lee *et al.*, 1980c). With the exception of the latter species all of the others were selective, feeding on diatoms and chlorophytes. In each, photosynthesis accounted for approximately 10% of the overall host/symbiont system's carbon budgets. *Heterostegina depressa* did not feed on algae, but Lee and co-workers (1980c) obtained some evidence that it might feed on bacteria. Although in Rottger's lab and our own, *H. depressa* survive for almost a year, the animals eventually change pigmentation and die after producing a single brood of offspring.

Because of their small size, abundance in marine sediments, and durability in the fossil record, foraminifera have been important as tools for micropaleontologists and paleoecologists. Demonstrations that the shells of some foraminifera and corals are out of equilibrium in C and O isotopes from the sea water they live in has intensified interest in the mechanisms by which disequilibria are brought about and in the usefulness of stable isotope fractionation in paleotemperature estimation. Disequilibrium between O and C isotopes is common in benthic foraminiferan shells (e.g., Duplessey *et al.*, 1970; Vinot-Bertouille and Duplessey, 1973; Buchardt and Hansen, 1977; Erez, 1978; Wefer and Berger, 1980). Disequilibrium is usually less for oxygen than for carbon, perhaps reflecting pool sizes (water, CO_2-bicarbonate) available for each element (Berger, 1981). Striking differences in carbon disequilibrium generally encountered between species depend on niche and habitat. The larger foraminifera, all of which are hosts for endosymbiotic algae, tend to have shells 2 to 2.5% lighter in carbon isotopes ($^{13}C/^{12}C$) than would be yielded by equilibrium values with seawater at the same site. Photosynthesis and rapid growth favor disequilibria. The effect suggests paleotemperatures up to 5°C higher than the actual temperatures (Vinot-Bertouille and Duplessey, 1973). In this respect the larger foraminifera are similar to the scleractinian reef-building corals. In a classical study Weber and Woodhead (1972) calculated the regression coefficients for 44 genera of corals from Australia, Polynesia, Micronesia, and Caribbean which suggested that skeletal carbonate derives from seawater and metabolic CO_2. The proportions contributed by each seem constant for a particular genus but differ among genera living in the same environment. Mollusc skeletons from the same locations had the same $^{18}O/^{16}O$ ratio as inorganic carbonate precipitated from solution at a given temperature. In general, corals of the suborder Faviina secrete skeletons with low $^{18}O/^{16}O$ ratios, Fungiina show intermediate ratios, and Astrocoeniina higher ratios. The larger foraminifera apparently follow this rule. Erez (1978) has shown in the Gulf of Elat that *Amphisorus hemprichii* and *Borelis* sp. (genera belonging to suborder Miliolina) cluster in ^{18}O and ^{13}C values whereas *Heterostegina depressa* and *Amphistegina lobifera* (suborder Rotaliina) had even lighter shells. Erez shows that the differences in isotopic compositions of the shells resulted from differing rates of photosynthesis of the endosymbionts in each foram species at the same depth. A common view among workers in the field (e.g., Berger, 1981) is that photosynthesis enriches the ^{12}C and ^{16}O isotopes in the skeleton by recycling metabolic CO_2 from the organism's internal pools.

Stable-isotope analysis in foraminifera has been much refined in recent years, e.g., Weber *et al.* (1981) examined portions of stable isotopes in recent larger foraminifera from Bermuda, Persian Gulf, and Philippines. Shells of these long-lived giant protozoa yielded information as to annual seasonal temperature ranges and life-history stages—all recorded as fluctuations in isotopic values of oxygen and carbon (Wefer *et al.*, 1981). In this respect larger foraminifera seem more valuable than planktonic foraminifera which sink from warm surface waters to deeper, cool zones where they undergo additional calcification before gametogenesis (Berger, 1981; Erez and Luz, submitted).

Various investigators have been tempted to regard light as the primary factor affecting the distribution of different species of larger foraminifera in the 150 m euphotic zone in the Gulf of Aqaba/Elat (Reiss, 1977; Hansen and Buchardt, 1977). Additional depletion of ^{18}O observed in symbiont bearing larger foraminifera tests in proportion to depth/light is interpreted by Buchardt and Hansen (1977) as further evidence of the significance of the symbionts for shell growth.

Several experimental studies demonstrate that light-mediated symbiont processes enhance calcification and growth in these larger foraminifera. Muller (1979) found no growth of *Amphistegina lessonii* in dark incubated cultures while those incubated in the light grew 9 μm a day. Her more recent work (Hallock, 1981), and that of the group in Kiel (Röttger *et al.*, 1980), has confirmed and expanded data on this animal and *Heterostegina depressa*. Within the range of 150–800 klx, growth and primary production ($^{14}CO_3$ incorporation) increased as a function or light intensity. Photoinhibition was observed at high levels of light. Photosynthetic and calcification rates of the *Archaias angulatus/Chlamydomonas hedleyi* association were directly proportional to light intensity in the range of 0–2000 Einsteins/m^2/second (Duguay and Taylor, 1978). DCMU completely inhibited photosynthesis and light enhanced calcification (Duguay and Taylor, 1978). Immature specimens grew at faster rates than adults, but on the average the calcification rate is 7 n*M*/Ca/mg dry wt/day (~ 1.5% growth/day) (Duguay and Taylor, 1978; Lee and Bock, 1976; Lee and Zucker, 1969).

Röttger (1972) found that the nummulitid *Heterostegina depressa*, a diatom-bearing foram, survives and grows well in the absence of any obvious concentration of food if it is incubated in the light. Steady but slow growth was observed at an irradiance of 45 lx; 300 lx was optimal. Less rapid growth was observed at higher irradiance (300–2000 lx).

Recently Röttger and his co-workers (Williams *et al.*, 1981) studied the effects on light on stable isotope fractionation in *Heterostegina depressa*. Their results were quite equivocal, we suspect because there were too many uncontrolled variables. Light also effects the behavior of larger foraminifera. *Amphistegina radiata* is positively phototaxic within a photonic flux range of 10^{11}–10^{15} photons/cm/second and unresponsive at lower intensities. The action spectrum peaked at 500 nm (Zmiri *et al.*, 1974). *A. lobifera* is negatively phototaxic at illumination above 10 klx and positively phototaxic at some point above 0.1 to > 1–10 klx. *Amphisorus hemprichii* seems to be positively phototaxic at some point between 11–22 and < 6 klx but > 1.7 klx. Respirometric studies of the same two animals in the field suggested that they were photoinhibited at light intensities > 20 klx (Lee *et al.*, 1980a).

The photoadaptive responses of axenic cultures of four diatom species isolated as endosymbionts from the same two species of larger foraminifera (*Amphistegina lessonii* and *Heterostegina depressa*) collected in the Red Sea and Hawaii have recently been studied in the laboratory (M. J. Lee *et al.*, 1982). The two endosymbiotic diatoms isolated from *Amphistegina lessonii, Fragilaria shiloi,* and *Nitzschia laevis* grew fastest at 312 $\mu W/cm^2$ whereas the endosymbiotic diatoms isolated from *Heterostegina depressa, Nitzschia panduriformis,* and *N. valdestriata* grew best at 19 $\mu W/cm^2$. In nature *Amphistegina lessonii* tends to be found in shallower depths (5–50 m, peak 25 m; Hansen and Buchardt, 1977) whereas *Heterostegina depressa* tends to be found in deeper (20–70 m or shaded) habitats (Hottinger, 1977; Rottger, 1976). Photosynthetic rates, measured by respirometry, suggested that all four species were photoinhibited at high light intensities (625 $\mu W/cm^2$) and did well at moderately low light intensities (175 $\mu W/cm^2$). The photocompensation points of all four species were ~ 2% of the light available at 1 m depth at Elat on the Red Sea in summer. In winter light levels are reduced particularly on cloudy days so that photocompensation for the algae on a year round basis will be reached at < 40–50 *M*. Attenuation of light by the shells will reduce compensation levels to less than that.

Nutritional studies of isolated symbionts have given some clues to one aspect of the possible benefits of the symbiotic relationships to the algae. All of the algal strains tested thus far require or are stimulated by thiamine. Many require biotin and a few vitamin B_{12} (Lee *et al.*, 1974, 1979, 1980b). The sources of these vitamins could be the bacteria eaten by their hosts. There were three orders of magnitude difference between the optimum $NaNO_3$ concentration of the various symbiotic algal clones tested. *N. panduriformis* had the lowest optimum value (2 μM) and several species, *N. laevis, N. frustulum, N. valdestriata,* and *F. shiloi* had higher (2 m*M*) optima. At the depths where the foraminifera were captured the concentration of NO_2^- and NO_3^- in the Gulf of Elat rarely exceeds 1 μg at/liter (Levanson-Spanier *et al.*, 1979), a condition which suggests that even the foraminiferal/algal symbiotic systems with the lowest N requirements must tightly recycle bound nitrogen or they are constantly nitrogen limited. The profile for PO_4^{2-} is similar. At the same depths levels of phosphate in the Gulf rarely exceeded 0.3 μg at/liter.

IV. Establishment and Regulation of the Association

As indicated in the introduction some species of the miliolinid larger foraminifera seem always to be the host for particular species of dinoflagellate, chlorophyte, or rhodophyte endosymbionts while the generally faster growing perforate species in the suborder Rotaliina tend to be the hosts for a number of small pennate diatoms. When we compare our knowledge of the specificity of association, establishment, and regulation of algal numbers to knowledge of the *Hydra viridis, Paramecium bursaria, Convoluta roscoffensis,* and various coral systems we have barely scratched the surface in larger foraminifera (see Taylor, 1973, 1980; Muscatine *et al.*, 1975; Karakashian, 1975; Smith, 1979, 1980; Cook, 1980; Pool and Muscatine, 1980; Schoenberg, 1980; Doonan *et al.*, 1980). Of particular interest are the diatom-bearing larger foraminiferal species because they seem to have a less rigid, but still selective, relationship between themselves and their algae than do the better known systems (e.g., *Hydra viridis, Paramecium bursaris,* and *Convoluta roscoffensis*). The finding of a second species of algal endosymbiont in *Convoluta roscoffensis* (Mcfarlane, 1982) and *Anthopleura elegantissima* (Muscatine, 1971) may portend that marine symbioses may not be as simple as originally believed.

Our knowledge of endosymbiotic diatoms is based on six collections of these animals: one each near the Steinetz marine laboratory of Hebrew University and Wadi Taba on the Gulf of Elat (Red Sea); one each from the Makapuu Tide Pool Oahu and Maui Island Hawaii, and the Coral Gardens in the northeast part of the Heron-Wistori Channel of the Great Barrier Reef, Australia. Approximately 20 individuals of three different animal species have been studied from all three geographic sites, *Heterostegina depressa, Amphistigina lessonii,* and *A. lobifera*. In addition small numbers (<10) of *Amphistegina bicirculata, A. papillosa,* and *Borelis schlumberger* were examined. While the total numbers of organisms examined at each site may seem relatively small to those unfamiliar with the subject, the process of cleaning and isolating endosymbionts is quite labor intensive and time consuming. Very briefly, the animals' surfaces are carefully picked, brushed, and aseptically washed over the course of several days to remove all external algae (Lee *et al.*, 1980). An antibiotic mix is added to the sea water to suppress bacterial growth. The animals' surfaces are next examined under a water immersion lens (500×) in order to ensure the removal of all adhering algae before they are gently crushed with sterile glass rods or broken open with sterile forceps. The algae released are then spread with glass rods over the surface of agar-solidified differential media (Lee *et al.*, 1975, 1980). A few washed and cleaned organisms from each experimental batch have been mounted on stubs so that their surfaces could be more closely examined to determine the effectiveness of the cleaning procedures. Ten colonies from each animal are picked and the diatoms are cleaned and examined in phase or scanning electron microscopy. While a number of workers in Tübingen (Bertold, 1977), Kiel (Elbrachter, 1971; Schmaljohann and Rottger, 1977), and Basel (Leutenegger, 1977a–c) obtained fine structural evidence that the endosymbionts of a number of Rotaliniids (e.g., *Heterostegina depressa, Amphistegina lobifera,* and *A. lessonii*) were most probably diatoms, the question remained unresolved because the diagnostic envelop (frustrule) is not formed in the host. Fortunately we were able to isolate the first diatom endosymbionts from *H. depressa, A. lobifera,* and *A. lessonii,* larger foraminifera from the Red Sea, and they formed normal frustrules (Lee *et al.*, 1979, 1980a,b). Subsequent isolation and identification of the diatom zooxanthellae from the same host species from Hawaii showed that some of the diatom species were the same as in the Red Sea but for the most part the diatom species were different (Lee *et al.*, 1980c,d). We have isolated the diatom endosymbionts from three additional collections, one from the Great Barrier Reef and two from the Red Sea. The results from these new collections tend to support Hansen and Larsen's (1977) observations and suggest that the diatom endosymbionts may not be the same in populations of the same foraminiferal species from the Red Sea in every season of the year. While present data led us to initiate additional field work, several points are beginning to emerge from the information presently at hand.

1. The larger foraminifera which harbor endosymbiotic diatoms do not have a rigid specificity for particular species of diatoms.
2. All of the endosymbiotic diatoms isolated tend to be small (< 10 μm) pennate species.
3. Most larger foraminifera ($\sim 80\%$ of the population) are hosts for only one species of endosymbiotic diatom at a time.
4. Two species of diatoms, *Fragilaria shiloi* and *Nitzschia panduriformis* var *continua,* were found commonly as endosymbionts in each of the five collections of larger foraminifera examined. *Amphora tenerrima* and *Nitzschia laevis* have been found in specimens from two of the three collection sites.
5. Many animals in the same field collection are hosts for the same diatom species.

As working hypotheses we speculate that the animals' life cycles are partially or entirely responsible for points 1 and 5. Zygotes may not be able to recruit the same endosymbiont species as their parents. Although there are two morphologically distinct generations of larger foraminifera, in nature and in the fossil record megalospheric and microspheric, the gamontic forms are much less common. Best available field evidence on *Amphisorus hemprichii* and *Amphistegina* spp. suggests that larger foraminifera have basically apogamic life cycles (Zohery *et al.*, 1980; Reiss and students, in progress). Thus while zygotes must acquire symbionts from their surroundings this is unlikely to be a major factor accounting for changes in endosymbionts. Schizozoites acquire their symbionts from their parents. With respect to points 2–4, we suspect that some diatom species are much better suited to be endosymbionts than others. Little is known about the physiological basis for endosymbiotic diatom selection or the pathways of carbon flow.

Along these lines two projects presently underway may be of interest to readers of this article. We have just finished some field work at the H. Steinetz Marine Laboratory on the Red Sea. One of these was aimed at studying the potential for recruitment of symbionts from among free-living diatoms in the habitat. We carefully collected individual animals and their substrates at different depths and different stations along the Gulf of Elat; the endosymbionts from some animals at each station were isolated. The rest of the material was fixed for examination with the light and scanning electron microscopes (see Figs. 18 and 19). The second project involved experiments aimed at determining if one host, *Amphistogina lessonii*, has a preference for one or more of the algae which have been isolated as endosymbionts from larger foraminifer (in preparation); zoochlorellae and small (< 10 μm) free-living isolates of diatoms from a United States salt marsh were also tested for their potential as endosymbionts. DCMU (1×10^{-5} *M*) was used to make the foraminifera nearly aposymbiotic. Controls included unbleached animals, animals which were not fed (''rebrown'' controls), and animals whose symbionts were isolated shortly after harvest from the sea.

The results were interesting. DCMU inhibits the endosymbiotic diatoms and apparently makes them vulnerable for replacement. Several new findings emerged from these experiments:

1. The endosymbiotic diatoms within the foraminifera can be replaced within a period of 7 days.
2. None of the free-living diatoms isolated from northeastern littoral environment which we tested in the experiment survived the digestive process to become endosymbionts.
3. *Chlamydomonas hedleyi* and *C. provasolii*, endosymbiotic chlorophytes from *Archaias angulatus* and *Cyclorbiculina compressa* (both Soritidae), were not retained as endosymbionts by the Red Sea foraminifera *Amphistegina*.
4. We found that under the conditions tested there was a consistent preference for some endosymbiotic diatoms over others by *Amphistegina lessonii*. At 10 m *Nitzschia valdestriata* and *N. laevis* were the most preferred endosymbiotic species, *Fragilaria shiloi* the least preferred. However at 20 m, *N. valdestriata* was preferred over *N. laevis*. While time and the great effort required to do the experiments prevented us from comparing every species with every other species, we were able to develop the following preferential relationships: *N. laevis* > *F. shiloi*, *N. valdestriata* ≈ *N. laevis*, *Amphora* sp. > *F. shiloi*, *N. val-*

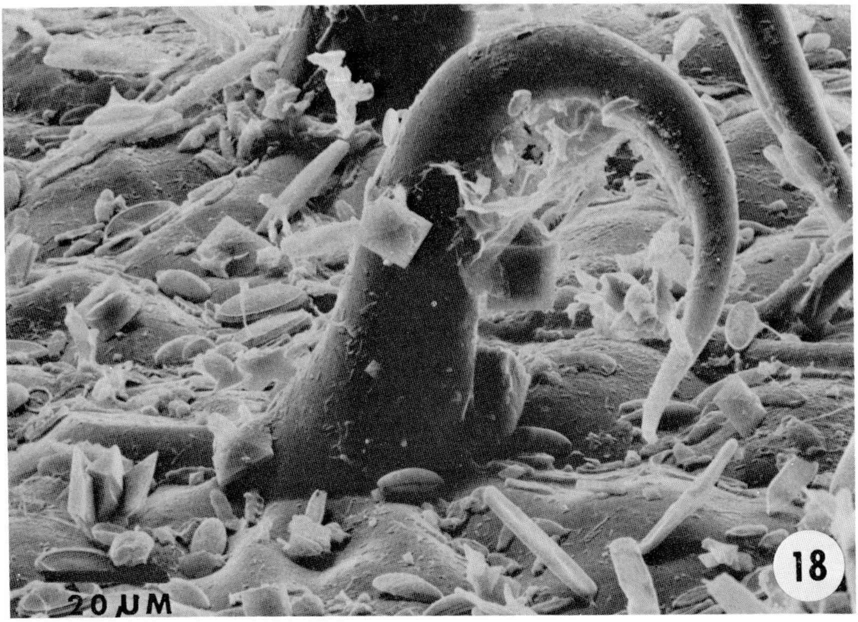

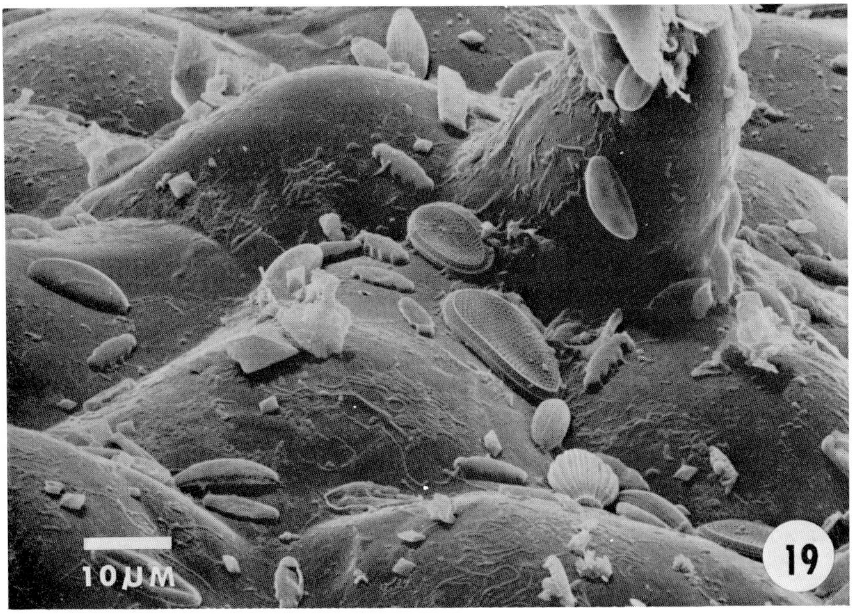

FIGS. 18 AND 19. The surface of a *Halophila* leaf showing pennate diatoms which serve as food for the animal and which potentially might be recruited as endosymbionts. The hook-like structure on the surface of the leaf is a unicellular leaf-hair (trichome). SEM. Figure 18. ×850; Figure 19. ×1200.

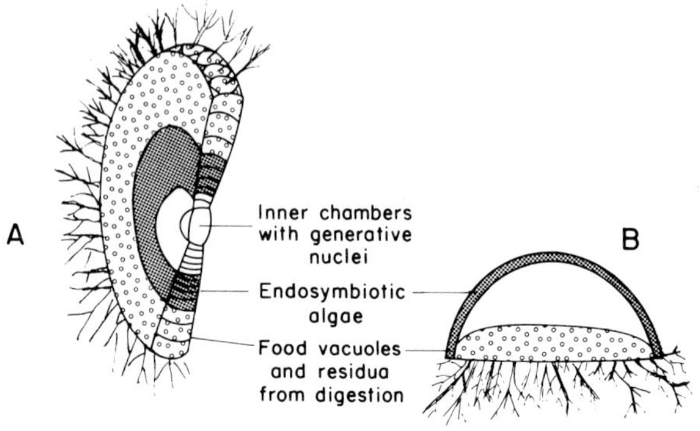

FIG. 20. Diagrammatic representation of regionalization of feeding and symbionts in (A) *Sorites marginalis* and (B) *Amphistegina lessonii*.

destriata > *Amphora* sp., *N. laevis* > *Amphora* sp., *N. laevis* > *F. shiloi, N. laevis* > *N. frustulum, N. valdestriata* > *N. frustulum*.

V. Unresolved Questions

At the present time it is easy to see that larger foraminifera have varied and interesting relationships with endosymbiotic algae. The subject is a fertile ground for research, and with some effort progress is being made to extend our knowledge of symbiosis in these rather numerous benthic marine animals. In the last few years our focus has been on the diatom-bearing species and in particular on questions relating to the identity and specificity of their endosymbionts and relating to symbiotic physiological and metabolic interactions with their hosts. Some of the questions we have thought about are listed below.

A. Taxonomic Specificity of the Diatom Endosymbionts

1. Do populations of diatom-bearing foraminifera have the same endosymbionts at (a) different depths, (b) different seasons, and (c) at moderate special separations (0.01–1 km)?

2. If populations at the same location change diatoms endosymbionts with time, is it accomplished by (a) feeding and survival of potential endosymbionts with the eventual dominance of the best adapted endosymbiont for the time and location, or (b) at sexual reproduction?

3. Do abiotic physiological conditions (temperature, light) favor the retention of some potential endosymbiont diatom species over others? Are diatom endosymbionts recruited from the natural epibenthic or epiphytic populations upon

which larger foraminifera feed? If so, what are their abundance relationships among the natural population?

B. Physiological and Metabolic Interactions between Hosts and Endosymbionts

1. What is the relative contribution of diatom photosynthesis to the C and O budget of the foraminiferal shell? How is this affected by light, nutrients, or temperature?

2. What is the relative contribution of ingested food to the C and O budget of the foraminiferal shell?

3. Is metabolic C recycled between the diatom endosymbionts and their hosts? (a) What is the nature of the exchange: (i) feeding on, or autolysis of, some endosymbionts; (ii) metabolite release; (iii) recycling of CO_2? (b) What are the quantitative aspects of the exchange in terms of C budget?

4. Are there differences in the above relationships between various hosts and the same endosymbionts and between the same host and various endosymbiotic species? Can abiotic physiological factors affect the relationships? (a) Is there in fact, a hierarchy among potential endosymbionts?

5. Are there any mechanisms for excluding or deterring the establishment of new symbionts? This may be accomplished by (a) inhibitory or protective substances produced by established symbionts; (b) special separation of established symbionts from regions of the cell with active digestive vacuoles.

My colleagues world wide and our students look forward with eager anticipation to answering some of these questions in the very near future.

Acknowledgments

The National Science Foundation, The U.S.–Israel BiNational Science Foundation, and the H. B. Cantor Foundation have all been generous in the support of various phases of the work done in our laboratory. Over the years, the author has worked in close collaboration with Drs. Marie McEnery, John H. Tietjen, and Norman Saks of City College of New York. In addition, collaboration with Drs. Moshe Shilo, Yehuda Cohen, Zeev Reiss, Yohanathan Erez, Rudolf Röttger, Edith Muller-Merz, Lucas Höttinger, Suzy Leutenegger, and H. J. Hansen is also acknowledged. The technical assistance of Mr. Carmine Mastropoalo, Mr. Robert Koestler, and Ms. Joan Whelan on various projects is also gratefully acknowledged.

References

Berger, W. H. (1981). *Paleogeogr. Paleoclimatol. Paleoecol.* **33,** 3–7.

Berthold, W. (1976). *J. Foram Res.* **6,** 22–29.

Berthold, W. (1978). *Arch. Protistenkunde* **120,** 16–62.
Buchardt, B., and Hanson, H. (1977). *Bull. Geol. Soc. Denm.* **26,** 185–194.
Cook, C. B. (1980) *In* "Endocytobiology" (W. Schwemmler and H. E. A. Schenk, eds.), Vol. I, pp. 249–277. de Gruyter, Berlin.
Dietz-Elbrachter, G. (1971). *Meteor Forschungsge Meinschaft Ergebnisse* **6,** 41–47.
Doonan, S. A., Douglas, A. E., and Gooday, G. W. (1980). *In* "Endocytobiology" (W. Schwemmler and H. E. A. Schenk, eds.), pp. 293–304. de Gruyter, Berlin.
Dupplessy, J. C., Lalow, C., and Vinot, A. C. (1970). *Science* **168,** 250–251.
Duquay, L., and Taylor, D. (1978). *J. Protozool.* **25,** 356–361.
Erez, J. (1978). *Nature (London)* **273,** 199–202.
Fermont, W. J. J. (1977). *Utrecht. Micropaleontogic Bull.* **15,** 149–164.
Freudenthal, H. D. (1962). *J. Protozool.* **9,** 45–52.
Hallock, P. (1979). *J. Foram. Res.* **9,** 61–69.
Hallock, P. (1981). *J. Foram. Res.* **11,** 40–46.
Hallock, P., and Hansen, H. (1979). *Bull. Geol. Soc. Denm.* **27,** 99–104.
Hansen, H. J. (1972). *Micropaleontology* **18,** 223–230
Hansen, H., and Dalberg, P. (1979). *Bull. Geol. Soc. Denm.* **28,** 47–55.
Haynes, J. (1965). *Cont. Cusham Found Foram Res.* **16,** 40–43.
Hottinger, L. (1977). *Utrecht. Micropaleont. Bull.* **15,** 35–109.
Hottinger, L. (1978). *In* "Foraminifera" (R. H. Hedley and C. G. Adams, eds.), Vol III, pp. 203–266. Academic Press, New York.
Hottinger, L., and Dreher, D. (1974) *Mar. Biol.* **25,** 41–61.
Hyman, L. H. (1940). "The Invertebrates: Protozoa through Ctenophora." McGraw-Hill, New York.
Karakashian, M. W. (1975). *Symp. Soc. Exp. Biol.* 145–173.
Karakashian, S. J., and Rudzinska, M. A. (1981). *Exp. Cell. Res.* **131,** 387–393.
Kremer, B., Schmazjohann, R., and Rottger, R. (1980). *Mar. Ecol. Prog. Ser.* **2,** 225–228.
Larsen, A. R. (1976). *Israel J. Earth Sci.* **25,** 1–26.
Larsen, A. R., and Drooger, L. W. (1977). *Utrecht Micropaleont. Bull.* **15,** 225–240.
Lee, J. (1974) *In* "Foraminifera" (R. H. Hendey and G. Adams, eds.), Vol. I. pp. 208–257. Academic Press, New York.
Lee, J., and Bock, W. (1976). *Bull. Mar. Sci.* **26,** 530–537.
Lee, J. J., and McEnery, M. E. (1982). *In* "Algal Symbiosis: A Continuum of Interaction Strategies" (L. I. Goff, ed.), pp. 105–128.
Lee, J., and Zucker, W. (1969). *J. Protozool.* **10,** 71–81.
Lee, J., McEnery, M., and Rubin, H. (1969). *J. Protozool.* **16,** 377–395.
Lee, J. Crockett, L, Hagen, J., and Stone, R. (1974). *Br. J. Phycol.* **9,** 407–422.
Lee, J., McEnery, M., Kennedy, E., and Rubin, H. (1975). *J. Phycol.* **11,** 14–49.
Lee, J. J., McEnery, M, Kahn, E., and Schuster, F. (1979a). *Micropaleontology* **25,** 118–140.
Lee, J. J., McEnery, M. E., Shilo, M., and Reiss, Z. (1979b). *Nature (London)* **280,** 57–58.
Lee, J., McEnery M., and Garrison, J. (1980a). *J. Foram Res.* **10,** 31–47.
Lee, J., McEnery, M., Reidy, J., Garrison, J., and Rottger, R. (1980b). *In* "Endocytobiology" (W. Schwemmler and H. E. A. Schenk, eds.), Vol. I, pp. 113–124. de Gruyter, Berlin.
Lee, J., McEnery, M., Rottger, R., and Reimer, L. (1980c). *Bot. Mar.* **23,** 297–302.
Lee, J., Reimer, C., and McEnery, M. (1980d). *Bot. Mar.* **23,** 41–48.
Lee, M. J., Ellis, R., and Lee, J. J. (1982). *Mar. Biol.* (in press).
Leutonegger, S. (1977a). *Utrecht Micropaleont. Bull.* **15,** 225–239.
Leutenegger, S. (1977b). *Cahiers Micropaleont.* **3,** 5–53.
Leutenegger, S. (1977c). *Mar. Biol.* **44,** 157–164.
Leutenegger, S., and Hansen, H. (1979). *Mar. Biol.* **54,** 11–16.

Levanson-Spanier, I., Padan, E., and Reiss, Z. (1979). *Deep Sea Res.* **26/6A,** 673–685.
Lopez, E. (1979). *Mar. Biol.* **53,** 201–211.
McAuley, P. J. (1981). *J. Cell. Sci.* **47,** 207–226.
McEnery, M., and Lee, J. (1981). *Micropaleontology* **27,** 71–83.
McFarlane, A. E. (1982). *J. Mar. Biol. Assoc. U.K.* **62,** 235.
Muller, P. (1978). *J. Foram Res.* **8,** 35–41.
Muller, P. (1974). *Limnol. Oceanogr.* **19,** 802–809.
Muller-Merz, E., and Lee, J. (1976). *J. Protozool.* **23,** 390–396.
Muscatine, L. (1971). *Pac. Sci.* **25,** 13–21.
Muscatine, L., and Pool, R. R. (1979). *Proc. R. Soc. London* **204,** 131–139.
Muscatine, L., Pool, R. R., and Trench, R. K. (1975). *Trans. Am. Micros. Soc.* **94,** 450–469.
Newell, R. (1949). *Evolution* **3,** 103–124.
Pardy, R. I., and Muscatine, L. (1973). *Biol. Bull.* **145,** 565–579.
Pool, R. R., and Muscatine, L. (1980). *In* "Endocytobiology" (W. Schwemmler and H. E. A. Schenk, eds.), Vol. I, pp. 225–240. de Gruyter, Berlin.
Reiss, Z. (1977). *Utrecht Micropaleont. Bull.* **1,** 7–25.
Reisser, W. (1980). *Arch. Microbiol.* **125,** 291–293.
Ross, C. (1974). *Proc. Sec. Int. Coral Reef Symp. I,* Brisbane, 1974.
Ross, C. A. (1972). *J. Protozool.* **19,** 181–192.
Röttger, R. (1972). *Mar. Biol.* **17,** 228–242.
Röttger, R. (1976). *Mar. Sed. Special Publ.* **1,** 75–79.
Röttger, R., Irwan, A., Schmaljohann, R. and Franzisket, C. (1980). *In* "Endocytobiology" (W. Schwemmler and H. E. A. Schenk, eds.), Vol. I, pp. 125–132. de Gruyter, Berlin.
Schmaljohann, R., and Rottger, R. (1978). *J. Mar. Biol. Assoc. U.K.* **58,** 227–237.
Schoenberg, D. (1980a). *In* "Endocytobiology" (W. Schwemmler and H. E. A. Schenk, eds.), Vol. I, pp. 146–154. de Gruyter, Berlin.
Schoenberg, D., and French, R. (1980a). *Proc. Soc. London* **207,** 405–427.
Schoenberg, D., and Trench, R. (1980b). *Proc. Soc. London* **204,** 429–444.
Schoenberg, D., and Trench, R. (1980c). *Proc. Soc. London* **207,** 445–460.
Smith, D. C. (1979). *Proc. R. Soc. London* **204,** 113.
Smith, D. C. (1980). *In* "Endocytobiology" (W. Schwemmler and H. E. A. Schenk, eds.), Vol. I, pp. 317–334. de Gruyter, Berlin.
Spindler, M. (1978a). *J. Foram. Res.* **8,** 258–261.
Spindler, M. (1978b). *J. Foram. Res.* **8,** 319–331.
Taylor, D. L. (1974). *In* "Symbiosis in the Sea" (W. B. Vernberg, ed.), pp. 245–262. Univ. of South Carolina Press, Columbia.
Taylor, D. L. (1980). *In* "Endocytobiology" (W. Schwemmler and H. E. A. Schenk, eds.), Vol. I. pp. 279–292. de Gruyter, Berlin.
Vinot-Bertowille, A., and Dupplessy, J. C. (1973). *Earth Plan. Sci. Lett.* **18,** 247–252.
Weber, J. N., and Woodhead, P. M. J. (1972). *J. Geophys. Res.* **77,** 463–473.
Wefer, G., and Berger, W. H. (1980). *Science* **209,** 803–805.
Wefer, G., Killingley, J. S., and Lutze, G. F. (1981). *Paleogr. Paleoclimatol. Paleoecol.* **33,** 253–270.
Weis, D. (1976). *J. Protozool.* **23,** 527–529.
Williams, D. F., Rottger, R. Schmaljohann, R., and Keigwin, L. (1981). *Paleogr. Paleoclimatol. Paleoecol.* **33,** 231–251.
Woodruff, F. W., Sarin, S. M., and Douglas, R. G. (1980). *Mar. Micropaleontol.* **5,** 3–11.
Zmiri, A., Kahan, D., Hochstein, S., and Reiss, Z. (1974). *J. Protozool.* **21,** 133–138.
Zohary, T., Reiss, Z., and Hottinger, L. (1980). *Ecologae Geol. Helv.* **73,** 1071–1094.

INTERNATIONAL REVIEW OF CYTOLOGY, SUPPLEMENT 14

The Biology of the Xenosome, an Intracellular Symbiont

A. T. SOLDO

Research Laboratories of the Veterans Administration Medical Center and the Department of Biochemistry, University of Miami School of Medicine, Miami, Florida

I. Introduction

Symbiotic associations between bacterium-like particles and eukaryotic host cells are widespread in nature. Intracellular bacteria-like particles have been found in the cells and eggs of insects (Ishikawa, 1978; Koerner, 1978; Wright *et al.*, 1978), root nodules of leguminous plants (Brewin *et al.*, 1980), blue green algae (Wyrek, 1979), a variety of protozoa including ciliates (Mouzarede, 1978), an ameba (Jeon and Ahn, 1978), a sporozoan (Desportes *et al.*, 1977), *Giardia* (Nemanic *et al.*, 1979), *Crithidia* (Mundion and Roitman, 1977), *Euplotes* (Heckmann, 1975; Rosati *et al.*, 1976), many strains of *Paramecium* (Sonneborn, 1959; Preer *et al.*, 1974), as well as strains of free-living ciliates devoid of mitochondria (Fenchell *et al.*, 1976).

A well-studied group of bacterium-like particles are the endosymbionts of *Paramecium aurelia*. These were originally designated by letters of the Greek alphabet (Sonneborn, 1959) but now have been given binominal names and classified as bacteria. In general, the symbionts of *Paramecium* are intracytoplasmic (Preer *et al.*, 1974) but may also be found in the nucleus (Preer,

ISBN 0-12-364375-9

1969; Ossipov and Podlipaev, 1978; Görtz, this volume). Some of the particle types, particularly kappa (*Caedibacter*) are capable of infecting homologous and heterologous symbiont-free strains. In these cases, infectivity may be observed by exposing particle-free animals to freshly prepared homogenates of kappa-bearing animals of the same genotype; lambda particles have been reported to be infective by a similar process (van Wagtendonk *et al.*, 1963). Mu particles are also infective but only through conjugation (Siegel, 1954). Kappa particles appear to be unique among the symbionts of *Paramecium* because of the presence of a so-called "R" or refractile body. Kappa-containing R bodies produce a "toxin" which, under certain conditions, can kill a susceptible symbiont-free *Paramecium* of suitable genetic constitution (Sonneborn, 1939). In no case, however, has it been demonstrated that kappa or any of the symbionts of *Paramecium* are essential for the growth and division of the host. Removal of the symbionts by treatment with chemical agents or antibiotics (Stock *et al.*, 1951; Soldo, 1961; Soldo *et al.*, 1970) or by genetic manipulation (Sonneborn, 1946) did not result in the death of the protozoan. In only one instance has it been shown that the particles contribute to nutritional economy of the host cell. Lambda particles produce sufficient quantities of biopterin and folic acid (both essential nutrients for the symbiont-free host protozoan) to permit lambda-bearing animals to grow and divide in the absence of these substances (Soldo, 1973).

In not all cases are bacterial symbionts "dispensable" with respect to growth and division of the host protozoan. "X" bacteria, cytoplasmic symbionts of certain amoeba, have become essential for the growth and survival of the host (Jeon and Jeon, 1976, and this volume). Similarly, the bacterial endosymbionts of *Pleomyxa palustris* (Andresen *et al.*, 1965) and omicron particles of the ciliate *Euplotes* appear to be essential for the survival of the protozoan (Heckmann, 1975, and this volume). In these cases, removal of the symbionts results in the ultimate demise of the host.

In recent years, we identified and introduced into axenic culture several strains of the marine protozoan *Parauronema acutum*, a number of which contain intracellular symbiotic bacteria-like particles (Soldo and Merlin, 1972). The particles, which we have termed xenosomes ("alien-bodies"), appear to be analogous to the symbiont particles of *P. aurelia*. They may be recovered from *P. acutum* at concentrations of $1-1.5 \times 10^{12}$ from 10 liters of culture containing $1-1.5 \times 10^{6}$ animals per ml. Xenosomes are examples of infectious, intracellular, symbiont particles and are useful models for the study of infective agents of bacterial origin (Soldo *et al.*, 1974). In the past, we employed the endosymbionts of the fresh water ciliate *Paramecium aurelia* in studies of this nature [e.g., kappa (*Caedibacter*); mu (*Pseudocaedibacter*), and lambda]. Because it has proved difficult, if not impossible, to regularly demonstrate infectivity with the *Paramecium* symbionts under nonbacterized conditions, we developed a methodology for handling large quantities of axenically cultured xenosome-bearing marine ciliates in which infectivity can readily and regularly be demonstrated. For this reason we adopted the marine xenosome:protozoan system as our experimental model.

The purpose of this article is to describe our studies concerning the biochemistry and molecular biology of these infectious intracellular particles.

II. Properties of Xenosomes

A. General

Xenosomes comprise a group of infectious intracellular bacterium-like particles found in certain strains of the marine hymenostome ciliate *Parauronema acutum*. The particles are rod-shaped, 0.3 μm in diameter, and 0.8 μm in length and occur exclusively in the cytoplasm in numbers ranging from 50 to 300 per protozoan. They contain DNA, RNA, protein, carbohydrate, and lipids in amounts comparable to those found in small bacteria. They are gram negative, divide by transverse fission in synchrony with the host protozoan, and are destroyed by the action of a number of antibiotics including penicillin, tetracycline, and chloramphenicol, but not cycloheximide, neomycin, or streptomycin. Other antimetabolites including mitomycin C, nalidixic acid, rifampcin, and tubercidin exert a selective effect on the particles and result in their irreversible loss from the cytoplasm. Xenosomes are also destroyed by irradiation with an electron beam (Fig. 1). The effective dosage (ED_{50}) was about 30,000 rad, comparable with values reported for X-irradiation of certain bacteria and members of the *Mycoplasma*. The ED_{50} for the protozoan itself was approximately 260,000 rad. Thus xenosomes could be removed from the cytoplasm of the ciliates by treatment with the electron beam using dosages of about 200,000 rad. The infectious nature of the particles may be regularly demonstrated by incubating freshly prepared homogenates of xenosome bearers with intact particle-free animals and observing the uptake of the particles into the cytoplasm (Soldo *et al.*, 1974). Particles thus taken up reproduce in the cytoplasm and are maintained indefinitely through serial subculture. Xenosomes are also capable of killing or inhibiting the growth of other marine strains, particularly those of the related genus *Uronema*. Isolated and purified particle preparations consume oxygen at a rate about 20 times lower than that of the bacterium *Escherichia coli*. A number of enzymes of the glycolytic pathway and those of the Krebs cycle have been identified, although it has not yet been possible to establish the principal means by which the particles derive energy. Highly purified xenosome DNA exhibits a sharp thermal transition suggesting that the molecule is a native duplex. The molecular size of the genome, determined from kinetic data, is comparable to that of some *Mycoplasma* and *Rickettsias*. Values obtained from analytical data

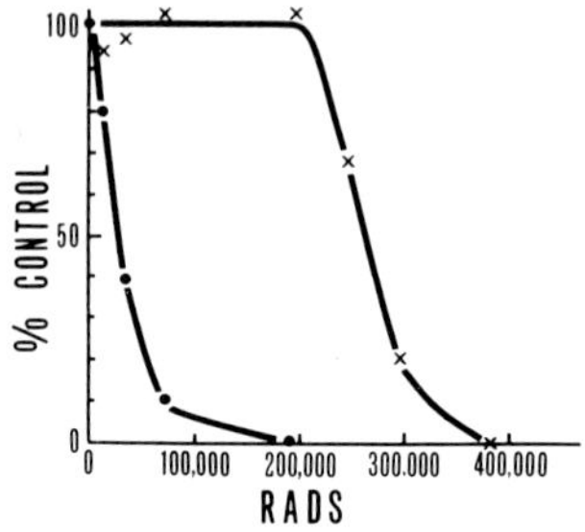

FIG. 1. Destruction of xenosomes by the electron beam. ●——●, Xenosomes; x——x, *P. acutum*

suggest a size several-fold greater than this. Hence, the particles possess multi-copy genomes. In addition, they contain extrachromosomal elements in the form of covalently closed circles of duplex DNA (plasmids).

Studies employing DNA–DNA hybridization techniques indicate that DNA isolated from xenosomes of several stocks of *P. acutum* possesses a high degree of base sequence homology and suggest that the symbionts are members of closely related species. At the present time we have on hand 12 xenosome-bearing stocks of *P. acutum*. These were isolated from various locations along the South Florida coast and have been maintained in axenic culture for several years. (Table I). Most of our work has been conducted with one strain of *P. acutum* designated as stock 110-3. Xenosomes in this stock are typical and serve to illustrate the features of xenosomes in general.

TABLE I
SOURCE OF XENOSOME-BEARING STOCKS OF *Paraurónema acutum*

Stock	Source	Year
108-7, 110-3	Backwater, Miami, Florida	1971
112-5	Tidal pool, Coral Gables, Florida	1971
263-2, 263-5, 263-11, 263-12, 263-13, 263-14, 263-18, 263-20	Tidal pools and sandy bottom, Coral Gables, Florida	1977
265-6	Backwater, Homstead, Florida	1977

B. Ultrastructure

Figure 2 is an electron micrograph of a xenosome-bearing *P. acutum*. The animals were fixed in a 1.3% OsO_4, 0.1 *M* collidine buffer pH 7.2, dehydrated through an alcohol series, embedded in Araldite, sectioned with a diamond knife, stained with uranyl acetate and lead citrate, and viewed in a Phillips EM 300. The xenosomes (arrows) are clearly distinguishable from other subcellular components such as the nucleus, mitochondria, and microbodies. Figure 3 illustrates a group of particles at higher magnification. The particles divide by a "pinching off" process; there is no evidence of the formation of a transverse septum as occurs in a number of gram-positive bacteria (xenosomes are gram-negative). Details of the ultrastructure of the particle are most readily seen in Fig. 4. The cell envelope is complex and consists of a well-defined double track outer membrane whose external surface is coated with minute particles of electron-dense material in direct contact with the cytoplasm of the host protozoan. The inner membrane appears to consist of a single electron-dense component. The inner and outer membranes are separated by an area which is largely electron transparent. The internal contents of the particle contain a number of areas of poorly defined electron-dense material enmeshed in a network of fibers (DNA). The fibrous material extends throughout the entire internal contents of the xenosome itself. Additional morphological features of the particles may be observed through negative staining techniques. When released from the protozoan (ruptured by gentle pressure from a glass coverslip) the particles can be seen spinning (propeller-like) and darting rapidly, suggesting a motile apparatus. Phosphotungstic acid stains of particles released by this technique reveal that many do, indeed, possess bacteria-like flagella. Figure 5 shows a xenosome with three flagella occurring at various points on the surface of the particle. The flagellum typically is wider at the point of attachment on the surface of the particle. Figure 6 compares the frequency of the occurrence of flagella with the particles themselves. Most particles contain two or more flagella although a substantial number (9%) have none.

C. Infectious Nature

Cell-free preparations of xenosomes may be obtained by filtering suspensions of xenosome-bearing *P. acutum* aseptically through Millipore filters of a pore size of 8 μm in diameter. Under these conditions the particles pass through the filter whereas the protozoans do not (Soldo and Brickson, 1978). The infectious nature of the particles could be demonstrated by adding aliquots of these cell-free preparations to whole-cell particle-free animals and observing the uptake of the particles into the cytoplasm. Particles taken up grow and divide in synchronism with the host protozoan and are maintained indefinitely through serial subculture.

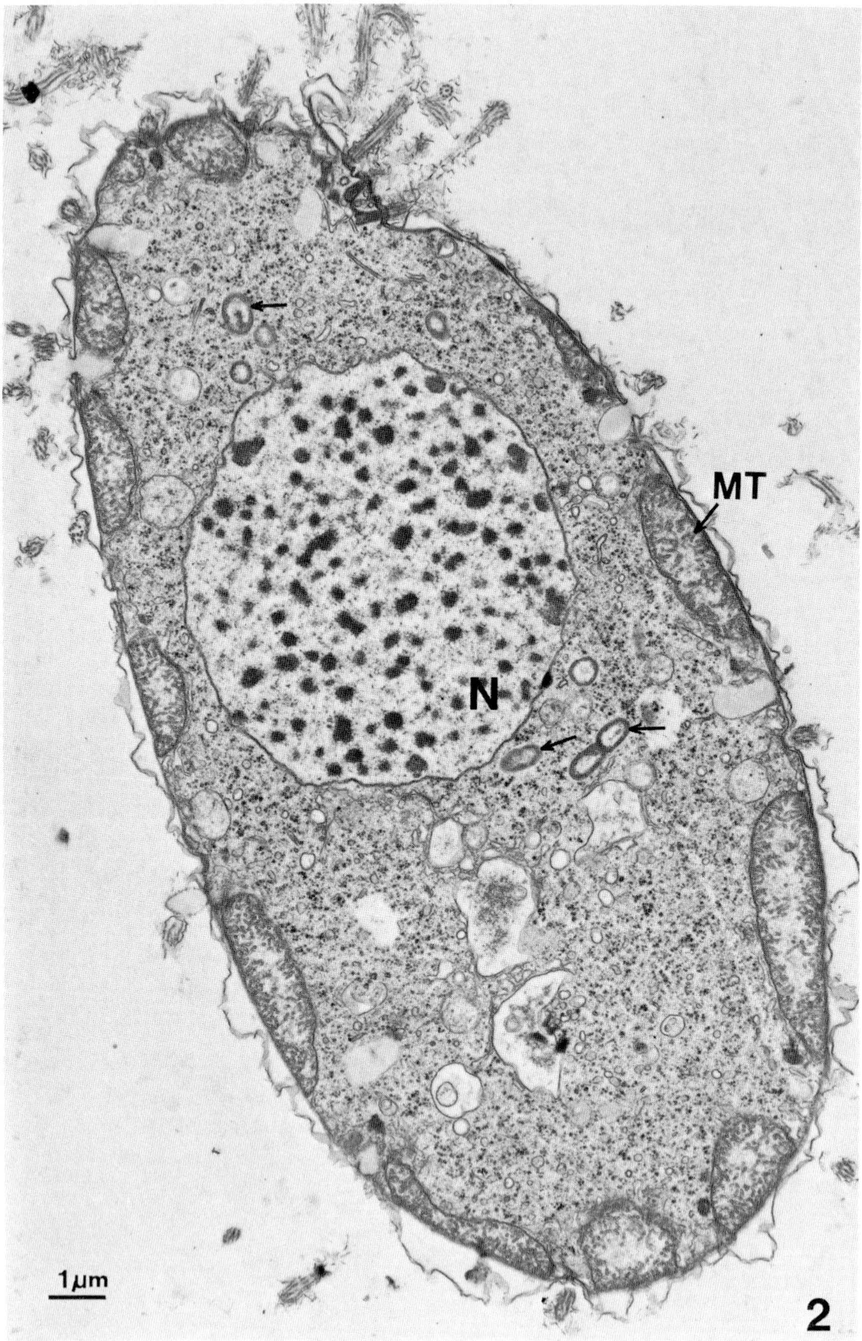

FIGS. 2–4. Electron micrographs of xenosome-bearing *P. acutum*. Arrows, xenosomes; MT, mitochondria; N, nucleus; OM, outer membrane; IM, inner membrane.

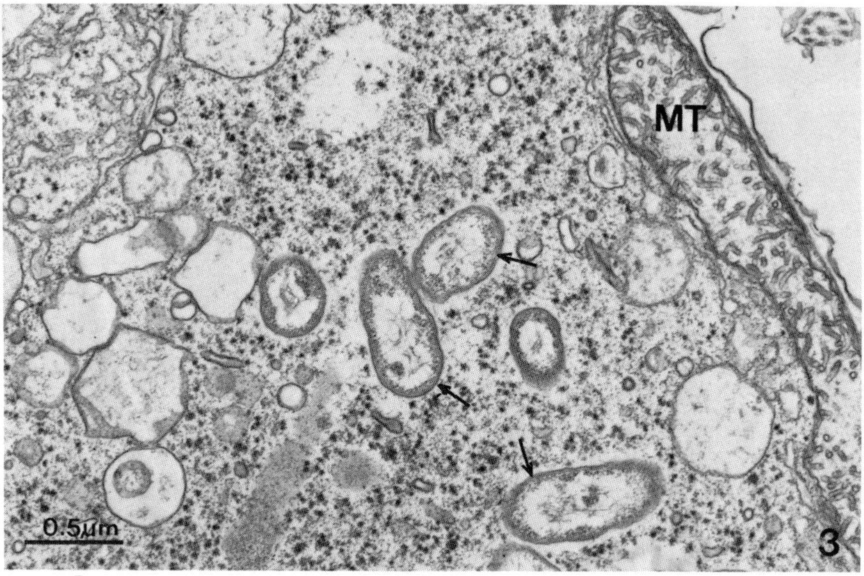

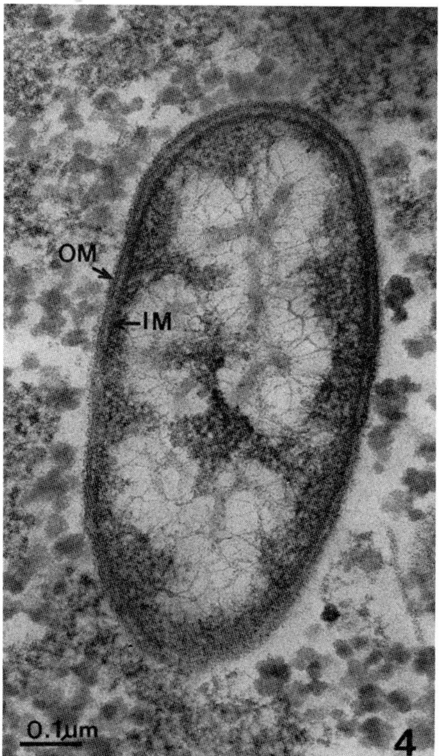

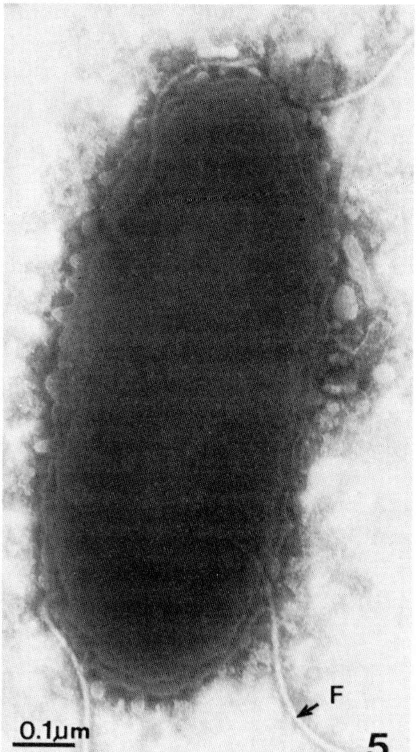

FIG. 5. Electron micrograph of a phosphotungstic acid stained xenosome. F, Flagellum.

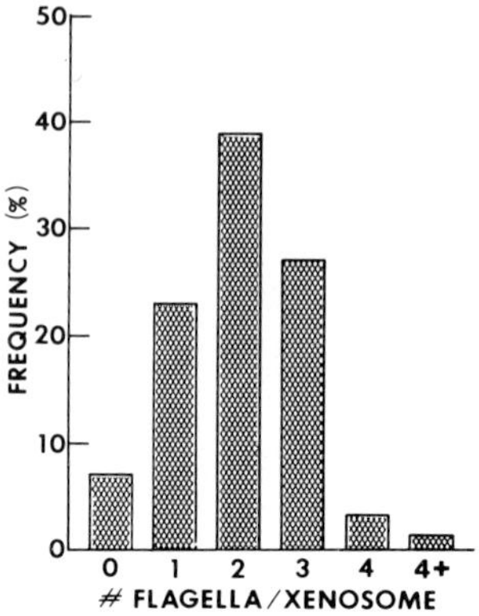

FIG. 6. Frequency of distribution of flagella in xenosomes.

Although the infective agent readily passes through Millipore filters with pores 8 μm in diameter, filters of 0.45 μm in diameter block its passage. When suction is used to filter suspensions of whole xenosome-bearing *P. acutum,* the protozoans rupture, thereby releasing their particles (Table II). If care is exercised during the process (i.e., slow filtration by gravity and allowing a small volume of the suspension to remain behind in the filter) it is possible to filter cell suspensions without disrupting the animals. Thus, we found that, during normal growth, intact protozoans actively release xenosomes into the ambient culture fluid (Fig. 7). The rate at which this process takes place parallels the rate of division of the protozoans during the first day of the logarithmic phase of growth, diminishes somewhat during the next 2 days, peaks after the fourth day, and then falls off. Net rate of production of xenosomes during the first 24 hours of growth is about 0.3 particle per protozoan per day and about 0.08 particle per protozoan per day during the 24-hour period from the third and fourth days of growth. Infectivity of the particle suspensions appears to decrease as a function of motility. Microscopic observation of xenosomes released from the protozoans

TABLE II
EFFECT OF FILTRATION ON RELEASE OF XENOSOME FROM XENOSOME-BEARING *Parauronema acutum*

Filtration	Viable cells recovered (%)	Xenosomes in filtrate (per ml)
Complete (full suction)	3	11.1×10^7
Complete (partial suction)	22	8.7×10^7
Complete (gravity)	94	0.7×10^7
Partial (gravity)	100	8.0×10^4

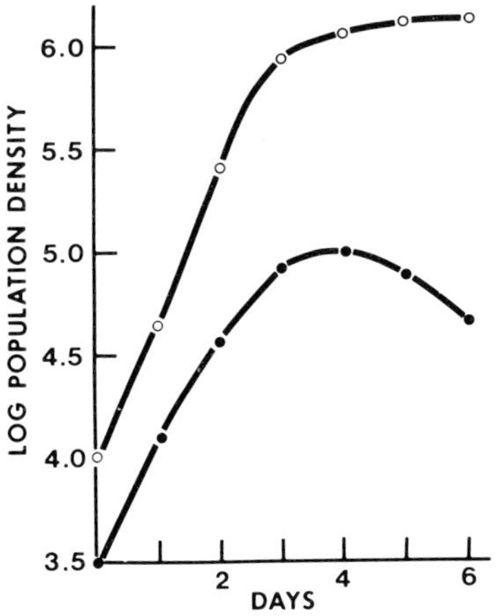

FIG. 7. Release of xenosomes during growth of *P. acutum*. ○——○, Host cells; ●——●, xenosomes.

shows large numbers of motile forms. Upon standing at room temperature for 2 hours, the number of motile forms decreases to about half those originally present (Fig. 8). Moreover, the rate of decrease parallels closely the loss of infectivity of the particle suspension (Fig. 9). In another experiment, we added xenosome-bearing animals to xenosome-free cultures. Despite the fact that xenosome bearers divide essentially at the same rate as xenosome-free protozoa (Fig. 10), addition of as few as one xenosome bearer for every 20 xenosome-free animals resulted in populations containing 100% xenosome-bearing cultures after two or three weekly subcultures (Table III). The infectious agent is relatively unstable to temperature (Fig. 11). Infectious activity is lost almost imme-

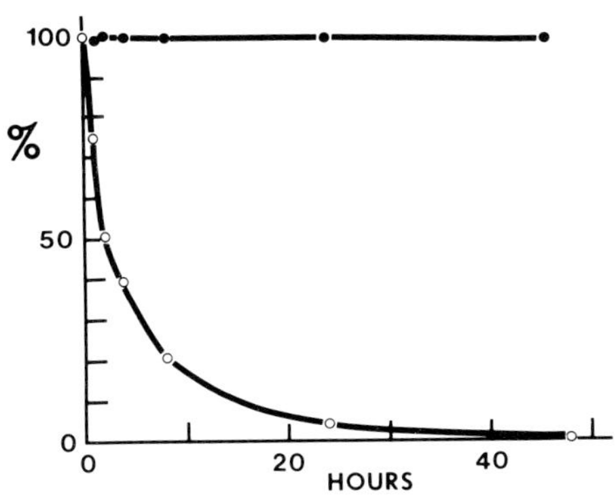

FIG. 8. Loss of motility of xenosomes released from *P. acutum*. ○——○, Motile; ●——●, total.

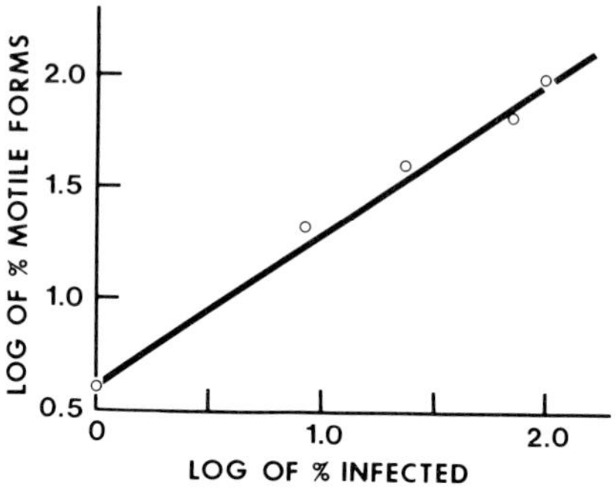

FIG. 9. Correlation of loss of motility of xenosomes and infectivity.

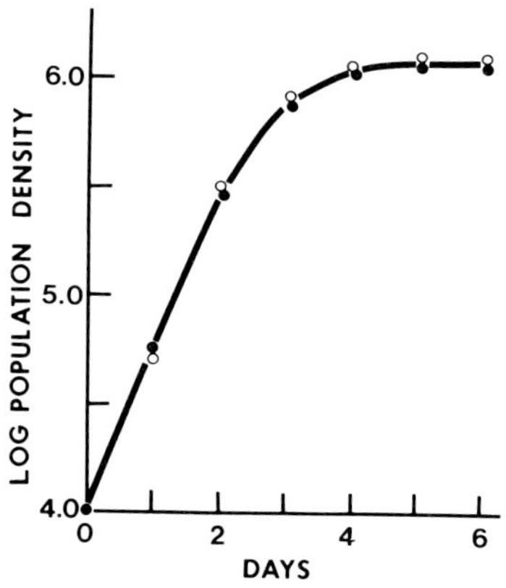

FIG. 10. Growth curves of whole cell xenosome-bearing and xenosome-free *P. acutum*. ○——○, Xenosome-bearing; ●——●, xenosome-free.

TABLE III
EFFECT OF ADDING XENOSOME-BEARING *P. acutum* TO CULTURES OF XENOSOME-FREE *P. acutum*[a]

Proportion of xenosome-bearing to xenosome-free *P. acutum* in the inoculum (%)	Proportion of *P. acutum* with xenosomes after growth at weekly subculture		
	1	2	3
100	100	100	100
50	62	100	100
25	35	97	100
5	7	34	100
0	0	0	0

[a]Inoculum size was 10,000/ml. Cultures were incubated at 27°C and transferred weekly to fresh medium.

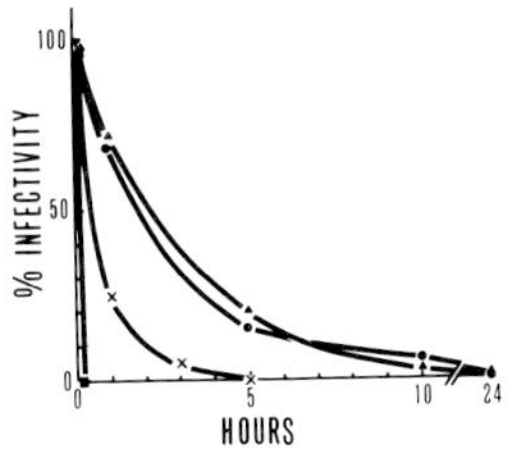

FIG. 11. Stability of infectious agent to temperature. ■, 65°C; x——x, 37°C; ●——●, 22°C; ▲——▲, 4°C.

diately at 65°C, in about 1 hour at 37°C, and 2 hours at 22 and 4°C. Sonication for periods as brief as 5 seconds completely destroys the infective activity of the particles. Attempts to stablize the infectious agent, by varying conditions or ionic strength, osmotic pressure, and pH, have proved ineffective. Statistical analysis shows that the mode of action of the infectious agent follows a "one-hit" phenomenon and suggests that a single particle is capable of infecting a susceptible particle-free animal (Fig. 12). However, actual titration using carefully counted xenosome suspensions reveal a "threshold" phenomenon, i.e., 100–200 xenosomes are required before a single xenosome could infect a protozoan. Host-range studies show that it is possible to establish infections, using xenosomes of stock 110-3, in *P. acutum* stocks 108-7, 110-3, and 112.5 *Parauronema* spp. 111-2, and *Miamiensis avidus,* Ma, but not with other marine ciliate strains tested. (Table IV).

D. Toxicity

Xenosomes of stock 110-3 markedly inhibited the growth of *Uronema nigricans,* Pc, *Uronema marinum* PW-2, and *Parauronema* spp. 34-7 (Table V). *Parauronema* spp. stocks 108-7, 111-2, and 114-6 were inhibited to a lesser extent. The growth inhibitory effect of 110-3 xenosomes on *Uronema nigricans* Pc was most dramatic. Suspensions containing as few as 100 xenosomes per protozoan blocked cell division and ultimately resulted in the death of the pro-

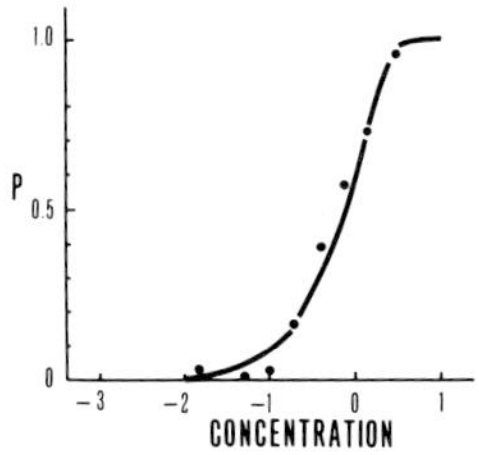

FIG. 12. Experimental frequency of infections. Points represent the average of four experiments. The solid line was derived from $P_{r>0} = 1 - e^{-s}$ for a "one-hit" phenomenon.

TABLE IV
HOST RANGE OF 110-3 XENOSOMES

Ciliate	Stock	Relative infectivity (%)
Parauronema acutum	108-7	42
Parauronema acutum	110-3	100
Parauronema acutum	112-5	10
Parauronema sp.	2-1	0
Parauronema sp.	34-7	Inhibitory
Parauronema sp.	108-6	0
Parauronema sp.	111-2	100
Parauronema sp.	114-6	Inhibitory
Miamiensis sp.	1-1	0
Miamiensis avidus	Ma	75
Uronema marinum	PW-2	Inhibitory
Uronema nigricans	Pc	Inhibitory

tozoan. The "toxic" agent exhibited characteristics that were strikingly similar to those observed for the "infective" agent. The "toxic" agent readily passed through 8-μm pore size Millipore filters, but not through 0.45- or 0.22-μm pore size filters. The toxic effect was lost after 2 hours when xenosome suspensions were maintained at 37°C, after 24 hours at 22°C, and after 3–4 days at 4°C. Treatment of the particles with pronase, trypsin, lysozyme, and RNase abolished the toxic effect whereas treatment with α-amylase, isoamylase, amyloglucosidase, lipase, or DNase had no effect. Attempts to block the toxic effect with a number of sugars including ribose, sorbose, xylose, glucose, fructose, rhamnose, mannose, maltose, sucrose, and raffinose were unsuccesful.

TABLE V
GROWTH INHIBITORY EFFECT OF 110-3 XENOSOMES ON XENOSOME-FREE MARINE CILIATES

		Xenosome/protozoan ratio						
Ciliates tested	Strain	10,000	2000	1000	500	250	100	50
Miamiensis avidus	Ma	40[a]	62	100	—	—	—	—
Parauronema sp.	34-7	0	0	0	0	0	32	83
Parauronema sp.	108-6	0	50	100	—	—	—	—
Parauronema sp.	111-2	0	20	68	100	—	—	—
Parauronema sp.	114-6	0	16	50	100	—	—	—
Uronema marinum	PW-2	0	0	0	0	0	47	82
Uronema nigricans	Pc	0	0	0	0	0	45	75
Parauronema acutum	110-3	100	—	—	—	—	—	—

[a]Percentage control.

Following addition of xenosomes to a suspension of *Uronema nigricans,* Pc in culture medium, the animals ceased to divide but remained active for periods up to several hours. After this period, swimming activity diminished rapidly; the animals became moribund and died shortly afterward, usually within 24 hours. Apparently, the intact particle is essential for both infectivity and the toxic effect. Ability of xenosomes to inhibit the growth or kill potential competitors may offer the host an advantage in its natural environment.

III. Genome

A. Size and Structure

Xenosomes were isolated and purified from three stocks of xenosome-bearing *P. acutum,* 108-7, 110-3, and 112-5 by procedures that were essentially the same as those used for the mu and pi symbionts of *P. aurelia* (Soldo and Godoy, 1974). Purified particle preparations were fractionated and the nucleic acids extracted from the dry, defatted residues with hot trichloroacetic acid. The extract was analyzed for DNA by the diphenylamine procedure (Burton, 1956) or by the method of Webb and Levy (1955) and for RNA by the orcinol procedure. (Ogur and Rosen, 1950). Protein was determined by solubilizing the TCA-extracted residue in 0.1 *N* NaOH and assaying the extract by the Lowry procedure (Lowry *et al.*, 1951). The DNA, RNA, and protein content of the isolated particles (fg per particle) of the three stocks are shown in Table VI. Highly purified chromosomal xenosome DNA exhibited a sharp thermal transition in SSC (0.15 *M* NaCl, 0.015 *M* Na citrate, pH 7.0). T_m for stock 110-3 xenosome DNA was 83.2°C, corresponding to a G + C content of 33.9 mole%. T_m of DNA isolated from the protozoan was 77.5°C, corresponding to a G + C content of 20.0 mole%, among the lowest ever reported for a eukaryotic organism. The size of the DNA genome of the xenosome was obtained kinetically from second-order

TABLE VI
Size, DNA, RNA, and Protein Content of Xenosomes of *Parauronema acutum*

Particle	Number of determinations	Size (μm) Length	Width	DNA (fg per particle)	RNA (fg per particle)	Protein (fg per particle)
108-7 Xp	5	1.0	0.3	5 ± 1[a]	10 ± 4	157 ± 20
110-3 Xp	7	1.0	0.3	5 ± 2	31 ± 7	41 ± 7
112-5 Xp	5	1.0	0.3	5 ± 2	11 ± 5	112 ± 15

[a] SD.

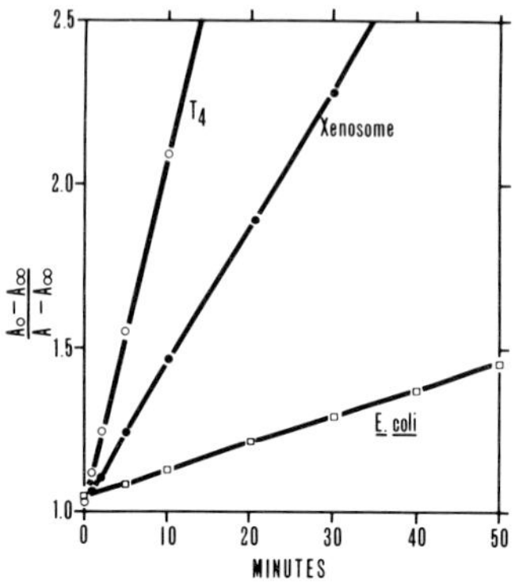

FIG. 13. Second order renaturation plots of bacteriophage T_4, xenosome, and *E. coli* DNA.

renaturation rate (Fig. 13) and C_0t plots (Fig. 14). The results show that xenosome DNA is much closer in size to that of bacteriophage DNA than it is to DNA of *E. coli*. Table VII presents data on the thermal transition, analytical, and kinetic complexity of xenosomes studied thus far, and Table VIII compares these data with similar data obtained from certain symbionts of *Paramecium*. In general, the size of the DNA genome of xenosomes and the symbionts of *Paramecium* are considerably smaller than that of the free-living bacterium *E. coli*. This suggests that the metabolic capabilies of the particles may be significantly reduced as compared to those of free-living bacteria. It also suggests that the particles, as a consequence of intracellular residence, may have lost a part of their original genomes. Consider, for example, that the size of the lambda genome, determined kinetically, is comparable to that of the xenosome, yet the volume of lambda particles (3.5 μm long and 0.6 μm wide is about 15-fold greater than that of xenosome particles (0.8 μm long and 0.3 μm wide). As a

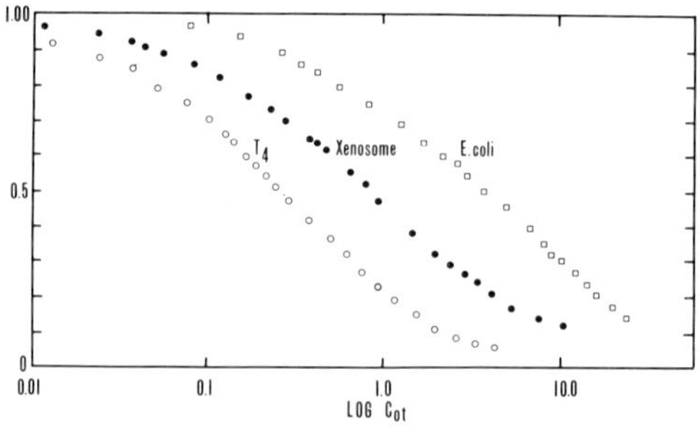

FIG. 14. C_0t plots of bacteriophage T_4, xenosome, and *E. coli* DNA.

TABLE VII
THERMAL TRANSITION, ANALYTICAL AND KINETIC COMPLEXITY OF XENOSOMES OF *Parauronema acutum*

	Xenosome		
	108-7	110-3	112-5
Thermal transition			
Hyperchromicity	1.35	1.39	1.37
T_m (C)	83.4	83.2	83.6
G + C content, from T_m (mole%)	34	34	35
Analytical complexity (daltons $\times$ 10^9)	3.0	3.0	3.0
Kinetic complexity (daltons $\times$ 10^9)			
Observed	0.49	0.50	0.55
Corrected for (G+C) content	0.34	0.35	0.38
Number of genomes	9	9	8

further consequence of prolonged intracellular existence, the particles may have evolved multicopy or "amplified" genomes (Table VIII).

Having established that chromosomal DNA of xenosomes of stock 110-3 was multicopy, we attempted to determine its arrangement within the genome by means of sedimentation rate data and electron microscopy. For sedimentation rate studies it was necessary to label xenosome DNA to provide a suitable marker. We were unable to label particle DNA to high specific activity with

TABLE VIII
KINETIC, ANALYTICAL COMPLEXITY AND GENOMIC SIZE OF XENOSOMES OF *P. acutum* AND SYMBIONT PARTICLES OF *Paramecium aurelia*

		Complexity ($\times$ 10^9 daltons)		
		Kinetic[a]	Analytic	Number of genomes
Xp	108-7	0.34	3.0	9
Xp	110-3	0.35	3.0	9
Xp	112-5	0.38	3.0	8
Lambda	299	0.39	7.5	19
Mu	138	0.56	5.1	9
Pi	139	0.59	4.5	8
T_4		0.13	0.12	1
E. coli	K-12	2.5	2.5	1

[a]Corrected for G+C content.

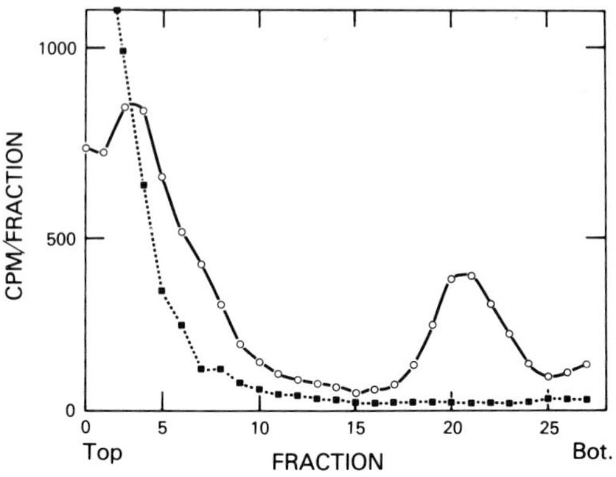

FIG. 15. Sucrose gradient of [^{14}C]adenine-labeled xenosomes. ○——○, No treatment; ■· · ·■, treatment with DNase.

radioactive thymine, thymidine, or thymidylic acid. [^{14}C]Adenine, however, proved to be an excellent precursor, and it became possible to prepare xenosome DNA at high specific activity (> 30,000 μg DNA). Xenosomes, isolated and purified from cultures containing [^{14}C]adenine, were carefully lysed on top of sucrose gradients and sedimentation rate patterns determined using the general procedures described by Kavenoff (1972). Figure 15 identifies the rapidly sedimenting labeled material as DNA by virtue of its destruction by the action of DNase. Treatment of lysates with RNase and/or proteases did not affect the

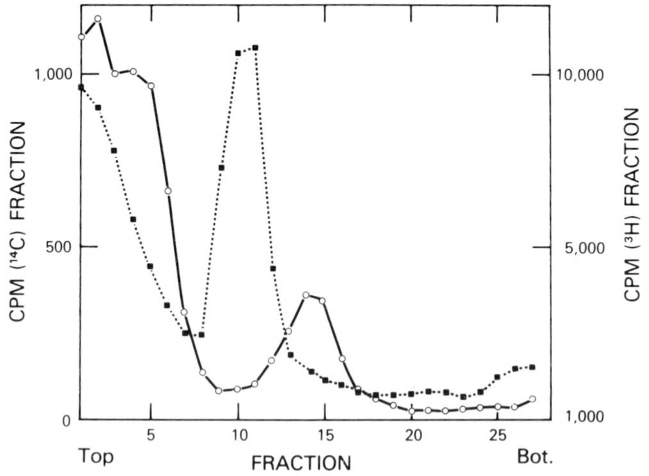

FIG. 16. Sucrose gradient of [^{14}C]adenine-labeled xenosomes and ^{3}H-labeledT_4 bacteriophage. ○——○, Xenosomes; ■· · ·■, T_4.

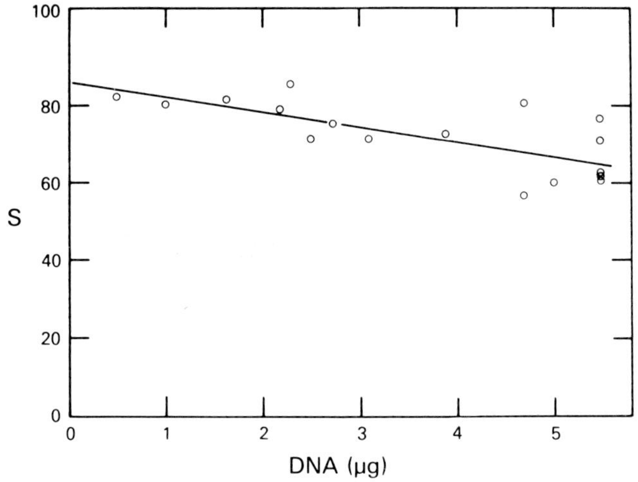

FIG. 17. Effect of DNA concentration of sedimentation rate coefficient.

sedimentation pattern (data not shown). Figure 16 shows the result of a typical run in which ^{3}H-labeled T_4 bacterophage was included with ^{14}C-labeled xenosome lysate as a marker. DNA peaks due to the presence of T_4 DNA and xenosome DNA are clearly shown. We found that the DNA concentration (Fig. 17) and speed of centrifugation (Fig. 18) exerted only a small effect on the sedimentation rate coefficient. Nevertheless, we calculated the sedimentation rate coefficient values from data extrapolated to zero. The size of the genome

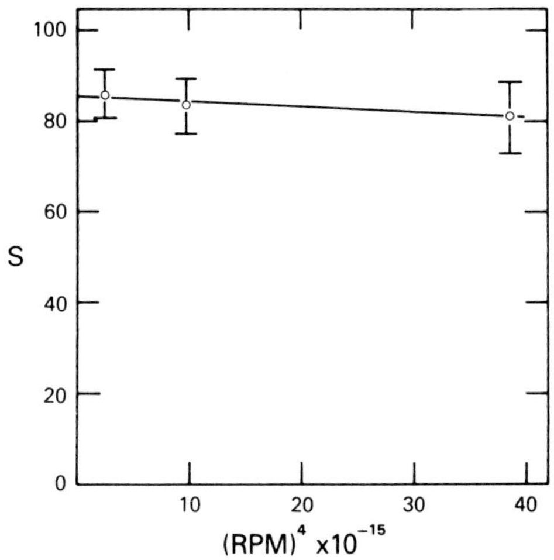

FIG. 18. Effect of speed of centrifugation on sedimentation rate coefficient.

TABLE IX
GENOMIC SIZE FROM SEDIMENTATION MEASUREMENTS

	S	Molecular complexity ($\times 10^9$)
T_4	57	0.13
Xp	85.7 ± 9	0.42 ± 0.13 (linear)
		0.30 ± 0.09 (circular)

calculated from sedimentation coefficient measurements yielded values ranging from 0.30×10^9 for circularly permuted molecules to 0.42×10^9 for linearly permuted molecules (Table IX). The observed value of 0.34×10^9 (from reassociation rate measurements) suggests that the DNA sedimented as a collection of linear and circular molecules. Conceivably linear molecules could have arisen as a result of breaks occurring during the lysis procedure. Electron microscopy of *gently* lysed xenosome preparations reveal linear molecules, the largest of which was 160 μm in length. This is equivalent to a molecular size of 0.32×10^9. Despite the fact that we have, as yet, been unable to obtain unequivocal evidence of large intact circles in EM, the most reasonable conclusion to be drawn from these data is that chromosomal DNA of the xenosome exists in the form of 8–14 circularly permuted molecules of molecular weight 0.34×10^9.

B. EXTRACHROMOSOMAL DNA

CsCl-ethidium bromide gradients of lysates of purified xenosomes revealed two bands when viewed in UV. Seen in the electron microscope, the upper band contained chromosomal DNA and the lower band covalently closed circular DNA molecules. Cleared lysates of xenosomes prepared by the method of Guerry *et al.* (1973) enabled us to isolate extrachromosomal DNA of sufficient purity to permit analysis by agarose gel electrophoresis (Meyers *et al.*, 1976) and by electron microscopy (Kleinschmidt *et al.*, 1976). Repeated freeze-thawing of purified extrachromosomal DNA (or treatment with small amounts of DNase) resulted in the formation of open circular and/or linear DNA molecules. Electrophoresis of this material yielded a band whose electrophoretic mobility was comparable to λ phage DNA (Fig. 19). Electron microscopy of xenosome extrachromosomal DNA revealed the presence of a large circular component whose molecular weight determined from contour length measurements was equal to 32×10^6 (Fig. 20a and b). A small circular DNA molecule whose molecular weight (from contour length measurements) was about 6×10^9 was observed in some preparations (Fig. 20c). This component, unlike the large component (32×10^9), was not readily observable on the gels (Fig. 19). It is not clear whether the small circular molecules are artifacts due to catenane formation. Occasionally very large molecules (MW $\sim 320 \times 10^6$) which *appear* to be circular were observed in lysates (Fig. 20d). These molecules are of sufficient size to be considered chromosomal in nature. At present we are in the process of isolating xenosomes from other *P. acutum* stocks and of mapping their extrachomosomal DNA using restriction endonucleases.

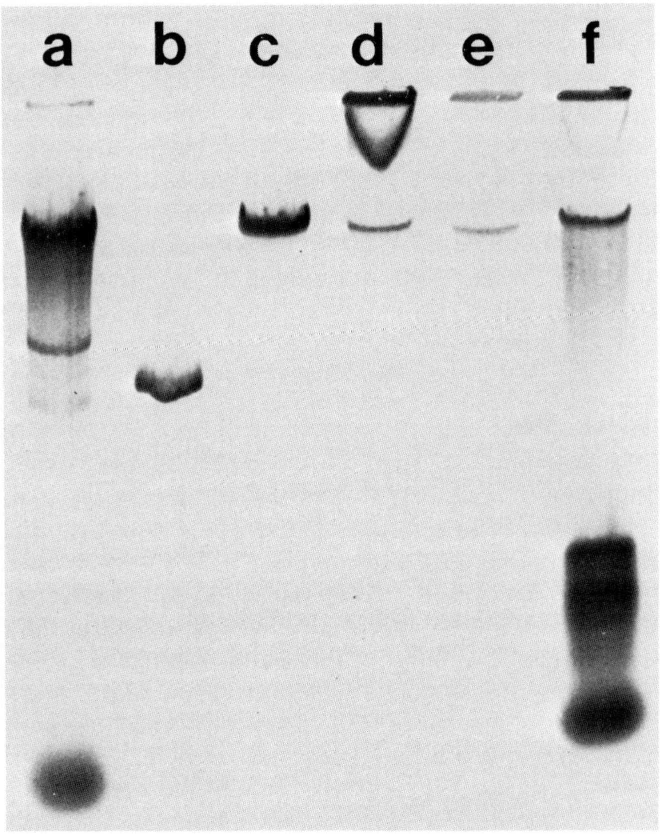

FIG. 19. Agarose gel electrophoresis of plasmid DNA. (a) *E. coli* HB 101, cleared lysate; (b) SV40; (c) Lambda phage; (d) xenosome, neutral extraction; (e) xenosome, alkaline extraction; (f) xenosome, cleared lysate

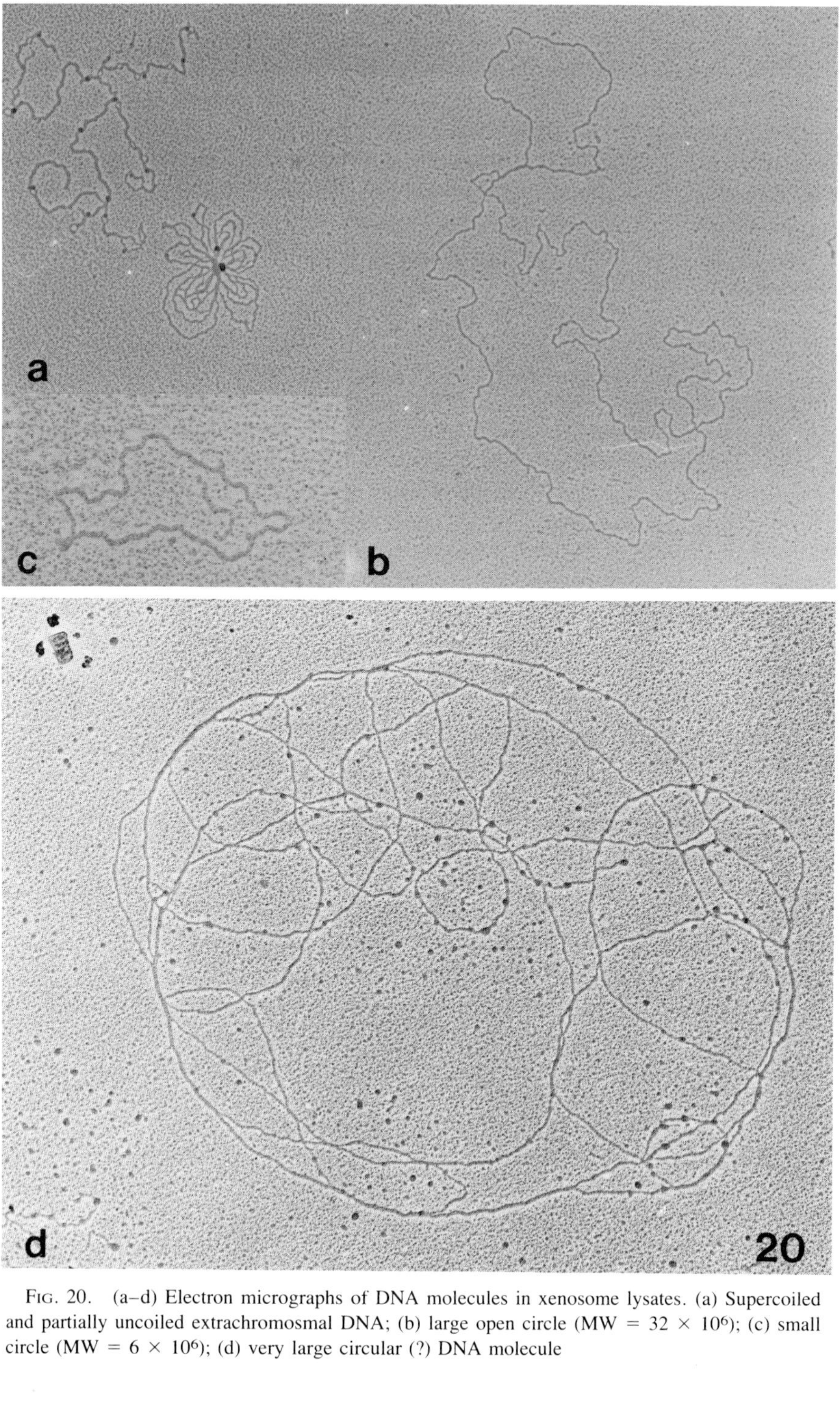

FIG. 20. (a–d) Electron micrographs of DNA molecules in xenosome lysates. (a) Supercoiled and partially uncoiled extrachromosmal DNA; (b) large open circle (MW = 32×10^6); (c) small circle (MW = 6×10^6); (d) very large circular (?) DNA molecule

C. DNA Base Sequence Homologies

To increase our understanding of the degree of relatedness that may exist among xenosome particles of *Parauronema* and those of the symbionts of *Paramecium,* we undertook a study of base sequence homologies of DNA extracted from several of these particles. Initially, DNA–DNA hybridization was determined by the membrane filter procedure of Gillespie and Spiegelman (1965) as modified by Denhardt (1966) for detection of complementary DNA. Results obtained using this procedure suggested that significant base sequence homologies existed among the DNA of xenosomes and the DNA of lambda, mu, and pi symbionts of *Paramecium.* We found, however, that "tracer" [^{3}H]- or [^{14}C]DNA used in our hybridization studies could be bound to *blank* filters depending upon conditions of ionic strength, temperature, and formamide concentration. For these reasons, we abandoned the procedure and repeated our hybridization studies in "solution." The procedure we adopted was as follows. "Tracer" DNA was labeled by the "nick-translation" procedure or by culturing the particle bearers in the presence of [^{3}H]adenine. Both procedures yielded

TABLE X
Base Sequence Homologies among DNAs of Xenosomes of *Parauronema acutum*, the Symbionts of *Paramecium aurelia*, and Selected Bacteria

Unlabeled DNA		^{3}H-labeled DNA		
Organism	Stock	110-3 xenosomes	139 pi	*E. coli*
P. acutum	110-3	100[a]	0	0
P. acutum	108-7	99		
P. acutum	112-5	96		
P. acutum	263-5	100		
P. acutum	263-11	98		
P. acutum	263-13	104		
P. acutum	263-18	84		
P. acutum	265-6	86		
P. acutum	Whole cell	0		
P. aurelia	299 lambda	2	4	2
P. aurelia	138 mu	1	68	1
P. aurelia	139 pi	1	100	2
S. faecalis		1		
S. aureus		1		
A. ugiliosum		2		
C. lytica		0		
Bacteriophage T_4		2	2	1
E. coli		1	1	100

[a]Percentage control.

DNA preparations of sufficient specific activity to effectively carry out the experiments. [^{3}H]DNA was mixed with 500- to 1000-fold excess of cold homologous or heterologous DNA, sheared by sonication to about 2×10^5 daltons, and, after heating at 100°C and quenching in ice, renatured in 0.12 *M* NaP, pH 7.0. Hybridization was carried out for a sufficient period to allow >95% reassociation of the cold DNA. Under these conditions hybridization of "tracer" DNA was insignificant. Unreacted single-stranded DNA was separated from duplex DNA molecules by hydroxyapatite chromatography and the degree of hybridization determined relative to a control containing homoduplex cold and "tracer" DNA. The results are shown in Table X. It can be seen that a high degree of homology exists among the DNAs of the xenosome stocks tested. Stocks 263-18 and 265-6 however showed a somewhat lower degree of base sequence homology. There was *no* significant base sequence homology between xenosome DNA and those of the symbionts of *Paramecium,* bacteriophage T_4 and bacteria including *E. coli, S. faecalis, S. aureus, A. ugiliosum* and *C. lytica.* The latter three bacterial species were included because their DNA base composition was similar to that of xenosome DNA (34 mole % guanine + cytosine). A significant degree of base sequence homology was observed between 139 *pi* and 138 *mu* particles of *Paramecium* using 139 *pi* DNA as tracer. These results confirm similar observations made earlier by Quackenbush (1978).

IV. Energy Metabolism and Nutrition

With respect to investigations concerning energy metabolism, we studied oxygen consumption in isolated and purified particles (H. A. Kareem and A. T. Soldo, unpublished results). Our studies showed that xenosomes consumed oxygen at a rate of about 1.3 nmoles/minute/mg protein at 29°C. This value compared favorably with those reported for *Rickettsiae* and kappa symbionts of *Paramecium aurelia.* It was higher than that reported for the *Bdellvibrio* bacteriovirus and much lower than that of the free-living bacterium *E. coli* K-12 (Table XI). The rate of endogenous oxygen consumption of freshly isolated xenosomes did not change significantly for periods up to 14 hours at room temperature and was unaffected by the presence of 0.005 *M* ethanol or methanol. A number of substances were tested to determine their effect on the rate of oxygen consumption of isolated and purified xenosomes. We found that the rate of oxygen consumption by xenosomes was stimulated in the presence of fatty acids such as acetate, butyrate, arachidonate, linoleate, stearate, and *n*-hexanoate. Acetoacetate, *n*-heptanoate, β-hydroxybutyrate, and *n*-octanoate inhibited oxygen uptake by xenosomes, whereas oleate and *n*-pentanoate had no effect. All the glycolytic pathway intermediates tested stimulated oxygen uptake of intact xenosomes. With the exception of citrate (which had no effect) all Kreb's

TABLE XI

COMPARISON OF THE ENDOGENOUS RATES OF OXYGEN UPTAKE OF SYMBIOTIC AND FREE-LIVING BACTERIA

Symbiont or bacterium	Endogenous rate (nmoles O_2/minute/mg) protein
Bdellovibrio (Bacteriovirus)	0.3
Xenosome (110-3)	1.3
Rickettsiae (Typhus)	2.2
Kappa (51, *P. aurelia*)	2.7
E. coli K-12	43, 51

Cycle intermediates stimulated oxygen consumption of intact xenosomes to some degree. Succinate was by far the most effective stimulator of oxygen consumption by xenosomes. At a concentration of 10^{-2} *M* succinate, the stimulation was more than 300% that of endogenous oxygen uptake (Fig. 21). These results were similar to those of Wink (1979), who found that succinate was a principal substrate for the O_2 respiration of the endosymbionts of *Glossina morsitans*. Glyoxylate and β-NADH yielded substantial stimulation (85 and 93%, respectively) over that of the endogenous rate. Inorganic phosphate was also a very effective stimulator of oxygen consumption by xenosomes (130%), second only to succinate in this respect. Ribose and phosphocreatine were less effective substrates; ADP inhibited oxygen uptake.

The effect of a number of respiratory inhibitors on oxygen consumption of xenosomes was also examined. The most potent inhibitor of oxygen consumption of xenosomes was cyanide, which showed 70 and 90% inhibition at concentrations of 10^{-3} and 10^{-2} *M*, respectively (Fig. 22). Other strong inhibitors were carbonyl-cyanide, *m*-chlorophenylhydrazone, 8-hydroxyquinoline, and salicyl-aldoxime. Antimycin, 2,2′-bipyridine, 2,4-dinitrophenol, and *o*-phe-

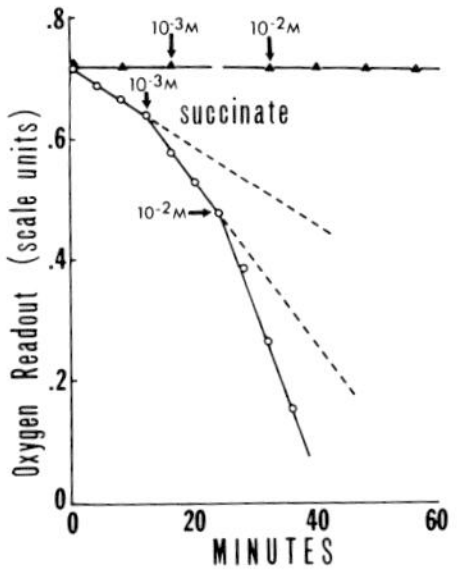

FIG. 21. Oxygen consumption of isolated xenosomes in the presence of succinate.

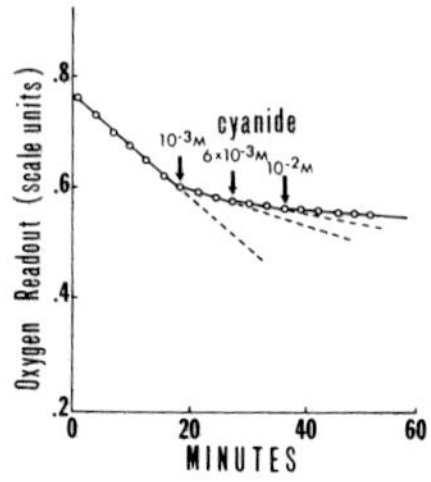

FIG. 22. Inhibition of oxygen consumption of isolated xenosomes by cyanide.

nanthroline were comparatively weak inhibitors and showed less than 50% inhibition of the endogenous rate. Arsenate, azide, rotenone, and valinomycin did not inhibit the rate of consumption by xenosomes. In other studies involving oxygen uptake of whole cell xenosome-bearing and xenosome-free *P. acutum* we observed that throughout the entire growth period there were significant differences in the rate of oxygen consumption. Xenosome-bearing animals consumed oxygen at a rate about 20 to 25% higher than xenosome-free host protozoans (Fig. 23). During the same period, we found that the level of cellular carbohydrate which accumulated during growth under these conditions was consistently higher in xenosome-free *P. acutum* than it was in xenosome-bearing protozoans (Fig. 24). We examined the nature of the carbohydrate present in *P. acutum* and found it to be indistinguishable from that of animal glycogen (Table XII) (Kareem and Soldo, 1978).

Previously, Soldo and Merlin (1977) described a synthetic medium which supported the growth of a number of ciliates of marine origin. This medium (Table XIII) also supported the growth of xenosome-bearing *P. acutum* 110-3. In order to establish the contribution xenosomes made (if any) to the overall nutrition of the host protozoan, we conducted growth experiments using both xenosome-bearing and xenosome-free animals. Employing the single-omission technique to determine the need for each component of the medium, we were unable to detect any significant differences in the growth response of cells containing

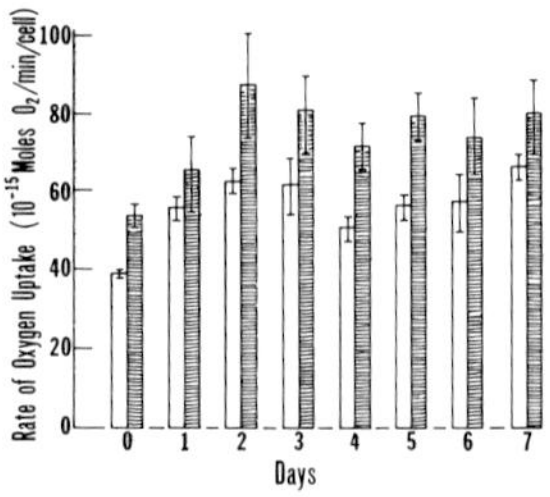

FIG. 23. Oxygen consumption of xenosome-free and xenosome-bearing *P. acutum* during growth. Open bars, xenosome-free *P. acutum;* hatched bars, xenosome-bearing *P. acutum.*

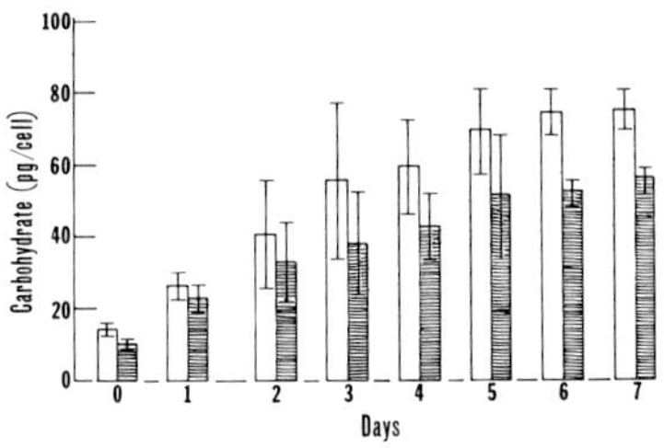

FIG. 24. Carbohydrate content of xenosome-bearing and xenosome-free *P. acutum* during growth. Open bars, xenosome-free *P. acutum;* hatched bars, xenosome-bearing *P. acutum.*

xenosomes compared to those freed of their particles by treatment with penicillin G. We did, however, conduct a detailed study of the nutrition of the xenosome-free strain. We found the following amino acids to be essential for growth: L-arginine, L-histidine, L-isoleucine, L-leucine, L-lysine–HCl, L-methionine, L-phenylalanine, L-proline, DL-threonine, L-tryptophan, and DL-valine. In the absence of *both* glycine and DL-serine the animals failed to grow; however, either glycine *or* DL-serine alone supported optimal growth when present in sufficient quantities. Guanine, guanosine, or guanylic acid satisfied the need for a purine for the growth. Pyrimidines were not required. The animals failed to grow beyond the first serial weekly transfer in the absence of any one of the following vitamins: folic acid, *d*-pantothenate, pyridoxal–HCl, riboflavin, or DL-6-thioctic acid. In the absence of nicotinamide, growth failed in the third weekly transfer. Lipids were essential for *optimal* growth of *P. acutum* 110-3. In the absence of a source of lipids, population densities were only about 3% (40,000 animals per ml after 5 days at 27°C), compared to that obtained in the presence of a mixture of asolectin, cephalin, and Tween 80. Sterols were not essential for the growth of the organisms.

TABLE XII
COMPARISON OF THE PROPERTIES OF PROTOZOAN CARBOHYDRATE WITH ANIMAL GLYCOGEN

Degradation by	Carbohydrate: Protozoan (%)	Carbohydrate: Glycogen (%)
Amyloglucosidase + glucose oxidase	102	100
α-Amylase	71	75–85
β-Amylase	45	45–55
Isoamylase	8.3	8.7
Isoamylase + β-amylase	105	109
Average chain length (CL)	12	12.7

TABLE XIII
COMPOSITION OF SYNTHETIC MEDIUM FOR THE CULTIVATION OF *Parauronema acutum* 110-3[a]

Component	μg/ml	Component	μg/ml
L-Alanine	900	Biotin	0.0005
LArginine	900	Folic acid	2.5
L-Asparagine	270	Nicotinamide	2.5
L-Aspartic acid	300	*d*-Pantothenate, Ca	5.0
L-Glutamic acid	1200	Pyridoxal–HCl	2.5
Glycine	225	Riboflavin	2.5
L-Histidine	300	Thiamine–HCl	7.5
L-Isoleucine	375	DL-6-Thioctic acid	0.05
L-Leucine	375		
L-Lysine–HCl	750	NaCl	16400
L-Methionine	450	$MgSO_4$	1980
L-Phenylalanine	375	$MgCl_2$	1390
L-Proline	225	$CaCl_2$	710
DL-Serine	900	KCl	430
DL-Threonine	1050	HCO_3	73
L-Tryptophan	225	Br	36
L-Tyrosine	225	B	5
DL-Valine	600	PO_4	2.7
		NO_3	1.8
Adenosine	90	Si	1.4
Guanosine 2′,3′-phosphate	150	Al	0.8
Asolectin	200	Li	0.2
Cephalin	200		
Tween 80	200		

[a]pH 7.2; density of sea water was 1.015 g/cm^3.

V. Comments on Methodology

In the course of developing xenosome-bearing *P. acutum* as models for studies of the infective process and for host–symbiont interactions it was necessary to generate a methodology that was sometimes unique to these organisms. For example, procedures for collecting and identifying the protozoa in sea water samples, screening for the presence of symbionts were required. In this respect we developed a new method for plating and cloning ciliates based on a principle differing from that traditionally used for plating and cloning microorganisms that greatly facilitated these procedures (Soldo and Brickson, 1980).

Methods were also needed for the axenic cultivation of these symbiont-containing organisms. This led to development of both crude (Soldo and Merlin, 1972) and chemically defined media (Soldo and Melin, 1977) which supported the luxurious growth of the protozoa and maintenance of the particles. In addition to these methods we also devised procedures for the isolation and purification of the symbionts so that sufficient material of high purity was available to conduct biochemical studies.

VI. Discussion

The morphological characteristics of xenosomes discerned from observations in both the phase-contrast and electron microscopes clearly establishes the symbionts as bacteria. As seen in ultrathin sections, the structure of the cell envelope, which consists of a double-track outer layer and an electron-dense inner layer (the plasma membrane), is typical of gram-negative bacteria (Murray, 1968). The presence of the outer layer, a structural characteristic not found in *Mycoplasma*, distinguishes xenosomes from this bacterial type. On the other hand, most gram-negative bacteria possess a well-defined "murein" layer inserted between the inner and outer layers of the cell envelope. This middle layer imparts rigidity and accounts for the shape and the strength of the cell wall. Lysis of gram-negative bacteria due to the action of the penicillin occurs as a consequence of the inhibition of a trans-peptidation reaction by the antibiotic and the resulting failure to complete the terminal step in the synthesis of a peptidylglycan present in this layer (Wise and Park, 1965). The cell envelope of the xenosome contains very little of this layer and probably accounts for the fact that optimal concentrations of the antibiotic required to effect lysis of xenosomes is about 50 times greater than that required for lysis of most free-living gram-negative bacteria. It appears that a "murein" layer is probably present as a functional component of the xenosome cell wall despite the fact that it is not readily observable in the electron microscope. Other morphological features of the xenosome include the arrangement of DNA fibers within the internal contents of the particles. Typical free-living bacteria, both gram-negative and gram-positive, contain DNA in the form of one or two densely packed masses of material, termed "nucleosomes" or nucleoids. The DNA of xenosomes does not occur in this form. Rather, it appears to be distributed throughout the particles as a network of fibrous material. A similar arrangement has been observed for mu (Beale and Jurand, 1960), kappa (Dippell, 1959), and lambda (Soldo and Godoy, 1973) and other symbionts of *P. aurelia* (Preer *et al.*, 1974). Xenosomes also possess bacterial flagella and are motile. With respect to size, shape, and general morphology they most closely resemble *Rickettsiae*. However *Rickettsiae* do not possess flagella and are nonmotile.

The infective nature of xenosomes is now well established. Particles released from the host have been shown to be infective both by *in vivo* and *in vitro* type experiments (see Tables III and IV). Once taken up by a susceptible host, the particles grow and divide and are maintained indefinitely. Little is known about the mechanism by which the particles gain entrance into the host cytoplasm. A likely possibility is through the food vacuole, perhaps through rupture of the food vacuole membrane. A process of this nature has been invoked to account for the uptake of kappa particles by *Paramecium* (Jurand *et al.*, 1971). In contrast to xenosomes which are flagellated and capable of moving about on their own, kappa particles do not possess flagella and are nonmotile. It is not clear whether motility provides xenosomes with advantage over that of kappa with respect to their ability to gain entrance into potential hosts and establish an infection. Of significance here is the fact that both xenosomes and kappa symbionts are actively released from the protozoans during normal growth and that the particles, once released, are taken up by non-particle-containing protozoan hosts. In this respect their ability to infect homologous as well as heterologous strains may provide the particles with a distinct advantage in nature. This advantage may be further enhanced by the fact that both xenosomes and kappa exert a toxic effect and are capable of killing or inhibiting the growth of a potential competitor. The mechanism by which the toxin exerts its effect on susceptible protozoans is not understood. In the case of kappa the killing action appears to be associated with the presence of refractile bodies, called R-bodies (Sonneborn, 1948). The refractile body has been shown to be a tightly wound ribbon of proteinaceous material envoloping virus-like particles within its coils (Preer and Preer, 1967). No such structure is present in xenosomes.

The molecular complexity of the xenosome genome (0.34×10^9) is comparable to that of some *Rickettsiae* (Kingsbury, 1969) and is intermediate in size between those of *Mycoplasma* (Bode and Morowitz, 1967) and bacteriophage T_4 whose sizes are 0.5×10^9 and 0.13×10^9, respectively. It is strikingly similar to that of lambda (0.39×10^9), a symbiont of *Paramecium* which is at least 15 times more massive than the xenosome (Soldo and Godoy, 1973). The molecular complexities of the genomes of mu and pi particles (Soldo and Godoy, 1974) are somewhat larger than that of the xenosome (0.63 and 0.59×10^9, respectively) but are much smaller than the free-living bacterium *E. coli* (2.5×10^9) (Cairns, 1963). The presence of multicopy genomes in xenosomes appears to be characteristic of a number of symbiotic bacteria studied thus far. Lambda, mu, and pi symbionts of *Paramecium* contain multicopy genomes containing 19, 9, and 8 copies, respectively. Omicron particles (bacterial symbionts of the fresh water *Euplotes*) were reported to contain a multicopy genome (Schmidt and Heckmann, 1980). The comparatively small size of the genome, which appears to be independent of the physical size of the symbiont itself, and the presence of multiple copies of DNA molecules supports the notion that the symbionts may have adapted to an intracellular environment. Free-living bacteria contain genomes much larger than those observed for symbionts and generally contain only a single copy of the genome. The only exception reported thus far is that of the bacterial genome of *Azotobacter* (Sadoff *et al.*, 1979). This bacterium contains 40 copies of the genome.

Extrachromosomal DNA in xenosomes must be evaluated with respect to the role this DNA may play in the infective process. It is well established that plasmid DNA of bacteria carries genes responsible for mating type, antibiotic resistance, and infectivity. Clearly, plasmid DNA plays a significant role in nitrogen fixation in symbiotic bacteria in root nodules of a number of plants (Bishop *et al.*, 1977). Although covalently closed circular DNA molecules have been observed in a number of kappa symbionts of *Paramecium* (Dilts, 1977; Quackenbush, 1978), it is uncertain whether they arise from a defective virus known to be present in these symbionts.

One clear consequence of the interaction of the symbiont with the host protozoan is the finding that xenosome-bearing *P. acutum* consumes oxygen at a rate significantly higher than symbiont-free protozoans. The increase is not due simply to the additive effect of the particle population within the host; oxygen consumption of the collective particles accounts for a relatively insignificant portion of the total increase. Similarly, kappa-bearing *Paramecium aurelia* consumed oxygen at a higher rate than that of kappa-free protozoans (Simonsen and van Wagtendonk, 1952). Further, the difference in the rate of oxygen consumption of kappa-bearing and kappa-free protozoans could not be accounted for by the number of kappa present in the cytoplasm. In these cases the *in vivo* and *in vitro* rates of oxygen consumption of the symbionts may differ. Alternatively, the presence of the symbionts within the host may affect the metabolic activities of the protozoan as a consequence of its interaction with the host cell.

VII. Concluding Remarks

Clearly, xenosomes are infective agents, bacterial in nature. The degree to which they have become integrated into the economy of the host is reflected in the fact that the symbiont–host association is a highly stable one. We have maintained several xenosome-bearing stocks of *P. acutum* in axenic culture for several years through thousands of generations without loss of the symbionts. Moreover, it is apparent that the symbionts are dependent upon the host for their survival, since all efforts to cultivate xenosomes extracellularly in artifical media have been unsuccessful. The ability of xenosomes to infect susceptible hosts and to exert a killing effect on others suggests a high degree of specificity with respect to symbiont–host recognition. Evidence we have accumulated suggests that the xenosome genome consists of nine copies of a comparatively small (MW $= 0.34 \times 10^9$) circularly permuted DNA molecule of unique sequence. We have postulated that the similarity (i.e., multicopy DNA) of the genomic structure of lambda, mu, and pi symbionts of *Paramecium* with those of other self-reproducing intracellular entities (e.g., mitochondria and chloroplasts) may be due to adaptation resulting from prolonged intracellular existence. A multicopy genome offers an obvious selective advantage in an intracellular environment especially where the infective agent is dependent upon the host for its growth and replication. The role of plasmid DNA is not yet understood, but its presence in the symbiont offers exciting future prospects for studies relating to infectivity and the nature and production of the toxic principle. The following postulates, to be used as working hypotheses, may serve as a guide for the further study of these infectious entities: (1) Xenosomes are examples of a new class of infectious intracellular entities that arose through adaptation of once free-living bacteria to an intracellular environment. (2) The nature of the interaction between the symbiotic bacteria and the host cell is an intimate one and is probably under genetic control. (3). Recognition between the various symbionts and susceptible hosts is effected by specific molecules located at or near the surface of the xenosome and within the protozoan, perhaps on the surface of the food vacuole membrane.

Because of the breadth and scope of the general problem of bacterial endosymbiosis and the fact that comparatively little is known about these associations, these postulates, of necessity, are broadly posited. It is anticipated that detailed answers will be forthcoming only through the application of a number and variety of techniques and experimental approaches.

Acknowledgments

I wish to express my sincere appreciation and thanks to Dr. Velia Figueroa de Soto and Mr. Hussein A. Kareem who contributed to some of the work presented here and to Ms. Sylvia A. Brickson and Frances Larin for their outstanding technical assistance.

References

Andresen, C. C., Chapman-Andresen, C., and Nilsson, J. R. (1965). *Progr. Protozool. Exerpa Med. Fed. Intern. Congr. Ser.* **91,** 258.
Beale, G. H., and Jurand, A. (1960). *J. Gen. Microbiol.* **23,** 243–252.
Bishop, P. E., Dazzo, F. B., and Applebaum, E. R. (1977). *Science* **198,** 938–940.
Bode, H. R., and Morowitz, H. J. (1967). *J. Mol. Biol.* **23,** 191–199.
Brewin, N. J., Beringer, J. E., and Buchanan-Wollaston, A. V. (1980). *J. Gen. Microbiol.* **116,** 261–270.
Burton, K. (1956). *Biochem. J.* **62,** 315–323.
Cairns, J. (1963). *Cold Spring Harbor Symp. Quant. Biol.* **28.** 43–45.
Denhardt, D. T. (1966). *Biochem. Biophys. Res. Commun.* **23,** 64–68.
Desportes, I., Theorides, J., and Vivares, C. (1977). *Protistologica* **13,** 611–619.
Dilts, J. A. (1977). *J. Bacteriol.* **129,** 888–894.
Dippell, R. V. (1959). *Science* **130,** 1415.
Fenchell, T., Perry, T., and Thane, A. (1976). *J. Protozool.* **24,** 154–163.
Gillespie, D., and Spiegelman, S. (1965). *J. Mol. Biol.* **12,** 829–842.
Guerry, P., LeBlanc, D. J., and Falkow, S. (1973). *J. Bacteriol.* **116,** 1064–1066.
Heckmann, K. (1975). *J. Protozool.* **22,** 94–104.
Ishikawa, H. (1978). *Biochem. Biophys. Res. Commun.* **81,** 993–999

Jeon, K. W., and Ahn, T. I. (1978). *Science* **202,** 637–8.
Jeon, K. W., and Jeon, M. S. (1976). *J. Cell. Physiol.* **89,** 337–344.
Jurand, A., Rudman, B. A., and Preer, J. R., Jr. (1971). *J. Exp. Zool.* **177,** 365–387.
Kareem, H. A., and Soldo, A. T. (1978). *J. Protozool.* **25,** 560–562.
Kavenoff, R. (1972). *J. Mol. Biol.* **72,** 801–806.
Kingsbury, D. T. (1969). *J. Bacteriol.* **98,** 1400–1401.
Kleinschmidt, A. K., Lang, D., and Jackerts, D. (1962). *Biochim. Biophys. Acta* **61,** 857–864.
Koerner, H. K. (1978). *Zool. Beitr.* **24,** 59–68.
Lowry, L. H., Rosebrough, N. S., Farr, A. L., and Randall, N. J. (1951). *J. Biol. Chem.* **193,** 265.
Meyers, J. A., Sanchez, D., and Elwell, L. R. (1976). *J. Bacteriol.* **127,** 1529–1537.
Mouzarede, M. (1978). *Protistologica* **14,** 177–182.
Mundion, M. H., and Roitman, I. R. (1977). *J. Protozool.* **24,** 329–331.
Murray, R. G. E. (1968). *In* "Microbial Protoplasts, Spheroplasts and L-forms" (L. B. Guze, ed.), pp. 1–18. Williams & Wilkins, Baltimore.
Nemanic, P. C., Owen, R. L., Stevens, D. P. (1979). *J. Infect. Dis.* **140,** 222–228.
Ogur, M., and Rosen, G. (1950). *Arch. Biochem.* **2,** 62–76.
Osipov, V., and Podlipaev, S. A. (1978). *Acta Protozool.* **16,** 289–308.
Preer, J. R., Jr., and Preer, L. B. (1967). *Proc. Natl. Acad. Sci. U.S.A.* **58,** 1774–1781.
Preer, J. R., Jr., Preer, L. B., and Jurand, A. (1974). *Bacteriol. Rev.* **38,** 113–163.
Preer, L. B. (1969). *J. Protozool.* **16,** 570–578.
Rosati, G., Verni, F., and Luporini, P. (1976). *Monot. Zool. Ital.* **10,** 446–460.
Quackenbush, R. L. (1978). *J. Gen. Microbiol.* **108,** 181–188.
Sadoff, H., Schimel, B., and Ellis, S. (1979). *J. Bacteriol.* **138,** 871–877.
Schmidt, H. J., and Heckmann, K. (1980). *In* "International Colloquium on Endosymbiosis and Cell Research" (W. Schwemmler and H. Schenk, eds.). de Gruyter, Berlin and New York.
Siegel, R. W. (1954). *Physiol. Zool.* **27,** 89–100.
Simonsen, D. H., and van Wagtendonk, W. J. (1952). *J. Gen. Microbiol.* **15,** 39–46.
Soldo, A. T. (1961). *Trans. N.Y. Acad. Sci.* **23,** 653–661.
Soldo, A. T., and Brickson, S. A. (1978). *Tissue Cell* **10,** 609–618.
Soldo, A. T., and Brickson, S. A. (1980). *J. Protozool.* **27,** 328–331.
Soldo, A. T., and Godoy, G. A. (1973). *J. Mol. Biol.* **73,** 93–108.
Soldo, A. T., and Godoy, G. A. (1973). *J. Protozool.* **20,** 502.
Soldo, A. T., and Godoy, G. A. (1974). *Nucleic Acid Res.* **1,** 387–396.
Soldo, A. T., and Merlin, E. J. (1972). *J. Protozool* **19,** 519–524.
Soldo, A. T., and Merlin, E. J. (1977). *J. Protozool.* **24,** 556–562.
Soldo, A. T., Musil, G., and Godoy, G. A. (1970). *J. Bacteriol.* **104,** 966–980.
Soldo, A. T., Godoy, G. A., and Brickson, S. A. (1974). *Nature (London)* **249,** 284–286.
Sonneborn, T. M. (1939). *Am. Naturalist* **73,** 390–413.
Sonneborn, T. M. (1946). *Cold Spring Harbor Symp. Quant. Biol.* **11,** 236–255.
Sonneborn, T. M. (1948). *Am. Naturalist* **82,** 26–34.
Sonneborn, T. M. (1959). *Adv. Virus. Res.* **6,** 229–356.
Stock, C. C., Jacobson, W. E., and Williamson, M. (1951). *Proc. Soc. Exp. Biol. Med.* **78,** 874–876.
Van Wagtendonk, W. J., Clark, J. A. D., and Godoy, G. A. (1963). *Proc. Natl. Acad. Sci. U.S.A.* **50,** 835–838.
Webb, J. M., and Levy, H. B. (1955). *J. Biol. Chem.* **213,** 107–111.
Wink, M. (1979). *Acta Trop.* **36,** 215–222.
Wise, E. M., Jr., and Park, J. T. (1965). *Proc. Natl. Acad. Sci. U.S.A.* **54,** 75–81.
Wright, J. D. Sjostrand, E. S., and Portaro, J. K. (1978). *J. Ultrastruct. Res.* **63,** 79–85.
Wryek, D. E. (1979). *Trans. Am. Microscop. Soc.* **98,** 143–145.

INTERNATIONAL REVIEW OF CYTOLOGY, SUPPLEMENT 14

Endosymbionts of *Euplotes*

KLAUS HECKMANN

Zoological Institute, University of Münster, Münster, Federal Republic of Germany

I. Introduction

The genus *Euplotes* belongs to a large, highly specialized and much studied order of ciliates, the hypotrichs. Currently some 51 species are recognized (Curds, 1975). Among them are marine and fresh water forms as well as a few brackish and euryhaline species. They are widely distributed throughout the world. Reviews and historical accounts have been given by Kahl (1932), Pierson (1943), Tuffrau (1960), Borror (1968), and others. Occasional reports in the older literature indicate the presence of endosymbionts in species of *Euplotes*, e.g., colored species of *Euplotes* have been described as far back as 1840 by Ehrenberg, who named a form from Berlin *E. viridis*, and by Stein (1859), who presented figures of forms he called *E. patella*, and noted that their cytoplasm frequently is more or less thickly filled with bright green chlorophyll bodies. However, it was not until 1952 that bacteria-like symbionts were noted in two

ISBN 0-12-364375-9

Euplotes species (Fauré-Fremiet, 1952), and only very recently has it become clear that the most common fresh water forms are dependent upon symbiotic bacteria (Heckmann *et al.*, 1982).

The object of this article is to survey the different kinds of endosymbionts which grow in the cytoplasm or macronucleus of *Euplotes*. The term "endosymbiont" is used simply to denote one organism living within another. It does not imply anything about the interactions between host and symbiont; the symbiont may be harmful, neutral, or beneficial. It is, however, assumed that the association is not just a transient one, but that the symbiont is maintained over long periods of time and generally is transmitted from one host generation to the next.

Most of the symbionts mentioned here are designated by Greek letters, as became customary for cytoplasmic hereditary units and symbionts of unclear taxonomic affiliation. When more is known about them eventually they should be given Latin binomial names.

Before we turn to the symbionts, a short general description of the animals in which the symbionts live will be given.

Euplotes (Fig. 1) has an inflexible ovoid body which is flattened on its ventral surface. The dorsal surface has a convex shape and bears longitudinal rows of short cilia, often called "bristles." They are not vibratile and play no direct role in cell movement. It has been repeatedly suggested that they may have a sensory function. The ventral surface bears compound ciliary organelles which bring about the swimming and creeping movements of the cell. If the cilia are arranged into plate-like structures, we speak of "membranelles." If they form thick bundles, they are called "cirri." The number and positions of these organelles are prescribed precisely and are of taxonomic importance. Among the cirri, frontoventrals, transversals, and caudals are distinguished according to their position on the ventral surface. Numerous membranelles form the left and anterior margin of a large buccal cavity, or peristome, located at the left anterior borders of the cell. These collectively make up the "adoral zone of membranelles" (AZM). In addition there is a "paroral membranelle" located to the right of the posterior end of the buccal cavity where it tapers to the cytostome, or mouth. The buccal structures are also of taxonomic importance, as is the size of the cell, its shape, and its "dargyrome," a system of lines on the dorsal surface which becomes visible after silver impregnating the cells.

Like all ciliates, *Euplotes* has two types of nuclei, a small generative "micronucleus" and a large somatic "macronucleus" which forms a long contorted, often C-shaped rod. Asexual reproduction is by transverse binary fission, and sexual reproduction is by conjugation, a process by which two cells unite and exchange gametic nuclei and then separate again. Autogamy, a mode of self-fertilization without cell pairing, occurs in some species.

Most of the *Euplotes* species can easily be cultivated if algae are provided as food. *Dunaliella salina* is accepted by most marine species and *Chlorogonium*

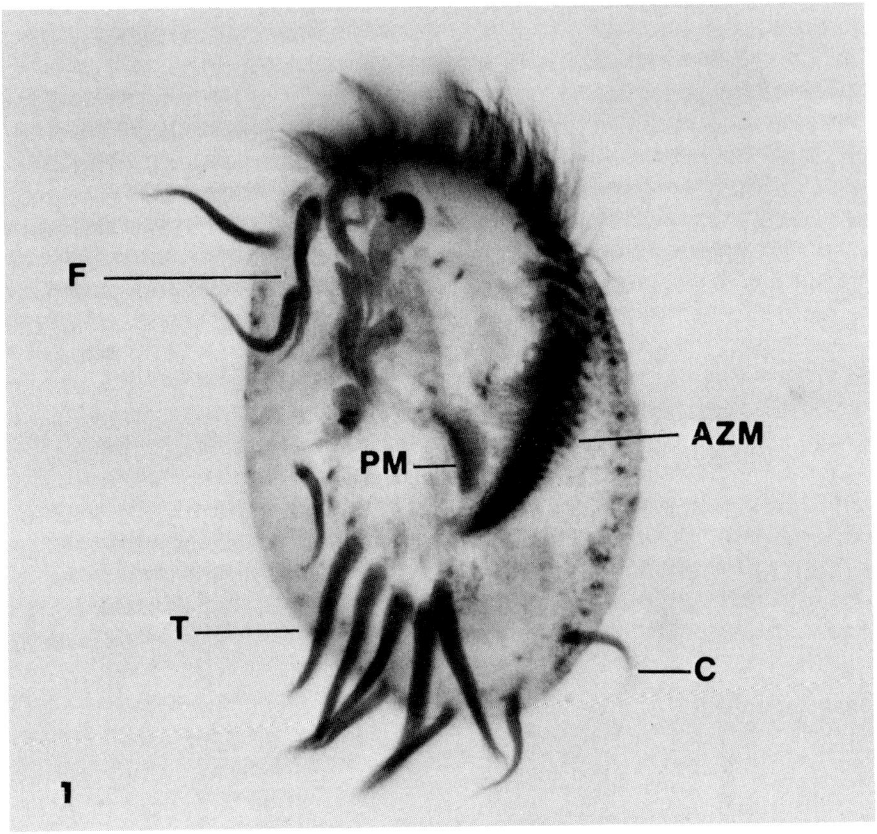

FIG. 1. Photograph of ink stained *Euplotes octocarinatus* depicting some of the distinctive characteristics of *Euplotes*. The micrograph shows the ventral surface of the cell with frontoventral (F), transversal (T), and caudal (C) cirri, the paroral membranelle (PM) and the adoral zone of membranelles (AZM). (Photograph prepared by H. W. Kuhlmann.)

elongatum by the fresh water forms. Some need in addition bacteria. Depending on the species, one to three fissions per day can be obtained.

The following terms, used throughout this article, need explanation:

Stock designates the progeny of a single *Euplotes* isolated from nature. The term does not exclude the possibility of sexual reproduction among the progeny of such a cell; in some stocks conjugation has frequently been observed.

Strain is used in a general way to designate a stock or mutant, hybrid or otherwise modified line of *Euplotes*.

Line has its usual ''dictionary'' meaning, individuals with a common ancestor.

Clone designates individuals derived from an ancestor by strictly asexual reproduction, i.e., a number of fissions.

Coconjugants are the two cells which form a pair and exchange nuclei during conjugation. They give rise to *coconjugants lines* which, in *Euplotes,* might or might not be genetically alike (see Heckmann, 1963).

Exconjugant simply denotes that a cell has undergone conjugation and as a consequence of this has renewed its nuclear apparatus.

Exautogamont refers to the same process after self-fertilization.

Syngen is a term used to designate protozoan sibling species when reproductive barriers have been recognized between groups of organisms but no criteria other than breeding tests, particularly no morphological differences, are at hand to distinguish between them. One group is then called syngen 1, another syngen 2, and so forth. Together they form a "species complex." The 14 syngens of the *P. aurelia* complex have recently been given separate binomial names (Sonneborn, 1975).

II. Omikron, an Essential Symbiont of *E. aediculatus*

Ever since bacteria and algae have been noticed inside ciliate cells, the question has been raised whether and to what extent they may benefit their hosts. The answer has proven to be exceedingly difficult. Some hold that the various cytoplasmic particles are just inhabitants of a specialized niche, the ciliate's cytoplasm, neither harming nor specifically benefiting their hosts. Others point out that large proportions of natural strains, and at some locations practically every isolated specimen, do bear symbionts when first brought into the laboratory and that therefore there must be some advantage to symbiont bearing cells. But on the other hand most of the symbionts are quickly lost when a specimen is brought into the laboratory and hence they are dispensable, at least under laboratory conditions. The best example of an indispensable symbiont of a ciliate is omikron, which shall be discussed in this section.

A. Morphology and Mode of Reproduction

Omikron particles were discovered by Heckmann (1975) in stocks of *E. aediculatus* Pierson, 1943, originally collected by Ammermann near Marseille (France). Heckmann found 900–1000 slightly curved rods evenly scattered in the cytoplasm of every animal. He named the symbiont "omikron."

Omikron can easily be observed with a phase contrast microscope if a cell is crushed (Fig. 2). The symbiont has a length of 2.5–7.5 μm—most commonly 5 μm—and a width of about 0.3 μm. It reacts gram-negatively. If omikron is stained with aceto-carmin or aceto-orcein, 3–9, in some case up to 12, inten-

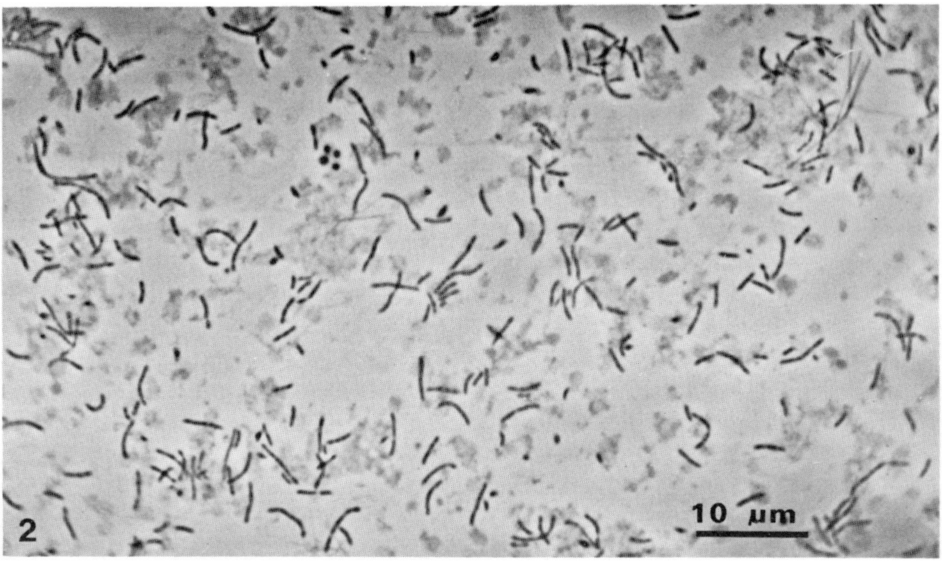

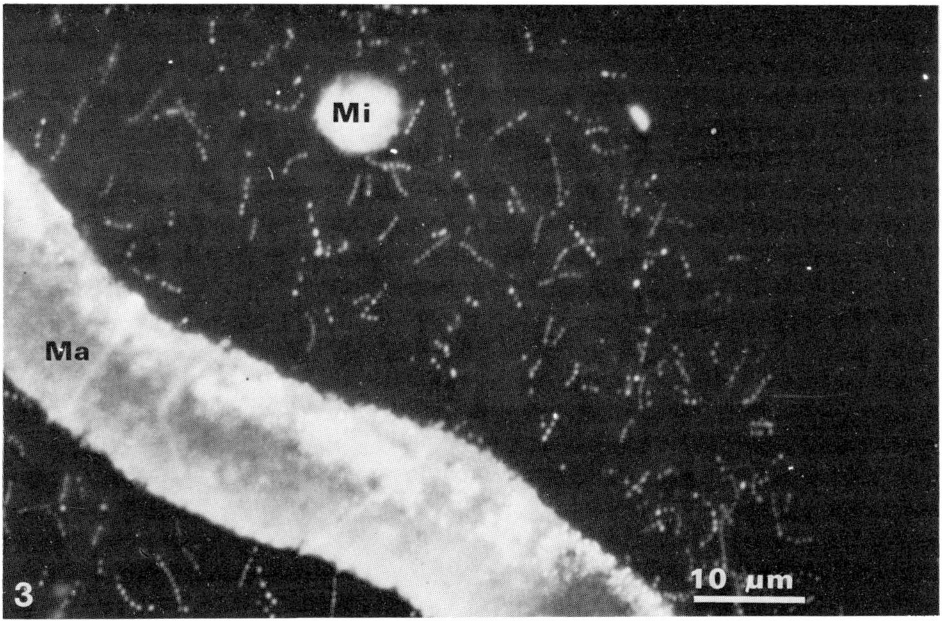

FIG. 2. Omikron particles in a crushed *E. aediculatus* cell; unfixed, unstained, phase-contrast. (From Heckmann, 1980.)

FIG. 3. Fluorescence micrograph of omikron particles in the cytoplasm of a slightly crushed *E. aediculatus* cell after staining with PIC. The omikron particles are revealed by the DNA-specific yellow fluorescence of their nucleoids. Ma, Macronucleus; Mi, micronucleus. (From Heckmann, 1975.)

sively stained and regularly spaced dots become visible. They are considered to be nucleoids. The same dots show a bright yellow fluorescence when the symbiont is stained with the fluorochrome PIC (*N,N'*-diethylpseudoisocyanin chloride) under conditions specific for DNA (Fig. 3). Multiple nucleoids have also been revealed in omikron by the electron microscope; they are one of the characteristics by which omikron differs from other symbiotic bacteria.

The nucleoids of omikron were found to look different from those of free living bacteria in that they regularly contain an electron-dense central core (Fig. 4). The material which forms this core resembles to some extent chromatin of eukaryotes. Whether it is formed by proteins associated with DNA or some other material is not clear.

Except for the nucleoids, the ultrastructure of omikron resembles that of gram-negative bacteria. Like them, it is bounded by two membranes, an inner one which is the plasma membrane and an outer one which is part of the cell wall (Fig. 5). The symbionts are contained in vesicles. The borders of the vesicles are formed by an electron-dense layer to which ribosomes are attached (Fig. 6). Heckmann (1975) reported that he has not been able to observe a trilaminar appearance of this border layer but found that it sometimes extended for some distance into the host's cytoplasm at the ends of a symbiont. The situation very much resembles that described for gamma particles of *P. biaurelia* (Beale *et al.*, 1969).

Omikron reproduces by transverse binary fission, the fission products often differing in size from each other. Frequently a 7-μm-long rod containing eight to nine nucleoids buds off a 2.5-μm-long piece that contains three nucleoids. This is preceded by a growing in of the symbiont's plasma membrane and cell wall in a typical bacteria-like manner.

From all of the above, Heckmann (1975) concluded that omikron is a gram-negative bacterium. In this respect it resembles kappa, lambda, mu, and other well investigated symbionts of the *P. aurelia* complex (Preer *et al.*, 1974). Omikron differs from these symbionts by possessing at least 3, often 5–9, and sometimes even 12, clearly visible nucleoids. It was not found that omikron confers any killing activity upon its host, as is the case for most of the symbionts of the *P. aurelia* complex and for those found in marine *Euplotes* species.

B. The Dependence of *E. aediculatus* on Omikron

Heckmann (1975) found that *E. aediculatus* can be freed of omikron by treating a rapidly growing culture with 100–500 units/ml penicillin for 5–6 days. He reported that cells which had lost their symbionts when transferred into penicillin-free medium may swim about, take in food, and undergo one or two divisions. Then, however, they stop multiplying. They gradually become smaller, stop taking in food, and eventually die, about 15–20 days after their last fission.

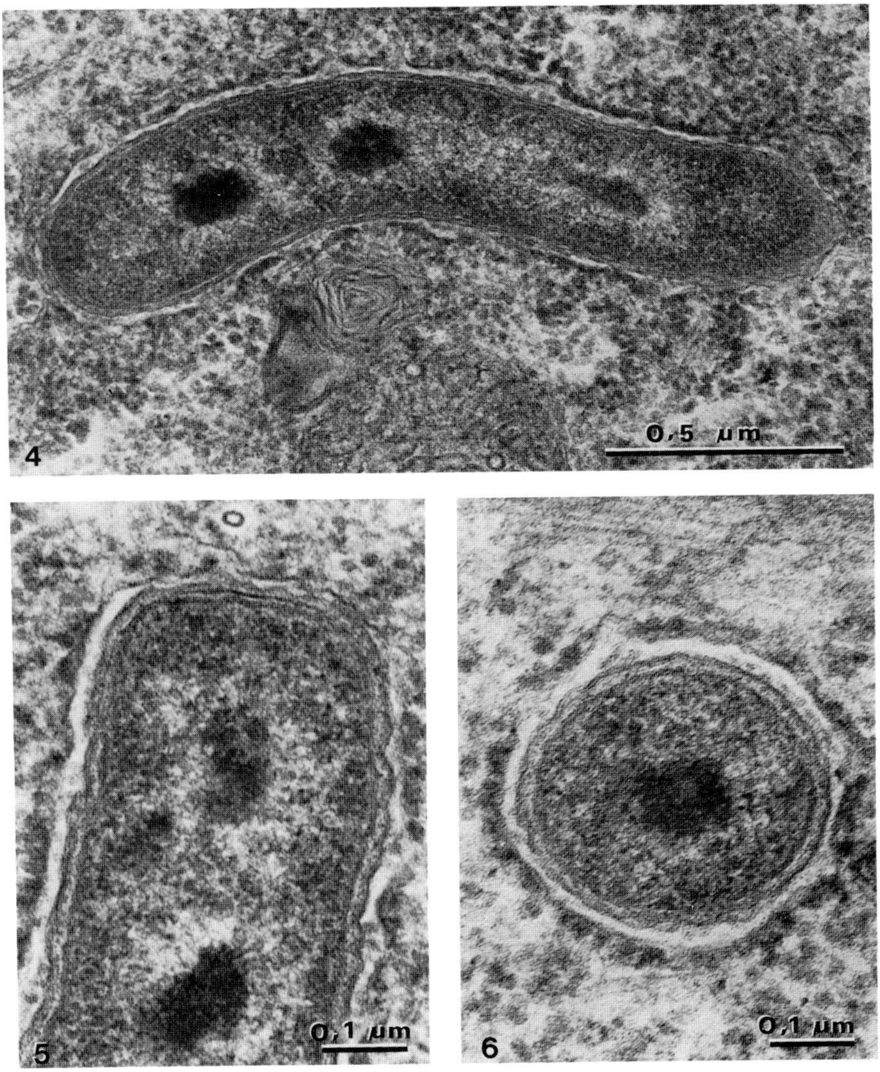

FIGS. 4–6. Electron micrographs of omikron particles. (From Heckmann, 1975.) Figure 4. Longitudinal section. The ultrastructure of omikron resembles that of gram-negative bacteria. The electron-dense areas inside the nucleoids remind one in their appearance of the chromatin of eukaryotes. Figure 5. Longitudinal section at higher magnification showing the inner and outer membranes of omikron. Figure 6. Transverse section revealing that omikron particles are contained in vesicles. The borders of the vesicles are formed by an electron-dense layer to which ribosomes are attached.

Basically the same was observed when *E. aediculatus* was treated with rifampicin, neomycin, or chloramphenicol or if the ciliate was irradiated with a sufficiently high dose of X rays (Heckmann, 1975) or γ rays (Kloetzel, 1981).

Heckmann (1975) reported that only after a treatment with penicillin was he able to obtain cells which were totally free of symbionts. In the other cases, he found at least a few symbionts still present when the ciliate had ceased dividing, although they were obviously severely harmed and had lost the ability to regrow a new population of symbionts. When *Euplotes* was X rayed with less than 50,000 R or the treatment with antibiotics was ended prematurely, *Euplotes* cells were found which started to divide again after lag periods of up to 15 days. Once *Euplotes* had started to multiply after such a lag period, it was found to do so with its previous fission rate. When cells were checked for symbionts, they were found to bear the usual number of about a thousand omikrons. Obviously the symbionts had been regrown during the lag period from one or a very few survivers. It was reported (Heckmann, 1975) that lower doses of X rays resulted in higher numbers of recovering *Euplotes* cells and in shorter periods of recovery. This is in good agreement with the assumption that a large number of omikron cells must first be regrown before the host can start to multiply again.

The observation that a dose of 50,000 R, which kills 100% of the symbionts in the ciliate, allows the animal still to undergo up to two fissions, is of interest. It probably means that a substance or substances produced by omikron, and needed by *Euplotes* for its multiplication, are rather stable, because the second division can take place as late as 2 days after the irradiation.

Aposymbiotic cells obtained by penicillin treatment can be rescued from death and may give rise to vigorous clones, if they are reinfected with omikron. Reinfection has been achieved in two ways: (1) by adding to isolated cells a few drops of a freshly prepared symbiont-containing cell homogenate (Heckmann, 1975); (2) by injecting symbiont-containing cytoplasm into aposymbiotic cells (Fujishima and Heckmann, 1982). Symbionts which enter the cytoplasm of an aposymbiotic cell start to reproduce and at the same time improve the vitality of their host. Reinfected cells increase their intake of food, grow, and eventually start multiplying at their previous fission rate. When they start to divide, about 900–1000 symbionts can be found in their cytoplasm.

It is unclear how the symbionts manage to enter the cytoplasm of an aposymbiotic cell if a symbiont-containing homogenate is added to their culture medium. It is likely that the ciliate takes up omikron into food vacuoles along with other particulate matter, and that most of the symbionts are digested. Probably only a few omikron cells manage to escape into the cytoplasm where they are saved and become able to establish new populations. It is also conceivable that the symbionts, after being taken up, actively change the membrane which encloses them and, in this way, prevent the fusion of lysosomes with the enclosing vacuole. Such an active interaction of a symbiont with the enclosing membrane has recently been proposed for algal symbionts of *P. bursaria* (Reisser, 1980; Karakashian and Rudzinska, 1981).

The proportion of cells which become reinfected by adding homogenate to their medium is small (less than 10%) and was found to vary from experiment to experiment for unknown reasons (Heckmann, 1975). Cells treated with homogenate needed an average of 5–6 days to recover. If symbionts were injected into aposymbiotic cells 50% of them recovered. Here, fissions have been observed in some cells as early as 2 days after an injection, although other cells required 3–4 days until they divided for the first time. The fact that recovery of *Euplotes* requires a much shorter period after microinjection than after homogenate feeding can be explained by the larger number of unharmed symbionts which reach the cytoplasm. It has been estimated (Fujishima and Heckmann, 1982) that the 10 pl of cytoplasm which were usually transferred during a microinjection experiment may contain as many as 60 omikron particles. Probably not all of them are capable of reproducing after such a transfer, since many may have become harmed by the procedure. The fact that only 50% of the injections were found to be successful, suggests that sometimes all of the transferred symbionts were irreversibly affected.

The observation that *E. aediculatus* stops multiplying when a large number of its symbionts is harmed or lost due to irradiation or treatment with antibiotics yet recovers and multiplies when a sufficiently large number of symbionts has become reestablished, either by regrowth from unharmed left overs, or by growth from symbionts which have been introduced by artificial means demonstrates that omikron serves some indispensable function for *E. aediculatus*. The exact nature of its function is not known.

C. The Dependence of Omikron on *E. aediculatus*

Heckmann (1975) has attempted to grow omikron free of *Euplotes*. He tried 12 different kinds of media but found none which allowed omikron to grow. The symbiont was introduced into these media either by breaking omikron-containing cells open and immediately plating the homogenate on the various agar plates, or by first isolating omikron with ECTEOLA-columns from homogenates and then plating them. Agar plates were then incubated for 5–6 days at 22°C and examined. Occasional growth of bacteria was found, but, judged by their morphology, none of them could be identified with omikron and they were therefore considered to be contaminants. Heckmann has also incubated omikron in *Euplotes* homogenate but could not obtain evidence that they continued to grow under these conditions. He therefore concluded that omikron requires very specific growth conditions, its reproduction probably depending on the intact cytoplasmic environment provided by *Euplotes*.

One could speculate that this dependence is due to a loss by the symbiont of genetic information which would be needed in a less favorable environment in which amino acids, nucleoids, etc. are not available but first have to be synthesized. Bacteria tend to be economical and to discard DNA no longer needed. One could also speculate that some of the information necessary for the growth of omikron is no longer located in the symbionts themselves but became transferred in the process of coevolution to the nucleus of the host, as is assumed for mitochondria and chloroplasts which are now held by many to have evolved from former free-living organisms. None of these speculations have been confirmed or refuted. However, it is clear that omikron has a genome size of approximately one-quarter of that usually found in free-living bacteria (Schmidt and Heckmann, 1980). In addition, there are arguments which speak in favor of the assumption that omikron is not a recent acquisition of *E. aediculatus,* but probably is an ancient symbiont. Similar-looking bacteria that also depend on their host have recently been found in a number of *Euplotes* species, all closely related to *E. aediculatus* (see Section III).

Since omikron depends on *E. aediculatus* as *E. aediculatus* depends on omikron, both organisms form a tight evolutionary unit which can no longer be resolved. One can express this thought also by calling omikron an organelle of *E. aediculatus.*

D. Genome Size of Omikron and Characterization of Its DNA

It is possible to isolate omikron from cell homogenates and to obtain pure preparations of the symbiont in quantities large enough for extraction and characterization of its DNA. Schmidt (1982) has done this, and determined the GC content of omikron DNA to be 47.7% if computed from the melting profile of the DNA. A value 2.8% lower was found if the GC content was calculated from the buoyant density of the DNA, which Schmidt found to be 1.7036 g/cm^3. Differences of the same magnitude have been reported by others and were found to be due to rare bases in the symbiont's DNA. It is at present not known whether or not omikron DNA has such rare bases.

The average DNA content of omikron was determined to be 5.8×10^{-3} pg per particle, which is equivalent to 3.5×10^9 daltons. It is known from microscopic observations that in omikron the DNA is compounded into several distinct nucleoids which can be counted with the help of a light microscope. If one assumes that each of these nucleoids comprises a genome, one can estimate the genome size of omikron by dividing 3.5×10^9 daltons by the average number of nucleoids of a symbiont. The genome size of omikron was in this way estimated to be 0.5×10^9 daltons. This value was found to be in good agreement with the one for the kinetic complexity of omikron DNA, which was determined from renaturation kinetics to be 0.57×10^9 daltons when corrected for GC content.

It is interesting to note here that similarly small genomes have been reported for lambda, mu, and pi, symbionts of the *P. aurelia* complex (Soldo and Godoy, 1973), for xenosomes of *Parauronema acutum* (Figueroa-de Soto and Soldo, 1977) and for chlamydias, rickettsias, and mycoplasmas (Bac *et al.*, 1969). The genomes of these organisms are the smallest known for cells. They code for only about 700 to 900 proteins.

III. Omikron-like Particles in Other Fresh Water *Euplotes* Species

A. The Bacterial Symbionts of *E. patella* and *E. eurystomus*

The occurrence of bacteria-like symbionts in members of the genus *Euplotes* was first noticed by Fauré-Fremiet (1952). In a short report to the French Academy of Science he described bacteria-like filaments which he found in the cytoplasm of 12 stocks of *E. patella* and *E. eurystomus* collected near Paris. They had a length of 5–12 μm and a width of 0.5 μm. After Feulgen staining he noticed many reddish colored granules in them, which must have been nucleoids. Fauré-Fremiet mentioned that the same symbionts had already been seen in *E. patella* by Chatton and Séguéla (1940), who had described them as "mitochondries bacilliformes." In a later paper, which is primarily concerned with the fine structure of *Euplotes* (Fauré-Fremiet and André, 1968), a short account of the ultrastructure of the symbionts is given.

Since Fauré-Fremiet found that small doses of penicillin not only led to a loss of symbionts within 4 days, but also to the death of the ciliates within 12 days, he suggested that the symbionts might be essential for the two *Euplotes* species. The possibility that the death was due to sensitivity to penicillin rather than to loss of symbionts, was, however, not excluded.

The symbionts described by Fauré-Fremiet were about twice as large as omikron described by Heckmann in *E. aediculatus*. Nevertheless, they showed several nucleoids per particle, a characteristic property of omikron, and therefore are probably related to omikron. Omikron-like symbionts with dimensions similar to those given by Fauré-Fremiet have later been found not only in stocks of *E. patella* and *E. eurystomus,* but also in some stocks of *E. aediculatus* (Heckmann *et al.*, 1982).

B. Do All Fresh Water *Euplotes* Species Depend upon Bacterial Symbionts?

Currently 22 fresh water *Euplotes* species are recognized. They can be subdivided into three major groups: one characterized by 10 frontoventral cirri, one by a 9 type 1 or *E. patella* cirrus pattern, and one by a 9 type 2 or *E. affinis* cirrus pattern. Gates (1978) has shown by morphometric measurement that the two cirrotype 9 subtypes can artificially be derived from the pattern of a species with 10 cirri by removing either cirrus 3/IV (9 type 1 pattern) or cirrus 2/IV (9 type 2 pattern). He suggested that the cirral pattern evolution in *Euplotes* might have occurred by the loss of these cirri.

When Heckmann had recognized that omikron was essential for *E. aediculatus* and symbionts of a similar type were apparently indispensable for *E. patella* and *E. eurystomus,* he started to wonder whether perhaps all fresh water *Euplotes* species depend upon symbionts. He knew that many marine *Euplotes* species can bear symbionts, but do not depend on them, a situation similar to that in the species of the *P. aurelia* complex (Sonneborn, 1959). Assuming that *Euplotes* has evolved in a marine environment and entered fresh water habitats only afterward, he was wondering whether fresh water habitats may lack certain organic compounds present in sea water, and therefore available to *Euplotes* in a marine environment which for fresh water forms had to be provided by symbionts. In an effort to clarify whether other fresh water *Euplotes* species also depend on symbionts, altogether 40 stocks were collected and carefully examined (Heckmann and ten Hagen, 1981; Heckmann *et al.*, 1982). Most of them were collected in Germany (near Münster/Westfalen) but 13 of them had come from France, the United States, and Japan. The results are summarized in Table I.

The 40 stocks are distributed among 10 species. Seven of them are characterized by a 9 type 1 cirrus pattern, one species has a 9 type 2 cirrus pattern, and 2 species are of cirrotype 10. The group of species with a 9 type 2 cirrus pattern is not well represented. On the other hand several species of this group have so far only been reported once and one cannot expect to find them easily.

TABLE I
OCCURRENCE OF OMIKRON-LIKE SYMBIONTS IN CURRENTLY RECOGNIZED FRESH WATER *Euplotes* SPECIES

Species	Number of stocks examined	Cirrotypes	Omikron-like symbionts
E. crenosus	2	10	No
E. palustris	1	10	No
E. aediculatus	9	9 type 1	Yes[a]
E. eurystomus	5	9 type 1	Yes
E. plumipes	1	9 type 1	Yes
E. daidaleos	5	9 type 1	Yes
E. octocarinatus	8	9 type 1	Yes
E. patella	3	9 type 1	Yes
E. woodruffi	3	9 type 1	Yes
E. muscicola	2	9 type 2	No

[a]Two stocks of *E. aediculatus* bear another type of symbiont which probably has displaced the original omikron-like symbiont.

Thirty-five of the 40 stocks bear bacterial symbionts. The symbionts are in appearance similar to omikron and are therefore called omikron-like with the exception of the symbionts of two *E. aediculatus* stocks which do not show nucleoids and will be discussed below. All symbiont-bearing stocks were found to belong to species characterized by a 9 type 1 cirrotype. The remaining five symbiont-free stocks belong to two species with a cirrotype of 10 and one species with the cirrotype 9 type 2.

Heckmann *et al.* (1982) have subjected their symbiont-bearing stocks to a penicillin treatment of the kind which had earlier been found to remove omikron. They reported that most of the stocks were affected in exactly the same way as had been observed for omikron-bearing strains earlier. Two stocks could not be freed of their symbionts, perhaps because their symbionts are penicillin resistant. They continued to grow unaffected. When the five symbiont-free stocks were treated with penicillin they also continued to multiply.

These findings are interpreted by Heckmann *et al.* (1982) as support for Gates' (1978) conclusion that species with a 9 type 1 cirrus pattern form a natural group which has evolved independently of the forms with a 9 type 2 cirrus pattern. They put forward the hypothesis that all *Euplotes* species with a 9 type 1 frontoventral cirrus pattern depend upon bacteria-like endosymbionts. They propose that these species have evolved from an ancestor with a cirrotype of 10 which already had become dependent upon bacteria-like symbionts.

In this connection it is of interest that Foissner (1978) has described a species with 10 frontoventral cirri (*E. moebiusi f. quadricirratus* Kahl, 1932) in which he has found omikron-like symbionts. He has not tested whether or not the ciliate depends upon the symbiont.

C. Are All Bacterial Symbionts Related to Omikron?

The bacterial symbionts of the various fresh water *Euplotes* stocks were found to differ in appearance from each other. In some stocks they were long and thin with a tendency to form filaments, whereas in other stocks short forms were found to prevail. Most of the symbionts were slightly curved but in some stocks they had the shape of straight rods. Figures 7–12 illustrate some of the forms encountered. Differences of this kind were not found to be species-specific. They were also found among stocks classified by the conventional morphological criteria as belonging to the same species. Another difference which was recognized was the average number of symbionts per host cell. For some stocks more than 1000 symbionts per cell seem to be normal, for other stocks only about half that number could be counted. At present it is not known to what extent such differences reflect differences in the physiology of the host. Heckmann *et al.* (1982) proposed that these differences are due, partially, at least, to differences in the symbionts themselves.

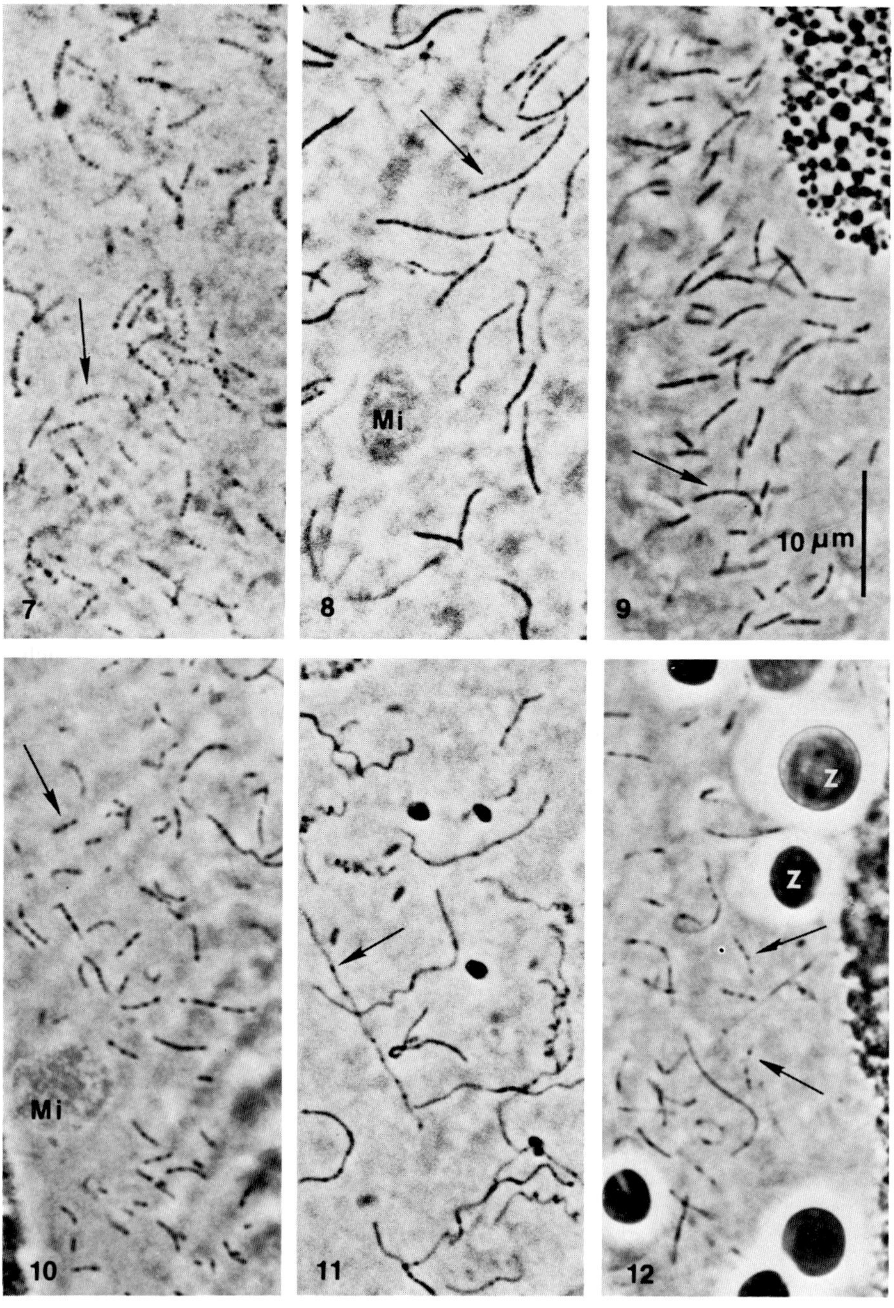
7
8
Mi
9
10 μm
10
Mi
11
12
Z
Z

Thirty-three of the 35 symbiont-bearing stocks were found to have symbionts which can be classified as omikron-like. They all showed several clearly defined nucleoids per particle although they differed considerably from each other in dimensions and other aspects. However, two stocks of *E. aediculatus* (stock 7 and 10) were found to bear bacteria which were of an entirely different type. They were wider than all the other symbionts, possessed tapered ends, and lacked nucleoids (Fig. 13). In both stocks the number of these bacteria varied between 400 and 600 per cell. Heckmann *et al.* (1982) found that this type of symbiont also can be removed with penicillin and that its host then also stopped multiplying in the same way as was observed for the other stocks. In order to find out whether this type of symbiont served the same function for its host as omikron, Heckmann *et al.* (1982) have tried to infect aposymbiotic cells of stock 15, the one which bears omikron, with the nontypical symbionts of stock 7. In an experiment in which 90 cells were provided with a homogenate that contained type-7 symbionts, they observed a resumption of growth in one cell definitely caused by a successful infection with type-7 symbionts. All of the other cells died. The experiment was later repeated, this time using a microinjection procedure for the transfer of type-7 symbionts, and a high percentage of recovery was achieved. Using the same method, omikron was established successfully also in stock 7 cytoplasm. These results show that both types of symbionts can restore to their hosts the ability to multiply, and therefore obviously provide to *E. aediculatus* the same kind of functions.

The fact that omikron-like symbionts were found in stocks belonging to different species of *Euplotes* and collected in different continents make Heckmann *et al.* (1982) argue that they are most probably the original inhabitants of the *Euplotes* species with a 9 type 1 cirrotype. They suggest that the various omikron-like symbionts evolved from a bacterium of this type which had become established in the ancestor of this group of ciliates. They consider the type-7 symbionts as a recent acquisition which may have entered the *Euplotes* stocks as a secondary infectant that displaced the original symbiont. They quote in favor of their view the fact that omikron has a very small genome which they interpret as

---

FIGS. 7–12. Omikron-like endosymbionts of various fresh water *Euplotes* species. Nucleoids can be recognized as dark dots (arrows). Ma, Macronucleus; Mi, micronucleus; Z, zoochlorellae. All figures are photomicrographs of aceto-orcein-stained *Euplotes* cells in phase contrast. All figures are of the same magnification (scale in Fig. 9). (From Heckmann *et al.*, 1982.) Figure 7. *E. aediculatus* stock 15. The symbionts of this stock have been named omikron. Figure 8. *E. aediculatus* stock 27. The symbionts are longer and have more nucleoids than omikron. Figure 9. *E. woodruffi* stock 22. The symbionts stain more intensely and appear less curved than omikron. Figure 10. *E. eurystomus* stock 25. The symbionts show dimensions very similar to omikron. Figure 11. *E. eurystomus* stock 40. The symbionts form in this stock long threads which often are twisted. Figure 12. *E. daidaleos* stock 13. The symbionts resemble those of *E. eurystomus* stock 40 but do not tend to curl. In addition zoochlorellae are present.

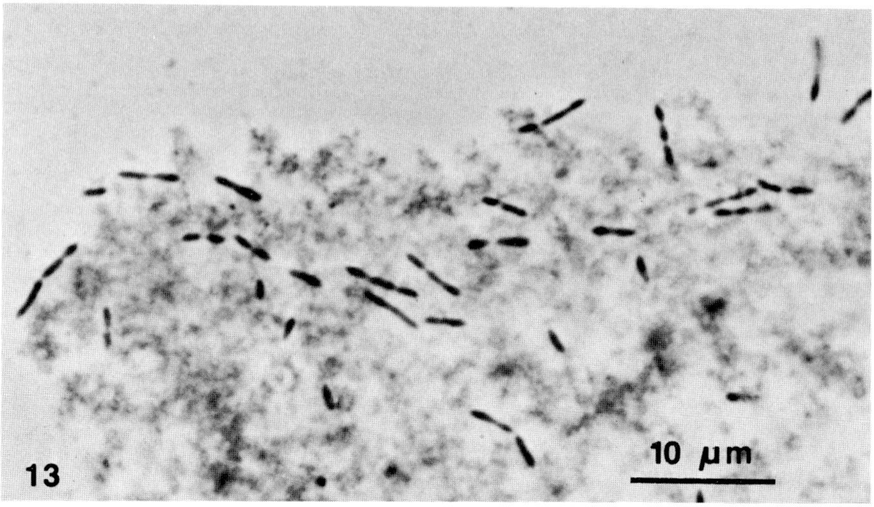

FIG. 13. Symbionts of *E. aediculatus* stock 7. They are not omikron-like, no nucleoids can be recognized in them, and in addition they have tapered ends. It is assumed that they are secondary symbionts which have displaced the original omikron-like symbionts. (From Heckmann *et al.*, 1982.)

a consequence of a long and well-adapted life in an intracellular environment. So far no data with respect to the genome size of the type-7 symbionts are available. It is also not known whether type-7 symbionts depend upon *Euplotes* for their reproduction, as was demonstrated for omikron, or whether they can reproduce in a less complex environment.

D. TRANSFER OF SYMBIONTS BETWEEN *Euplotes* SPECIES

Recently Fujishima and Heckmann (1982) have started to investigate the question of whether the omikron-like symbionts encountered in the seven species with a 9 type 1 cirrus pattern can replace each other. They have produced aposymbiotic cells by treating them with penicillin and have then tried to infect such cells by microinjecting into them symbiont-containing cytoplasm from other species. So far only a limited number of combinations has been investigated and most of them did not lead to a resumption of growth of the recipients. Restoration of growth and the ability to divide was, however, observed when symbionts of *E. woodruffi*, of a stock collected in Japan, were introduced into a stock of *E. aediculatus*, which had been collected in France. About half of the injected cells started to divide again and gave rise to healthy looking lines. The reverse combination, *E. woodruffi* as the recipient and *E. aediculatur* as the donor, did not work.

These interesting results are consistent with the notion of Heckmann *et al.* (1982) that the omikron-like particles are related to each other and restore a common inherited defect in all the species with a 9 type 1 cirrus pattern, but with secondary dependencies that have developed along with their integration in some cases. Thus all symbionts are no longer able to grow in all species, or support all the needs which have evolved. It might very well be that by testing the various possible combinations one may learn more about the process by which symbionts become integrated in an intracellular habitat and at the same time obtain hints of how the species with a 9 type 1 cirrus pattern are related to each other.

IV. Killers and Mate-killers in *Euplotes*

A. The Killers

Certain stocks of ciliates are called "killers" because they liberate toxic substances into the medium in which they live which kill cells from certain other stocks, which are therefore called "sensitives." Killers and sensitives were first discovered in *Paramecium aurelia* by Sonneborn (1938) at about the same time that he discovered mating types. They have been studied intensively by him and his associates, and it is now clear that the killing phenomenon results from toxins produced by bacterial endosymbionts living within the paramecia (reviewed in Preer *et al.*, 1974). Killers liberate bacteria into the culture medium which release toxins if they are taken up into food vacuoles. The toxins are generally harmless to ciliates which bear this type of bacterium, but kill sensitives. Many different kinds of bacteria have been discovered to be associated with the killer phenomenon. Although often referred to only by common names, kappa, lambda, etc., they have more recently been classified and given binomial names (Preer *et al.*, 1974; Quackenbush, 1978; Preer and Preer, 1982).

Killer stocks have been reported not only for various species of the *P. aurelia* complex but also for *P. bursaria* (Chen, 1945, 1955) and *P. polycaryum* (Takayanagi and Hayashi, 1964; Takayanagi *et al.*, 1966) and outside the genus *Paramecium* for *E. minuta* (Siegel and Heckmann, 1966) and *E. crassus* (Luporini, 1974). The particles which cause the killing in the two *Euplotes* species and the phenotypes bestowed by them upon their hosts are the subject of this section.

1. *Epsilon Particles*

a. *The Cytoplasmic Particles.* These symbionts were described by Heckmann *et al.* (1967) from killer strains K_1, K_3, K_7, and VF_{17} of *E. minuta* collected from the Mediterranean Sea at Villefranche sur Mer (France). The symbionts are rather small—about 0.8–2.5 μm long and 0.4 μm wide—and appear, when observed in freshly crushed cells with the help of bright phase optics, somewhat darker than free-living bacteria (Fig. 14). The same is characteristic of the cytoplasmic symbionts of *Paramecium* (J. R. Preer, Jr. personal communication). When the killer strain K_3 was fixed in Schaudinn's fluid, hydrolyzed in 1 *N* HCl, and stained with basic fuchsin as described by Dippell and Chao (Sonneborn, 1950), several dozen darkly staining bodies were found scattered in the cytoplasm (Fig. 15). Their staining properties, size, shape, and cellular distribution resemble that found for kappa, pi, mu, nu, sigma, and lambda in *Paramecium* (Beale *et al.*, 1969). Since they are only found in killer strains and not in nonkillers, it is very likely that they are the agent responsible for the killer phenotype in *E. minuta*.

Heckmann *et al.* (1967) were not able to distinguish between the particles in strain VF_{17} and those of the other killer strains (K_1, K_3, K_7), in spite of the fact that VF_{17} was found to kill sensitives in a somewhat different way (see below). No evidence of R bodies, such as are present in kappa in *P. aurelia*, was observed.

b. *The Killer Trait.* That certain strains of *E. minuta* kill certain other strains

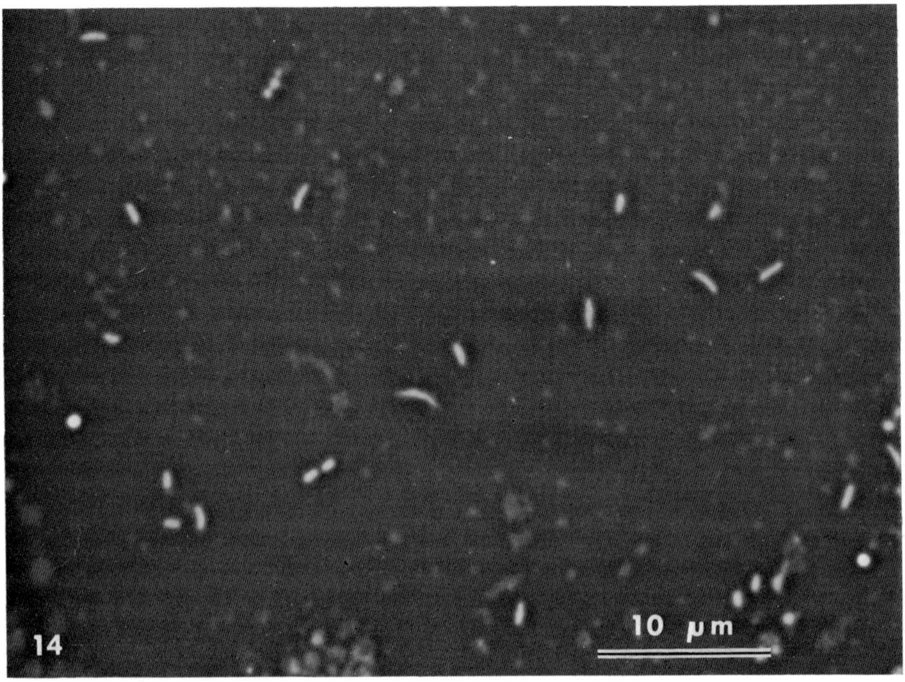

FIG. 14. Epsilon particles of a freshly crushed cell of *E. minuta* stock K_3. Bright phase. (From Heckmann *et al.*, 1967.)

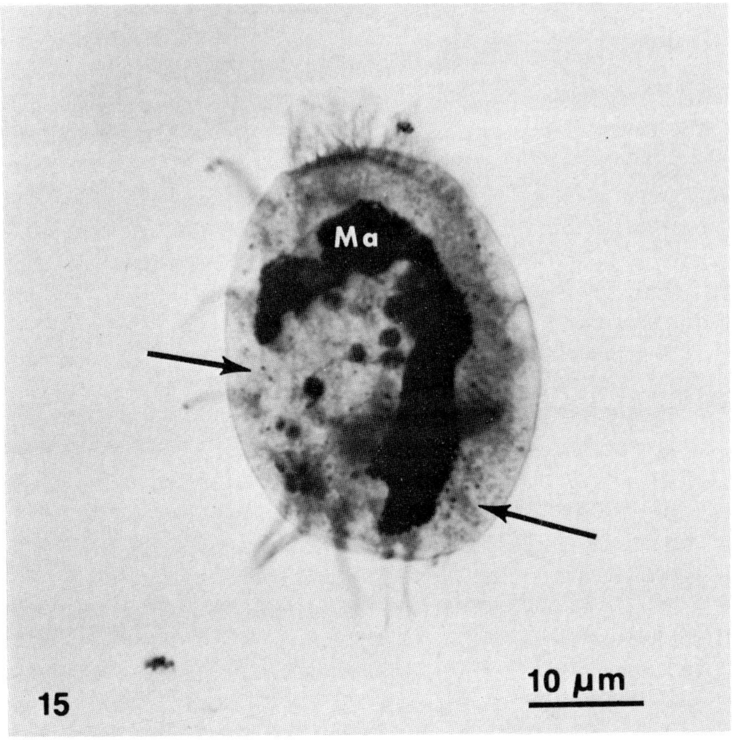

FIG. 15. Whole stained cell of *E. minuta* stock K_3 with epsilon particles (small darkly stained bodies) scattered randomly about the cytoplasm. Ma, Macronucleus. (From Heckmann *et al.*, 1967.)

of this species was first noticed when Siegel and Heckmann (1966) combined various strains and observed the mixtures for conjugating pairs. At first three killer strains, designated K_1, K_2, and K_3 were distinguished. They were found to cause the death of certain other strains named S_1, S_2, S_3, S_4, and A_{23} within 1 to 4 days. No killing occurred when cells of the three killer strains were combined nor when those of the sensitive strains were mixed. It was later found that the killers of *E. minuta* also have an effect on *E. crassus* and that the symptoms of killing were particularly clear and easy to read in cells of strain D of *E. crassus*, syngen 2, which therefore was adopted as the standard sensitive tester. Cells of this strain are markedly larger and distinctly lighter in appearance than cells of *E. minuta* and in addition are particularly sensitive toward the toxin liberated by the *E. minuta* killers. As a response they form hundreds of tiny vacuoles which later enlarge and coalesce (Fig. 16). Such vacuolated individuals cease their normal swimming movements, lie quiescent on the bottom of the culture vessel, and will not become active even when the vessel is shaken. They finally lyse. A some-

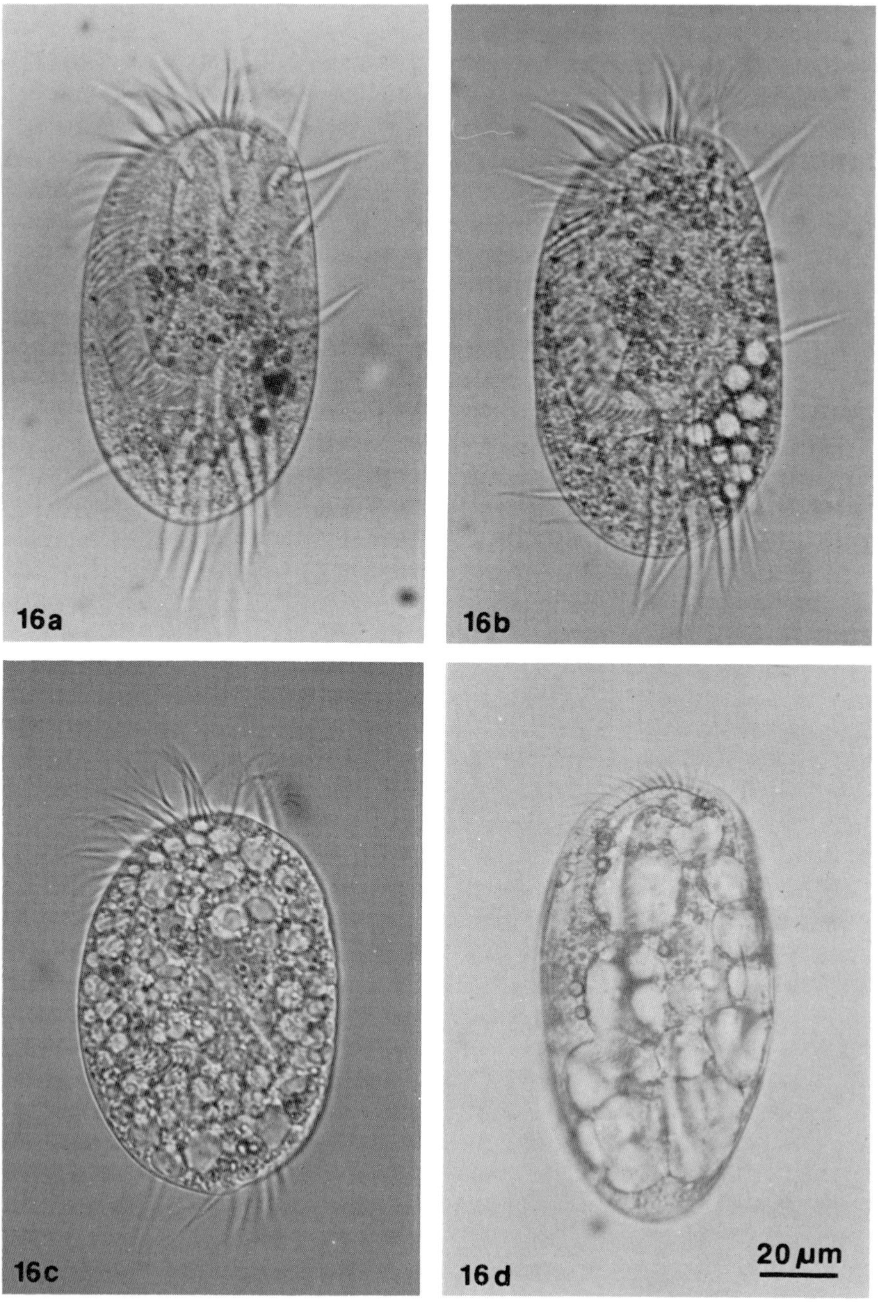
16a
16b
16c
16d
20 µm

what different killing action is produced by VF_5 and VF_{17}. These are killer strains from the same location as the other K strains, but collected 2 years later. Here vacuole coalescence in the *E. crassus* testers generally proceeds to the point where only one large vacuole is present, that comes to lie in the posterior region of the sensitive, giving the whole affected organism a pear-shaped form (Heckmann *et al.*, 1967).

The above described events also occur when sensitive cells are transferred to the cell-free fluid from cultures of the killer cells.

c. *Properties of the Killer Substance.* The fluid-stable poison liberated by the killer cells into the surrounding medium was called "euplotein" (Siegel and Heckmann, 1966). It is probably bound to epsilon particles liberated by the killers into the medium, because it can easily be sedimented by centrifugation with a speed which would be expected for particles of the size of epsilon (Heckmann *et al.*, 1967). Strong killing activity was obtained after killer cells were concentrated and homogenized. The activity derived from the homogenate precipitated with the same speed as the one observed in culture fluid. On the other hand it was lost when a homogenate was passed through a French pressure cell so that all visible particles were fragmented. These facts support the view that epsilon particles are the bearers of the toxic activity.

The killer substance of *E. minuta* is rather heat-labile, like the other killer substances which have been studied: kappa killer substance of *P. aurelia* (Sonneborn, 1959), a lambda killer substance of *P. aurelia* (Butzel and van Wagtendonk, 1963), and a killer substance of *P. bursaria* (Dorner, 1957). At 50°C all killing activity of a filtrate is lost within a few minutes.

Filtrates were also exposed to a number of enzymes (Heckmann *et al.*, 1967). Trypsin, papain, ribonuclease, and deoxyribonuclease had no effect on killing activity. Pepsin destroyed about 80% of the activity, while chymotrypsin destroyed all detectable activity.

Proteolytic enzymes have been found to inactivate killing particles of stock 51 of *P. aurelia* (van Wagtendonk *et al.*, 1950; Setlow and Doyle, 1956) just as they inactivate the killing activity of *E. minuta*. However, caution must be used in interpreting the effect of enzymes on the large complex particles which bear killing activity. The fact that chymotrypsin and pepsin destroy killing activity may mean only that these enzymes disrupt an outer membrane of the particles and thereby indirectly cause inactivation of the toxic principle or interfere with the uptake of the particles by sensitives. The chemical nature of the toxin itself is unknown.

d. *The Inheritance of the Killer Trait.* As in *P. aurelia*, the killer trait is

FIG. 16. Consecutive stages in animals (*E. crassus* strain D) affected by the killer activity of stock K_1 of *E. minuta*. (a) Before the exposure to the toxin; (b) several hours later; (c) about 1 day later; (d) final stage, shortly before the cell disintegrates. (From Heckmann *et al.*, 1967.)

always transmitted in *E. minuta* to both daughter cells. No special precautions were required to avoid loss of symbionts from stocks. In fact, lines were observed to lose their symbionts only in old age, generally first becoming weak killers with low fission rates (Heckmann, unpublished). Conjugation between killers and sensitives always yielded killer clones from the killer mate and sensitive clones from the sensitive mates. Such crosses were fruitful if paired cells were isolated into fresh culture medium after they had formed and if the coconjugants were again isolated to fresh medium soon after they separated. No indications for special genotypes necessary for the maintenance of epsilon particles could be obtained; all the autogamous F_2 clones derived from F_1 killer exconjugants were found to be killers, nor did the trait segregate following subsequent autogamous reorganizations or back crosses with nonautogamous sensitive strains (Siegel and Heckmann, 1966; Heckmann and Frankel, 1968).

2. *Eta Particles*

a. *Morphology of the Symbiont.* Recently protozoologists at the University of Pisa (Italy) have described endosymbionts in stocks of *E. crassus* collected by Luporini (1974) from the Indian Ocean at beaches along the coast of Somalia. One of the symbionts, first designated type B (Rosati *et al.*, 1976) and then named eta (Nobili *et al.*, 1976) or B_2 (Rosati and Verni, 1977), was found to be associated with a killer trait. The symbiont, observed only in the electron microscope, is round or oval shaped with an average diameter of 0.9 μm. It is bounded by three juxtaposed membranes. The outermost membrane has ribosomes attached to it. The interior of the symbiont is granular with a few scattered fibrous strands.

b. *The Killer Trait.* Nobili *et al.* (1976) reported that cells of stock $21A_7$ are killed when mixed with those of stock M or C, or when exposed to their culture fluid or to homogenate prepared from cells of these stocks. Within 6 to 24 hours the sensitive testers develop a large vacuole in the posterior part of the body which progressively increases in size. Affected cells are gradually transformed into transparent spheres which ultimately "explode." These changes were only rarely observed in mixtures of cells of stock Mog with those of the two killer strains M and C and they were never found in mixtures of Mog cells with those of the sensitive strain $21A_7$. Strain Mog, therefore, is neither a sensitive nor a killer but constitutes a third, "resistant" phenotype which combines properties of the sensitive stock (not to kill) with those of the killer lines (to be resistant).

The four stocks ($21A_7$, M, C, Mog) were followed over a long period of time and found not to change their phenotypes. It is not reported whether the two killer stocks were tested against each other and found to be resistant to the toxin produced by the other stock.

Since the killing activity was found not to pass through a membrane with a pore diameter of 0.4 μm, cells of the three phenotypes were fixed and investigated for cytoplasmic symbionts with the help of an electron microscope. In the cells of the two killer stocks, 500–1000 eta particles per cell were found (the number estimated from the density of symbionts counted in randomly chosen EM sections). In cells of the Mog strain 150–200 similar looking particles per cell were found, which were assumed by Nobili *et al.* (1976) to be eta symbionts. No particles were found in cells of stock $21A_7$.

That the loss of killing activity is associated with a loss of eta particles was shown in two ways: first, by treating killer cells of stock M with penicillin, and second, by treating them with UV (Nobili *et al.,* 1976). In both experiments about one-third of the treated cells gave rise to sensitive lines which were found to be free of eta, one-third remained killers, and one-third showed the resistant (Mog-like) phenotype. It is not reported whether those with a resistant phenotype maintained this character or reverted to the killer phenotype. It is reported that the sensitive lines remained sensitives.

c. *Reinfection Experiments.* Reinfection of eta-free M cells was achieved by placing them into eta-containing cell homogenate (Nobili *et al.,* 1976). Although most of the cells were killed by this treatment, a few survived and continued to reproduce when isolated into fresh culture medium. After 1 week their phenotype was tested by mixing them with cells of the killer strain M and those of the sensitive strain $21A_7$, and it was found that about one-half of them had become killers and the other half did not kill, but was resistant against the eta toxin. Killers had at the average the same number of symbionts as the original killer stocks and the resistant lines were found to have a low number of symbionts like cells of the Mog strain. Both types of reinfected cells were, however, found to be sensitive again when they were tested a month later, i.e., they had not maintained their killer or resistant phenotypes.

A similar experiment was carried out with cells of strain $21A_7$ as the recipients (Nobili *et al.,* 1976). When lines of successfully infected $21A_7$ cells were tested a week after they had been exposed to eta-containing homogenate, about half of them were found to be resistant and to bear eta symbionts. One month later when the test was repeated they also were found to be sensitives.

The results of the reinfection experiments are difficult to interpret. First of all, it is surprising that 100% of the cells which survived an exposure to homogenate gave rise to lines of cells which a week later had hundreds of symbionts in their cytoplasm and therefore not only had been successfully infected, but also maintained the symbionts. Unfortunately, it is not reported how long the cells were treated with homogenate and how many of the aposymbiotic cells were killed by this procedure and it is also not stated whether the reinfection was carried out in mass cultures or whether single cell isolates were treated. The results imply that the uptake of a very few symbionts, if they do not kill, immediately makes the host cell resistant and that the symbionts must then rapidly reproduce to large numbers in such cells. That eta particles are quickly taken up by *E. crassus* into food vacuoles was demonstrated by Verni *et al.* (1977). They also showed that unaffected sensitives had particles in their cytoplasm 6 hours after feeding with eta-containing homogenate. Second, it is stated that all infected lines had lost their symbionts 30 days later. Why should cells which after infection had allowed the symbionts to reproduce rapidly in their cytoplasm later cease doing so? This is especially difficult to understand for reinfected M cells, because M cells which had not been deprived of their symbionts for a short period have never been found to lose their symbionts under conditions that were otherwise exactly the same. Verni *et al.* (1977) reported that eta particles, when found after an infection in the cytoplasm of cells of strain $21A_7$, did not show the outermost third membrane which is present in M cells. They interpreted this as a sign for an incompatibility between eta particles and the $21A_7$ cells. They did not report whether this membrane is also missing in reinfected M cells which also lose eta.

d. *Crosses between Killer and Nonkiller Strains.* Nobili *et al.* (1976) reported a number of crosses between killer and nonkiller strains which, however, are also difficult to interpret. In a cross of M (killer) × Mog (resistant) all progenies, including those which derived their cytoplasm from Mog cells, were killers and maintained this character for at least 2 months. An F_1 clone was backcrossed with Mog and here, killers, sensitives, and resistants were found among the progenies. Unfortunately, it is not mentioned which phenotypic classes derived their cytoplasm from which parent or whether cytoplasmic ancestry was irrelevant to the outcome. Some of the resistant clones later became killers. A third cross is reported in which a killer progeny clone of the backcross was crossed with the sensitive stock $21A_7$. The majority of the progeny was found to be resistants, which implies that the symbiont-free cells of stock $21A_7$ had become infected during conjugation. A few were found to be sensitives and a few others to be killers. When the clones were retested after 30 days two of the original sensitive clones had become resistant and two of the original killer clones had become resistant nonkillers. From the clones found to be resistants after 7 days, many had become sensitives when tested again after 30 days. Nobili *et al.* (1976) concluded from these results that nuclear genes are involved in the determination of the three phenotypes. Although this possibility cannot be ruled out, it is certainly not yet established. What can be concluded from their experiments is that the resistant and the killer phenotypes are associated with the respective presence of low and high numbers of eta symbionts, and that sensitive cells lack symbionts. Understanding how the number of eta symbionts is regulated in *E. crassus* requires further investigation.

B. The Mate-killers

Mate-killers were discovered by Siegel (1950, 1953, 1954) in a species of the *P. aurelia* complex now called *P. octaurelia,* and have subsequently been found in *P. primaurelia* (Beale, 1957; Beale and Jurand, 1966) and *P. biaurelia* (Beale *et al.*, 1969). Siegel found that following conjugation between mate-killer and normal paramecia, exconjugants or subsequent fission products deriving cytoplasm from the normal parents were killed, while exconjugants deriving cytoplasm from mate-killer parents yielded viable progeny. No toxin has been detected in the medium in which mate-killers live, nor do homogenates of mate-killers kill. Cell-to-cell contact, but not nuclear transfer, seems to be necessary for mate-killing. In the *P. aurelia* complex, mate-killing is due to bacterial symbionts called mu. They look very much like kappa without R bodies (Beale *et al.*, 1969).

In the 1960s mate-killing was found for *E. patella* and *E. minuta* and more recently for *E. crassus*. The mate-killer stocks of the three *Euplotes* species and their symbionts are the topic of this section.

1. *Mate-killing in E. patella*

Two stocks, named 25 and K_{13}, both collected from ponds in Hiroshima, were reported by Katashima (1965) to be mate-killers, because they kill their mates during conjugation in a way similar to that described by Siegel for stock 138 of *P. octaurelia*. Both *Euplotes* stocks belong to syngen 1 of *E. patella* of Japan and are of different mating type. When the mate-killing was recognized and the stocks were examined cytologically, symbionts were found in them. They are described as spherical with a diameter of 4–7 μm. They stain with Giemsa more deeply than with the Feulgen technique. Katashima reported that they are maintained through all stages of the life cycle including conjugation and that their number per cell varies from 7 to 28. Similar symbionts were not found in other stocks of *E. patella* which when crossed to each other gave rise to high numbers of viable exconjugants from both parents. These normal stocks of *E. patella* proved with one exception to be sensitives when crossed to the mate-killers.

In his paper, Katashima was concerned primarily with various kinds of damage to the nuclear apparatus of sensitives when they are mated with stock K_{13}. He distinguished and described in detail six types of nuclear abnormalities, five of them affecting the macronuclear anlagen, which may show complete failure of formation, disintegration, excessive swelling, alteration in position, and delayed differentiation following multiple fissions of the synkaryon which results in formation of multiple anlagen. In addition he observed that the degeneration of old macronuclear fragments is markedly changed in sensitives. They become vacuolized and show condensation of large chromatic aggregates. These abnormalities already become visible when the coconjugants separate. The relative frequency of the events seems to depend on the sensitive stocks mated with the mate-killer K_{13}. In all cases, however, damage to the nuclear apparatus of the sensitive conjugant was presumed to be the cause which led finally to its death. Katashima pointed out that a similar conclusion had been reached for *P. oct-*

aurelia by Siegel (1953) and Levine (1953), who were convinced that sensitives die as a result of nuclear damage after conjugation with mate-killers.

Although the mate-killer system of *E. patella* shows great resemblance to the mate-killer system in *P. octaurelia* it must be kept in mind that the symbionts in the two systems do not resemble each other. In *E. patella,* they are unusually large and look more like yeasts than bacteria. Since no information on their ultrastructure is available, the question as to which group of organisms they belong must stay open.

Although not explicitly stated by Katashima, it is clear from data in an earlier paper (Katashima, 1961) that at least one of his strains (named clone 3) was not affected when crossed to the mate-killer stock 25. Since strain 3 did not kill when crossed to other sensitive stocks, its phenotype must be classified as resistant non-mate-killing. The strain has apparently not been investigated cytologically. It is therefore not clear whether its resistance was associated with a low number of symbionts in its cytoplasm or whether it was due to other properties.

Katashima observed bacterial symbionts of the omikron-like type in his sensitive lines as well as in the mate-killers. He did not think that they had anything to do with the mate-killing phenomenon and therefore did not study them further. We know now that they are essential for *E. patella* but do not confer any killing properties on them (see Section III).

2. *Mate-killing in E. minuta*

Stock VF_{17} was described by Heckmann *et al.* (1967) as a killer stock which showed somewhat different killer properties than the other killer stocks of *E. minuta* (see Section IV,A,1). When stock VF_{17} was employed for cross-breeding experiments in a study concerned with the genetic control of cortical pattern in *E. minuta* (Heckmann and Frankel, 1968), it was found that this stock not only killed sensitive strains via epsilon particles liberated into the medium, but that it also acted as a mate-killer. From two coconjugants, one generally developed into a healthy looking exconjugant, whereas the other one failed to develop a large macronuclear anlage, became quiescent, and finally died. The surviving exconjugant clones had the corticotype of stock VF_{17} and were all found to act as killers and therefore were cytoplasmic descendents of the mate-killer parent. That they were true hybrids, i.e., that they had received a pronucleus from the sensitive partner, is revealed from genetic markers which were followed in these crosses.

The cytoplasmic particles found in VF_{17} were of the epsilon type. No differences to those found in stocks K_1 and K_3 could be recognized (Heckmann *et al.*, 1967). Whether VF_{17} was the host of a mixed population of symbionts, the one type responsible for the killer trait, and the other one for the mate-killer trait, with both types so similar that they could not be distinguished from each other, or whether a single class of particles present in VF_{17} determined both traits, cannot be decided. Both the killer trait and the mate-killer trait were lost concurrently with aging of the stock (Frankel, 1973); loss of cytoplasmic symbionts in aging lines has also been observed in *P. aurelia* (Sonneborn *et al.*, 1956).

The stock VF_{17} is remarkable in that it combined the capacities to act both as a killer and as a mate-killer. When it was found to have lost its symbionts it had lost both capacities and had become sensitive toward the epsilon particles liberated by other killer stocks into the culture medium. More recently Demar-Gervais and Genermont (1976) have reported a *E. vannus* stock which also acts simultaneously as a killer and as a mate-killer.

3. *Mate-killing in E. crassus*

A symbiont very similar in morphology to the eta particles associated with the killer trait in *E. crassus* seems to be responsible for the mate-killer trait in stock C_8 described by Dini and Luporini (1976). The symbiont can be seen only with the help of an electron microscope and was later named B_1 by Rosati and Verni (1977). It is a spherical or oval-shaped particle with an average diameter of 1 μm (Fig. 17). The particle is bounded by three membranes, the outermost one appearing granular. It is assumed that the granules attached to the outer membrane are ribosomes and that this membrane is derived from the endoplasmatic reticulum which envelops the symbiont. The symbiont is considered to be a bacterium. Dini and Luporini (1976) estimated that a cell of the mate-killer stock C_8 contains 500–800 symbionts. They found the same type of symbionts also in the sensitive stock W_2, at a concentration, however, about one-third of that observed in the mate-killer. The symbiont was not found in another sensitive stock designated G_{10}.

That the mate-killer trait is associated with the presence of the symbiotic bacterium B_1 is concluded from the observation that mate-killing and the symbiont disappear when the killer stock C_8 is treated with 250 units/ml penicillin for a week (Dini and Luporini, 1976).

The effect which the mate-killers have on sensitives becomes recognizable shortly after the coconjugants have separated. The sensitive exconjugant then appears weak, usually lying on the bottom of the culture vessel. Six to seven hours later, small vacuoles are formed in the posterior part of the cell. They gradually merge into a large vacuole which fills up the posterior half of the animal. By now the sensitive is practically immobile and dies. The sequence of events is strikingly similar to the one reported by Nobili *et al.* (1976) for the killer trait in *E. crassus* caused by eta symbionts which have almost the same morphology.

Like other mate-killers, stock C_8 also seems to act on the nuclei of the sensitives. Dini and Luporini (1976) reported that in 35% of the sensitives they found the gametic nuclei not to have fused into a synkaryon. By labeling the mate-killer with [^{3}H]thymidine, they were able to state that an exchange of the

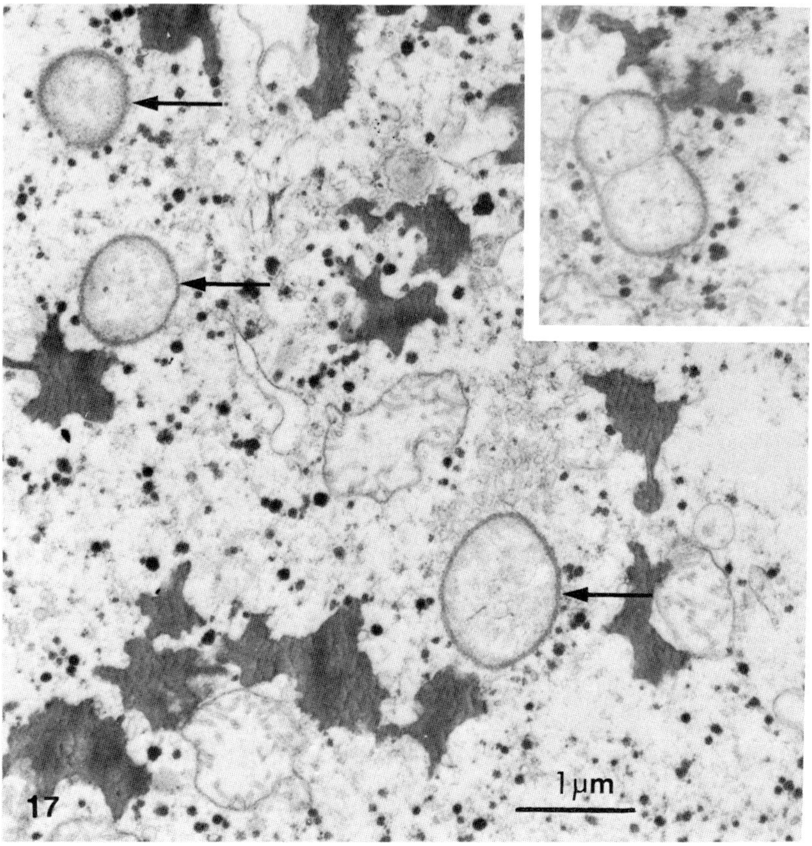

FIG. 17. B_1 particles in a section of the mate-killer stock C_8 of *E. crassus*. The inset shows a dividing B_1 symbiont. (From Dini and Luporini, 1976.)

pronuclei had, however, taken place. If a synkaryon was formed, which occurs in 65% of the sensitives only, very small macronuclear anlagen are developed. Another observation is that fragments of the old macronucleus do not gather in the anterior part of the sensitives, as is the case in uneffected cells, but remain scattered throughout the cytoplasm.

In a search for genes involved in the maintenance of the symbionts in their mate-killer stock, Luporini and Dini (1975) have carried out a number of crosses which gave rise to mate-killers, sensitives, and resistant non-matekillers. They interpreted their data as evidence for the existence of two such genes at two unlinked loci and assumed that the mate-killer C_8 is heterozygous at both loci. They argued that the products of a single gene may not suffice to keep enough symbionts in a cell to let it act as a mate-killer, but would perhaps change it into a resistant line with a low number of symbionts in the cytoplasm. The data are so far preliminary and need to be substantiated.

V. Other Symbionts

A number of endosymbionts which are not as well described as the ones we have discussed so far are the subject of this section.

A. Zoochlorellae

Symbiotic algae are called zoochlorellae if they are green (Brandt, 1882). They are well known to occur in *P. bursaria, Stentor polymorphus,* and a number of other protozoa and have been mentioned repeatedly to occur in *E. patella* or forms which have later been considered conspecific with *E. patella* (Tuffrau, 1960). They are certainly not a constant feature of *E. patella,* and it is possible that the putative *E. patella* stocks in which they were observed in the past in fact belong to *E. daidaleos,* a form closely related to *E. patella* and only relatively recently recognized as a separate species (Diller and Kounaris, 1966). Diller and Kounaris collected *E. daidaleos* for a period of 7 years from a pond at the campus of the University of Pennsylvania and found that the ciliates always contained at least a few algae. Normally, however, they were filled with algae (up to 100 were counted per ciliate) and therefore appeared bright green. The algae are of spherical shape, their size ranging from 3 to 5 μm in diameter.

Heckmann *et al.* (1982) reported that of the five stocks of *E. daidaleos* which they have collected in Germany two were free of algae. This shows that *E. daidaleos* does not depend on algae. They also noted that the two stocks could easily be infected with algae and that upon infection the number of omikron-like symbionts, on which *E. daidaleos* depends, dropped from approximately 400 to roughly 200 per cell. They observed that cells with low numbers of zoochlorellae tend to have numbers of omikron-like symbionts and vice versa. This suggests that the algae can to some extent compensate for the bacteria which have been found to be indispensable for *E. daidaleos* (see Section III,B).

The zoochlorellae of *E. daidaleos* have so far not been investigated carefully. It is therefore not clear whether they are related to the ones present in *P. bursaria* and whether or not they serve similar functions for their host as the zoochlorellae do for *P. bursaria* (Karakashian, 1975; Reisser, 1980).

B. Macronuclear Symbionts

In connection with a search for symbionts in killer-strains of *E. crassus,* Rosati and Verni (1976) came across symbiotic bacteria that apparently have chosen the macronucleus as their habitat. They are described as rod shaped with an average diameter of 0.5 μm and a length of about 1.5 μm. The symbionts are surrounded by two membranes, the outermost one probably being part of the cell wall.

The symbionts were found in two strains, collected at widely separate places. One was from the coast of Somalia and bore in addition eta particles in its cytoplasm (see Section IV,A,2); the other stock was collected near Pisa from the Tyrrhenian Sea and served as a sensitive in killer tests. No effects of the symbionts on their hosts are reported.

Macronucleus-specific symbionts are well known from *P. caudatum* and *P. biaurelia* (see Görtz, this volume). Whether the macronuclear symbionts of *E. crassus* are related to them is not known.

C. Inconspicuous Symbionts

A number of symbionts have been recognized by chance when certain stocks were investigated with the electron microscope. No special functions of these symbionts are known, nor are effects reported which the symbionts exert on their hosts.

Symbiont type A was found in nonautogamous stocks of *E. crassus* collected from the Somalian and Tuscan coasts (Rosati *et al.*, 1976; Rosati and Verni, 1977). It is spherical in appearance, and has a diameter of about 1 μm. A set of three membranes encloses the symbiont. The outermost membrane is probably formed by the host cytoplasm and resembles endoplasmic reticulum.

Symbiont type B_3 is ultrastructurally indistinguishable from either B_1, which causes mate-killing (see Section IV,B,3) or B_2, which transforms *E. crassus* into a killer and has been named eta by Nobili *et al.* (1976) (see Section IV,A,2). Symbionts with a B-type appearance were found in all nonautogamous stocks collected in Somalia. The symbionts of type B are characterized by the fact that the three membranes which bind them are closely juxtaposed. Otherwise they are spherical and have a diameter of about 1 μm like the type A symbionts with which they might share the cytoplasm of *E. crassus* (Rosati *et al.*, 1976).

Symbiont type C is described as "roundish with an irregular cell boundary silhouette" (Rosati *et al.*, 1976). It has a diameter of about 2 μm. It was recognized in an autogamous stock of *E. crassus* collected at the Somalian coast.

Symbiont type D is the largest of all these *E. crassus* symbionts. It has a curved rod-shaped body and is 2 μm wide and 6 μm long. It was found in another autogamous stock collected in Somalia (Rosati *et al.*, 1976).

No names have been given to some tiny bacteria of the coccobacilli type which have been observed in addition to omikron-like symbionts in one stock of *E. eurystomus*, one stock of *E. patella*, and four stocks of *E. octocarinatus* (Heckmann *et al.*, 1982). They are considered to be dispensable symbionts.

They illustrate that mixed populations of symbionts do occur occasionally. Two types of bacteria side by side have also been noticed by Ruffolo (1980) in a stock of *E. eurystomus* and a stock of *E. daidaleos*.

VI. Summary and Concluding Remarks

The endosymbionts of *Euplotes* comprise a heterogeneous assortment of microorganisms. Most of them are considered to be bacteria and were found to be gram-negative; one is an alga, and one might be a yeast. They may lie freely in the cytoplasm or macronucleus, but more commonly, they are enclosed in membrane-bound vacuoles of various types. Symbionts occur in fresh water species as well as in marine ones, and one obtains the impression that only a limited number of them has been observed so far, mainly because only a small number of stocks has been investigated. Probably many more symbionts would be found if a systematic search for them were undertaken.

The onikron particle and its relatives, designated here collectively as omikron-like particles, are remarkable because they are necessary for the life of their hosts. A whole group of fresh water *Euplotes* species, all characterized by the same 9 type 1 cirrus pattern and therefore believed to be closely related to each other, depend upon these symbionts. If they are removed, which can be achieved with the help of penicillin, or by other means, then these ciliates stop multiplying and eventually die. If the symbionts are reintroduced into aposymbiotic cells, achieved most elegantly by injecting them into the cytoplasm, then the symbionts quickly reproduce to their usual number per cell, at the same time restoring the ability of their hosts to multiply.

Which kinds of substances are provided by omikron and its relatives is unknown. It could be something as simple as a vitamin, essential for growth of *Euplotes,* and the trigger of cell division could be a consequence of this; it could also be that something more directly concerned with fission is provided by the symbionts. Whatever it is, the molecules needed by these *Euplotes* species in order to divide appear to be rather stable, because it was observed that the substances exerted their function for another 2 days after a massive irradiation which killed all the symbionts. Up to two fissions were usually carried out after such a treatment.

It is in keeping with the findings for other bacterial endosymbionts that omikron was found not to grow outside of *Euplotes,* and that is true most probably also for the omikron-like symbionts. It is assumed that this is due to evolutionary modifications brought about by the intracellular habitat. The small genome size of omikron, which is with 0.5×10^9 daltons about one-quarter the size of that of a free-living bacterium, and approaches the minimum for a cell, is explained in the same way.

Omikron and its relatives form a group of symbionts which have been under the influence of cytoplasmic habitats in the species with a 9 type 1 pattern probably for a long time. This may offer a unique opportunity to study adaptive processes brought about by such conditions. To learn more about them is of the utmost importance in the light of the present debate on the descent of mitochondria and chloroplasts and on the evolution of the eukaryotic cell.

The epsilon particles found to be responsible for the killer phenotype of *E. minuta* very much resemble the kappas, mus, lambdas, and nus of the *P. aurelia* complex. Epsilon was the first example of the fact that symbionts with toxic properties do not occur only in ciliates of fresh water habitats but also in forms which live in marine environments. It is not yet clear what advantage killer strains have over sensitives. However, it does not appear to be just a laboratory effect, which has no biological meaning in natural habitats. Landis (1981) argues that his collections of *P. tetraurelia* from ponds in Indiana, carried on for several years, clearly speak in favor of a selective advantage of killers over sensitives. The findings that killers exist also in marine forms like *E. minuta* and *E. crassus* support this view. A search for killers among other groups of ciliates probably would furnish a large group of additional examples.

Mate-killing is another interesting phenomenon, of unknown biological significance. But again the fact that it occurs not only in species of the *P. aurelia* complex but also in several *Euplotes* species, and the finding that very different types of symbionts are able to bring it about—different bacteria as well as perhaps a yeast—argues in favor of the view that there must exist selective pressure in natural habitats which favors this trait. It is of interest that in all cases of mate-killing known so far, damage to the nuclear apparatus of the sensitive conjugant appears to be the cause which eventually leads to the death of the sensitive.

The finding that some of the symbionts of *Euplotes* confer upon their hosts killer and mate-killer properties, phenomena discovered much earlier by Sonneborn and his associates for species of the *P. aurelia* complex, nicely illustrates the point that similar selective pressures act upon very different types of ciliates. We do not yet understand the forces which bring about these phenotypes, nor do we really understand the laws that govern the establishment of symbiotic complexes, the increasing mutual dependence of their components, and, in some cases, their complete integration to form new units of a higher order. We need to learn more about these laws if we want to understand how the eukaryotic cell has evolved.

Acknowledgments

I am indebted to the National Science Foundation and the Deutsche Forschungsgemeinschaft for support of those portions of my work which are reported in this article. The constructive comments of Dr. J. Frankel and Dr. J. R. Preer, Jr. are gratefully acknowledged.

REFERENCES

Bac, A. L., Black, F. T., Christiansen, C., and Freundt, E. A. (1969). *Nature (London)* **224,** 1209–1210.
Beale, G. H. (1957). *Proc. R. Phys. Soc. Edinburgh* **26,** 11–14.
Beale, G. H., and Jurand, A. (1966). *J. Cell Sci.* **1,** 31–34.
Beale, G. H., Jurand, A., and Preer, J. R. (1969). *J. Cell Sci.* **5,** 65–91.
Borror, A. C. (1968). *Stain Technol.* **43,** 293–294.
Brandt, K. (1882). *Arch. Anat. Phys., Abt. Phys.* pp. 125–151.
Butzel, H. M., Jr., and van Wagtendonk, W. J. (1963). *J. Protozool.* **10,** 250–252.
Chatton, E., and Séguéla, J. (1940). *Bull. Biol.* **74,** 349–442.
Chen, T. T. (1945). *Proc. Natl. Acad. Sci. U.S.A.* **31,** 404–410.
Chen, T. T. (1955). *Proc. Soc. Exp. Biol. Med.* **88,** 541–543.
Curds, C. R. (1975). *Bull. Br. Mus. Nat. Hist. (Zool.)* **28,** 1–61.
Demar-Gervais, C., and Genermont, J. (1976). *J. Protozool.* **23** (Suppl.), 8 A.
Diller, W. G., and Kounaris, D. (1966). *Biol. Bull. Mar. Biol. Lab. Woods Hole* **131,** 437–445.
Dini, F., and Luporini, P. (1976). *Monit. Zool. Ital.* (N.S.) **10,** 15–24.
Dorner, R. W. (1957). *Science* **126,** 1243–1244.
Ehrenberg, C. G. (1840). *Monatsber. Preuss. Akad. Wiss. Berlin* pp. 197–219.
Fauré-Fremiet, E. (1952). *C. R. Acad. Sci.* **235,** 402–403.
Fauré-Fremiet, E., and André, J. (1968). *Arch. Anat. Microsc.* **57,** 53–78.
Figueroa-de Soto, V. E., and Soldo, A. T. (1977). *Int. Congr. Protozool., 5th New York* 339.
Foissner, W. (1978). *Naturkd. Jahrb. Stadt Linz* **23** (year 1977), 17–24.
Frankel, J. (1973). *J. Exp. Zool.* **183,** 71–94.
Fujishima, M., and Heckmann, K. (1982). *J. Exp. Zool.* (in press).
Gates, M. A. (1978). *Protistologica* **14,** 125–132.
Heckmann, K. (1963). *Arch. Protistenkd.* **106,** 393–421.
Heckmann, K. (1975). *J. Protozool.* **22,** 97–104.
Heckmann, K. (1980). *In* "Endocytobiology, Endosymbiosis, and Cell Biology" (W. Schwemmler and H. E. A. Schenk, eds.), Vol. I, pp. 393–400. De Gruyter, Berlin.
Heckmann, K., and Frankel, J. (1968). *J. Exp. Zool.* **168,** 11–38.
Heckmann, K., and ten Hagen, R. (1981). *Int. Congr. Protozool., 6th, Warsaw* **134.**
Heckmann, K., Preer, J. R., Jr., and Straetling, W. H. (1967). *J. Protozool.* **14,** 360–363.
Heckmann, K., ten Hagen, R., and Görtz, H.-D. (1982). *J. Protozool.*
Kahl, A. (1932). *In* "Die Tierwelt Deutschlands" (F. Dahl, ed.), Teil 25, pp. 399–650. Fischer, Jena.
Karakashian, M. W. (1975). *Symp. Soc. Exp. Biol.* **29,** 145–173.
Karakashian, S. J., and Rudzinska, M. A. (1981). *Exp. Cell Res.* **131,** 387–393.
Katashima, R. (1961). *Jpn. J. Zool.* **13,** 39–61.
Katashima, R. (1965). *Annot. Zool. Jpn.* **38,** 207–215.
Kloetzel, J. A. (1981). *Int. Congr. Protozool., 6th, Warsaw* 187.
Landis, W. G. (1981). *Can. J. Zool.* **59,** 1734–1743.
Levine, M. (1953). *Genetics* **38,** 561–578.
Luporini, P. (1974). *Monit. Zool. Ital. Suppl.* (N.S.) **V,** 129–132.
Luporini, P., and Dini, F. (1975). *Boll. Zool.* **42,** 475–476.
Nobili, R., Rosati, G., and Verni, F. (1976). *Boll. Zool.* **43,** 251–258.
Pierson, B. F. (1943). *J. Morphol.* **72,** 125–165.
Preer, J. R., Jr., and Preer, L. B. (1982). *Int. J. Syst. Bacteriol.* **32,** 140–141.
Preer, J. R., Jr., Preer, L. B., and Jurand, A. (1974). *Bacteriol. Rev.* **38,** 113–163.
Quackenbush, R. L. (1978). *J. Gen. Microbiol.* **108,** 181–187.
Reisser, W. (1980). *In* "Endocytobiology, Endosymbiosis and Cell Biology" (W. Schwemmler and H. E. A. Schenk, eds.), Vol. I, pp. 97–104. De Gruyter, Berlin.

Rosati, G., and Verni, F. (1976). *Boll. Zool.* **42,** 231–232.
Rosati, G., and Verni, F. (1977). *Int. Congr. Protozool., 5th, New York* **443.**
Rosati, G., Verni, F., and Luporini, P. (1976). *Monit. Zool. Ital.* (N.S.) **10,** 449–460.
Ruffolo, J. J., Jr., (1980). *J. Protozool.* **27** (Suppl.), 27 A - 28 A.
Schmidt, H. J. (1982). *Exp. Cell Res.* **140,** 417–425.
Schmidt, H. J., and Heckmann, K. (1980). *In* "Endocytobiology, Endosymbiosis and Cell Biology" (W. Schwemmler and H. E. A. Schenk, eds.), Vol. I, pp. 401–407. De Gruyter, Berlin.
Setlow, R., and Doyle, B. (1956). *Biochim. Biophys. Acta* **22,** 15–20.
Siegel, R. W. (1950). *Microbiol. Genet. Bull.* **3,** 12.
Siegel, R. W. (1953). *Genetics* **38,** 550–560.
Siegel, R. W. (1954). *Physiol. Zool.* **27,** 89–100.
Siegel, R. W., and Heckmann, K. (1966). *J. Protozool.* **13,** 34–38.
Soldo, A. T., and Godoy, G. A. (1973). *J. Mol. Biol.* **73,** 93–108.
Sonneborn, T. M. (1938). *Science* **88,** 503.
Sonneborn, T. M. (1950). *J. Exp. Zool.* **113,** 87–143.
Sonneborn, T. M. (1959). *Adv. Virus Res.* **6,** 229–356.
Sonneborn, T. M. (1975). *Trans. Am. Microsc. Soc.* **94,** 155–178.
Sonneborn, T. M., Schneller, M. V., and Craig, M. F. (1956). *J. Protozool.* **3** (Suppl.), 8.
Stein, F. (1859). "Der Organismus der Infusionsthiere. I. Allgemeiner Theil und Naturgeschichte der Hypotrichen Infusionsthiere." Leipzig.
Takayanagi, T., and Hayashi, S. (1964). *J. Protozool.* **11,** 128–132.
Takayanagi, T., Inoki, S., and Yoshikawa, K. (1966). *J. Genet.* **41,** 241–246.
Tuffrau, M. (1960). *Hydrobiologia* **15,** 1–77.
van Wagtendonk, W. J., Zill, L. P., and Simonsen, D. H. (1950). *Proc. Indian Acad. Sci.* **60,** 64–66.
Verni, F., Rosati, G., and Nobili, R. (1977). *Trans. Am. Microsc. Soc.* **96,** 363–369.

Endonuclear Symbionts in Ciliates

HANS-DIETER GÖRTZ

Zoological Institute, University of Münster, Münster, Federal Republic of Germany

I. Introduction

Intracellular symbionts living in protozoa have been investigated in a variety of associations. The observations have been thoroughly reviewed several times (Bütschli, 1889; Kirby, 1941; Ball, 1969). Since then many findings have been made and it seems advisable now to subdivide the topic, permitting discussion of the data in more detail. Accordingly, in recent reviews the symbioses of only a few host species or genera are treated comprehensively, e.g., *Paramecium* (Preer *et al.*, 1974; Soldo, 1974), *Euplotes* (Heckmann, this volume), and *Amoeba* (Jeon, 1980).

The aim of this article is to show the diversity of endonuclear symbioses in ciliates and to review what is known about the relationships of symbionts and hosts. Ciliates have a nuclear dimorphism with a diploid, germinal micronucleus and a polyploid, somatic macronucleus. The latter is functionally active throughout the vegetative life of the cell. During sexual reproduction the old macronucleus is replaced by a new one which is generated by the micronucleus (see Raikov, 1972; Grell, 1973; Vivier, 1974; Nanney, 1980; Ossipov, 1981).

Endonuclear symbionts in ciliates are situated either in the micro- or the

ISBN 0-12-364375-9

macronucleus. The influence of nucleus-specific symbionts on development, morphological differences, and the specific functions of the different types of nuclei at various stages of the cell cycle has been studied by several authors.

Older reports being reviewed by Bütschli (1889), Kirby (1941), and Ball (1969) will not be discussed in detail, but some results of the papers can be better appreciated now that new data are available. Moreover, some of the symbionts described before have been reisolated and studied.

Throughout this article the term symbiosis denotes an intimate association of different organisms including parasitism, commensalism, and mutualism according to the definition of De Bary (1879). Accordingly, the organisms living within nuclei are termed endonuclear symbionts and those living within the cytoplasm, endocytoplasmic symbionts. These terms do not define the kind of relationships of host cells and symbionts. Although this may be unsatisfactory in some cases, the nature of most of the associations is not really known.

II. Eukaryotic Organisms in Ciliate Nuclei

Eukaryotic symbionts belong to various taxonomic groups. Rees (1878) observed infections of the macronuclei in *Oxytricha fallax* by parasitic algae, which were later thought to be fungi or flagellates by Bütschli (1889). Flagellates occasionally invading the macronucleus of *Stentor* were also observed by Balbiani (1881). They caused a strong hypertrophy of the nucleus. In the macronucleus of *Paramecium calkinsi* Diller (1948) frequently found fungoid symbionts. Hovasse (1950) observed a microsporidian, which he named *Gurleya nova,* in the cytoplasm and in the macronucleus of the apostome ciliate *Spirobütschliella chattoni*.

Two leptomonad flagellates infecting the macronuclei of ciliates have been investigated in detail. Gillies and Hanson (1962, 1963) described *Leptomonas karyophilus* from the macronucleus of *Paramecium trichium*. The flagellate is spindle-shaped with a body length of 5 μm, an average width of 1.5 μm, and a flagellum 12 μm long. The aflagellar end sometimes shows an extremely fine thread-like projection, being more slender than a flagellum with no mobility. The ciliate is infected by the mature leptomonad form and when the host dies mature flagellates are set free. Gillies and Hanson (1963) assumed that the life cycle of *L. karyophilus* is completed within one host, because the flagellate shows a leishmanial form after the initial infection of a new host. The flagellate was found to multiply in the leptomonad form. Freedman and Hanson (1970) have reisolated *L. karyophilus* from nature and then observed the reproduction of the leishmanial form. The leishmanial form matures into the leptomonad form before being extruded into the medium.

Division of the flagellate is observed only in the macronucleus and the micronucleus remains uninfected. Besides this nucleus specificity the flagellate also shows some host specificity: It cannot infect *Paramecium bursaria* but is able to infect *P. aurelia* and *P. caudatum*. In *P. aurelia*, the symbionts even can divide. A cultivation of *L. karyophilus* was also possible (Freedman and Hanson, 1970) and an additional elongated, slender form was discovered *in vitro* (21 μm long, 1.5 μm wide, the flagellum being 30 μm long).

Another *Leptomonas* species was observed by Wille *et al.* (1981) in the macronucleus of a *Euplotes* species (Fig. 1). It is about the same size as *L. karyophilus*. Its fine structure resembles that of other trypanosomatids concerning the flagellar apparatus, flagellar sheath, subpellicular microtubules, and kinetoplast. The flagellate divides within the macronucleus exclusively. Infected macronuclei may contain up to several hundred totally quiescent flagellates.

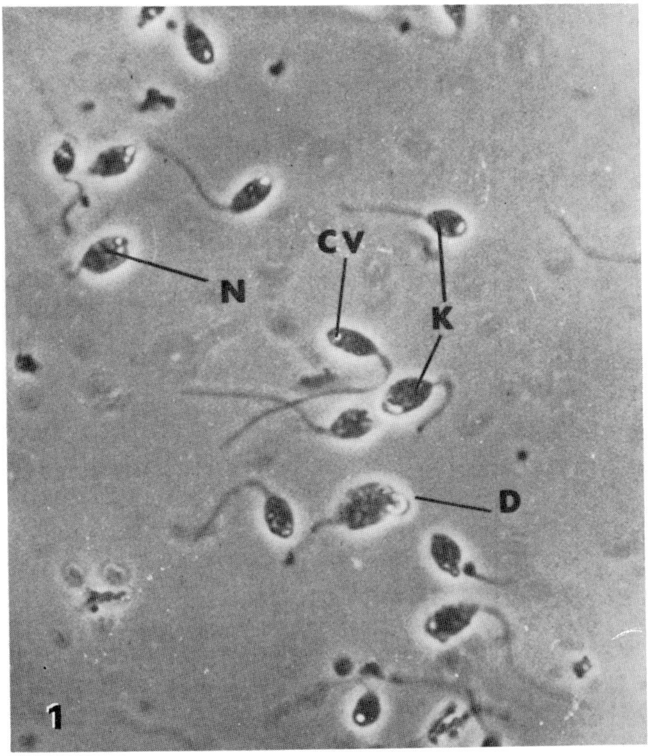

FIG. 1. Flagellates isolated from a symbiont-bearing macronucleus of *Euplotes*, fixed and stained for light microscopy, showing kinetoplast (K), nucleus (N), and posterior contractile vacuole (CV). Note also the biflagellated dividing cell (D). ×1300. (From Wille *et al.*, 1981, with permission of the *Journal of Protozoology*.)

When released from a ruptured host cell, the flagellates develop a vigorous but ineffective motility.

Leptomonads released from *Euplotes* could infect symbiont-free cells, but infestation of symbiont-free cells by coculturing with symbiont-bearing cells was unsuccessful. Besides *Euplotes, Didinium nasutum* was also infected, but not *Paramecium*. This and the fact that no leishmanial form has been observed in this flagellate are taken as arguments against its identity with *L. karyophilus* (Wille *et al.*, 1981).

The flagellates investigated by Gillies and Hanson (1963), Freedman and Hanson (1970), and Wille *et al.* (1981) are the only eukaryotes that have been observed in the nuclei of ciliates recently. In spite of their relative host specificity these rare observations might indicate that ciliates are not the normal hosts. In nature the flagellates may well be maintained in other hosts and the infection of ciliates may be rather accidental. Freedman and Hanson (1970) suggested that the paramecia might serve as a vector or reservoir of *L. karyophilus* for infections of metazoan hosts. In a reinfection experiment not all of the lines of symbiont-free *Euplotes* exposed to liberated leptomonads became infected (Wille *et al.*, 1981). The estimated high infection rate of the wild *Euplotes* population (above 90%) may not be due to a high infectivity of the flagellates but at least to some degree may be the result of the multiplication of symbiont-bearing *Euplotes*. In fact, both *P. trichium* and *Euplotes* are able to divide if they are not heavily infected.

III. Bacteria in Ciliate Nuclei

Proof that bacteria-like structures in ciliate nuclei are bacteria appears in many cases. As summarized for the various symbionts of the *P. aurelia* complex (Preer, 1975; Preer *et al.*, 1974) there is now convincing evidence for their bacterial nature. In this section I will include structures which appear to be bacteria from their fine structure, together with those proven to be bacteria. Most of the bacteria are gram-negative or at least appear gram-negative by the fine structure of their cell wall.

The associations of endonuclear bacteria and ciliate hosts differ in their degree of adaptation. Kirby (1941) has distinguished between associations of a constant and obligatory character and those of an occasional character. None of the endonuclear symbionts proved to be essential for the ciliates, and the kind of relationships is unknown in most cases. Therefore, a classification must be done in a different way. Some of the bacteria have been found only once, and others repeatedly even at various locations. Some have been observed in the same ciliate over several years. I will therefore classify the bacterial symbionts as accidentally observed and repeatedly observed. This may be arbitrary and in some cases will have to be revised in the future, but seems helpful for the moment. Again, this does not imply a classification into parasitic and mutual relationships.

A. Accidentally Observed Bacterial Symbionts

An endonuclear bacterium being cylindrical or biscuit-shaped (1.5 to 3 μm long) is reported by Engelmann (1876) from *Stylonychia mytilus*. The bacteria are harbored in the macronuclei or nuclear fragments.

A small rod-shaped bacterium was observed by Bozler (1924) in the macronucleus of *P. caudatum*. Infected macronuclei did not swell and the bacteria were seen only if the macronucleus was ruptured. The micronuclei appeared dislocated to the anterior part of the cell.

Kirby (1942) described a bacterium from the macronucleus of *Vorticella*, probably *V. similis* Stokes. The bacteria were straight to slightly curved rods (0.5 to 1.5 μm long), being spindle-shaped with pointed ends or rounded at one or both ends. Symbiont-bearing macronuclei swell to varying extents.

Bretschneider and van Vorstenbosch (1964) observed bacteria in the macronucleus of the rumen ciliate *Isotricha intestinalis*. The bacteria were only 0.4 to 0.8 μm in diameter, and were considered to be rickettsiales or anaplasmales. It must be emphasized that the bacteria are situated within the chromatin bodies either singly or in small groups. An electron-transparent area between the bacteria and the surrounding host chromatin has been taken as an indication of digestion of chromatin. Endonuclear symbionts have also been observed in the macronuclei of other rumen ciliates (Grain, 1966). Grain suggested digestion of chromatin by the symbionts, too. Albaret and Grain (1973) observed bacteria in the macronucleus of *Plagiotoma lumbrici*. The bacteria were juxtaposed to the chromatin bodies which appeared to be grooved out at the contact sites.

Hausmann (1972) reported the presence of bacteria in the macronucleus of *Loxophyllum meleagris*. These bacteria are 1 to 1.5 μm long and 0.3 to 0.4 μm wide. They are arranged in small groups of up to 10, distributed in the karyoplasm of the macronuclear nodes. The nuclei of the host cells appear normal in shape and size.

Rosati and Verni (1975) found a rod-shaped bacterium (about 0.5 μm wide and 1.5 μm long) in the macronucleus of the marine *Euplotes crassus* in a strain collected on the Somali coast and in a strain from the Tyrrhenian Sea near Pisa. The strain from the Somali coast harbors an additional cytoplasmic symbiont responsible for a killer trait of this strain. Cytoplasmic and endonuclear symbionts differ in morphology and are regarded as independent species occasionally coexisting in the same strain. Similar bacteria were also observed in the macronucleus of *Pseudoprorodon arenicola* (Kattar, 1972), *Helicoprorodon multinucleatum* (Puytorac and Kattar, 1969), *Metaradiophrya gigas*, and *Meslinella trispicula* (Puytorac, 1959, 1963).

A kappa-like (now Caedibacter, Preer and Preer, 1982a) bacterium in the macronucleus of *P. caudatum* was observed by Estève (1978). One form of the bacterium is 1.5 to 3.0 μm long with a diameter of 0.3 to 0.4 μm. The diameter of a second form is up to 0.8 μm. This form contains a multilamellar structure (Fig. 2) similar to the R-body in kappa (see also Preer *et al.*, 1974). As in kappa the R-body is associated with phage-like particles. *P. caudatum* harboring this bacterium in the macronucleus shows a typical killer trait when mixed with symbiont-free paramecia. Preer and Preer (1982b) assume that the bacterium described by Estève belongs to the genus *Caedibacter*.

Unusual structures observed in infected nuclei as isolated multilamellar structures, membrane fragments close to the nuclear envelope, and areas of granu-

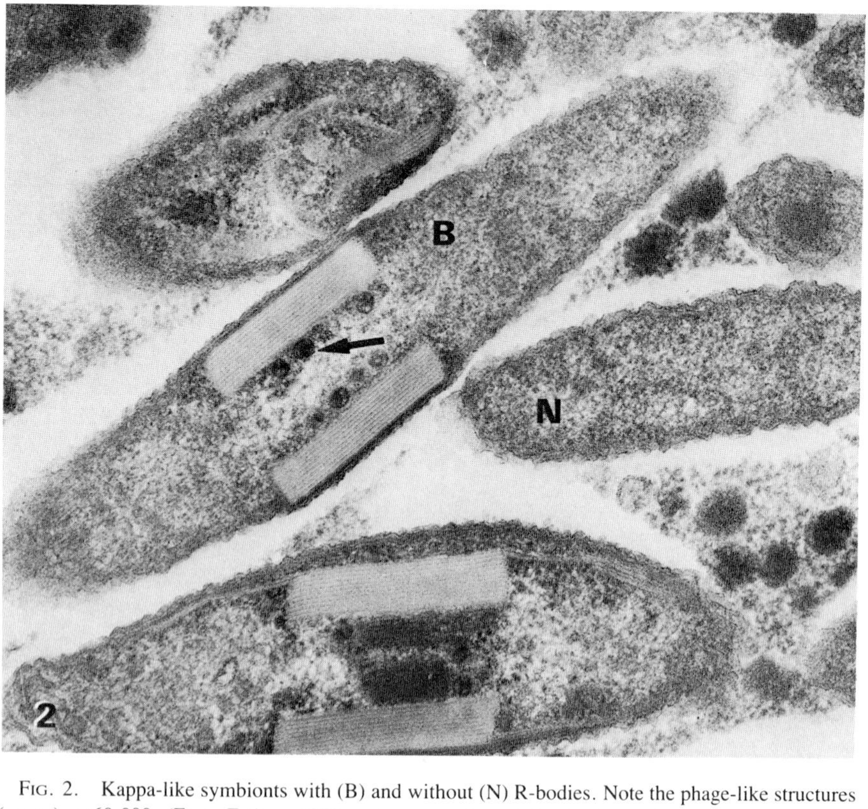

FIG. 2. Kappa-like symbionts with (B) and without (N) R-bodies. Note the phage-like structures (arrow). ×60,000. (From Estève, 1978, with permission of *Protistologica*.)

lar–fibrillar material (Estève 1978) are suggested to be remnants of bacteria lysed by the action of the phages. For the killer trait it must be assumed that the symbionts are released from the macronucleus into the culture medium. Whether this is achieved only after lysis of the host cells or also by some other mechanism is unknown.

A small bacterium called MS 2 has been observed in the macronucleus of *P. caudatum* by Görtz (1980). The bacterium (0.6 to 3.0 μm long) is found in clusters within the macronucleus, the size and shape of which are not affected. The symbionts are transmitted regularly during divisions of the host cells and are not outgrown during fast multiplication of the host. Like most other endonuclear bacteria, MS 2 is only weakly infectious when homogenate of symbiont-bearing paramecia is added to symbiont-free cultures.

A bacterium designated epsilon was described by Jenkins (1970) in *P. multimicronucleatum*. Epsilon is not found within the nuclei but is included in this article because it is situated in bulbous distensions (loculi) of the outer membrane of the micronuclear envelope. Loculi with bacteria are also formed by the outer membrane of the macronuclear envelope, but exclusively proximate to the micronucleus. The space of the loculi is always continuous with the perinuclear space, sometimes with both, the macronuclear and the micronuclear envelopes. Connections to the rough endoplasmic reticulum are observed, the membranes of the loculi being free of ribosomes. The bacteria are rod-shaped, sometimes coccoid (about 0.35 μm in diameter and up to 0.7 μm long). They contain closely packed granules, regions possibly representing nuclear areas and mesosome-like membranous whorls.

B. Repeatedly Observed Bacterial Symbionts

There are two reasons for the repeated observations of these symbioses. (1) The bacteria live permanently within the nuclei, being regularly transmitted with nuclear divisions. Consequently these bacteria need not be highly infectious. However, they must not be clearly parasitic. (2) The symbionts are highly infectious. Infectious symbionts need not be transmitted with the nuclear divisions. They may even be clearly parasitic. Both types of associations are found in ciliates.

Nilsson (1969) observed bacteria (about 1 μm long and 0.7 μm wide) in the macronucleus of *Neobrusaridium gigas*. The bacterial cytoplasm has a granular appearance with ribosome-like bodies and electron-dense networks. The symbionts have been observed ever since the isolation of *N. gigas* over a period of 14 years, which is taken as evidence that they can scarcely be pathogenic. This was supported by Kattar's (1975) observation of similar endonuclear bacteria in a strain of *N. gigas* from Brazil. The density of bacteria in the Brazil strain was lower than that in the Uganda strain used by Nilsson.

Inaba (1960) discovered a bacterium in the macronucleus of *Spirostomum ambiguum*, and the association has been further studied by Harrison *et al.* (1976a,b). The symbiont is of irregular pleomorphic shape (0.3 to 3.5 μm long and 0.3 to 2.5 μm in diameter). It has been isolated by lysing the host in KOH, triturating, and passing through a 0.45-μm pore size filter. The filtrates generated bacterial growth on a blood agar. Fluorochromated antibodies obtained against the cultured bacteria reacted with the symbionts in the macronuclei, which confirmed their identity in addition to fine structural comparison. Although many properties of the bacteria have been analyzed the taxonomic position is still unknown (Harrison *et al.*, 1976a).

Harrison *et al.* (1976a,b) have not observed harmful or mutual effects of the symbionts on the hosts. No symbiont-free *S. ambiguum* have been obtained by the authors. As in *N. gigas* the constant character of endonuclear symbiosis in *S. ambiguum* is further supported by an observation of endonuclear bacteria in a strain of this species from Germany (Moritz and Storch, 1971). An attempt should be made to liberate the ciliates of their symbionts by means of antibiotics and to examine whether they retain the ability to grow.

Highly infectious endonuclear bacteria were first observed by Müller (1856) who described thread-like elements in the macronucleus of *P. caudatum*. Similar bacteria were reported in *Paramecium* and other ciliates by various earlier authors (see for review Bütschli, 1889). It may be of historical interest that Müller (1856), Stein (1861), Claparède and Lachmann (1857), and Engelmann (1862) first regarded the bacteria as sperms located in a receptaculum. Balbiani (1860, 1861), Kölliker (1864), and Bütschli (1876) were able to clarify the nature of the symbionts as being bacteria.

The bacteria infecting nuclei of *P. caudatum* were reisolated by Hafkine (1890). He designated the bacteria *Holospora obtusa* being limited to the macronucleus, and *H. elegans* and *H. undulata*, both occurring exclusively in the micronucleus. Hafkine described short fusiform bacteria and long rod-shaped ones. He considered the long form to be infectious and called it a spore since he did not observe any division stages. He was aware that this was not a typical bacterial endospore. In order to emphasize this difference he called the bacterium *Holospora* (whole spore) (see also Kirby, 1957). Hafkine reported a budding of the long form, which must be regarded as a misinterpretation (Görtz and Dieckmann, 1980).

Hafkine regarded the bacteria as parasites as did Dehorne (1911) and Fiveiskaja (1929). Wichterman (1940, 1953) also emphasized the hypertrophy of *H. obtusa*-bearing macronuclei. He observed infected cells dying at the bottom of the culture vessel.

Petschenko (1911) reported an infection of *P. caudatum*, the bacterium lying in the cytoplasm. He named it *Drepanospira mülleri* (see also Kirby, 1957). Petschenko could not detect a micronucleus in infected paramecia and the bacteria from his descriptions were much like *H. undulata*. They may in fact be identical with this species, especially since they are always found in clusters (presumably within the micronucleus). This was previously suggested by Kirby (1941).

Recently all three of the *Holospora* species described by Hafkine (1890) have been reisolated (Ossipov and Ivakhnyuk, 1972; Ossipov *et al.*, 1975a,b; Ossipov, 1975, 1981; Görtz and Dieckmann, 1980; Görtz 1980). In addition, two new species of the genus have been described: *H. caryophila* (formerly alpha) in the macronucleus of *P. biaurelia* (Preer 1969; Preer and Preer, 1982a) and *H. acuminata* in the micronucleus of *P. bursaria* (Ossipov *et al.*, 1980a). I will first describe *H. elegans* in more detail showing the features common to all *Holospora* species.

H. elegans specifically infects the micronucleus of *P. caudatum*, and infected micronuclei are visible *in vivo* (Fig. 3). The symbionts are released into the medium, when a host cell is ruptured. Then morphologically different forms of the bacterium can be observed (Figs. 4, 5, and 6). The two main forms are a long rod-shaped one (8 to 12 μm long and 0.8 to 1 μm wide) and a short fusiform stage (2.5 to 3.5 μm long and 0.5 to 0.8 μm wide). The long form shows a part appearing dark when observed with the dark phase contrast microscope and a part appearing refractile. At the end of the refractile part a nonrefractile tip is visible which is clearer than the dark part. Orcein, Feulgen, or the fluorochrome 4,6-diamidino-phenylindole (DAPI) only stain the part appearing dark, indicating the presence of DNA.

When homogenate of *H. elegans*-bearing cells is added to a culture of symbiont-free cells, the long form of the bacterium is taken up into the food vacuoles. Later long forms are found in the cytoplasm, and after 1 hour the first bacteria appear within the micronucleus. Since the micronucleus is often smaller than the long form of *H. elegans*, micronuclei are stretched after infection or the symbionts appear to protrude from the nucleus. Later on, the micronucleus swells almost reaching the size of the macronucleus.

The long form of *H. elegans* is the only form capable of infections and therefore is called the infectious form. During the process of infection it undergoes a differentiation. The refractile part shortens while the dark part elongates, the overall length of the bacterium remaining constant. This development starts in the food vacuole and is finished in the nucleus, when the refractile part has disappeared completely. Sometimes the tip is still visible after invasion of the micronucleus. Some hours to 1 day later the infectious form constricts at four to six points and undergoes a multiple fission, by which the short form is developed.

The short form multiplies by binary fission and is called the reproductive form. Occasionally it stops dividing, grows, and develops into the infectious form. The small tip becomes visible and the refractile part gradually grows to

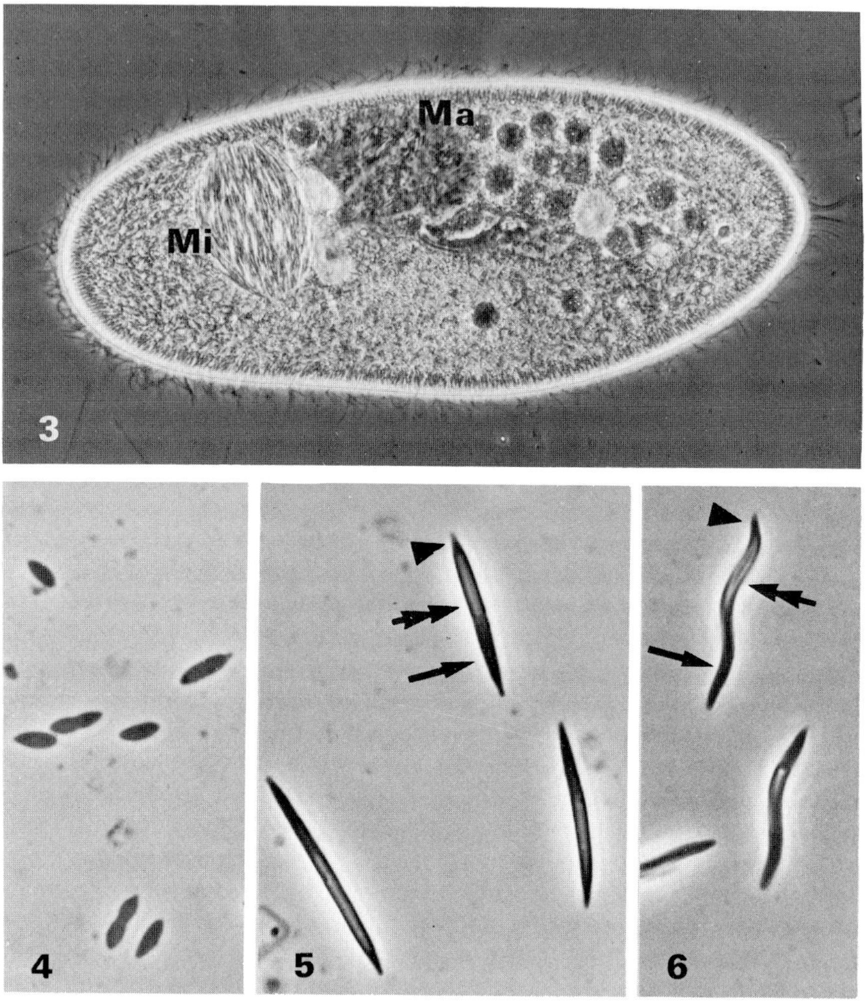

Fig. 3. *P. caudatum* harboring *H. elegans* in the micronucleus (Mi), unfixed, microflash. Ma, Macronucleus. ×400.

Fig. 4. The reproductive form of *H. elegans*. ×1800.

Fig. 5. The infectious form of *H. elegans*. Note the dark part (arrow), the refractile part (double arrow), and the pale tip (arrowhead). ×1800.

Fig. 6. The infectious form of *H. undulata*. The different parts can be seen as in *H. elegans* (see Fig. 5). ×1800.

about half the length of the bacterium. The regulation of this process and the physiology of the infectious form are not yet clear. Sometimes the infectious forms of *H. elegans* (Görtz and Dieckmann, 1980) and other *Holospora* species (Görtz, unpublished) have constrictions. The significance of these constrictions is unknown. They may be taken as signs of rare divisions of the infectious form. However, in contrast to the reproductive form, the infectious form is resistent to penicillin possibly indicating that it does not grow. Therefore, it appears unlikely that the infectious forms can divide. Rather, the infectious forms with constrictions will have developed from division stages of the reproductive form. This would mean that a reproductive form of *Holospora* can be triggered at any stage, even during division, to differentiate into the infectious form.

The infectious form cannot be regarded as a spore. A "hatching" as suggested by Gromov *et al.* (1974) and Ossipov *et al.* (1976) to occur within the food vacuole was not observed in *H. elegans* nor in *H. obtusa* or *H. undulata* by Görtz and Dieckmann (1980, and Görtz, unpublished). Moreover, a "hatching" seems unlikely because the overall length of the bacterium does not change during the differentiation process and the pale tip remains visible throughout the course of infection even after invasion of the nucleus. Structures suggesting a hatching may be induced by treating infectious forms with weak acids (Gromov *et al.*, 1974).

The fine structure of the infectious form corresponds to the light microscopic aspect (Figs. 7, 8, and 10). The dark part is a dense bacterial cytoplasm sometimes with mesosome-like structures (Preer, 1969; Gromov *et al.*, 1975, 1976; Görtz, 1980). The refractile part contains an electron-dense material of fine granular appearance. This material is located within the periplasmic space of the bacterium. The pale tip at the end of the refractile part appears fine granular and rather electron translucent. The fine granular material of the refractile part disappears a few hours after the symbiont has entered the nucleus (Fig. 9). It is not known whether the material is taken up into the cytoplasm of the bacterium and metabolized, or whether it is excreted, possibly sheltering the infectious form against being digested inside the food vacuole.

During binary fission of the host cell the infectious forms of *H. elegans* remain within the separation spindle (stem body) and are later released into the medium (see Section VI,B). The reproductive forms are distributed into the daughter micronuclei. The life cycle of *H. elegans* is summarized in Fig. 11.

Morphologically different forms representing stages of a life cycle exist in all *Holospora* species and the long form always is the infectious form (Rautian *et al.*, 1978, 1979; Preer, 1969; Görtz, 1980; Ossipov, 1981). Aside from morphological differences (Figs. 5 and 6) the differences in host specificity and nucleus specificity are characteristic for the *Holospora* species (Table I). These differences are therefore taken as main taxonomic features. The names of *H. elegans, H. undulata,* and *H. obtusa* have been revived recently and must be

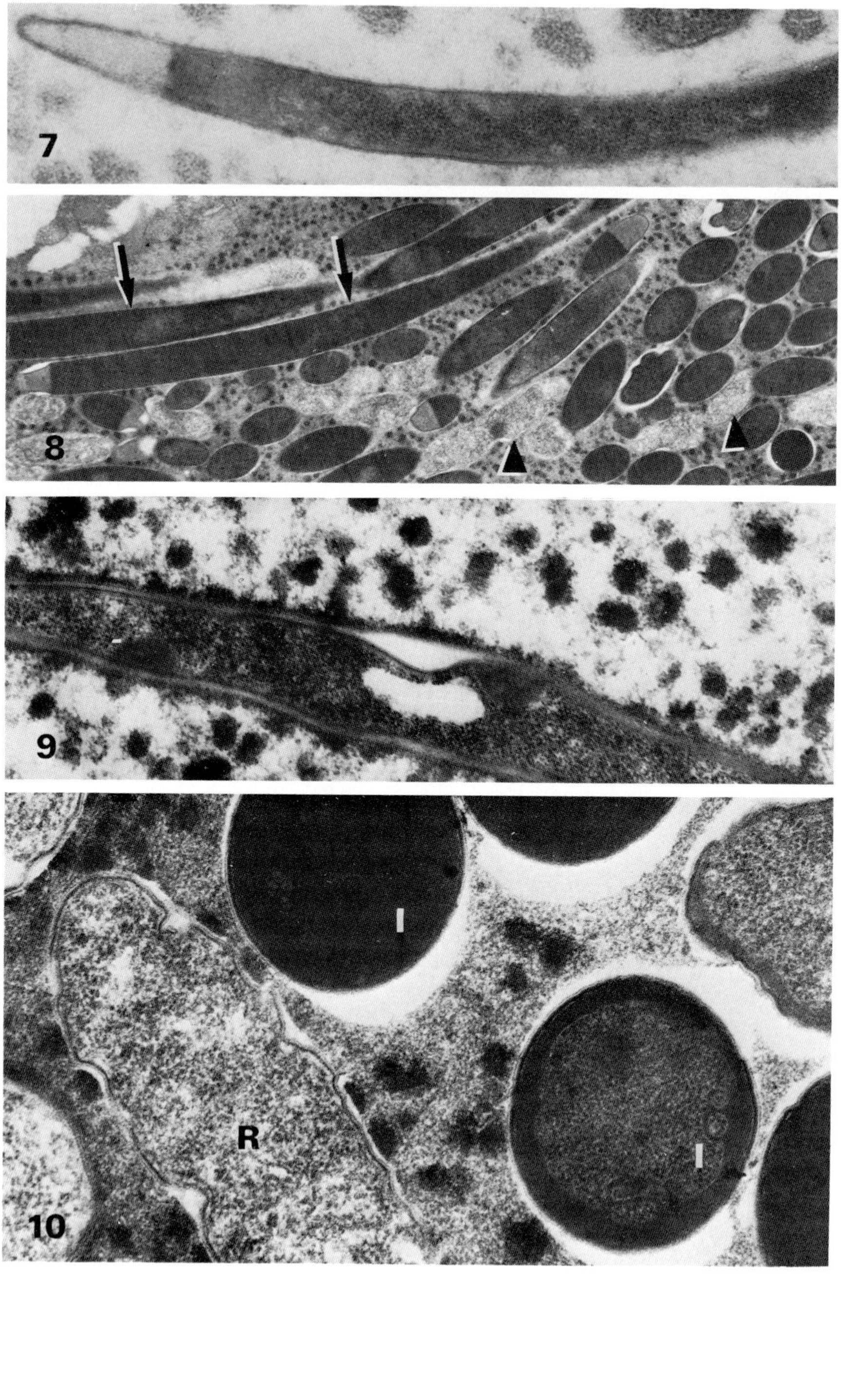
7
8
9
10
R
I
I

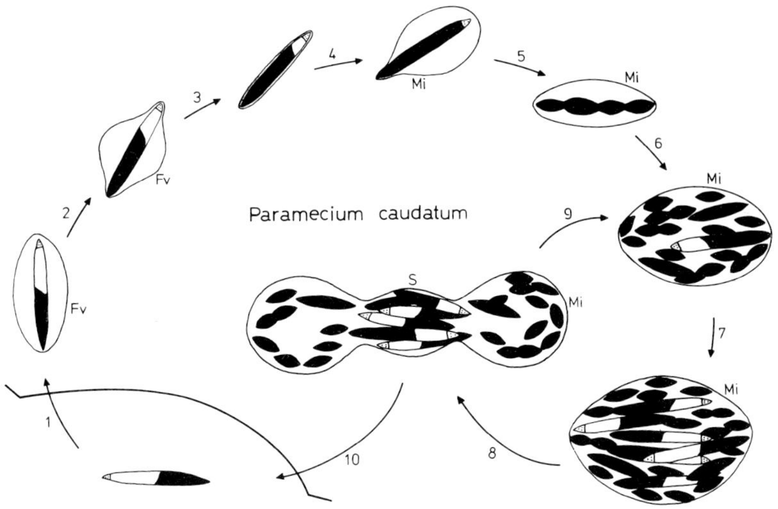

FIG. 11. Life cycle of *H. elegans*. (1) The infectious form is ingested into a food vacuole (Fv). (2) The food vacuole shrinks and evaginates. The bacterium starts a differentiation process finally leading to the development of the reproductive form. (3) The symbiont separates from the food vacuole but remains surrounded by host membranes. (4) The symbiont enters the micronucleus (Mi). (5) Inside the micronucleus the symbiont constricts at several points, divides, and so develops into the reproductive form. (6) The reproductive form multiplies by binary fission. (7) Some bacteria start to grow and develop into the infectious form again. (8) The micronucleus divides. The reproductive forms are distributed on the daughter nuclei, while the infectious forms remain within the separation spindle (S). Later, the infectious forms are released into the external medium (see text). (Modified from Görtz and Dieckmann, 1980.)

regarded as nomina revicta (Gromov and Ossipov, 1981; Preer and Preer, 1982a).

Ossipov (1973) tested the host specificity of *H. undulata* using 174 clones of *P. caudatum* from various places of the USSR and two clones from Kita (Mali). Of these 130 clones could be infected. Ten of the 46 clones which could not be infected were amicronucleate. Among the others were the two clones coming

---

FIG. 7. The infectious form of *H. caryophila* with clear area visible at the left end. The clear area presumably corresponds to the pale tip of the other *Holospora* species. ×40,000. (From Preer, 1969, with permission of the *Journal of Protozoology*.)

FIG. 8. *H. obtusa* in the macronucleus of *P. caudatum*. Infectious forms (arrows) and reproductive forms (arrowheads) are visible. ×9500.

FIG. 9. *H. obtusa* some hours after having entered the macronucleus. The inner appearance of the infectious form has changed. Note the sheath of chromatin-like material surrounding the bacterium. ×30,000.

FIG. 10. Infectious forms in cross sections (I). The reproductive form (R) is associated with chromatin bodies (arrows). ×51,000.

TABLE I
CHARACTERISTICS OF THE *Holospora* SPECIES

	H. elegans	*H. undulata*	*H. acuminata*	*H. caryophila*	*H. obtusa*
Host	*P. caudatum*	*P. caudatum*	*P. bursaria*	*P. biaurelia*	*P. caudatum*
Nucleus	Micro-	Micro-	Micro-	Macro-	Macro-
Length (infectious form) (μm)	10–20	7–15	Up to 6	5–6	7–20
Shape (infectious form)	Rod	Spiral	Rod	Spiral	Rod
Ends (infectious form)	Tapered	Tapered	Slightly tapered	Tapered	Rounded

from Africa. A number of other ciliates and *Amoeba proteus* could not be infected. Görtz and Dieckmann (1980) tested the host specificity of *H. elegans*, finding that all strains tested (from syngens 1, 3, 12, 13) could be infected. However, the symbionts appear to be maintained in the strains to a different degree (Görtz, unpublished). Preer (1969) reported that only strains of *P. biaurelia* (formerly *P. aurelia* syngen 2, for renomination of the *P. aurelia* complex see Sonneborn, 1975) could be infected by *H. caryophila*. The recent description of *H. acuminata* (Golikova, 1978b; Ossipov *et al.*, 1980a,b) from the micronucleus of *P. bursaria* shows that the *Holospora* species are host-specific and may have coevolved with the ciliates. It should be interesting to look for *Holospora* species in ciliates other than *Paramecium*, some of which may have been observed long ago by Stein (1861), Engelmann (1876), and Claparède and Lachmann (1857).

IV. Viruses in Ciliate Nuclei

Viruses cannot be regarded as symbionts. Still, they will be included because they may cause effects similar to bacterial or eukaryotic symbionts. The universal occurrence of submicroscopic organisms in eukaryotic cells benefiting their hosts had been postulated by Altenburg (1946). Intranuclear viruses, the major components of which are formed or assembled inside the nucleus, have been found in different wild-type, mutant, and neoplastic tissues of various plants and animals (see for reviews Heine, 1974; Gatteff, 1978).

Viruses have been rarely observed in ciliates. Phage-like structures present in kappa particles (Preer *et al.*, 1974) and in the kappa-like bacterium in the macronucleus of *P. caudatum* (Estève, 1978) may not be regarded as viruses in ciliates but rather as phages in bacterial symbionts of ciliates.

Other reports do not deal with the localization but with the propagation of viruses in ciliates. Brutsaert *et al.* (1946) and Groupé *et al.* (1955) could not induce a propagation of viruses in ciliates. Kesa and Teras (1981) also observed that myxovirus is inactivated by *Tetrahymena pyriformis* while picorna-virus persists, eventually bringing about a pathogenicity of *T. pyriformis*. The propagation of other viruses could also be induced in various ciliates (Kesa and Teras, 1977; Teras, 1981; Moevus-Kobb, 1965; Kovács and Bucz, 1967).

Indirect evidence of a parasitic microorganism "perhaps being a virus" in the macronucleus of *Paramecium* was presented by Wichterman (1953). He observed large refractile areas free of chromatin in the macronuclei and unstable micronuclear behavior causing variations in the number of micronuclei.

Vivier and André (1961, see also Raikov, 1969) observed structures within the macronucleus of *P. caudatum* which possibly were viruses during many generations. Unusual Feulgen-negative zones were observed in the nuclei, which contained large bundles (1 to 2 μm in diameter) of parallel solid fibers (approximately 25 nm thick). In cross-sections the fibers were arranged in centered hexagons. A pyroninophilia of the fibers is RNase-resistant. They contained nonbasic proteins.

Virus-like particles have also been observed in protozoa other than ciliates (e.g., Schuster, 1969). Thus, in spite of few reports viruses may be of common occurrence in the nuclei of various organisms.

V. Mode of Infection

Infectious organisms invade eukaryotic cells by two different mechanisms: (1) active penetration by local destruction of the membrane; and (2) uptake by the host without destruction of membranes, e.g., phagocytosis. In all cases of endonuclear or endocytoplasmic symbiosis in ciliates the symbionts are found to be taken up into the food vacuoles. An exception may be the infection of the apostome *S. chattoni* by *G. nova* (Hovasse, 1950). Once the symbionts have entered the host cell, the invasion of the nuclei seems to be achieved differently by various symbionts.

During the infection of *P. trichium, L. karyophilus* penetrates the food vacuole membrane and actively swims in the cytoplasm of the host cell (Gillies and Hanson, 1963). The parasite approaches the macronucleus touching it with the aflagellar end. This causes an indentation of the macronuclear membrane at the point of contact. The penetration into the nucleus requires 20 to 30 minutes. Whether the nuclear membranes are locally destroyed remains unknown. It is also unknown if the fine thread-like protrusion at the aflagellar end of the leptomonad has any significance for the penetration process.

The mode of infection by nucleus-specific bacteria is unknown in most cases. Information is available only for *Holospora* species. The infectious forms of these bacteria are taken up into the food vacuoles (Ossipov and Ivakhnyuk, 1972; Ossipov *et al.*, 1975b; Görtz and Dieckmann, 1980; Ossipov, 1981). There the infectious form starts the differentiation process, finally leading to the development of the reproductive form (see Section III,B).

H. elegans and *H. obtusa* cause an evagination of the food vacuole membrane and thereby become enclosed in a vesicle when leaving the food vacuole. During the transport toward the nucleus these bacteria remain surrounded by host membranes (Ossipov and Podlipaev, 1977; Görtz, 1980, 1981b). *H. undulata* leaves the food vacuole in an unknown way (Ossipov, 1981). In the cytoplasm *H. undulata* is surrounded by a layer (up to 28 nm thick) of electron-dense material. Later, before entering the micronucleus, it is also wrapped in host membranes (Podlipaev and Ossipov, 1979). The authors presented a model for the uptake of *H. undulata* into the micronucleus. Approaching the micronucleus a zone (200 to 400 nm wide) of fibrillar material appears as an additional cover around the bacterium. Connected to this, cylindrical strings composed of periodic fibrillar structures are visible. These strings (up to 300 nm long) appear like rope-ladders from the pictures (Podlipaev and Ossipov, 1979). Additional network-like structures and the fibrillar strings sometimes appear to have contact with the micronuclear envelope. The significance of the fibrillar structures and their nature is unknown. It is tempting to suggest that they draw the symbiont to the nucleus. So the presence of actin-like components in these structures should be no surprise.

The micronuclear envelope reacts to the approaching symbiont by forming invaginations and projections (Podlipaev and Ossipov, 1979). When the symbiont touches the nuclear envelope, it is surrounded by the membranous projections and thus enclosed within a compartment. Inside this compartment a second set of nuclear membrane projections surrounds the bacterium before it is passed into the nucleus. During this process many intranuclear microtubules appear in the area of penetration. Microtubules might function to transport symbiont-containing vesicles from the food vacuoles to the nucleus in all *Holospora* species. Such transport seems possible, since microtubules are known to be involved in the transport of phagosomes and other vesicles in ciliates (e.g., Bardele, 1972, 1974; Allen, 1975; Hausmann and Peck, 1979).

For the invasion of *H. obtusa* into the macronucleus of *P. caudatum* a similar model is proposed by Ossipov and Podlipaev (1977). When the transport vesicle with *H. obtusa* approaches the macronucleus, membrane projections of the nuclear envelope are established, which surround the symbiont and partly fuse with the membranes of the transport vesicle. So, as in the case of *H. undulata* invading the micronucleus, the karyoplasm never comes into contact with the cytoplasm. Neither microtubules nor fibrillar elements such as those observed within the micronucleus during infection by *H. undulata* have been seen in the macronucleus. It is not known what induces the fusion of transport vesicle membranes with nuclear membranes, both of which have different origins, just as it is not known why the membranes of pronuclei fuse forming the synkaryon during conjugation (Franke and Scheer, 1974).

During the infection process the pale tip of the infectious form might have a special function. The pale tip of *H. elegans* is found in close contact with the food vacuole membrane when this membrane evaginates (Görtz, 1981b). Later, thin fibrils are observed between the pale tip and the membranes of the transport vesicle, and the bacterium enters the micronucleus with the pale tip first. The nature of the material within the pale tip is unknown. From morphological observations the pale tip must be regarded as a distinct organelle. It is tempting to compare the pale tip with the acrosome of sperms or with organelles of sporozoa possibly serving a function for the infection of host cells, e.g. the roptries of *Plasmodium* (Aikawa, 1981). Accordingly the material of the pale tip might be responsible for the nucleus specificity of the *Holospora* species and for the host specificity, too.

The high infectivity of *Holospora* is an exception among endosymbiotic bacteria. Moreover, for endocytoplasmic symbionts the nuclear envelope appears to be a barrier, which is overcome only in rare events, e.g., the endocytoplasmic bacteria in *Urostyla grandis* have been observed once in the macronucleus (Altmann and Ruthmann, 1979). An exception is a bacterium in the cytoplasm of *P. sexaurelia* (Görtz, 1981a), which invades the macronucleus in up to 10%. There it multiplies without causing a hypertrophy of the nucleus. After autogamy the bacteria are not observed in the new macronuclei but are restricted to the cytoplasm.

In connection with the infectivity of *Holospora* it must be questioned whether the infectious form being 5 to 10 times longer than the food organisms (e.g., *Enterobacter aerogenes*) is ingested just by chance or whether it has a particular attractivity for the ciliates. The latter might be true, as it has been suggested that paramecia can discriminate food particles by screening their size (Vivier, 1974).

VI. Effects on Host Nuclei

Endonuclear symbionts cause some morphological and functional changes in the infected nuclei. These changes are small for some symbionts, but for others they end in a complete destruction of the infected nuclei. Besides the changes in interphase nuclei, the micronuclear mitosis, macronuclear amitosis, and the nuclear behavior during sexual processes in endonuclear symbiont-bearing cells are of special interest.

A. Symbiont-Bearing Interphase Nuclei

Structural changes caused by endonuclear symbionts are observed at different levels. Gross changes affect the shape and size of the nuclei and their location within the cell.

Shape and size of infected macronuclei are not changed in *I. intestinalis* (Bretschneider and van Vorstenbosch, 1964), *P. lumbrici* (Albaret and Grain, 1973), and *L. meleagris* (Hausmann, 1972). All these symbionts are small and do not completely fill the nuclei. Fine structural changes and the location of these bacteria inside the nuclei are different. The endonuclear bacteria in *I. intestinalis* are the only ones located within the chromatin bodies. The symbionts in *P. lumbrici,* and to some degree those in *L. meleagris* are found close to the chromatin but not surrounded by it. The bacterium MS 2 in *P. caudatum* (Görtz, 1980) is found in clusters of tens to hundreds having contact with the chromatin only at the margins of the clusters. The significance of the close association of some of the symbionts with the chromatin is unknown. Digestion of chromatin by the bacteria as suggested by Bretschneider and van Vorstenbosch (1964) has not been proven (see also Section VIII).

The *Leptomonas* species in *P. trichium* and *Euplotes* may completely destroy the chromatin (Gillies and Hanson, 1963; Wille *et al.*, 1981). Heavily infected nuclei swell and lose their normal shape. Also the kappa-like bacteria in the macronucleus of *P. caudatum* cause hypertrophy (Estève, 1978). Heavily infected macronuclei can occupy almost the whole cell.

We do not know whether the deformations of the nuclei are caused indirectly by the overpopulation with symbionts or also directly by destroying the nuclear skeleton of microtubules and microfilaments. Both might occur. The macronucleus of *P. caudatum* being heavily infected with the small bacteria observed by Bozler (1924) is not deformed, although Bozler suggests that the entire chromatin disappears. In *Vorticella* the macronucleus swells after infection with bacteria (Kirby, 1942), at an extreme of hypertrophy losing its elongated form, becoming a large ellipsoidal body. The chromatin granules disappear.

The symbiont-bearing macronuclei in *N. gigas* (Nilsson, 1969) and *S. ambiguum* (Harrison *et al.*, 1976a,b) are not swollen. However, in both species a comparison with uninfected nuclei was not possible, as symbiont-free individuals have not been found. For the same reason one cannot speak of fine structural changes but rather of fine structural peculiarities. In *N. gigas,* the macronuclear regions occupied by the symbiotic bacteria lack chromatin and stain heavily with pyronin. In these areas unusual chains of granules are observed. The granules with a diameter of about 30 nm have less dense centers.

The extent of hypertrophy of *H. obtusa*-bearing macronuclei in *P. caudatum* depends on the physiological conditions of the host (Ossipov *et al.*, 1976; Görtz and Dieckmann, 1977a,b). An extreme hypertrophy is observed in starved cells. In fast growing cells the macronuclei are of normal size. The effect of *H. obtusa* on the macronuclei is not the same in different strains of *P. caudatum.* In some strains the symbionts cause an extreme hypertrophy at least during starvation. In others the symbionts are well maintained at different conditions and the macronuclei keep their normal shape and size. In some host strains *H. obtusa* even appears to be destroyed within the macronuclei under starvation conditions (Görtz, unpublished).

After infecting the macronucleus, the infectious form of *H. obtusa* is surrounded by chromatin-like material (Fig. 9) (Ossipov and Podlipaev, 1977; Görtz and Dieckmann, 1977a,b). This material can be stained with the DNA-specific fluorochrome 4,6-diamidino-phenylindol (Görtz, 1982a). Within this fibrillar sheath the infectious form divides by multiple fission. Thereafter the fibrillar sheath disappears. The reproductive form also shows some affinity to chromatin as indicated by the close contact with chromatin bodies (Ossipov and Podlipaev, 1977). The shape of chromatin bodies in contact with symbionts appears flattened and of smaller than normal size (Fig. 10). When the infectious form is developed again, it is not associated with chromatin.

Micronuclei are apparently more affected by their symbionts than macronuclei. They always swell, sometimes reaching the size of macronuclei. The swelling follows the multiplication of symbionts. However, an infection with high numbers of *H. undulata* may cause the immediate destruction of a micronucleus (Skoblo *et al.*, 1978). Symbiont-bearing micronuclei usually become dislocated. They may be found before or behind the macronuclei. Within a micronucleus the infectious forms of *H. elegans* and *H. undulata* are arranged parallel to each other along the axis of the micronucleus. So, infected micronuclei have an appearance similar to a mitotic spindle (Görtz and Dieckmann, 1980).

The invasion of *H. undulata* is mostly observed at the achromatic cap of the micronucleus (Podlipaev and Ossipov, 1979). The chromatin fine structure of the micronuclei is considerably affected by *H. undulata* in *P. caudatum* and *H. acuminata* in *P. bursaria* (Ossipov *et al.*, 1980a). Parts of the chromatin aggregate. In other areas thin microfibrills (10 to 13 nm) are observed, which may represent decondensed chromatin. It is assumed by the authors that the fine structural changes indicate a functional activity of symbiont-bearing micronuclei. Other changes observed in infected micronuclei of *P. bursaria* are the appearance of microtubules, microfilaments, and kinosomes (Ossipov *et al.*, 1980a) which are typically observed during mitosis (Lewis *et al.*, 1976; Schwartz, 1976). Invaginations and projections of the micronuclear envelope are observed as well as multimembrane structures (Ossipov *et al.*, 1980a,b).

An argument for metabolic activity of infected micronuclei is that the density of micronuclear pores increases to that of the macronuclear pores, as calculated from transmission electron micrographs (Ossipov *et al.*, 1980a). The nature of the assumed micronuclear activity is unknown. This should be clarified, especially since a transcriptional activity in symbiont-free micronuclei does not exist for those genes tested in *Tetrahymena* (Mayo and Orias, 1981), although there are indications for morphogenetic activity of the micronucleus (Diller, 1965; Ossipov and Skoblo, 1973; Mikami, 1979; Fujishima and Watanabe, 1981; Ng and Mikami, 1981).

Like macronuclei, micronuclei appear to be affected by endonuclear symbionts to different extents according to the strains used. Whether these differences exist only for changes of size and shape or also for fine structural changes and changes of the nuclear metabolism is unknown.

B. Mitosis of Symbiont-Bearing Micronuclei

Holospora species are the only micronucleus-specific bacteria reported. They do not prevent mitosis (Ossipov, 1981; Görtz and Dieckmann, 1980). This indicates that the elements necessary for mitosis, e.g., microtubules and microfilaments, are neither digested nor is their assembly prevented. Moreover, the chromatin appears not to be destroyed and a normal distribution of host genome must be assumed, because *Paramecium* after being liberated from the micronucleus-specific symbionts can conjugate and has viable offspring (Ossipov *et al.*, 1974).

Besides the distribution of genetic material, the mitosis achieves the distribution of the symbionts. The distribution of the reproductive and the infectious forms of *Holospora* is different. The reproductive form is observed at the poles of dividing micronuclei at early anaphase, and it is found within the daughter micronuclei at the end of mitosis. The infectious form remains within the separation spindle of the dividing micronucleus (Ossipov and Ivakhnyuk, 1972; Ossipov *et al.*, 1976; Görtz, 1980, 1982a). At the end of the fission process in some cases the separation spindle is drawn into one of the daughter cells. There it is retained or disappears within 2 hours of division. In the latter case it is suggested that the symbionts are excreted via the cytoproct (Ossipov, 1981). In *P. caudatum* infected with *H. elegans* the separation spindle often bursts before the daughter cells separate, releasing the infectious forms into the cytoplasm. Again, 2 hours later all symbionts disappear, probably being excreted.

Thus, the mechanism by which the symbionts are distributed during mitosis can distinguish the reproductive and the infectious forms. The controlled maintainance of reproductive forms ensures that symbiont-bearing paramecia remain infected, even if the distribution of symbionts in the daughter nuclei may not be equal. The liberation of infectious forms during mitosis enables the symbionts to infect symbiont-free paramecia. The mechanism of mitosis is thus utilized by *Holospora* for its propagation.

It is still unknown how the different distribution of the two forms is achieved. Four mechanisms seem possible: (1) The reproductive forms actively move into the daughter nuclei while the infectious forms do not. However, motility of *Holospora* has not been observed. (2) The reproductive form is attached to and distributed with the chromatin, while the infectious form is not. This seems unlikely, since the amount of chromatin in the micronucleus is low, while several hundred symbionts are distributed. (3) The reproductive form is attached to the nuclear envelope. However, symbionts located in the midst of the micronucleus become distributed. (4) The symbionts are distributed by microtubules of the spindle apparatus. Electron micrographs of *Holospora*-bearing micronuclei show many microtubules even at interphase. In *P. bursaria* structures regarded as microtubule-organizing centers are observed (Ossipov *et al.*, 1980a). In symbiont-free nuclei these structures are seen especially at mitosis (Lewis *et al.*, 1976; Schwartz, 1976). During mitosis of *H. elegans*-bearing micronuclei microtubules are observed close to the symbionts (Görtz, unpublished). If they serve a function for the distribution of symbionts, which is not proven, it must be postulated that infectious forms either are not recognized by the spindle components, or are drawn with equal force to the opposite poles. If microtubules or microfilaments really attach to the symbionts, it should be interesting to look for the substances on the bacterial surface serving as the attachment sites.

Despite the normal mitosis of symbiont-bearing micronuclei fission anomalies occur. Individual cells with heavily infected micronuclei sometimes divide incompletely and form "monsters" (Görtz and Dieckmann, 1980). Also, the direction of mitosis may be wrong and both daughter nuclei remain within one of the daughter cells, the other becoming amicronucleate. In certain strains this is not a rare event and may be used to obtain amicronucleate lines. Amicronucleate cells can also result from new infections with high densities of infectious forms (10^6 to 10^7 infectious forms/ml) (Rautian *et al.*, 1978; Skoblo *et al.*, 1978).

C. Amitosis of Symbiont-Bearing Macronuclei

Symbiont-bearing macronuclei can divide and the symbionts are transmitted into the daughter macronuclei. Even macronuclei infected with organisms harmful to the host cells, e.g., *Leptomonas,* can divide, when infection is slight (Gillies and Hanson, 1963; Wille *et al.*, 1981). Wichterman (1953) and earlier authors reported that macronuclei which are heavily infected cannot divide, whereas Ossipov *et al.* (1975b, 1976) and Görtz (1980) observed that even macronuclei filled with symbionts undergo normal amitosis. This may depend on culture conditions and on the host strains used. In cases where heavily infected macronuclei can no longer divide, it is unknown whether this is due to any effect on the nuclei themselves or to harmful effects on the host metabolism.

The normal stages of macronuclear amitosis are swelling, elongation, karyokinsesis, and division (Raikov, 1969; Grell, 1973; Vivier, 1974), showing many differences in the various ciliate species. In the first stage of amitosis in *P. caudatum, H. obtusa* often shows a remarkable circular arrangement (Görtz, 1982a), while it appears disordered in interphase nuclei. The circular arrangement is reminiscent of the swirling of chromatin strands in the early amitosis of *Tokophrya* (Heckmann, 1965, 1967; Millecchia and Rudzinska, 1971), *Paracineta,* and *Ichthyophthirius* (Hauser, 1973). However, a movement of symbionts in the nucleus has not been observed in *Paramecium in vivo.* Yet, the peculiar arrangement of the symbionts at this stage might indicate a corresponding behavior of the chromatin.

During amitosis the reproductive forms of *H. obtusa* are distributed to daughter macronuclei as is the chromatin. The infectious forms are collected at the fission line in a body resembling the separation spindle of the micronucleus (Ossipov *et al.,* 1975a,b; Ossipov, 1981; Görtz, 1980, 1982a). Before this stage, during the elongation stage the infectious forms appear to spread over the macronucleus. Therefore it is assumed that the infectious forms are transported into this "separation spindle" (stem body) during the course of karyokinesis. It is unknown whether this separation spindle contains chromatin, but it cannot be excluded at present. Certainly, the separation spindles of symbiont-bearing macronuclei cannot be regarded as extrusion bodies observed in the macronuclei of *Tetrahymena* (Cleffmann, 1968, 1980).

As in the micronucleus the mechanism by which the two forms of *H. obtusa* become differently distributed is unknown. A role of microtubules and/or microfilaments seems most likely for the same reasons as in the micronucleus. The infectious forms are released after binary fission of the host cells, bringing about an optimal propagation of the symbionts.

D. Effects of Endonuclear Symbionts on Sexual Reproduction

Paramecium biaurelia bearing *H. caryophila* in the macronucleus are capable of autogamy (Preer, 1969). In slow growing cultures some infectious forms may get into the new macronuclei of the exautogamonts and thus be maintained in the cells. In starved cultures infectious forms have been observed in the macronuclear anlagen. However, in rapidly growing cultures after autogamy all of the macronuclear fragments containing *H. caryophila* disappear before the new macronucleus can be infected, resulting in the loss of symbionts.

Ossipov *et al.* (1974) tested the capability of conjugation in *P. caudatum* infected to 100% with either *H. obtusa* or *H. undulata,* using mating reactive clones infected with symbionts. All clones lost the ability to conjugate and intraclonal conjugation could not be induced by chemical methods described by Hiwatashi (1959) and Miyake (1968). After spontaneous loss of *Holospora* the cells regained the capacity to conjugate and showed the same mating type as before.

Görtz and Fujishima (1982) tested the effect of *H. elegans* on the conjugation of *P. caudatum.* A strong agglutination reaction was observed upon mixing infected and noninfected mating types. After agglutination, pairs were formed with one partner bearing *H. elegans.* This was observed in heavily infected cells with micronuclei having almost the size of macronuclei. After pair formation the micronuclei of *H. elegans*-bearing cells divided meiotically. All three of the pregamous divisions described in *P. caudatum* (Calkins, 1907; Wichterman, 1953; Vivier, 1974) were observed. Due to the presence of bacteria in the nuclei the crescent stage and all following stages of meiosis could be examined *in vivo* unstained. As during mitosis the *Holospora* symbionts are oriented parallel to the spindle axis. The symbionts are distributed among division products and most of the infectious forms are lost in the separation spindles after each of the pregamic divisions. Thus, the pronuclei contain few and often no symbionts.

Using the same strains the effects of the macronucleus-specific *H. obtusa* on conjugation have been tested. Again, agglutination and pair formation have been observed. However, the mating reactivity appears considerably lower than in symbiont-free cells (Görtz, unpublished).

These observations are contradictory to those of Ossipov *et al.* (1974). The explanation might be that different strains of *P. caudatum,* possibly belonging to different syngens, behave differently. In addition, the various *Holospora* strains may also behave differently. The influence of micronucleus-specific symbionts on the mating reactivity is interesting, since even amicronucleate strains of *P. caudatum* and other paramecia are known to be mating reactive and capable of conjugation (Diller, 1936; Schwartz, 1939; Miyake, 1957; Golikova, 1978a). So endonuclear symbionts appear to have some indirect effect on the host cell, reducing the mating reactivity.

Besides the general significance of conjugation and autogamy these processes have additional consequences for ciliates bearing endonuclear symbionts. In *H. elegans*-bearing paramecia the pronuclei are often free of symbionts, thus producing symbiont-free exconjugants. In *H. caryophila*-bearing *P. biaurelia* autogamous lines more frequently lose their symbionts than do nonautogamous lines (Preer, 1969). The same consequences are observed for the bacterium in *P. sexaurelia* (Görtz, 1981a), which occasionally invades the macronucleus (Section V). During and after autogamy these symbionts are found within the macronuclear fragments but not in the macronuclear anlagen. Thus, sexual processes may be a means for the ciliates to get rid of endonuclear symbionts. This may be significant especially for inbreeders and it might explain why nucleus-specific symbionts are not observed in the *P. aurelia* complex with the exception of the highly infectious *H. caryophila*. Endonuclear symbionts are often observed in *P. caudatum,* a typical outbreeder with a long life cycle (Takagi and Yoshida, 1980).

VII. Effects on Host Viability

Many bacterial symbionts do not have an obvious effect on host viability. The endonuclear symbionts in *N. gigas* (Nilsson, 1969) and *S. ambiguum* (Harrison *et al.*, 1976a) are permanently maintained in the macronuclei. The permanent character of these symbioses is taken as an indication that the symbionts are not parasitic. Otherwise symbiont-free lines would be present under some conditions. The limited destruction of host chromatin appears not to be serious. Other endonuclear bacteria clearly harm their hosts, e.g., the kappa-like symbionts in the macronucleus of *P. caudatum*. These bacteria may cause a hypertrophy of the nucleus, leading to fission anomalies or to death of the host (Estève, 1978). During starvation the number of symbiont-bearing cells decreases faster than the number of symbiont-free cells (Fig. 12). However, like the killer strains in *P. aurelia* (Beale *et al.*, 1969; Preer *et al.*, 1974), *P. caudatum* bearing killer particles might have a slight advantage over symbiont-free paramecia, which may be killed.

The leptomonads in *P. trichium* and *Euplotes* (Gillies and Hanson, 1963; Wille *et al.*, 1981) exhibit parasitic effects. The ciliates survive only under optimal conditions and then may even outgrow the flagellates. The observation of Freedman and Hanson (1970) that paramecia freed of the flagellates show a higher resistance to new infections is probably the first observation of a quasi-immunization of a ciliate against a parasite.

The cytoplasm of paramecia-bearing endonuclear symbionts often appears

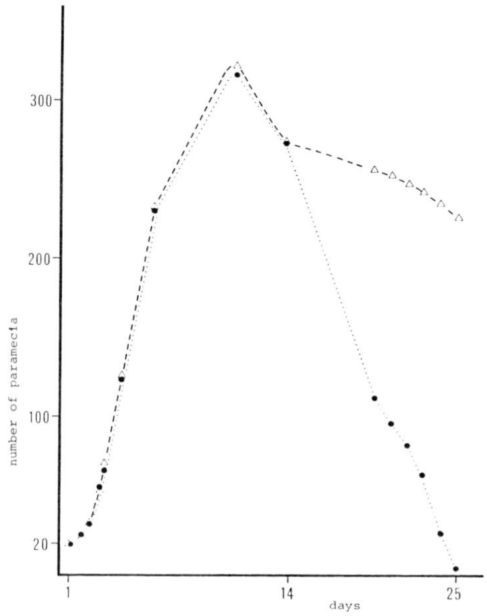

FIG. 12. Comparison of the development of symbiont-free (open triangles) and populations of *P. caudatum* infected with kappa-like bacteria (closed circles). No food has been added after the beginning of the experiment. (From Estève, 1978, with permission of *Protistologica*.)

vacuolized (Hafkine, 1890; Bozler, 1924; Fiveiskaja, 1929; Wichterman, 1953). The nature and significance of the vacuoles are unknown. However, the extent of damage depends on the culture conditions and some of the host strains do not appear affected. The decrease of fission rate in some *H. elegans*-bearing strains might indicate a function of the micronucleus during the vegetative life. A morphogenetic role of the micronucleus and an influence on the fission rate have been observed several times in studies using amicronucleate cells (Ossipov and Skoblo, 1973; Fokin *et al.*, 1979; Fujishima and Watanabe, 1981; Fokin and Ossipov, 1981; Ng and Mikami, 1981). *H. elegans*-bearing cells, however, keep their micronuclei, but they are swollen and dislocated. It is not known whether the decrease of fission rates in some *H. elegans*-bearing strains is due to the dislocation of the nucleus or to any other effects on the micronucleus. The macronucleus-specific *H. obtusa* does not influence the fission rate of host strains (Ossipov *et al.* 1975a,b; Görtz and Dieckmann, 1977a,b; Fujishima and Görtz 1982).

On the other hand, no clear benefit of *Holospora* on its hosts has been observed. Jenkins (1970) suggested that the close association of symbionts with the nucleus facilitates an uptake of nutrients excreted by symbionts into the nucleus. In ciliates an example of a nutrient produced by a symbiont is folic acid, which is essential for the purine metabolism of the ciliates. *Lyticum flagellatum* Preer and Preer 1982 (formerly lambda) supplies its host with folic acid (Soldo, 1963; Soldo and Godoy, 1973). So the host cells can grow even if the food does not contain folic acid. It might be mentioned that mitochondria are sometimes found within the nuclei of human heart cells (e.g., Jensen *et al.*, 1976). This may be due to an exceptional situation of this organ, requiring a maximum of energy. However, the type of mutualistic effects which are expressed by certain endocytoplasmic symbionts in *Euplotes* (Heckmann, 1975, 1977, and this volume) or other protozoa, e.g., *A proteus* (Jeon, 1972, Jeon and Jeon, 1976, 1980) can hardly be expected in endonuclear symbiosis. The organelle nucleus may be best supplied with any metabolites from the cytoplasm. Thus, if there are mutualistic effects, they might not be founded on nucleus specificity.

The presence of one *Holospora* species impedes a secondary infection by another *Holospora* (Ossipov *et al.*, 1975a; Görtz and Dieckmann, 1977b). It is unknown whether *Holospora* also provides the host cells with resistance against pathogenic germs. Certainly, this would be an advantage for *Holospora*-bearing paramecia (Görtz and Dieckmann, 1977b; Heckmann, 1977). Preer (1969) and Preer and Preer (1977) noted that *P. aurelia* normally harbors only one symbiont species. In *P. bursaria* zoochlorellae prevent infections of host paramecia with bacteria or yeast (Görtz and Dieckmann, 1982; Görtz, 1982b), which, if symbiotic in *Chlorella*-free paramecia, prevent infections with other bacteria or yeasts. So, the inhibition of a secondary infection may be a general feature of endocytoplasmic as well as endonuclear symbionts. However, not every symbiont may prevent other secondary infections. So the omikron-like bacteria in *Euplotes daidaleos* coexist with zoochlorellae (Heckmann *et al.*, 1982), and the *Euplotes* infected by leptomonads also harbor omikron-like symbionts (see Fig. 3 in Wille *et al.*, 1981).

Hitherto, the influence of endonuclear symbionts on the aging of host clones has not been investigated. Two phenomena of aging in ciliates must be distinguished: germinal aging, leading to defective micronuclei resulting in the genetic death of a clone, and somatic aging, leading to somatic death (for reviews see Nanney, 1980; Smith-Sonneborn, 1981). It is tempting to investigate germinal aging in clones bearing micronucleus-specific symbionts and somatic aging in both micro- and macroncleus-specific symbionts. The effects of the micronucleus on germinal aging have been shown in several investigations (e.g., Sonneborn and Schneller, 1960; Golikova, 1978a; Karino and Hiwatashi, 1981). Changes of macronuclear chromatin structure and DNA content are correlated with somatic aging (Schwartz and Meister, 1975; Sundararaman and Cummings, 1976; Smith-Sonneborn, 1979; Takagi and Kanazawa, 1982). Alterations in the structure of chromatin, possibly leading to altered transcriptional control of genes, might be involved in aging processes generally (for review see Smith-Sonneborn, 1981). Endonuclear symbionts cause changes in the micro- or macronuclei. Even if the DNA might not be affected directly by the symbionts, e.g., *Holospora,* the fine structural changes of chromatin (Gromov *et al.*, 1975, 1976; Ossipov *et al.*, 1980a; Görtz, 1980, 1982a) suggest some influence of endonuclear symbionts on the process of aging in the host organisms.

Several effects on host cells caused by endonuclear symbionts depend on culture conditions. The same is known for some endocytoplasmic symbionts. The most impressive example may be that certain paramecia die when transferred into axenic medium, presumably because their kappa symbionts multiply without restraint under these conditions (van Wagtendonk *et al.*, 1953) So, culturing the ciliates under laboratory conditions may never show whether endosymbionts are beneficial or parasitic. It will be difficult if not impossible to imitate the natural environments of the symbiont-bearing ciliates in the laboratory. Therefore, it seems necessary to acurately characterize the conditions under which parasitic or mutualistic effects of symbionts to their hosts are observed. Certainly Soldo (1974) is right in postulating the use of axenic media for such experiments.

VIII. Benefit to Symbionts: The Nuclei as Habitats

The nucleus might be regarded as either a favorable environment or as a trap for the symbionts. For those bacteria permanently maintained within the nuclei we cannot distinguish between the two possibilities. For most symbionts, however, the nucleus is an ecological niche as is the host cell (Trager, 1960) and in this respect endonuclear symbionts do not differ from endocytoplasmic symbionts.

Some symbionts can live in either cytoplasm or nucleus (Hovasse, 1950; Görtz, 1981a), but only one endonuclear symbiont has been successfully cultivated outside its host (Harrison *et al.*, 1976a).

The growth rate of symbiotic bacteria is very low (for those growing in balance with their hosts about one to three fissions per day) as compared to free living bacteria. In some cases the number of symbionts per cell is regulated by the host genome (Sonneborn 1943, 1959; Chao, 1953). Such regulation has not been proven for endonuclear symbionts. However, host genes may be involved in the maintenance of *H. caryophila* (Preer, 1969; Beale *et al.*, 1969). Moreover, *H. caryophila* was found to infect only kappa-bearing strains of *P. biaurelia* or strains derived from them (Preer, 1969). Many endonuclear symbionts grow in balance with the hosts (Nilsson, 1969; Harrison *et al.*, 1976a; Görtz, 1980), while others multiply until the host dies (Gillies and Hanson, 1963; Estève, 1978; Wille *et al.*, 1981).

Obviously, the symbionts are provided with nutrients by the host. It is unknown which types of substrates are taken up by endonuclear symbionts. Some authors assume that they digest chromatin. However, the significance of the association of symbionts with host chromatin certainly cannot be clarified by fine structural investigations alone. Also without digesting chromatin, metabolites of normal nuclear pools are available. Even if these pools were small, they might be preferentially refilled at the expense of cytoplasmic pools. This could explain why endocytoplasmic symbionts sometimes cluster at the macronuclear envelope (Wichterman, 1945). Kappa-like symbionts in a strain of *P. aurelia* are observed close to the macro- and micronuclei, sometimes within invaginations of the nuclear envelopes (Stevenson, 1970). The preferential supply of the nuclei with certain metabolites could also be the reason why the number of endocytoplasmic symbionts in *P. sexaurelia* strain 801 is reduced after some symbionts have invaded the macronucleus (Görtz, 1981a). Moreover, it could be a simple explanation for the impediment of double infections with different *Holospora* species: once symbiosis is established, the pools of metabolites could be too small for the establishment of a second symbiosis, if the second symbiont requires the same nutrients. The shortage of certain metabolites might be a trigger for the development of the infectious form of *Holospora,* which is chiefly observed in slow growing cultures and under starvation conditions (Ossipov *et al.*, 1976; Görtz and Dieckmann, 1980).

Besides the general advantage of nutrient supply and the advantage of a motile host, the association with nuclei provides regular distribution to the daughter cells (Jenkins, 1970). This may be true for binary fission but not for sexual reproduction (see Section VI,C).

Endonuclear symbionts generally are not enclosed by host membranes (Smith, 1979). The reason is unknown. It might be speculated that the nucleus has no pools of membrane components like fatty acids, which would be necessary for the formation and maintenance of host membranes surrounding the symbionts.

Symbionts surrounded by membranes during the course of infection lose the membranes when invading the nuclei (see Section V).

Finally, inside the nuclei, symbionts may be sheltered from digestion by lysosomal enzymes. Starving cells digest whole parts of their cytoplasm including organelles and probably endocytoplasmic symbionts, but not the nuclei. Also, living within nuclei may keep the symbionts from being excreted. It has been shown that *Holospora* which have left the nuclei during mitosis or meiosis normally do not remain within the host cells. So, once the symbionts leave the nucleus, they are no longer safe. As reported in Section VI,A, *Holospora* appears to be digested within the nuclei of some host strains under starvation conditions. It is not known whether this is done by autolysis of the bacteria, or by host enzymes. Only a few lytic enzymes have been proven inside nuclei (Vorbrodt, 1974).

IX. Perspectives

Endonuclear symbiosis commonly occurs in ciliates and we shall find more examples of this in the future. It will be of special interest to look for endonuclear symbionts in well-investigated species such as *Tetrahymena thermophila, P. aurelia,* and *Stylonychia mytilus*. In these species the consequences of symbiosis might be better analyzed. Also, in *Euplotes aediculatus* with its essential endocytoplasmic omikron particles (Heckmann, 1975), the search for endonuclear symbionts which are less parasitic than the leptomonads described by Wille *et al.* (1981) should be promising.

Besides the further elucidation of symbiosis itself, the study of endonuclear symbiosis may help to clarify general phenomena of ciliate biology, e.g., the role of micro- and macronuclei during conjugation and autogamy, and the sorting and distribution of chromatin during amitosis. Moreover, the infection with *Holospora* can be used as a model system for endocytosis and intracellular transport. The species specificity and the nucleus specificity of several endonuclear symbionts are attractive phenomena for parasitologists and cytologists.

In order to gain further insight into endonuclear symbiosis and its consequences it will be necessary to use biochemical and genetic methods. It is hoped that this article will encourage other biologists to study the fascinating manifestations of endonuclear symbiosis.

Acknowledgments

I should like to thank Dr. K. Heckmann for stimulating discussions and Dr. M. Freiburg for critically reading the manuscript. I also want to thank Mrs. G. Newels for her help in preparing the manuscript and Mr. D. Bell for some help with the English text.

References

Aikawa, M. (1980). *In* "Cellular Interactions in Symbiosis and Parasitism" (C. B. Cook, P. W. Pappas, and E. D. Rudolph, eds.), pp. 31–46. Ohio State University Press, Columbus.

Albaret, J.-J., and Grain, J. (1973). *Protistologica* **9,** 221–234.

Allen, R. D. (1975). *J. Cell Biol.* **64,** 497–503.

Altenburg, E. (1946). *Am. Nat.* **80,** 559–567.

Altmann, G., and Ruthmann, A. (1979). *Differentiation* **15,** 107–117.

Balbiani, E. G. (1860). *C. R. Acad. Sci.* **51,** 319–322.

Balbiani, E. G. (1861). *J. Physiol. Homme Animal* **4,** 465–520.

Balbiani, E. G. (1881). *J. Micrograph.* **5,** 63.

Ball, G. H. (1969). *In* "Research in Protozoology" (T. T. Chen, ed.), Vol. 3, pp. 565–718. Pergamon, Oxford.

Bardele, C. F. (1972). *Z. Zellforsch.* **126,** 116–134.

Bardele, C. F. (1974). *Symp. Soc. Exp. Biol.* **28,** 191–208.

Beale, G. H., Jurand, A., and Preer, J. R., Jr. (1969). *J. Cell Sci.* **5,** 65–91.

Bozler, E. (1924). *Arch. Protistenk.* **49,** 163–215.

Bretschneider, L. H., and Van Vorstenbosch, C. J. A. H. V. (1964). *Proc. K. Nederl. Akad. Wetensch., Ser. C* **67,** 313–319.

Brutsaert, P., Jungblut, C. W., and Knox, A. (1946). *Proc. Soc. Exp. Biol. Med.* **61,** 163–168.

Bütschli, O. (1876). *Abh. Senckenb. Naturf. Ges.* **10,** 231–464.

Bütschli, O. (1889). *In* "Klassen und Ordnungen des ThierReichs" (H. G. Bronn, ed.), Vol. 1, Protozoa 3. Leipzig.

Calkins, G. N., and Cull, S. W. (1907). *Arch. Protistenk.* **10,** 375–440.

Chao, P. K. (1953). *Proc. Natl. Acad. Sci. U.S.A.* **39,** 103–113.

Claparède, E., and Lachmann, J. (1857). *Ann. Sci. Nat. Zool.* **4,** 221–244.

Cleffmann, G. (1968). *Exp. Cell Res.* **50,** 193–207.

Cleffmann, G. (1980). *Chromosoma* **78,** 313–325.

DeBary, A. (1879). "Die Erscheinungen der Symbiose." Trübner, Strassburg.

Dehorne, A. (1911). *Arch. Zool. Exp.* **60,** 47.

Diller, W. F. (1936). *J. Morphol.* **59,** 11–67.

Diller, W. F. (1948). *Biol. Bull.* **95,** 265.

Diller, W. F. (1965). *J. Morphol.* **116,** 51–64.

Engelmann, T. W. (1862). *Z. Wiss. Zool.* **11,** 347–393.

Engelmann, T. W. (1876). *Morph. Jahrb.* **1,** 573–635.

Estève, J.-C. (1978). *Protistologica* **14,** 201–207.

Fiveiskaja, A. (1929). *Arch. Protistenk.* **65,** 275–298.

Fokin, S. I., and Ossipov, D. V. (1981). *Acta Protozool.* **20,** 51–73.

Fokin, S. I., Borchsenius, O. N., and Ossipov, D. V. (1979). *J. Protozool.* **26,** (Suppl.), 37 A.

Franke, W. W., and Scheer, U. (1974). *In* "The Cell Nucleus" (H. Busch, ed.), Vol. 1, pp. 219–347. Academic Press, New York.

Freedman, F. J., and Hanson, E. D. (1970). *J. Protozool.* **17** (Suppl.), 18 A.

Fujishima, M., and Görtz, H.-D. (1982). *Proc. Chugoku-Shikoku Branch Zool. Soc. Jpn.* **34,** 14.

Fujishima, M., and Watanabe, T. (1981). *Exp. Cell Res.* **132,** 47–56.

Gatteff, E. (1978). *In* "The Genetics and Biology of *Drosophila*" (M. Ashburner and T. R. F. Wright, eds.), Vol. 2, pp. 182–275. Academic Press, New York.

Gillies, C., and Hanson, E. D. (1962). *J. Protozool.* **9** (Suppl.) 15 A.

Gillies, C., and Hanson, E. D. (1963). *J. Protozool.* **10,** 467–473.

Golikova, M. N. (1978a). *Acta Protozool.* **17,** 89–108.

Golikova, M. N. (1978b). *Tsitologiya* **20,** 576–580.

Görtz, H.-D. (1980). *In* "Endocytobiology" (W. Schwemmler and H. E. A. Schenk, eds.), Vol. 1, pp. 381–392. de Gruyter, Berlin.

Görtz, H.-D. (1981a). *Verh. Dtsch. Zool. Ges.* **1981,** 227.
Görtz, H.-D. (1981b). *Abstr. VI. Int. Congr. Protozool.*, Warszawa, 1981, 122.
Görtz, H.-D. (1982a). *J. Protozool.* **29** (Suppl.), 224.
Görtz, H.-D. (1982b). *J. Cell Sci.*, in press.
Görtz, H.-D., and Dieckmann, J. (1977a). *Abstr. V. Int. Congr. Protozool.*, New York, 1977, 441.
Görtz, H.-D., and Dieckmann, J. (1977b). *Verh. Dtsch. Zool. Ges.* **1977,** 267.
Görtz, H.-D., and Dieckmann, J. (1980). *Protistologica* **16,** 591–603.
Görtz, H.-D., and Dieckmann, J. (1982). *J. Protozool.* **29** (Suppl.), 225.
Görtz, H.-D., and Fujishima, M. (1982). *Verh. Dtsch. Zool. Ges.* **1982,** in press.
Grain, J. (1966). *Protistologica* **2,** 5–51.
Grell, K. G. (1973). "Protozoology." Springer Verlag, Berlin and New York.
Gromov, B. V. (1978). *Uspekhi Mikrobiol. AN. SSSR* **13,** 50–73, 215–220.
Gromov, B. V., and Ossipov, D. V. (1981). *Int. J. Syst. Bacteriol.* **31,** 348–352.
Gromov, B. V., Ossipov, D. V., and Mamkaeva, K. A. (1974). *Dokl, Akad. Nauk. SSSR* **217,** 712–714.
Gromov, B. V., Mamkaeva, K. A., and Ossipov, D. V. (1975). *Microbiologija* **44,** 97–102.
Gromov, B. V., Mamkaeva, K. A., and Ossipov, D. V. (1976). *Izv. Akad. Nauk. SSSR Ser. Biol.* **3,** 399–409.
Groupé, V., Herrmann, E. C., and Rauscher, F. J. (1955). *Proc. Soc. Exp. Biol. Med.* **88,** 479–482.
Hafkine, W. M. (1890). *Ann. Inst. Pasteur* **4,** 148–162.
Harrison, D. N., Dorsey, C. H., and Finley, H. E. (1976a). *Trans. Am. Microsc. Soc.* **95,** 560–564.
Harrison, D. N., Dorsey, C. H., and Brown, C. A. (1976b). *Trans. Am. Microsc. Soc.* **95,** 565–568.
Hausmann, K. (1972). *Z. Naturforsch.* **27b,** 1572.
Hausmann, K., and Peck, R. K. (1979). *Differentiation* **14,** 147–158.
Hauser, M. (1973). *Chromosoma* **44,** 49–71.
Heckmann, K. (1965). *Inst. Wiss. Film Göttingen,* Film E 913.
Heckmann, K. (1967). *J. Protozool.* **14** (Suppl.), 115.
Heckmann, K. (1975). *J. Protozool.* **22,** 97–104.
Heckmann, K. (1977). *Proc. 5th. Int. Congr. Protozool.* 160–164.
Heckmann, K., TenHagen, R., and Görtz, H.-D. (1983). *J. Protozool.*, in press.
Heine, U. J. (1974). *In* "The Cell Nucleus" (H. Busch, ed.), Vol. III, pp. 489–563. Academic Press, New York.
Hiwatashi, K. (1959). *Sci. Rep. Tohoku Univ. (Biol.)* **25,** 81–90.
Hovasse, R. (1950). *Bull. Inst. Océanograph. Monaco* **962,** 1–10.
Inaba, F. (1960). *Biol. J. Nara Women's Univ.* **10,** 26–29.
Jenkins, R. A. (1970). *J. Gen. Microbiol.* **61,** 355–359.
Jensen, H., Engedal, H., and Saetersdal, T. S. (1976). *Virchows Arch. B Cell Pathol.* **21,** 1–12.
Jeon, K. W. 1972). *Science* **176,** 1122–1123.
Jeon, K. W. (1980). *In* "Cellular Interactions in Symbiosis and Parasitism" (C. B. Cook, P. W. Pappas, and E. D. Rudolph, eds.), pp. 31–46. Ohio State University Press, Columbus.
Jeon, K. W., and Jeon, M. S. (1976). *J. Cell Physiol.* **89,** 337–344.
Karino, S., and Hiwatashi, K. (1981). *Exp. Cell. Res.* **136,** 407–415.
Kattar, M. R. (1972). *Protistologica* **8,** 135–141.
Kattar, M. R. (1975). *Acta Protozool.* **14,** 179–184.
Kirby, H., Jr. (1941). *In* "Protozoa in Biological Research" (G. N. Calkins and F. M. Summers, eds.), pp. 1009–1113. Columbia University Press, New York.
Kirby, H., Jr. (1942). *J. Parasitol.* **28,** 311–314.
Kirby, H., Jr. (1957). *In* "Bergey's Manual of Determinative Bacteriology" (R. S. Breed, E. G. D. Murray, and N. R. Smith, eds.), 7th ed., pp. 927–929. Williams & Wilkins, Baltimore.
Kesa, L., and Teras, J. H. (1977). *Abstr. V. Int. Congr. Protozool.*, New York, 1977, 126.

Kesa, L., and Teras, J. H. (1981). *Abstr. VI. Int. Congr. Protozool.*, Warszawa, 1981, 181.
Kölliker, A. (1864). "Icones Histiologicae." 1. Abth. Der feinere Bau der Protozoen. Leipzig.
Kovàcs, E., and Bucz, B. (1967). *Life Sci.* **6,** 347–358.
Lewis, L. M., Witkus, E. R., and Vernon, G. M. (1976). *Protoplasma* **89,** 203–219.
Mayo, K. A., and Orias, E. (1981). *Genetics* **98,** 747–762.
Mikami, K. (1979). *J. Exp. Zool.* **208,** 121–128.
Millecchia, L. L., and Rudzinska, M. A. (1971). *Z. Zellforsch.* **115,** 149–164.
Miyake, A. (1957). *J. Inst. Polytech., Osaka City Univ. Ser. D* **8,** 1–10.
Miyake, A. (1968). *J. Exp. Zool.* **167,** 359–380.
Moevus-Kobb, L. (1965). *Ann. N.Y. Acad. Sci.* **126,** 328–342.
Moritz, K., and Storch, V. (1971). *Mikrokosmos* **60,** 77–80.
Müller, J. (1856). *Monatsberichte Berliner Acad. Wissenschaften* **1856,** 390–392.
Nanney, D. L. (1980). "Experimental Ciliatology." Wiley, New York.
Ng, S. F., and Mikami, K. (1981). *Protistologica* **17,** 497–509.
Nilsson, J. R. (1969). *C. R. Trav. Lab. Carlsberg* **37,** 49–76.
Ossipov, D. V. (1973). *Tsitologiya* **15,** 211–216.
Ossipov, D. V. (1975). *Acta Protozool.* **14,** 43–57.
Ossipov, D. V. (1981). "Problems of Nuclear Heteromorphism of the Unicellular Organisms." Nauka, Leningrad.
Ossipov, D. V., and Ivakhnuyk, I. S. (1972). *Tsitologiya* **14,** 1414–1419.
Ossipov, D. V., and Podlipaev, S. A. (1977). *Acta Protozool.* **16,** 289–308.
Ossipov, D. V., and Skoblo, I. I. (1973). *Acta Protozool.* **12,** 151–164.
Ossipov, D. V., Rautian, M. S., and Skoblo, I. I. (1974). *Genetica (USSR)* **10,** 62–70.
Ossipov, D. V., Skoblo, I. I., and Rautian, M. S. (1975a). *Acta Protozool.* **14,** 263–280.
Ossipov, D. V., Skoblo, I. I., and Rautian, M. S. (1975b). *Tsitologiya* **17,** 95–97.
Ossipov, D. V., Gromov, B. V., Mamkaeva, K. A., Rautian, M. S., Skoblo, I. I., and Borchsenius, O. N. (1976). *Protozoologia (USSR)* **1,** 101–139.
Ossipov, D. V., Borchsenius, O. N., and Podlipaev, S. A. (1980a). *Acta Protozool.* **19,** 315–326.
Ossipov, D. V., Skoblo, I. I., Borchsenius, O. N., Rautian, M. S., and Podlipaev, S. A. (1980b). *Tsitologiya* **22,** 922–929.
Petschenko, B. de (1911). *Arch. Protistenk.* **22,** 248–298.
Podlipaev, S. A., and Ossipov, D. V. (1979). *Acta Protozool.* **18,** 465–480.
Preer, J. R., Jr. (1975). *Symp. Soc. Exp. Biol.* **29,** 125–144.
Preer, J. R., Jr., and Preer, L. B. (1982a). *Int. J. Syst. Bacteriol.* **32,** 140–141.
Preer, J. R., Jr., and Preer, L. B. (1982b). In preparation.
Preer, J. R., Jr., Preer, L. B., and Jurand, A. (1974). *Bacteriol. Rev.* **38,** 113–163.
Preer, L. B. (1969). *J. Protozool.* **16,** 570–578.
Preer, L. B., and Preer, J. R., Jr. (1977). *In* "Cell Biology: A Comprehensive Treatise" (D. M. Prescott and L. Goldstein, eds.), Vol. 1, pp. 319–373. Academic Press, New York.
Puytorac, P. de (1959). *Arch. Anat. Microsc. Morphol. Exp.* **48,** 49–62.
Puytorac, P. de (1963). *J. Microsc.* **2,** 189–196.
Puytorac, P. de, and Kattar, M. R. (1969). *Protistologica* **5,** 549–560.
Raikov, I. B. (1969). *In* "Research in Protozoology" (T. T. Chen, ed.), Vol. 3, pp. 1–128. Pergamon, Oxford.
Raikov, I. B. (1972). *In* "Research in Protozoology" (T. T. Chen, ed.), Vol. 4, pp. 147–289. Pergamon, Oxford.
Rautian, M. S., Ossipov, D. V., and Skoblo, I. I. (1975). *Tsitologiya* **17,** 1200–1207.
Rautian, M. S., Skoblo, I. I., and Ossipov, D. V. (1978). *Tsitologiya* **20,** 455–459.
Rautian, M. S., Skoblo, I. I., and Ossipov, D. V. (1979). *J. Protozool.* **26** (Suppl) 34A.
Rees, J. van (1878). *Z. Wiss. Zool.* **31,** 473–481.

Rosati, G., and Verni, F. (1975). *Boll. Zool.* **42,** 231–232.
Schuster, F. L. (1969). *J. Protozool.* **16,** 724–727.
Schwartz, V. (1939). *Naturwissenschaften* **27.**
Schwartz, V. (1976). *Arch. Protistenk.* **120,** 255–277.
Schwartz, V., and Meister, H. (1975). *Arch. Protistenk.* **117,** 85–109.
Skoblo, I. I., Rautian, M. S., and Ossipov, D. V. (1978). *Acta Protozool.* **17,** 321–330.
Smith, C. C. (1979). *Proc. R. Soc. London Ser. B* **204,** 115–130.
Smith-Sonneborn, J. (1979). *Science* **203,** 1115–1117.
Smith-Sonneborn, J. (1981). *Int. Rev. Cytol.* **73,** 319–354.
Soldo, A. T. (1963). *Ann. N.Y. Acad. Sci.* **108,** 380–388.
Soldo, A. T. (1974). *In* "Paramecium—A Current Survey" (W. J. van Wagtendonk, ed.), pp. 377–442. Elsevier, Amsterdam.
Soldo, A. T., and Godoy, G. A. (1973). *J. Protozool.* **20** (suppl.), 502.
Sonneborn, T. M. (1943). *Proc. Natl. Acad. Sci. U.S.A.* **29,** 329–343.
Sonneborn, T. M. (1959). *Advan. Virus Res.* **6,** 229–356.
Sonneborn, T. M. (1975). *Trans. Am. Microsc. Soc.* **94,** 155–178.
Sonneborn, T. M., and Schneller, M. (1960). *In* "Biology of Ageing" (B. L. Strehler, ed.), pp. 286–287. Waverly Press, Baltimore, Maryland.
Stein, F. (1861). *Sitzungsberichte königl. böhm. Ges. W. zu Prag.* **1861,** 50–57.
Stevenson, I. (1970). *Cytobios* **2,** 207–224.
Sundararaman, B., and Cummings, D. J. (1976). *Mech. Ageing Dev.* **5,** 325–338.
Takagi, Y., and Kanazawa, N. (1982). *J. Cell Sci.* **54,** 137–147.
Takagi, Y., and Yoshida, M. (1980). *J. Cell Sci.* **41,** 177–191.
Teras, J. H. (1981). *Protozoologia (USSR)* **6,** 31–55.
Trager, W. (1960). *In* "The Cell" (J. Brachet and A. E. Mirsky, eds.), Vol. 4, pp. 151–213. Academic Press, New York.
van Wagtendonk, W. J., Conner, R. L., Miller, C. A., and Rao, M. R. R. (1953). *Ann. N.Y. Acad. Sci.* **56,** 929–937.
Vivier, E. (1974). *In* "Paramecium—A Current Survey" (W. J. van Wagtendonk, ed.), pp. 1–89. Elsevier, Amsterdam.
Vivier, E., and André, J. (1961). *C. R. Seances Acad. Sci.* **252,** 1848–1850.
Vorbrodt, A. (1974). *In* "The Cell Nucleus" (H. Busch, ed.), Vol. III, pp. 309–344. Academic Press, New York.
Wichterman, R. (1940). *J. Parasitol.* **26** (Suppl.), 29.
Wichterman, R. (1945). *J. Parasitol.* **31** (Suppl.), 25.
Wichterman, R. (1953). "The Biology of Paramecium." Blakiston, New York.
Wille, J. J., Jr., Weidner, E., and Steffens, W. L. (1981). *J. Protozool.* **28,** 223–227.

Metabolic Interchange in Algae–Invertebrate Symbiosis

CLAYTON B. COOK

Marine Science Institute, Northeastern University, Nahant, Massachusetts

I. Introduction

The past decade has seen a proliferation of research on the biology of algae–invertebrate symbioses. A number of metabolic features characterize these associations. Paramount among these is the movement of large amounts of carbohydrate from algae to host cells (Smith *et al.*, 1969). This phenomenon occurs in a variety of mutualistic associations, and forms the basis of biotrophy (Lewis, 1973). Other metabolic features of algae–invertebrate symbioses include the uptake, retention, and recycling of dissolved organic carbon and nitrogen (Lewis and Smith, 1971; Lewis, 1973; Muscatine and Porter, 1977) and the uptake and metabolism of dissolved inorganic nutrients (Muscatine, 1980a).

This article examines the interflow of metabolites in these associations and its consequences. Earlier work on biotrophic transport has been reviewed by Smith

ISBN 0-12-364375-9

et al. (1969) and Smith (1974); general cellular interactions have been reviewed by Taylor (1973a) and Trench (1979). Symbioses with chloroplasts (reviewed by Taylor, 1970; Muscatine and Greene, 1972; Trench, 1974, 1979, 1980) and cyanobacteria (blue-green algae; Trench, 1979, 1980, 1981) are not considered, except for recent work on *Prochloron*. The work considered is limited to non-protistan invertebrates, especially coelenterates. Muscatine (1974) has reviewed earlier work on symbioses within this group.

II. Diversity of Algal Symbionts

A number of algal groups enter into symbiosis with invertebrate cells. The taxonomy and distribution of endosymbiotic algae have been reviewed elsewhere (McLaughlin and Zahl, 1966; Taylor, 1973b, 1974; Muscatine, 1974). What follows is a brief discussion of the algae and hosts which will be considered in subsequent sections.

Two points may be raised about terminology. *In situ* generally refers to algae in association with hosts. *In vitro* usually means in culture outside of the host, but this term has often been used in reference to experiments using algae freshly isolated from host tissue, rather than algae which have been grown in culture. In this article, reference will be made to "cultured algae" or to "isolated symbionts" to make this distinction. With respect to hosts, reference will be made to "aposymbiotic" and "nonsymbiotic." The former refers to normally symbiotic animals which do not have symbionts; the latter refers to species which never possess algae.

A. *Chlorella*-LIKE ALGAE

The common symbionts of freshwater sponges and metazoans are unicellular algae of the order Chlorococcales, generally known as "zoochlorellae." These algae morphologically resemble free-living *Chlorella,* and are usually considered to be members of this genus. Specific names have not been assigned; algae which have been maintained in axenic culture bear strain designations of specific isolates, while symbionts which have not been grown *in vitro* are identified by strain designations of the host (e.g., Jolley and Smith, 1980). Precise characterizations of strains are difficult due to plastic phenotypic responses to different host environments (Pardy, 1976; Pool, 1979), and even cultured isolates with the same strain designation may have different physiological characteristics (e.g., Jolley and Smith, 1980).

Green hydra (*Hydra viridis,* = *Chlorohydra viridissima*) have been the most intensely studied of these associations. As is true of a great variety of algal–invertebrate associations, the algae are restricted to endodermal cells which have a primarily digestive function, and reside in individual membrane-limited vacuoles (Fig. 1). The question of "digestion" of algal symbionts has been a vexed one, but it is now clear that in *Hydra* symbionts actively interfere with host lysosomal activity (Section V). Research on these hydra has involved a number of different host and symbiont strains (e.g., "Florida" and "European") and it is clear that there are metabolic and physiological differences between these strains (Jolley and Smith, 1980; Mews, 1980; McAuley, 1981a).

Other symbioses with *Chlorella* which will be briefly considered include those with the freshwater sponge *Spongilla lacustris* (Williamson, 1979), the sea anemone *Anthopleura xanthogrammica* (O'Brien, 1978), turbellarian flatworms, and the freshwater bivalve *Anadonta* (Pardy, 1980). Reference will be made to the association between *Chlorella* and *Paramecium bursaria* where appropriate.

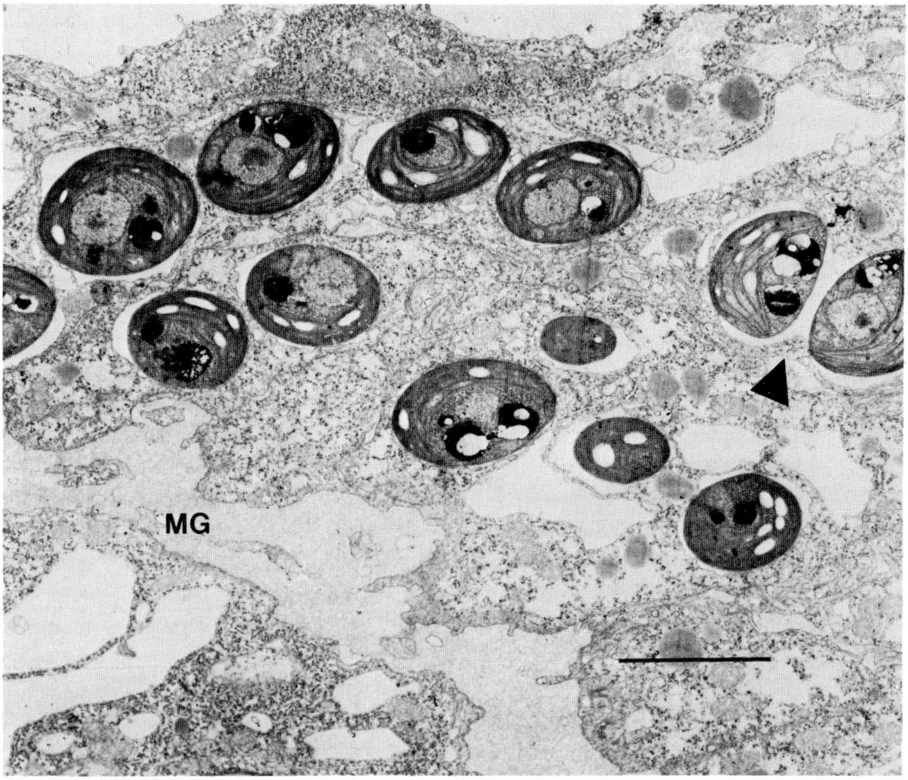

FIG. 1. Symbiotic zoochlorellae in gastrodermal cells of *Hydra viridis* (Florida strain). Except for the daughter cells indicated by the arrow, each alga is in an individual vacuole. The mesoglea (MG) separates this layer from the ectoderm. Scale bar = 5 μm. (Original micrograph.)

B. Dinoflagellate Symbionts

A number of dinoflagellates are symbiotic with marine invertebrates. Those which have been classically identified as *Symbiodinium microadriaticum* (Freudenthal, 1962; placed by some in the genus *Gymnodinium;* Taylor, 1971) will be the only ones considered here. A recent suggestion that these algae be transferred to the genus *Zooxanthella* (Loeblich and Sherley, 1979) has not yet gained wide acceptance. For the time being, it may be best to refer to them in terms of the hosts with which they are associated (or from which they have been isolated), following the terminology of Schoenberg and Trench (1980a–c). These authors have shown that what were formerly considered to be more or less similar algae have genetic, morphological, and physiological differences, and that symbionts isolated from the same host species have consistently similar characteristics.

Whatever their taxonomy, the symbionts in question are the familiar zooxanthellae which are associated with marine cnidarians such as hydroids, sea anemones, and corals, and with the "giant clams" (*Tridacna* and *Hippopus*). Typically these algae *in situ* are coccoid vegetative cells which lack flagella and the "gymnodinoid" arrangement of epicone and hypocone, and occur in membrane-bound vacuoles (Fig. 2). In some cases the algae are extracellular, as in *Tridacna* (Fankboner, 1971; Trench *et al.,* 1981a). Cells with flagella have been reported *in situ* (Schoenberg and Trench, 1980b), and it is clear that coccoid cells give rise to flagellated zoospores which are probably dispersive forms (Trench, 1980). Many strains are in culture, and regularly produce zoospores; zoosporogenesis is a process which follows a circadian rhythm (Fitt *et al.,* 1981; Lerch and Cook, 1982).

C. Other Algae

Also considered in this article are the prasinophycean flagellates in the genus *Platymonas* which occur with acoelous flatworms in the genus *Convoluta* (Parke and Manton, 1967). *Prochloron* is a genus of prokaryotic algae which has been proposed as a new division, the Prochlorophyta (Lewin, 1976, 1977). These algae are extracellular symbionts of didemnid ascidians. While there is an indication that translocation may occur (Pardy and Lewin, 1981), studies on isolated algae have shown that the release of photosynthetic products is very much less than that of other symbiotic algae, and is more similar to release by free-living phytoplankters (Fisher and Trench, 1980).

FIG. 2. A zooxanthella (*Symbiodinium microadriaticum*) from a digestive cell of the hydroid *Myrionema amboinense*, showing the typical morphology of the *in situ* vegetative cell. Scale bar = 1.5 μm. (Courtesy of Dr. R. K. Trench.)

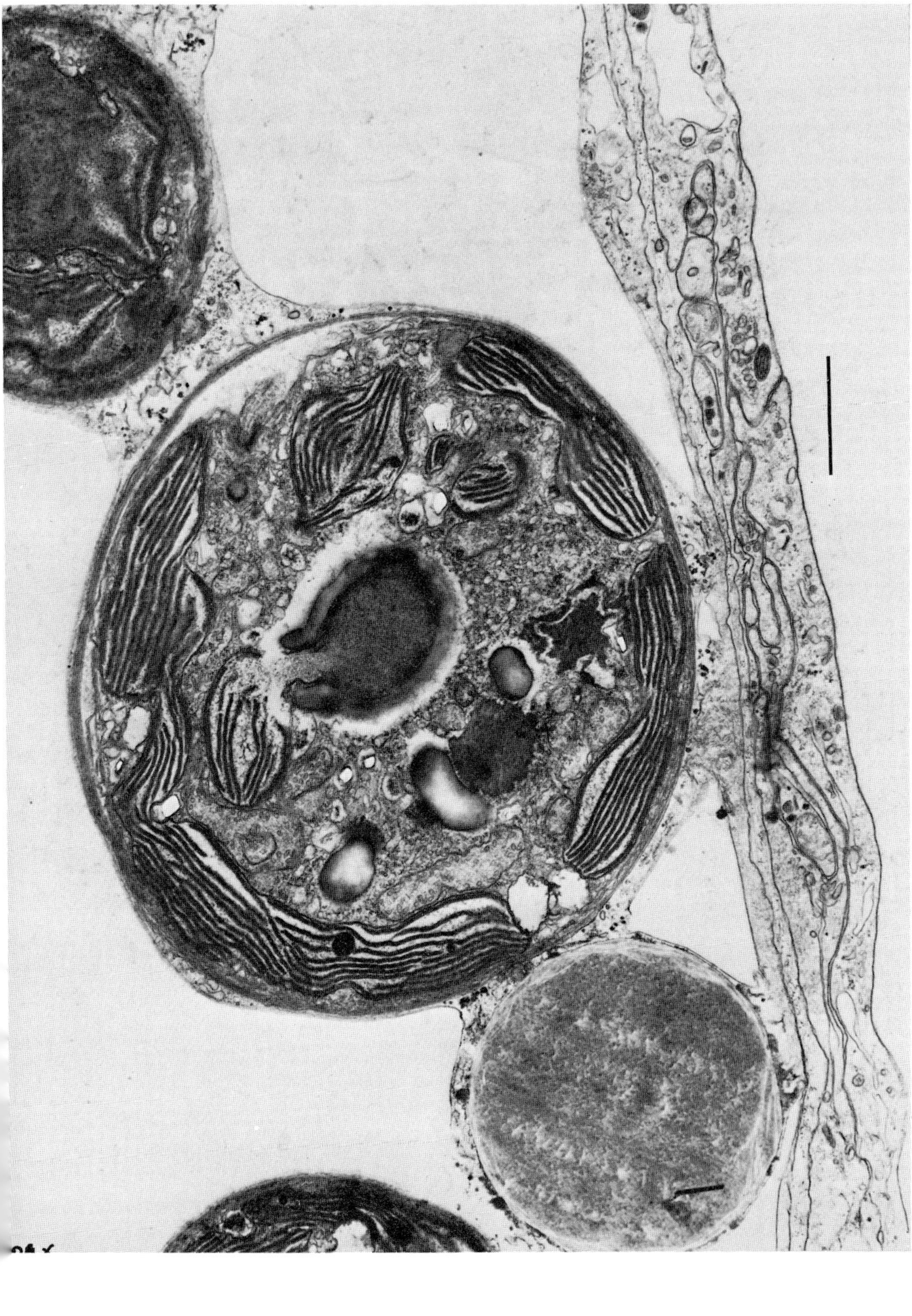

III. Photosynthetic Products and Translocation

A. Methods of Study

The photosynthetic products translocated by symbiotic algae to host tissue have been identified in two general sorts of experiments. The usual approach has been to identify extracellular and intracellular products of freshly isolated symbionts (*in vitro*), and then compare these with photosynthetic products which occur in host tissue. Isolated algae are incubated with $^{14}CO_2$ in an appropriate medium; labeled compounds in the medium ("extracellular") are then compared with soluble compounds extracted from the algae with ethanol ("intracellular"). The absence of intracellular compounds in medium samples is evidence that the extracellular compounds result from specific release processes, rather than from cell lysis (Muscatine, 1965; Trench, 1971b). Translocation of the extracellular compounds *in situ* may be confirmed by the identification of either the compounds or their derivatives in host tissue (Section IV).

The "inhibition" technique was developed to identify translocated carbohydrates in lichens (Drew and Smith, 1967). Lichens labeled with $^{14}CO_2$ were incubated with large amounts of ^{12}C compounds suspected to be translocated by the alga; these compounds subsequently appeared labeled in the medium. The phenomenon was originally explained as due to saturation of fungal uptake sites (Drew and Smith, 1967; Smith *et al.*, 1969). More recent observations suggest that, at least in intracellular symbioses, the effect is partly due to exchange diffusion processes between host cells. Thus, a successful "inhibition" probably indicates the existence of the compound in question within host cells, rather than its actual transport from algal symbionts to the medium (Smith, 1974, 1980; Trench *et al.*, 1974).

B. Extracellular Products

1. *Isolated Chlorella Symbionts*

Maltose is the principal extracellular product of most strains of symbiotic *Chlorella*. Originally discovered in the "Florida" strain of green hydra, where it may comprise more than 80% of the carbon fixed by isolated algae (Muscatine, 1965; Cernichiari *et al.*, 1969; Fig. 3), it is also produced by symbionts of other hydra strains (Mews, 1980) as well as by a variety of cultured isolates from *Paramecium bursaria* (Muscatine *et al.*, 1967; Weis, 1979, 1981). Typically it is detected only in fleetingly small amounts as an intracellular product, an indication that it is released soon after synthesis (Section III,D). As studies of symbiotic *Chlorella* are virtually the only reports of maltose production by algae, maltose synthesis and transport are evident specializations of symbiotic *Chlorella* for symbiosis.

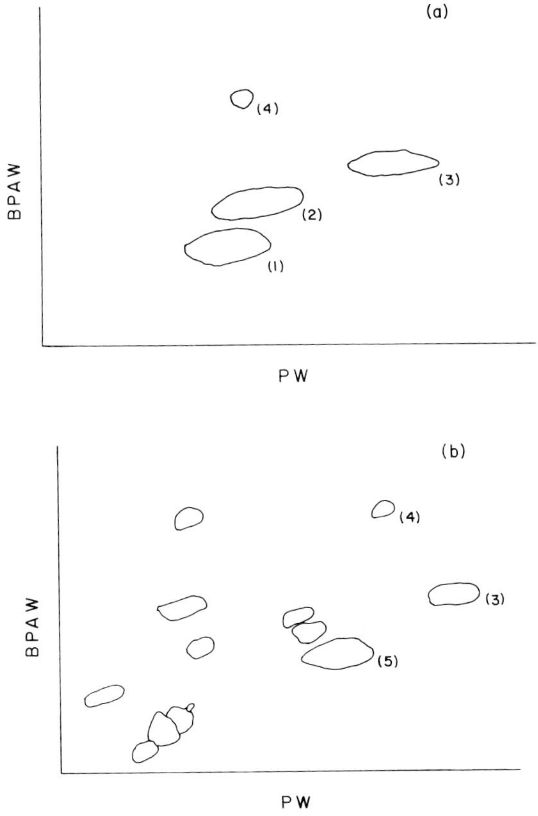

FIG. 3. Distribution of photosynthetic products by isolated algae of *Hydra viridis* (Florida) after 60 minutes of photosynthesis, as separated by paper chromatography. Solvent systems: PW, phenol/water, 100:40, w/v; BPAW, butanol/proprionic acid/water, 142:71:100,v/v/v. (a) Extracellular products. (b) Ethanol extracts of algae (intracellular products). (1) Maltose; (2) glucose; (3) alanine; (4) glycollate; (5) sucrose. [Data of Muscatine (1965) with permission of Pergamon Press and L. Muscatine.]

Glucose also occurs as an extracellular product, although to some extent it is produced by the action of animal maltase when freshly isolated algae are examined (Muscatine *et al.*, 1967; Mews, 1980). In two strains it is the major product: it represents more than 95% of the photosynthate released from algae of a stolonizing hydra, and it is the only extracellular product produced by cultured symbionts of *Spongilla lacustris* (Muscatine *et al.*, 1967). Kremer (1980) found that glucose was the major photosynthetic product in the intact sponge *Ephydatia fluviatilis*. Maltose was not found in either of these studies.

Other compounds may occur as extracellular products. Alanine and glycollic acid usually appear in trace amounts. Since glycollic acid can be released by free-living *Chlorella*, the significance of glycollate to the symbiosis is not clear. In all of these experiments, sucrose (the major intracellular product of both symbiotic and free-living *Chlorella*) does not occur among these products (Fig. 3).

2. *Studies of Isolated Zooxanthellae*

Parallel studies led to the discovery of glycerol as a major extracellular product of zooxanthellae freshly isolated from corals and *Tridacna* (Muscatine, 1967). Since that initial report, glycerol has been identified as a major extracellular photosynthetic product of freshly isolated zooxanthellae from virtually every host which has been examined (Muscatine and Cernichiari, 1969; Trench, 1971b, 1974; Muscatine *et al.*, 1972). [One exception to this is a study of products of zooxanthellae from several *Zoanthus* species (von Holt and von Holt, 1968b). However, the analysis of "neutral" sugars did not specifically examine glycerol, and Trench has clearly demonstrated its release by zooxanthellae from *Z. pacificus* (Trench, 1971b) and by *Z. sociatus* (Trench, 1974).]

A typical distribution of extracellular and intracellular photosynthetic products from such experiments is shown in Fig. 4 and Table I (Trench, 1971b). Under the usual conditions of such experiments (30–60 minute incubations in seawater plus some amount of host tissue; Section III,E), zooxanthellae release up to 40% of their fixed carbon as glycerol, although the proportions vary from strain to strain (Trench, 1971b). The highest reported percentage appears to be from the reef coral *Pocillopora damicornis;* glycerol constituted 82% of the ^{14}C-labeled extracellular products of zooxanthellae isolated from this coral after a 2 hour incubation (Muscatine and Cernichiari, 1969).

In addition to glycerol, freshly isolated zooxanthellae release a variety of other photosynthetic compounds (Fig. 4; Trench, 1971b), although there are likely to be significant differences between algae from different hosts. Notable among these are alanine and a number of other organic acids. Glucose is also generally found, although never in the amounts in which it appears as an intracellular product. Traces of glycollic acid are usually reported, and may become more significant in prolonged experiments. Schmitz and Kremer (1977) reported a variety of other photosynthetic products in an intact coral, including mannose and a number of sugar phosphates. However, their experiments did not distinguish between intracellular and extracellular algal products, or metabolites of algal products by host tissue.

It should be noted that zooxanthellae are more susceptible to physical damage than are zoochlorellae, so it is important that intracellular products are not produced by cell damage. The nature of released products also changes with time after isolation. For example, zooxanthellae from the sea anemone *Anthopleura elegantissima* show a striking increase in the proportion of alanine released, with

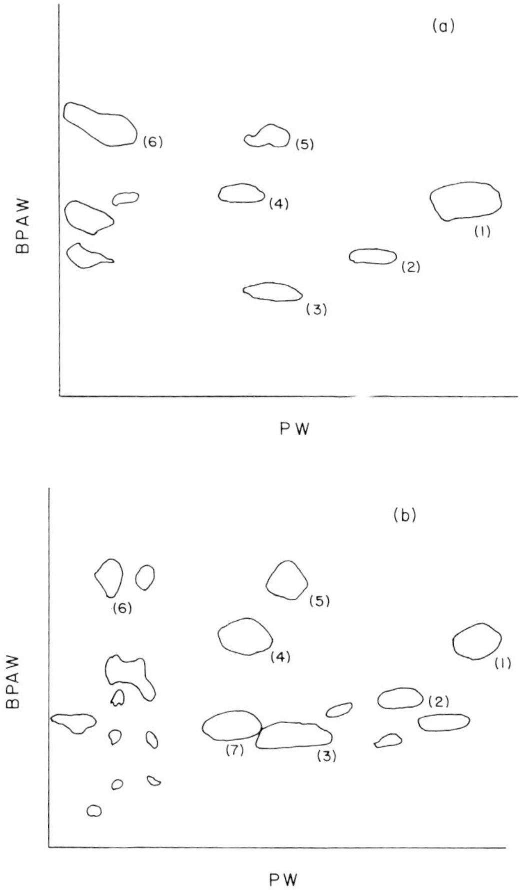

FIG. 4. Distribution of photosynthetic compounds of zooxanthellae freshly isolated from *Anthopleura elegantissima*. Chromatographic solvents as in Fig. 2. (a) Extracellular products. (b) Ethanol extracts of algae (intracellular products). (1) Glycerol; (2) alanine; (3) glucose; (4) glycollate; (5) succinate; (6) fumarate; (7) glutamic acid. Compounds along the left origin are probably non-amino organic acids. [Reproduced from Trench (1971b) with permission of the Royal Society and R. K. Trench.]

a concomitant decrease in glycerol, within 10 hours of isolation (Trench, 1971c). The difference between intracellular and extracellular products during short-term photosynthesis experiments can be seen in Fig. 4 and Table I. The absence of labeled glutamate as an extracellular product indicates the selective release of particular products. However, it should be emphasized that the total release of photosynthate may include compounds which are not synthesized during short periods of photosynthesis.

TABLE I
THE DISTRIBUTION OF ^{14}C IN EXTRACELLULAR AND INTRACELLULAR PRODUCTS AFTER 1 HOUR OF PHOTOSYNTHESIS BY ZOOXANTHELLAE ISOLATED FROM *Anthopleura elegantissima*[a,b]

Compound	Percentage of intracellular products	Percentage of extracellular products
Glycerol (1)	3.6	43.0
Alanine (2)	[c]	1.2
Glucose (3)	32.5	0.5
Glycollic acid (4)	[c]	2.3
Succinate (5) and fumarate (6)	[c]	28.0
Malate and citrate	[d]	25.0
Glutamate	6.7	0.0

[a]Data of Trench (1971b).
[b]Numbers refer to compounds in Fig. 4a and b.
[c]Total intracellular amino acids, excluding glutamic acid, = 6.7% of intracellular ^{14}C. Total non-amino organic acids, including glycollate, = 32.8%.
[d]Provisional identification, not noted in Fig. 4.

3. *Extracellular Products of Platymonas convolutae*

In culture, the symbionts of the *Convoluta* system produce mannitol as the principal photosynthetic product (Craigie *et al.*, 1966; Gooday, 1970). This appears to be the major intracellular product of the algae *in situ* (Muscatine *et al.*, 1974; Boyle and Smith, 1975; Holligan and Gooday, 1975). Because of the fragility of these cells, it has proved rather difficult to work with symbionts isolated from homogenates of tissue. Using this technique, Muscatine *et al.* (1974) demonstrated alanine as the major extracellular product, although it represented a smaller proportion (4–5%) of total fixed carbon than do the major extracellular products of symbiotic *Chlorella* or zooxanthellae. By using high CO_2 tensions, Boyle and Smith (1975) were able to isolate symbionts without physical damage. They found higher percentages of release (suggesting that damage had occurred during physical isolation), and verified alanine and glutamate as major extracellular compounds. There has been no indication that mannitol is released. No evidence to date bears on the release of compounds which might be used for lipid synthesis (Meyer *et al.*, 1979; Section V).

C. CHANGES AFTER ISOLATION FROM THE HOST

It is important to note that the physiology of symbiotic algae in culture is often different from that *in situ*, and to some extent these differences may reflect responses of the algae to the host cell habitat (Smith, 1974, 1979).

The pattern of release of extracellular compounds by zooxanthellae changes following isolation (Trench, 1971c; Muscatine *et al.*, 1972). Zooxanthellae in culture undergo a number of morphological and physiological changes, including the production of flagellated zoospores (McLaughlin and Zahl, 1957) and the synthesis of thick cell walls (Schoenberg and Trench, 1980b); the amount of released photosynthate decreases (Trench *et al.*, 1981b), and "mucoid" substances may be produced (McLaughlin *et al.*, 1963). More detailled studies are needed on the changes in synthetic patterns which occur when these algae are grown *in vitro*.

Strains of symbiotic *Chlorella* retain the ability to release maltose or glucose in culture, but the percentage of photosynthate which is released may decrease. Despite the numerous reports of maltose and glucose release by *Chlorella* cultured from *P. bursaria,* there seems to be no analysis of extracellular products of algae freshly isolated from these ciliates; this situation also applies to symbionts cultured from freshwater sponges. In the latter instance, the percentage of released photosynthate is particularly low. Successful culture of the symbionts from *Hydra* has been reported only once (Jolley and Smith, 1978). The cultured symbiont has a striking decrease in the amount of maltose released, but normal levels are restored following reestablishment of the symbiosis (Jolley and Smith, 1978). It is interesting that the ability to release maltose is maintained by freshly isolated "Florida" symbionts for at least 24 hours (Cernichiari *et al.*, 1969).

D. Synthesis of Extracellular Products

The chief translocated compounds of symbiotic *Chlorella* and zooxanthellae—maltose and glycerol, respectively—are products which are unusual both as algal products and in the quantities in which they are synthesized and released. Other extracellular compounds such as alanine and glucose are normally synthesized by algae, but their release by symbiotic algae presumably represents specific transport mechanisms (Smith *et al.*, 1969; Smith, 1974). Studies of maltose and glycerol synthesis by symbiotic algae have been limited to kinetic studies of photosynthetic products by intact algae. Cell-free systems have not been examined in these algae, although glycerol synthesis has been examined in cell-free extracts of the free-living flagellate *Dunaliella* (Mutschler and Wegmann, 1974). Such an approach may prove difficult for those symbiotic algae which have structurally resistant cell walls, such as *Chlorella* (Atkinson *et al.*, 1972).

1. *Maltose Synthesis by Chlorella*

Cernichiari *et al.* (1969) have provided the best evidence for what may be a unique synthetic pathway for an extracellular compound in an alga symbiotic with an invertebrate. They exploited the striking pH dependence of maltose synthesis and release by algae isolated from *Hydra* (Fig. 5; Cernichiari *et al.*, 1969). At pH 4.5 these algae may release up to 80% of fixed ^{14}C as maltose,

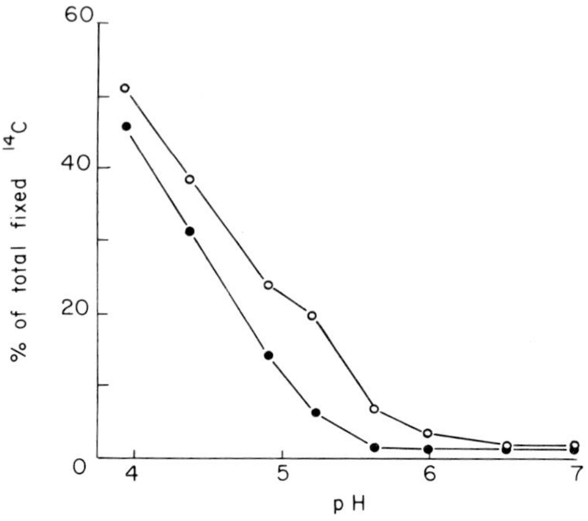

FIG. 5. The influence of pH on release of extracellular products of isolated *H. viridis* algae. Open circles, total release; closed circles, amount released as maltose. [Reproduced from Cernichiari *et al.* (1969) by permission of the Royal Society and L. Muscatine.]

while at neutral pH this release becomes negligible. By incubating symbionts with $H^{14}CO_3^-$ at either acidic or neutral pH, comparisons could be made of photosynthesizing cells which were or were not synthesizing and releasing maltose. Their experiments strongly suggested that the maltose-synthesizing system was located near the algal surface, and rapidly converted small-molecular-weight precursors to maltose, rather than hydrolyzing stored starch or other polysaccharides. Studies of short-term photosynthesis and dark metabolism of [^{14}C]glucose, suggested the pathways outlined in Fig. 6. Cernichiari *et al.* (1969) considered that maltose was synthesized by the condensation of UDPglucose (UDPG) and glucose 6-phosphate, analogous to other synthetic mechanisms for disaccharides in plants. Because of the sensitivity of this system to pH, and because maltose apparently is transported out of cells very soon after synthesis, these authors suggested that the synthesis of UDPG and the subsequent conversion to maltose occurred at the cell surface.

Mews (1980) has provided support for this general scheme. She interpreted the rapid equilibration of translocated products in host tissue as an indication that the algal maltose pool was small, and turned over rapidly. However, she demonstrated maltose production in darkness and by cells in which photosynthesis had been inhibited by DCMU. The source of the maltose appeared to be stored polysaccharide, although it is not clear if this maltose was a direct hydrolysis product, or synthesized via a mechanism as in Fig. 6. In either case these observations are consistent with the finding that the starch reserves of *Hydra* symbionts are depleted when hydra are grown in darkness (Pardy, 1976).

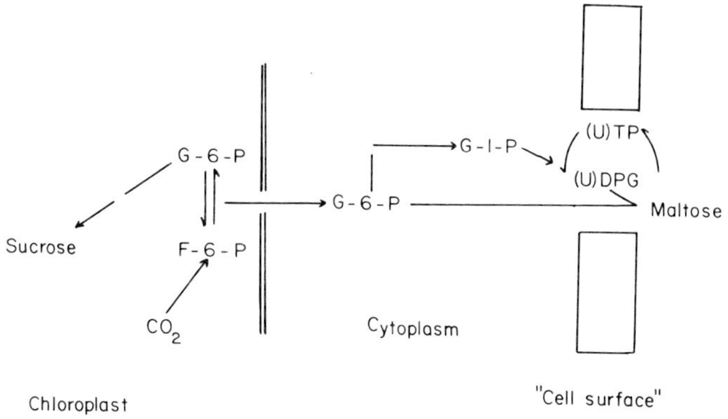

FIG. 6. Proposed scheme of Cernichiari *et al.* (1969) for the synthesis and release of maltose by *H. viridis* zoochlorellae. F-6-P, Fructose 6-phosphate; G-6-P, glucose 6-phosphate; G-1-P, glucose 1-phosphate; (U)TP, uridine triphosphate; (U)DPG, UDPglucose. UTP is considered to be the most likely nucleotide to be involved in maltose synthesis. (Reproduced by permission of the Royal Society and L. Muscatine.)

2. *Glycerol Synthesis by Zooxanthellae*

Trench (1979) has summarized what is known about glycerol synthesis by zooxanthellae: virtually nothing. Glycerol appears as a photosynthetic product within 3 minutes of photosynthesis by an intact coral (Schmitz and Kremer, 1977), but shorter term incubations of either intact associations or isolated zooxanthellae have not been reported. Schmitz and Kremer found 3-phosphoglyceric acid and various hexose phosphates to occur during these incubations, but the role of these compounds in glycerol synthesis is not known. In fact, there is some debate as to whether these algae are "C-3" or "C-4" plants; this question is discussed by Trench (1979) and Muscatine (1980a).

E. TRANSLOCATION TO HOST TISSUE

The actual translocation of extracellular products has been confirmed by either identifying products or their metabolites in host tissue, or in "inhibition" experiments (Section III,A). Unambiguous demonstration of translocation products in host tissue may be made difficult by the techniques used to separate host tissue and algae. In the case of zooxanthellae, these preparations (homogenates of host tissue plus algae, which are subsequently centrifuged to sediment algae) tend to favor the release of extracellular products from the algae. Hence, host metabolites of translocated compounds are taken as more conclusive evidence (Section IV).

1. *"Inhibition" Experiments*

"Inhibition" experiments provide the most convincing demonstrations that extracellular products found in isolated algae are in fact translocated to host tissue. This technique has been applied to two algae–invertebrate systems: marine coelenterates with zooxanthellae, and to *Convoluta roscoffensis.* (It has also been applied to marine slugs with symbiotic chloroplasts; Trench *et al.*, 1974.) There appear to be no reports of its having been used with respect to the symbioses with zoochlorellae.

In general, these experiments have confirmed such translocation. Lewis and Smith (1971) investigated 12 species of reef-dwelling coelenterates and found that in all cases glycerol, alanine, and glucose stimulated the release of [^{14}C]glycerol, -alanine, or -glucose.

Muscatine *et al.* (1974) applied the technique to *C. roscoffensis,* and found that alanine stimulated the release of not only [^{14}C]alanine, but also several other amino acids which were labeled. They interpreted this observation to indicate that the "exchange diffusion" process involved a transport system which was effective for structurally similar amino acids. As discussed above, these experiments probably indicate the existence of translocated compounds (or their metabolites) within host cells, rather than the actual transport of compounds from symbiont to host during the experiments.

2. *Factors Influencing Release and Translocation*

Several factors influence the release of extracellular products by isolated algae, but the extent to which these operate in intact associations is unclear. The effects of pH on zoochlorellae have been discussed; Pardy (1980) has also shown that the zoochlorellae of the freshwater bivalve *Anadonta* also show a similar pattern of pH-dependent release, although the levels of release are notably less than in *Hydra,* and the identity of the released product(s) is not known. It has been suggested that the host could regulate the translocation of maltose by controlling the pH of the algal environment. There are difficulties in directly testing this supposition, and there is no evidence that the host can control maltose release. For instance, maltose translocation continues when green hydra starve, and rates of translocation appear to be similar to those of well-fed hosts (Mews, 1980). The host must exert some influence over algal translocation, since maltose release by cultured algae is stimulated by reintroduction into host tissue (Jolley and Smith, 1978). This may be a long-term adaptation process, similar in time scale to the changes in morphology and cell surface characteristics which occur in heterologous hydra–algae combinations (Pardy, 1976; Pool, 1979).

Glycerol release by isolated zooxanthellae is stimulated by the addition of host homogenate (Muscatine, 1967; Trench, 1971c; Muscatine *et al.*, 1972). This discovery suggested a "host factor" which might mediate translocation in intact associations. Little is known about this putative factor, or how it exerts its stimulatory effect. It is heat-labile, and apparently is nonspecific, since tissue from *Tridacna* stimulates release by coral zooxanthellae, and vice versa (Muscatine, 1967). Trench (1971c) observed that tissue from aposymbiotic (algae-free) sea anemones was not stimulatory, but if these anemones were infected with zooxanthellae, the tissue would stimulate release.

"Host factors" may also stimulate release by *Convoluta* symbionts. Muscatine *et al.* (1974) and Boyle and Smith (1975) observed that algae in tissue homogenates released more amino acids than did isolated symbionts. Host homogenates do not stimulate maltose release by *Hydra* symbionts; if anything, release is reduced in its presence (Cernichiari *et al.*, 1969; Mews, 1980).

IV. Translocation and Its Significance for Host Tissue

A. Extent of Translocation

The amount of fixed carbon which is translocated from algal symbionts to host tissue is usually estimated by the distribution of ^{14}C between algae and host samples in photosynthesis experiments. Because of the loss of soluble compounds during processing, autoradiographic studies are less preferable than direct assays of host tissue homogenates and sedimented algal fractions. Muscatine and Cernichiari (1969), Trench (1971a), and Mews (1980) have pointed out some of the pitfalls of these experiments. These include contamination between animal and algal fractions, translocation which may occur during separation procedures, and the effects of heterotrophic fixation and changes in the specific activity of ^{14}C compounds during the course of an experiment.

Eisenstadt (1971) and Mews (1980) have used this approach to estimate translocation in *Hydra viridis*. Eisenstadt (1971) estimated that in the "Florida" strain of green hydra 30% of the fixed carbon moved to the animal within 1 to 3 hours, and 40% within 5 hours. Mews (1980) repeated these observations with similar results (Fig. 7). Over a 2 hour incubation with $^{14}CO_2$ the animal fraction contained 40–50% of the label found in algae, or 25–30% of the incorporated label.

The amount of translocation in other associations involving *Chlorella*-like algae may be substantially less than in green hydra. Of total fixed carbon 9–17% may be translocated to host tissue in the freshwater sponge *Ephydatia fluviatilis* (Wilkinson, 1980). Measurements of translocation to algae-free ectoderm in tentacles of the sea anemone *Anthopleura xanthogrammica* indicate that less than

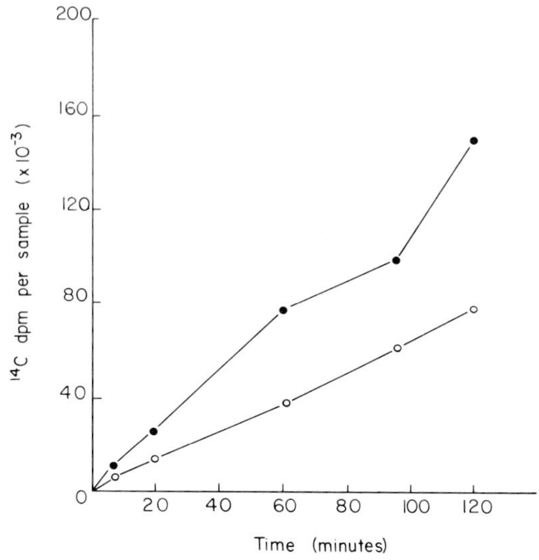

FIG. 7. Distribution of ^{14}C between animal and algal fractions of *H. viridis*, based on analyses of host supernatants and algal pellets. Closed circles, algal fraction; open circles, animal fraction. [Reproduced from Mews (1980) by permission of the Royal Society and L. K. Mews.]

4% of total fixed carbon is involved (Muscatine, 1971; O'Brien, 1980). These zoochlorellae release less than 2% of their fixed carbon in isolation experiments (Muscatine, 1971). There appear to be no other estimates of translocation in other *Chlorella* associations, although it apparently occurs in *Anadonta* (Pardy, 1980).

Similar studies have been carried out with associations with zooxanthellae. Muscatine and Cernichiari (1969) carried out 24 hour incubations of the coral *Pocillopora damicornis* and found that 25–40% of the incorporated label had been translocated to host tissue. Trench (1971a) reported values of 20–25% for the zoanthid *Palythoa townsleyi* and as high as 50% in *Anthopleura elegantissima* in 24 hour experiments. Von Holt and von Holt (1968a) found a range of values of 25–40% in four species of reef coelenterates following 3 hour incubations.

Boyle and Smith (1975) used another approach to measure translocation in *Convoluta roscoffensis*. Because of the fragility of the algae, physical separation of symbiont and host tissue would not produce reliable estimates. They labeled intact worms with $^{14}CO_2$, and then at intervals induced the loss of symbionts by increasing CO_2 tensions. By measuring the amount of label lost in this fashion, they found that more than 50% of the original label had been translocated with 4 days. A similar approach might be applied to symbioses in which algae-free hosts can be produced in a short time, with little damage to the algae.

B. Metabolism of Translocated Products by Host Tissue

1. *Metabolism of Maltose and Other products by H. viridis*

Despite the number of invertebrates which harbor zoochlorellae, the metabolism of translocated carbon has been studied extensively only in green hydra, perhaps reflecting the ease of experimentation with this organism. The results verify the translocation of maltose and glucose, and probably alanine, in the intact association.

The classic observations of the phenomenon were made by Muscatine and Lenhoff (1963). They demonstrated that translocated carbon became distributed in the major chemical fractions of algae-free ectoderm: after 48 hours of labeling, they found that 41% of the label was associated with the insoluble protein fraction.

It might be expected that glycogen synthesis would be a major use of translocated maltose. Roffman and Lenhoff analyzed the distribution of ^{14}C in green hydra samples which included gastrodermal tissue. Following a 10-minute pulse, label left the small molecule pool and became mainly associated with the polysaccharide fraction, comprising 30–40% of the incorporated label for 0.5 to 4 hours after the pulse. Lesser amounts of label occurred in lipid, protein, and nucleic acid fractions. A preliminary analysis indicated glycogen in the polysaccharide fraction.

Mews (1980) has provided the most detailed analysis of the fate of photosynthetic carbon in this strain of *H. viridis,* employing an enymatic assay specific for glycogen. The results of glycogen synthesis for 4 hours after a 20-minute pulse are shown in Fig. 8. Glycogen was synthesized rapidly from small-molecular-weight precursors, presumably mostly maltose; by 4 hours, half of the translocated label had been incorporated. Label in protein and nucleic acid amounted to less than 10% of the translocated ^{14}C. Of interest was the finding that the "European" hydra incorporated less label into glycogen.

The greater incorporation into protein which Muscatine and Lenhoff (1963) reported may have resulted from the analysis of only ectodermal tissue, which probably contains a higher proportion of protein, and the extended period of incubation. During the 2 days of the experiment these hydra possibly mobilized translocated carbohydrate for amino acid synthesis.

2. *Metabolism of Products of Zooxanthellae: Lipid Synthesis*

The metabolism of glycerol has been the major focus on the fate of translocated products in zooxanthellae associations. The metabolism of alanine will be

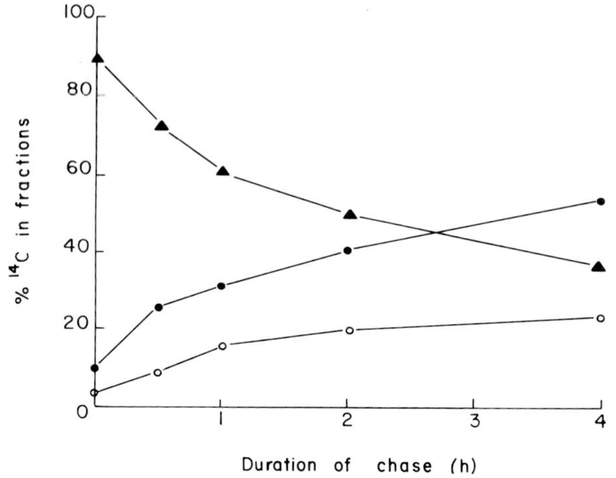

FIG. 8. Synthesis of glycogen from the small-molecular-weight pool (principally maltose) by animal tissue in *H. viridis* following a 20 minute pulse with $^{14}CO_2$. Triangles, soluble fraction; closed circles, glycogen; open circles, proteins and nucleic acids. [Reproduced from Mews (1980) by permission of the Royal Society and L. K. Mews.]

considered in Section V. Metabolism of non-amino acids has received little attention, and it is perhaps relevant that Lewis and Smith (1971) could not demonstrate the translocation of these compounds, although they did not examine the associations from which Trench (1971b) described these compounds as extracellular products.

The role of glycerol in lipid synthesis has been emphasized in numerous studies. Von Holt and von Holt (1968a) reported that radioactive lipids comprised one-half of the label in ethanol-soluble fractions of host tissue after a 3 hour incubation of the coral *Scolymia lacera*. Muscatine and Cernichiari (1969) incubated *Pocillopora damicornis* for 24 hours and found that 27–43% of the label in host tissue was ethanol-soluble; the lipid in this fraction yielded labeled glycerol upon deacylation. Trench (1971a) obtained similar results after 24 hour incubations of *Anthopleura elegantissima* and *Palythoa townsleyi*, with virtually all of the label in lipid fractions associated with glycerol moities. Schmitz and Kremer (1977) found large amounts of glycerol in lipid fractions of whole corals within 90 minutes, but their procedure did not discriminate between label in algae and in host tissue. Crossland *et al.* (1980b) measured the diurnal pattern of photosynthetic incorporation into coral lipids and demonstrated that lipids were rapidly synthesized during periods of photosynthesis (Fig. 9).

Studies of lipid metabolism in corals raise several questions about the role of zooxanthellae. Short-term lipid synthesis from glycerol involves triglyceride formation, but the fatty acid moieties of triglycerides are not labeled in short-

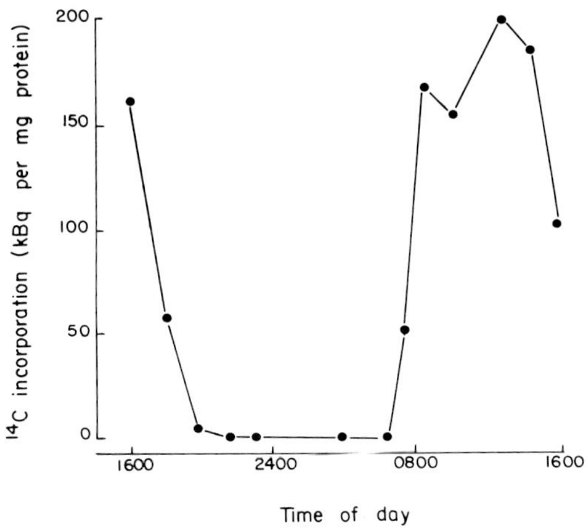

FIG. 9. The diurnal pattern of ^{14}C incorporation into tissue lipids of the coral *Acropora acuminata*. At each sampling time corals were incubated with $H^{14}CO_3^-$ for 60 minutes, and the chloroform-extractable components assayed. [Data of Crossland *et al.* (1980b) with permission of Springer-Verlag and C. J. Crossland.]

term photosynthesis experiments. Benson and Muscatine (1974) found that most of the photosynthetic incorporation into mucolipids involved triglyceride synthesis. Only 16% of the label occurred in wax esters, although they comprised 80% of this fraction.

Patton *et al.* (1977) investigated the role of symbiont acetate metabolism in this system. In contrast to bicarbonate incorporation, they found that both isolated zooxanthellae and intact corals would synthesize fatty acids from acetate in short-term (<5 hour) incubations, and that light incorporation was three times that in the dark. Blanquet *et al.* (1979) reported very similar findings for acetate metabolism by *Anthopleura elegantissima*. Based on their results, Patton *et al.* proposed an acetate recycling scheme in which host acetate, resulting from the catabolism of previously translocated carbohydrate, is recycled back to the algae. The algae utilize the acetate for fatty acid synthesis, and the fatty acids are then translocated back to the host. Wax ester and triglyceride synthesis using these acids would take place within host cells. Evidence for active lipid synthesis by zooxanthellae includes not only the biochemical studies cited above, but also ultrastructural observations of a variety of lipid inclusions within the algae. Figures 10 and 11 illustrate some of these inclusions, and also lipid inclusions in the host cell which are in close proximity to the algae. Crossland *et al.* (1980b) have interpreted some of these inclusions as wax ester deposits which may have crystallized during processing for microscopy.

This hypothesis raises several testable questions. If algae utilize previously

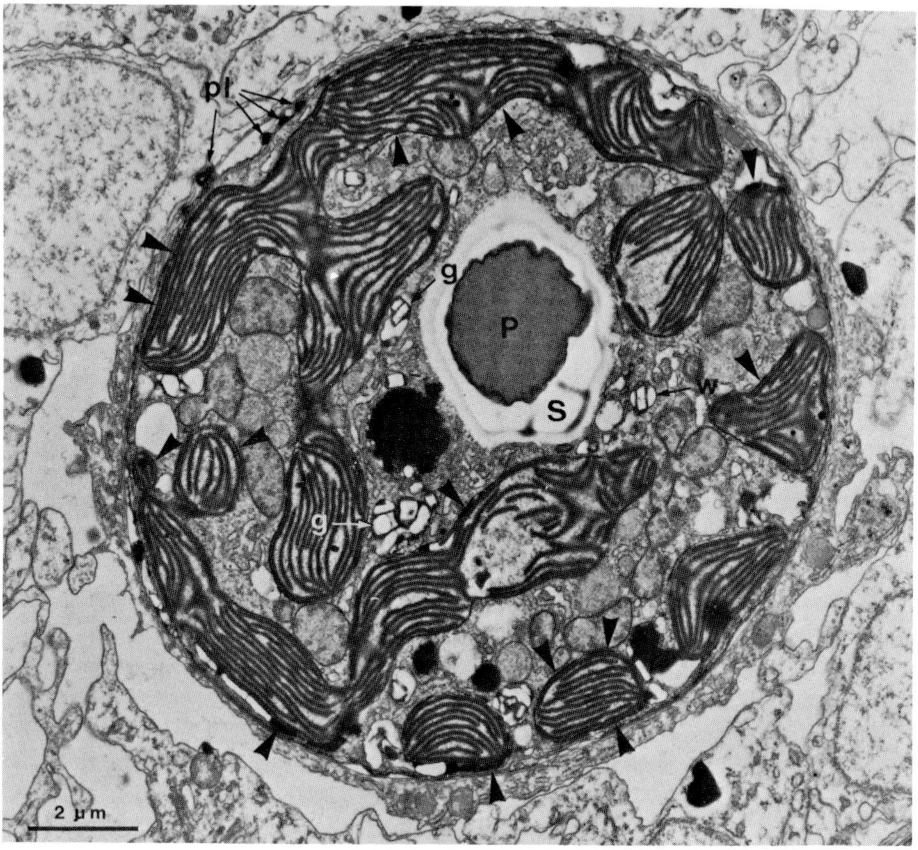

FIG. 10. A zooxanthella in the coral *Acropora acuminata*, showing lipid deposits on the chloroplast enveolpe (arrows) and near the plasmalemma (pl). P, Pyrenoid, S, starch; g, guanine crystals (note arrows). Scale bar = 2 µm. [Reproduced from Crossland *et al.* (1980) by permission of Springer-Verlag and C. J. Crossland.]

fixed carbon in fatty acid synthesis, it might be feasible to label corals with $H^{14}CO_3^-$ of high specific activity, and then subsequently test the algae for the translocation of labeled fatty acids. Studies of the extracellular products of [1-^{14}C]acetate metabolism should indicate whether these compounds are actually released by the algae. One unanswered question is the extent to which fatty acid synthesis occurs in host tissue; neither Patton *et al.* (1977) nor Blanquet *et al.* (1979) distinguished between host and symbiont synthesis, except by the observation of light enhancement.

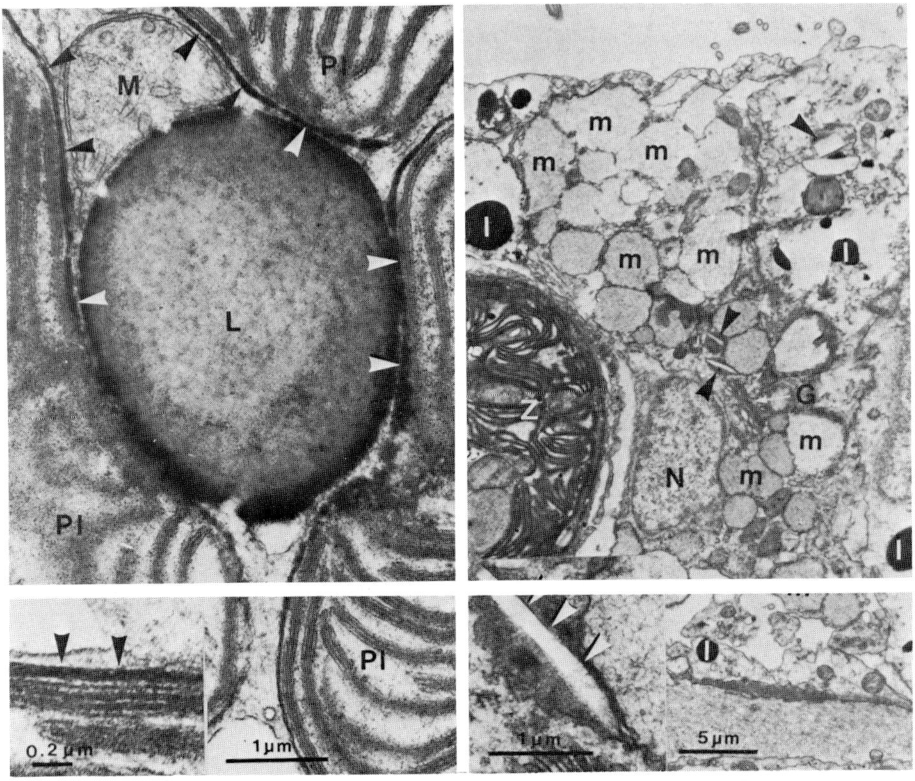

FIG. 11. Lipid deposits in zooxanthellae and coral tissue, from *A. acuminata*. Left micrograph: intrachloroplastic lipid droplet (L) in a zooxanthella; arrows indicate lipid deposits on chloroplast membrane, details in inset; Pl, plastid. Right micrograph: mucus gland of *A. acuminata* showing lipid droplets (l), mucus droplets (m), and cavities where wax ester crystals were extracted during processing (arrows); Z, Zooxanthella. Inset shows the detail of a wax ester crystal. [Reproduced from Crossland *et al.* (1980b) by permission of Springer-Verlag and C. J. Crossland.]

3. *Metabolism of Translocated Compounds in Convoluta roscoffensis*

The major products translocated by the symbionts of *C. roscoffensis* are amino acids. Muscatine *et al.* (1974) demonstrated the incorporation of these amino acids into host tissue by two sorts of experiments. They obtained algae-free eggs from adults which had been labeled with $^{14}CO_2$; acid hydrolysis of the eggs produced a variety of radioactive amino acids. A second line of evidence was obtained by the collection of mucus from worms which were allowed to photosynthesize for 1 hour. At the end of the incubation worm mucus was highly labeled; incorporation of photosynthetic products into mucus declined by 90% within 4 hours, indicating that translocated products were rapidly converted into mucus. Hydrolysis of the mucus yielded labeled amino acids and glucose.

The role of the symbionts in host lipid metabolism is similar to that of zooxanthellae in corals, except that it appears that the worms depend upon their symbionts to supply fatty acids and cholesterol. Meyer *et al.* (1979) incubated larval (algae-free) worms with [1-^{14}C]acetate and found that the larvae could not synthesize fatty acids, although they could lengthen existing fatty acid chains. (Presumably the larvae obtain sufficient fatty acids in the diet.) Adult worms and cultured symbionts do synthesize fatty acids from acetate, but only in the light. Fatty acids which are made in the intact association are primarily those which are synthesized by the isolated algae; ''worm'' fatty acids appear to be derived from these. Similar results were obtained with cholesterol, which is converted by worm tissue to 24-methylene chlolesterol, the major sterol in the adult. As with corals, it would be interesting to know if isolated algae translocate products of acetate metabolism.

C. Physiological Consequences of Translocation for Hosts

Numerous studies have demonstrated the benefit of symbionts to hosts. This section focuses on the use of translocated products during periods of reduced food supply, the fate of lipids and mucus in corals, and the role of translocated products in the establishment of associations. The roles of products of zooxanthellae in enhancing coral calcification have been reviewed by Vandermeulen and Muscatine (1974) and by Johnston (1980); Taylor (1973a) and Trench (1979, 1980) have summarized data on the involvement of symbionts in host developmental processes.

It has been difficult to evaluate precisely the nutritional contribution of algal symbionts to their hosts except under conditions of limited food or starvation; few instances exist such as that of adult *C. roscoffensis,* which do not feed and presumably are completely dependent upon their algae. Recent measurements have been made of O_2 exchange in various symbiotic invertebrates (*Hydra viridis:* Pardy and Dieckmann, 1975; Pardy and White, 1977; *Spongilla lacustris,* Gilbert and Allen, 1973; *Dalyellia viridis:* Heitkamp, 1979; *Convoluta roscoffensis:* Nozawa *et al.*, 1972; *Tridacna maxima:* Trench *et al.*, 1981a), particularly in corals (reviewed by Muscatine, 1980a). Muscatine (1980a) has discussed the use of such data in estimates of algal contribution. He and his colleagues (Muscatine *et al.*, 1981) have derived equations which utilize respirometry and translocation data to yield *in situ* measurements of symbiont contribution to host metabolism. Such an approach offers promise for the evaluation of the effects of nutritional state and environmental factors.

1. *Benefit during Starvation*

The most unambiguous demonstrations of the advantages of symbiont translocation to hosts are studies of survival or maintenance during periods of reduced food supply, although these have investigated a relatively small number of associations. Unequivocal evidence for enhanced survival during starvation has come from work on green hydra (Muscatine and Lenhoff, 1965; Kelty and Cook, 1976; Rahat and Reich, 1980), reduced weight loss during starvation of symbiotic sea anemones (Muscatine, 1961; Sebens, 1980) and zoanthids (Sebens, 1977). The accepted interpretation of these findings is that translocated photosynthate serves as an energy source in lieu of host tissue reserves.

In *Hydra viridis,* maltose translocation continues during starvation at rates which appear to be similar to those in well-fed animals (Eisenstadt, 1971; Mews, 1980), even though host energy demands are probably lower during starvation. Glycogen stores in these animals decrease for the first week of starvation (Mews, 1980; Cook and Kelty, 1982), but subsequently increase. After 2 weeks, green hydra may actually contain more glycogen than at the onset of starvation (Fig. 12). Translocation thus appears to provide starving animals with more carbon than they can utilize. The increase in glycogen apparently results from decreased energy demand coupled with the lack of additional inorganic nutrients.

Symbiotic hydra have lower rates of protein catabolism during starvation than aposymbionts (Muscatine and Lenhoff, 1965); aposymbionts rapidly catabolize all available tissue reserves (Fig. 13; Cook and Kelty, 1982). The consumption of tissue protein by starving aposymbionts has rather drastic consequences. These animals begin to lose the ability to feed within 10 days, while starving

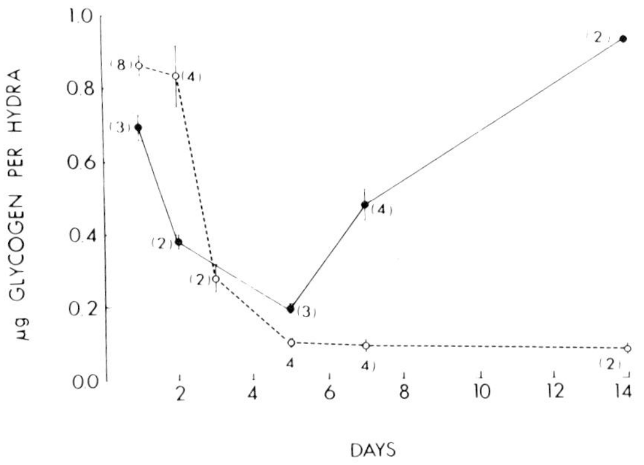

FIG. 12. Glycogen levels in symbiotic (closed circles) and aposymbiotic (open circles) *Hydra viridis* during starvation in the light. Sample sizes in parentheses. [Reproduced from Cook and Kelty (1982) with permission of Alan R. Liss.]

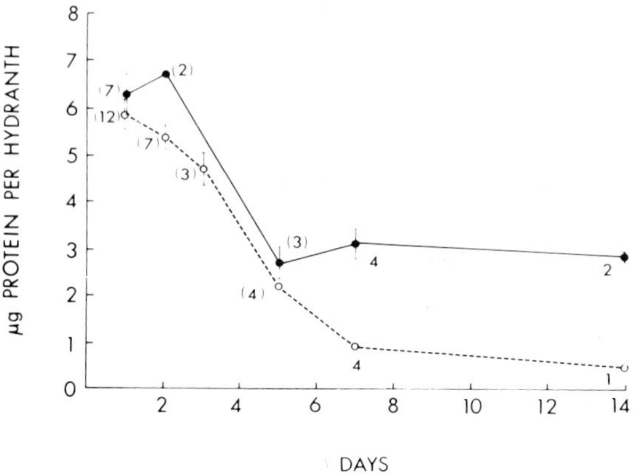

FIG. 13. Protein levels in starving symbiotic and aposymbiotic *H. viridis*. Conventions as in Fig. 12. [Reproduced from Cook and Kelty (1982).]

green hydra retain this ability for at least 6 weeks; interestingly, the effects of starvation are much less severe for nonsymbiotic hydra (Kelty and Cook, 1976).

There have not been comparable studies of the levels of storage compounds during starvation in other symbiotic invertebrates. The major energy stores of sea anemones and corals which have zooxanthellae are lipids, principally wax esters (Benson and Muscatine, 1974; Benson *et al.*, 1978). The available data suggest that lipid stores in starving sea anemones (Fitt and Pardy, 1981) and corals (Szmant-Froelich and Pilson, 1980) are analogous to glycogen stores in green hydra.

2. *Lipid Release by Corals*

Perhaps the most striking aspect of lipid metabolism in corals is that a sizable percentage of the total carbon fixed—as much as 60%—is exported as mucolipid and mucopolysaccharide (Crossland *et al.*, 1980a). Crossland *et al.* (1980b) followed this release for 33 hours in a continuous labeling experiment, and showed that the release follows a strong diurnal pattern (Fig. 14) which correlated with the pattern of lipid synthesis (Fig. 9). Wax esters comprised between 40 and 80% of this material, and the energy lost from the coral may be as much as 300 μg/mg protein per day. Such a luxury—used in feeding, for reducing overgrowth by epibionts and bacteria, and for hygienic purposes—is made possible by the extensive translocation of glycerol and other products.

3. *Role in "Recognition" by Host Cells*

The interactions which occur when symbioses between algae and invertebrates are established have received much recent attention (Smith, 1979, 1981; Trench,

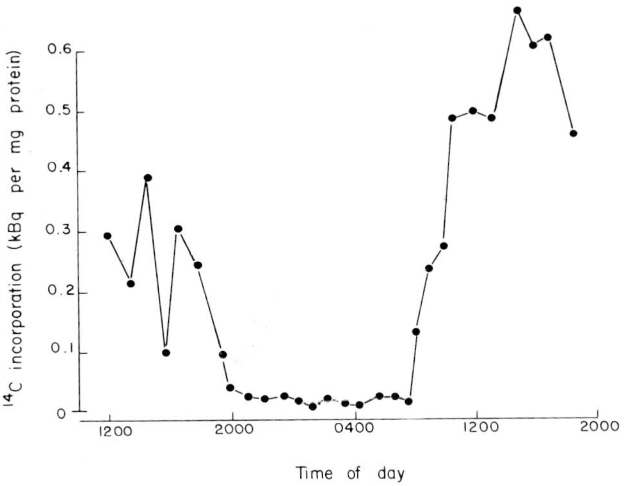

FIG. 14. The diurnal pattern of release of lipids (as mucolipids) from the coral *Acropora acuminata* during continuous labeling with $^{14}CO_2$. The mucolipids were collected from the seawater incubation medium in a flow-through system. [Data of Crossland *et al.* (1980b) with permission of Springer-Verlag and C. J. Crossland.]

1979, 1980; Cook, 1980a; Trench *et al.*, 1981b). It now appears that translocated carbohydrate (and other products?) are probably involved in the suite of events which lead to the formation of a functional symbiosis. Smith (1981) has discussed the more general aspects of this problem.

Hydra viridis has proved to be the most convenient experimental system for such studies. Among a range of potentially infective particles, zoochlorellae which release maltose are the most "acceptable" to host cells (Pardy and Muscatine, 1973; Jolley and Smith, 1980; McAuley and Smith, 1982). Muscatine *et al.* (1975) and Smith (1979) have summarized the various stages of the infective process; these include the interaction with host cell lysosomes and the transport of algae within host cells. Maltose release appears to be related to both phenomena. Treatment of symbionts with polyamines or the photosynthetic inhibitor DCMU reduces or abolishes maltose release by the algae, and interferes with the normal perturbation of phagosome–lysosome fusion (Hohman *et al.*, 1982). The transport of algae within host cells is also reduced if maltose release is affected (Hohman *et al.*, 1982; McAuley and Smith, 1982). How maltose affects these processes is not clear. The transport system is vinblastine-sensitive, and thus appears to involve microtubule arrays (Cooper and Margulis, 1978; Jolley and Smith, 1980; Hohman *et al.*, 1982; McAuley and Smith, 1982). It has been suggested (Cook *et al.*, 1978) that translocated carbohydrate might stabilize these arrays.

Translocated products may be important in the establishment of other symbioses with algae. Weis (1979, 1981) has suggested that the persistence of ingested algae in *Paramecium bursaria* is related to sugar release by the algae. Perhaps sugar release is related to phagosome–lysosome interactions in this system as well (Karakashian and Rudzinska, 1981). In experiments involving the infection of polyps of the jellyfish *Cassiopeia xamachana* with zooxanthellae, Trench *et al.* (1981b) found that infectivity decreased with the time that algae had been isolated. While this would correlate with the decrease in glycerol release with isolation time, it seems more likely that host tissue attached to isolated algae is involved in recognition in this system. Since a variety of zooxanthellae strains which do release glycerol do not infect polyps of *Cassiopeia,* glycerol release per se does not appear to be a factor in recognition (Trench *et al.,* 1981b; R. K. Trench, personal communication).

V. Other Metabolic Interactions

A. Inorganic Nutrients

The recycling of nutrients between hosts and symbiotic algae is one of the major features of algae–invertebrate symbioses (Lewis and Smith, 1971; Smith, 1977, 1981). Emphasis has been placed on the flux and metabolism of inorganic nutrients. This question has ecological importance, as associations with zooxanthellae flourish in nutient-poor tropical waters. Typical schemes involve the use of host excretory products by the algae, which subsequently translocate these nutrients to host tissue in the form of other compounds (e.g., Lewis and Smith, 1971; Smith, 1979). The acetate recycling model (Patton *et al.,* 1977), cited in the previous section, is an example.

1. *Uptake and Retention of Dissolved Inorganic Nitrogen, Phosphorus, and Sulfur*

Symbiotic invertebrates are notable for the ability to take up dissolved phosphate, ammonium, and nitrate from low concentrations, and to retain these nutrients rather than excrete them (reviewed by Muscatine, 1980b). The uptake by corals of both orthophosphate (OPO_3^{3-}; ''reactive phosphorus''; D'Elia, 1977) and ammonium (Muscatine and D'Elia, 1978) is light dependent, indicating the role of the algae in the uptake process. The role of host cells in uptake is not clear. Most studies utilize nonsymbiotic species as controls, rather than aposymbiotic individuals of symbiotic species; thus, uptake mechanisms of host cells are not examined directly. In one comparison of ammonium uptake by symbiotic and aposymbiotic corals (*Madracis mirabilis*), no uptake was detected by aposymbionts, while the uptake of ammonium by the intact association was greater than that by isolated algae (Muscatine *et al.*, 1979). These observations suggest that coral tissue lacks specific uptake mechanisms for ammonium, but the coral–algae association has a synergistic effect on ammonium uptake by the algae.

Green hydra tissue does have the capacity to take up sulfate from solution (Cook, 1976, 1981). Following a 24 hour incubation with $^{35}SO_4{}^{2-}$, aposymbiotic *H. viridis* had 15 times the tissue-specific ^{35}S as did nonsymbiotic hydra, while the intact association contained even more label. Analogous to the retention of nitrogen and phosphorus by marine symbioses, symbiotic hydra retain most of this label; aposymbionts gradually excrete it (Fig. 15). Sulfate is rarely taken up from solution by animal cells, and is rarely excreted. Both algae and symbiotic bacteria may enhance the uptake of phosphate from solution in green hydra (Wilkerson, 1980a,b).

2. *Metabolism and Recycling of Nitrogen and Sulfur*

Nitrate, ammonium, and sulfate are primarily metabolized by plant cells and not by animal tissue. Symbiotic algae assimilate these compounds into forms which can be utilized by the host. This pattern is the basis for schemes for the recycling of nitrogen in corals (Lewis and Smith, 1971) and *Convoluta roscoffensis* (Smith, 1979), and of sulfur in green hydra (Cook, 1980b).

Nitrogen recycling in corals has received the most attention. Crossland and Barnes (1977) reported nitrate reductase activity in zooxanthellae, and the work

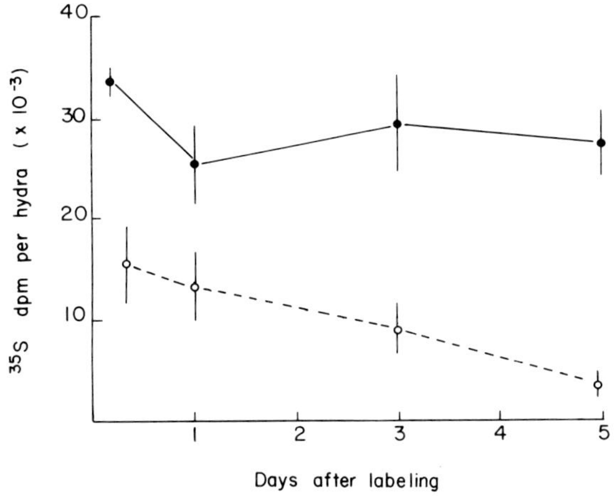

FIG. 15. The retention of ^{35}S by symbiotic (solid circles) and aposymbiotic (open circles) *Hydra viridis* (Florida strain). The hydra were labeled for 24 hours with radiosulfate, then transferred to sulfate-free medium (0 time). [Data of Cook (1981) by permission of the New York Academy of Sciences.]

of Marian (1982) provides direct evidence for the incorporation of this reduced nitrogen into host tissue. Lewis and Smith (1971) originally proposed that translocated alanine could serve as a vehicle by which ammonium nitrogen could be recycled to host tissue. Muscatine (1980b) has discussed how this model relates to current views of ammonium assimilation in plants. Glutamate is one of the photosynthetic products of zooxanthellae which is not translocated in short-term experiments. It is thought that ammonium assimilation occurs via glutamine synthetase (GS) and glutamate synthase (GOGAT) pathways. Glutamate serves as an ammonium acceptor, producing glutamine via GS. GOGAT catalyzes the transamination of α-ketoglutarate from glutamine, yielding more acceptor and amino acids available for transamination. Of the latter, only alanine is translocated to host tissue. Support for the GS–GOGAT pathway is provided by observations that increasing ammonium concentrations increase both the amount of ^{14}C incorporated into glutamine during photosynthesis (Muscatine, 1980b) and the amount of alanine which is translocated (Lewis and Smith, 1971). The data of Boyle and Smith (1975) indicate that a similar system of ammonium assimilation may operate in *Convoluta*.

In green hydra, there is reasonable evidence that sulfate reduction by zoochlorellae provides host tissue with sulfur amino acids. As with nitrogen assimilation, the enzymes for assimilatory sulfate reduction are not found in animal cells. Comparison of the incorporation of sulfur administered as sulfate indicates that sulfate reduction products of the algae are utilized by hydra tissue for protein synthesis (Figure 16; Cook, 1980b). However, translocation of sulfate reduction products (presumably amino acids) has not yet been confirmed by work with isolated algae, or in "inhibition" experiments. However, work with sulfate metabolism in hydra must proceed with the caveat that bacterial symbionts, which are associated with a number of green hydra strains (e.g., Wilkerson, 1980a,b), will influence sulfate metabolism. How these bacteria affect metabolism of sulfur (or other elements) in either hydra or algae is a question which will require future investigation.

B. "Reverse Translocation" from Host to Algae

Implicit in the recycling schemes discussed above is the assumption that metabolites flow from host to symbionts. Intracellular algae are isolated from the external environment by at least two layers of animal membranes (vacuole and plasmalemma), in addition to host cell cytoplasm. It seems evident that the host to some extent must supply nutrients such as nitrogen, phosphorus, and sulfur to the algae. This is particularly so for green hydra and others which are routinely cultured in media which lack these nutrients. What is perhaps an unanswerable question is whether the compounds which are passed to the algae are normal excretory products, or are other compounds. Unlike the situation of algal prod-

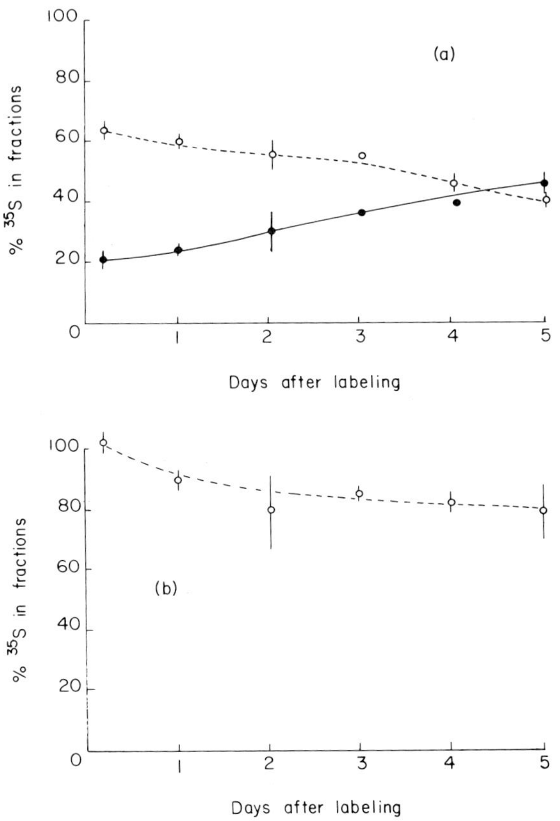

FIG. 16. The fate of ^{35}S, administered as sulfate, in animal tissue of (a) symbiotic and (b) aposymbiotic *H. viridis*. Open circles, soluble (small-molecular-weight) fraction; closed circles, protein (not shown in b; incorporation < 10%). [From Cook (1980b) by permission of Walter de Gruyter.]

ucts, the identification of putative "reverse translocation" products would appear to be a difficult problem.

Various estimates have been made of the extent of "reverse translocation," but precise measurement depends on the purity of "algal" samples. Such pellets can contain significant amounts of host tissue even after repeated washing (Mews, 1980). Cook (1971) estimated that perhaps 20% of the ^{35}S ingested by sea anemones was associated with algae, based on analyses of tentacles; the observation that most of this label was soluble was an indication that contamination by insoluble animal material was negligible. Preliminary observations on the heterotrophic use of carbon by green hydra symbionts indicated that the algae utilize perhaps 10% of the ingested carbon (Cook, 1972). More recent studies indicate that *Hydra* symbionts can utilize amino acids and nucleotides in ingested food (Thorington and Margulis, 1981). Particularly interesting in this work is the finding that ingested thymidine can be used for DNA synthesis by the algae, since there is evidence that *Chlorella* may lack thymidine kinases. It has been suggested that the zooxanthellae of a zoanthid may be dependent on either host feeding or the environment to supply glycine, since the symbionts did not incorporate $^{14}CO_2$ into glycine during a 3 hour experiment (von Holt, 1968). To my knowledge, this supposition has not been rigorously tested.

C. Other Host Influences on Algal Physiology

In addition to nutrient supply, hosts may influence their symbionts in other ways. The nutritional state of the host influences algal storage products (Pardy, 1976), and a number of morphological features vary according to the host cell environment (Pardy, 1976; Pool, 1979). Control of symbiont cell division is currently a topic of much interest. In some strains of hydra, it appears that nutrient levels in host cells influence algal populations (Muscatine and Pool, 1979; Muscatine and Neckelmann, 1981). The mitotic activity of host cells and the availability of "space" are also factors which influence algal cell division in *H. viridis* (McAuley, 1981b).

Phipps and Pardy (1982) have provided evidence that hydra "enhances" symbiont photosynthesis by removing O_2 which might otherwise induce photorespiration, although photorespiration (particularly *in situ*) has not been demonstrated, and these authors did not consider the possibility that symbiotic algae might be adapted to the O_2 conditions which may exist within a host cell. Boyle and Smith (1975) found that O_2 tensions influenced the products released by *Convoluta* symbionts during "inhibition" experiments, but their results appear to be related to host lactate metabolism.

One suggestion which has repeatedly occurred in the literature is that hosts "digest" their symbionts. The evidence for this appears to be at best equivocal, and in green hydra (Hohman *et al.*, 1982) and *Paramecium bursaria* (Karakashian and Rudzinska, 1981) it is established that symbionts normally perturb the fusion of lysosomes with vacuoles containing algae. The possibility that such digestion occurs in *Tridacna* (e.g., Fankboner, 1971) appears to be unlikely (Trench *et al.*, 1981a).

VI. Conclusions

It is apparent that much is known about the interactions which occur between invertebrates and symbiotic algae, at least on organismal and cellular levels. In this article I have indicated some of the directions which future research may take along these lines. The larger questions are molecular and ecological ones. Our lack of knowledge of the enzyme systems involved in the synthesis of extracellular products of symbiotic algae indicates the need for such studies, particularly in comparison to free-living algae. At the other end of the spectrum the ecological implications of symbiosis, such as the fate and significance of mucus and lipid exported from corals, remain to be critically examined.

Acknowledgments

I am grateful to R. A. Shepard and the staff of the Marine Science Institute, Northeastern University, for providing facilities, and the to Harvard University librarians for their assistance and cooperation. I am indebted to L. Muscatine and R. K. Trench for their many suggestions which have improved the manuscript. This is contribution number 111 of the Marine Science Institute, Northeastern University.

References

Atkinson, A. W., Gunning, B. E. S., and John, P. C. L. (1972). *Planta* **197,** 1.
Benson, A. A., and Muscatine, L. (1974). *Limnol. Oceanogr.* **19,** 810.
Benson, A. A., Patton, J. S., and Abraham, S. (1978). *Atoll Res. Bull.* **220,** 33.
Blanquet, R. S., Nevezel, J. C., and Benson, A. A. (1979). *Mar. Biol.* **54,** 185.
Boyle, J. E., and Smith, D. C. (1975). *Proc. R. Soc. London Ser. B* **189,** 121.
Cernichiari, L., Muscatine, L., and Smith, D. C. (1969). *Proc. R. Soc. London Ser. B* **173,** 557.
Cook, C. B. (1971). *In* "Experimental Coelenterate Biology" (H. M. Lenhoff, L. Muscatine, and L. V. Davis, eds.), pp. 218–224. Univ. of Hawaii Press, Honolulu.
Cook, C. B. (1972). *Biol. Bull.* **142,** 236.
Cook, C. B. (1976). *In* "Coelenterate Ecology and Behavior" (G. O. Mackie, ed.), pp. 415–422. Plenum, New York.
Cook, C. B. (1980a). *In* "Cellular Interactions in Symbiosis and Parasitism" (C. B. Cook, P. W. Pappas, and E. D. Rudolph, eds.), pp. 47–74. Ohio State Univ. Press, Columbus.
Cook, C. B. (1980b). *In* "Endocytobiology: Endosymbiosis and Cell Research" (W. Schwemmler and H. E. A. Schenk, eds.), pp. 249–257. De Gruyter, Berlin.
Cook, C. B. (1981). *Ann. N.Y. Acad. Sci.* **361,** 273.
Cook, C. B., and Kelty, M. D. (1982). *J. Exp. Zool.* **222,** 1.
Cook, C. B., D'Elia, C. F., and Muscatine, L. (1978). *Cytobios* **23,** 17.
Cooper, G., and Margulis, L. (1978). *Cytobios* **19,** 7.
Craigie, J. S., McLachlan, J., and Majak, R. C. (1966). *Can. J. Bot.* **44,** 1247.
Crossland, C. J., and Barnes, D. J. (1977). *Comp. Biochem. Physiol.* **57B,** 151.
Crossland, C. J., Barnes, D. J., Cox, T., and Devereux, M. (1980a). *Mar. Biol.* **59,** 181.
Crossland, C. J., Barnes, D. J., and Borowitzka, M. A. (1980b). *Mar. Biol.* **60,** 81.
D'Elia, C. F. (1977). *Limnol. Oceanogr.* **22,** 301.
Drew, E. A., and Smith, D. C. (1967). *New Phytol.* **66,** 389.
Eisenstadt, E. (1971). *In* "Experimental Coelenterate Biology" (H. M. Lenhoff, L. Muscatine, and L. V. Davis, eds.), pp. 202–208. Univ. of Hawaii Press, Honolulu.
Fankboner, P. V. (1971). *Biol. Bull.* **141,** 222.

Fisher, C. R., and Trench, R. K. (1980). *Biol. Bull.* **159,** 636.
Fitt, W. K., and Pardy, R. L. (1981). *Mar. Biol.* **61,** 199.
Fitt, W. K., Chang, S. S., and Trench, R. K. (1981). *Bull. Mar. Sci.* **31,** 436.
Freudenthal, H. D. (1962). *J. Protozool.* **9,** 45.
Gilbert, J. J., and Allen, H. L. (1973). *Verh. Int. Verein. Limnol.* **18,** 1413.
Gooday, G. W. (1970). *J. Mar. Biol. Assoc. U.K.* **50,** 199.
Heitkamp, U. (1979). *Arch. Hydrobiol.* **86,** 499.
Hohman, T. C., McNeil, P. L., and Muscatine, L. (1982). *J. Cell Biol.* (in press).
Holligan, P. M., and Gooday, G. W. (1975). *Symp. Soc. Exp. Biol.* **29,** 205.
Johnston, I. S. (1980). *Int. Rev. Cytol.* **67,** 171.
Jolley, E., and Smith, D. C. (1978). *New Phytol.* **81,** 637.
Jolley, E., and Smith, D. C. (1980). *Proc. R. Soc. London Ser B* **207,** 311.
Karakashian, S. J., and Rudzinska, M. A. (1981). *Exp. Cell Res.* **131,** 387.
Kelty, M. O., and Cook, C. B. (1976). *In* "Coelenterate Ecology and Behavior" (G. O. Mackie, ed.), pp. 409–414. Plenum, New York.
Kremer, B. P. (1980). *In* "Endocytobiology: Endosymbiosis and Cell Research" (W. Schwemmler and H. E. A. Schenk, eds.), pp. 89–96. De Gruyter, Berlin.
Lerch, K. A., and Cook, C. B. (1982). *Bull. Mar. Sci.* (Submitted).
Lewin, R. A. (1976). *Nature (London)* **261,** 697.
Lewin, R. A. (1977). *Phycologia* **16,** 217.
Lewis, D. H. (1973). *Biol. Rev.* **48,** 261.
Lewis, D. H., and Smith, D. C. (1971). *Proc. R. Soc. London Ser. B* **178,** 111.
Loeblich, A. R., III, and Sherley, J. L. (1979). *J. Mar. Biol. Assoc. U.K.* **59,** 195.
McAuley, P. J. (1981a). *J. Exp. Zool.* **217,** 23.
McAuley, P. J. (1981b). *J. Cell Sci.* **47,** 197.
McAuley, P. J., and Smith, D. C. (1982). In preparation.
McLaughlin, J. J. A., and Zahl, P. A. (1957). *Proc. Soc. Exp. Biol. Med.* **95,** 115.
McLaughlin, J. J. A., and Zahl, P. A. (1966). *In* "Symbiosis" (S. Henry, ed.), Vol. 1, pp. 257–297. Academic Press, New York.
McLaughlin, J. J. A., Zahl, P. A., Nowak, A., and Marchisotto, J. (1963). *Proc. Int. Congr. Protozool., 1st, Prague* pp. 205–206.
Marian, R. E. (1982). In preparation.
Mews, L. K. (1980). *Proc. R. Soc. London* **209,** 377.
Meyer, H., Provasoli, L., and Meyer, F. (1979) *Biochim. Biophys. Acta* **573,** 464.
Muscatine, L. (1961). *In* "The Biology of Hydra" (H. M. Lenhoff and W. F. Loomis, eds.), pp. 255–268. Univ. of Miami Press.
Muscatine, L. (1965). *Comp. Biochem. Physiol.* **16,** 77.
Muscatine, L. (1967). *Science* **156,** 516.
Muscatine, L. (1971). *Pac. Sci.* **25,** 13.
Muscatine, L. (1974). *In* "Coelenterate Biology: Reviews and New Perspectives" (L. Muscatine and H. M. Lenhoff, eds.), pp. 359–395. Academic Press, New York.
Muscatine, L. (1980a). *In* "Primary Productivity in the Sea" (P. G. Falkowski, ed.), pp. 381–402. Plenum, New York.
Muscatine, L. (1980b). *In* "Cellular Interactions in Symbiosis and Parasitism" (C. B. Cook, P. W. Pappas, and E. D. Rudolph, eds.), pp. 229–244. Ohio State Univ. Press, Columbus.
Muscatine, L., and Cernichiari, E. (1969). *Biol. Bull.* **137,** 506.
Muscatine, L., and D'Elia, C. F. (1978). *Limnol. Oceanogr.* **23,** 725.
Muscatine, L., and Greene, R. W. (1972). *Int. Rev. Cytol.* **36,** 137.
Muscatine, L., and Lenhoff, H. M. (1963). *Science* **142,** 956.
Muscatine, L., and Lenhoff, H. M. (1965). *Biol. Bull.* **129,** 316.
Muscatine, L., and Neckelmann, N. (1981). *Ber. Dtsch. Bot. Ges.* **94,** 571.

Muscatine, L., and Pool, R. R. (1979). *Proc. R. Soc. London Ser. B* **204,** 131.
Muscatine, L., and Porter, J. W. (1977). *Bioscience* **27,** 454.
Muscatine, L., Karakashian, S. J., and Karakashian, M. W. (1967). *Comp. Biochem. Physiol.* **20,** 1.
Muscatine, L., Pool, R. R., and Cernichiari, E. (1972). *Mar. Biol.* **13,** 298.
Muscatine, L., Boyle, J. E., and Smith, D. C. (1974). *Proc. R. Soc. London Ser. B* **187,** 221.
Muscatine, L., Cook, C. B., Pardy, R. L., and Pool, R. R. (1975). *Symp. Soc. Exp. Biol.* **29,** 175.
Muscatine, L., Matsuda, H., and Burnap, R. (1979). *Bull. Mar. Sci.* **29,** 572.
Muscatine, L., McCloskey, L. R., and Marian, R. (1981). *Limnol. Oceanogr.* **26,** 601.
Mutschler, P., and Wegmann, A. (1974). *Biol. Zentralbl.* **93,** 725.
Nozawa, K., Taylor, D. L., and Provasoli, L. (1972). *Biol. Bull.* **143,** 420.
O'Brien, T. L. (1978). *Trans. Am. Microsc. Soc.* **97,** 328.
O'Brien, T. L. (1980). *J. Exp. Zool.* **211,** 343.
Pardy, R. L. (1976). *J. Cell Sci.* **29,** 655.
Pardy, R. L. (1980). *Biol. Bull.* **158,** 349.
Pardy, R. L., and Dieckmann, C. L. (1975). *J. Exp. Zool.* **194,** 373.
Pardy, R. L., and Lewin, R. A. (1981). *Bull. Mar. Sci.* **31,** 817.
Pardy, R. L., and Muscatine, L. (1973). *Biol. Bull.* **145,** 565.
Pardy, R. L., and White, B. N. (1977). *Biol. Bull.* **153,** 228.
Parke, M., and Manton, I. (1967). *J. Mar. Biol. Assoc. U.K.* **47,** 445.
Patton, J. S., Abraham, S., and Benson, A. A. (1977). *Mar. Biol.* **44,** 235.
Phipps, D. W., and Pardy, R. L. (1982). *Biol. Bull.* **162,** 83.
Pool, R. R. (1979). *J. Cell Sci.* **35,** 367.
Rahat, M., and Reich, V. (1980). *In* "Developmental and Cellular Biology of Coelenterates" (P. Tardent and R. Tardent, eds.), pp. 465–469. Elsevier, Amsterdam.
Roffman, B., and Lenhoff, H. M. (1969). *Nature (London)* **221,** 381.
Schmitz, K., and Kremer, B. P. (1977). *Mar. Biol.* **42,** 305.
Schoenberg, D. A., and Trench, R. K. (1980a). *Proc. R. Soc. London Ser. B* **207,** 405.
Schoenberg, D. A., and Trench, R. K. (1980b). *Proc. R. Soc. London Ser. B* **207,** 429.
Schoenberg, D. A., and Trench, R. K. (1980c). *Proc. R. Soc. London Ser. B* **207,** 445.
Sebens, K. P. (1977). *Proc. Int. Coral Reef, Symp., 3rd* pp. 397–404.
Sebens, K. P. (1980). *Biol. Bull.* **158,** 370.
Smith, D. C. (1974). *Symp. Soc. Exp. Biol.* **28,** 437.
Smith, D. C. (1979). *Proc. R. Soc. London Ser. B* **204,** 115.
Smith, D. C. (1980). *In* "Cellular Interactions in Symbiosis and Parasitism" (C. B. Cook, P. W. Pappas, and E. D. Rudolph, eds.), pp. 197–227. Ohio State Univ. Press, Columbus.
Smith, D. C. (1981). *Ber. Dtsch. Bot. Ges.* **94,** 517.
Smith, D. C., Muscatine, L., and Lewis, D. (1969). *Biol. Rev.* **44,** 17.
Szmant-Froelich, A., and Pilson, M. E. Q. (1980). *J. Exp. Mar. Biol. Ecol.* **48,** 85.
Taylor, D. L. (1970). *Int. Rev. Cytol.* **27,** 29.
Taylor, D. L. (1971). *J. Mar. Biol. Assoc. U.K.* **51,** 301.
Taylor, D. L. (1973a). *Adv. Mar. Biol.* **11,** 1.
Taylor, D. L. (1973b). *Annu. Rev. Microbiol.* **27,** 171.
Taylor, D. L. (1974). *In* "Symbiosis in the Sea" (W. B. Vernberg, ed.), pp. 245–262. Univ. of South Carolina Press, Columbia.
Thorington, G., and Margulis, L. (1981). *Biol. Bull.* **160,** 175.
Trench, R. K. (1971a). *Proc. R. Soc. London Ser. B* **177,** 225.
Trench, R. K. (1971b). *Proc. R. Soc. London Ser. B* **177,** 237.
Trench, R. K. (1971c). *Proc. R. Soc. London Ser B* **177,** 251.
Trench, R. K. (1974). *Helgol. Wiss. Meeresunters.* **26,** 174.
Trench, R. K. (1975). *Symp. Soc. Exp. Biol.* **29,** 229.
Trench, R. K. (1979). *Annu. Rev. Plant Physiol.* **30,** 485.

Trench, R. K. (1980). *In* "Cellular Interactions in Symbiosis and Parasitism" (C. B. Cook, P. W. Pappas, and E. D. Rudolph, eds.), pp. 275–297. Ohio State Univ. Press, Columbus.
Trench, R. K. (1981). *Ann. N.Y. Acad. Sci.* **361,** 341.
Trench, R. K., Boyle, J. E., and Smith, D. C. (1974). *Proc. R. Soc. London Ser B.* **184,** 63.
Trench, R. K., Wethey, D. S., and Porter, J. W. (1981a). *Biol. Bull.* **161,** 180.
Trench, R. K., Colley, N. J., and Fitt, W. K. (1981b). *Ber. Dtsch. Bot. Ges.* **94,** 529.
Vandermeulen, J. H., and Muscatine, L. (1974). *In* "Symbiosis in the Sea" (W. B. Vernberg, ed.), pp. 1–19. Univ. of South Carolina Press, Columbia.
Von Holt, C. (1968). *Comp. Biochem. Physiol.* **26,** 1071.
Von Holt, C., and von Holt, M. (1968a). *Comp. Biochem. Physiol.* **24,** 83.
Von Holt, C., and von Holt, M. (1968b). *Comp. Biochem. Physiol.* **24,** 73.
Weis, D. (1979). *J. Protozool.* **26,** 117.
Weis, D. (1981). *Ber. Dtsch. Bot. Ges.* **94,** 547.
Wilkerson, F. P. (1980a). *In* "Endocytobiosis: Endosymbiosis and Cell Research" (W. Schwemmler and H. E. A. Schenk, eds.), pp. 269–277. De Gruyter, Berlin.
Wilkerson, F. P. (1980b). *In* "Nutrition of the Lower Metazoa" (D. C. Smith and Y. Tiffon, eds.), pp. 149–156. Pergamon, Oxford.
Wilkinson, C. R. (1980). *Hydrobiologia* **75,** 241.
Williamson, C. E. (1979). *Trans. Am. Microsc. Soc.* **98,** 59.

INTERNATIONAL REVIEW OF CYTOLOGY, SUPPLEMENT 14

The Molecular Biology of *Rhizobium*–Legume Symbiosis

DESH PAL S. VERMA* AND SHARON LONG†

**Department of Biology, McGill University, Montreal, Quebec, Canada, and †Department of Biological Sciences, Stanford University, Stanford, California*

I. Introduction

Plant–microbe interactions are common in nature. Most of these are pathogenic; however, some microbes, such as diazotrophic bacteria, form beneficial symbiotic associations with plants. The interaction of *Rhizobium* bacteria and host plants in the legume family results in the development of specialized structures, root nodules. Within these organs, some host cells are inhabited by endosymbiotic forms of the bacterial partner; the bacteria reduce (fix) molecular nitrogen (N_2) into ammonia (NH_3), which the host plant can use as a nutrient. The resulting ability to grow without large amounts of soil nitrogen makes this an important process agriculturally, and has probably contributed to the ecological and evolutionary success of the family Leguminosae.

Biological nitrogen fixation is catalyzed by the enzyme nitrogenase. This enzyme, and consequently the ability to fix nitrogen, is restricted to certain prokaryotes, bacteria and cyanobacteria. Many groups of plants have evolved

ISBN 0-12-364375-9

relationships with nitrogen-fixing prokaryotes; for more detailed discussions of some of these systems, see Vose and Ruschel (1981). Some plants in the cereal family associate with soil bacteria such as *Azospirillum* (Jordan, 1981). The plant rhizosphere appears to be colonized by the bacteria, and an exchange of metabolites is thought to take place. No specific invasion, or morphogenesis of new structures, is known to occur. Another type of nitrogen-fixing symbiosis is that between the water-fern *Azolla* and an associated cyanobacterium, *Anabaena*. The plant develops cavities in the dorsal lobes of its leaves, in which the cyanobacteria live and reproduce. The association remains extracellular (Peters *et al.*, 1980). In the case of nitrogen-fixing Actinomycetes and *Rhizobium*, the microbe–host associations are intracellular in nature. These different types of association, represented in Fig. 1, show increase in complexity, from loose interaction to the development of an intracellular state.

Rhizobium and its legume host interact at many levels: as distinct organisms; as adjoining cell types within a complex developing organ; as inhabitants within a common cell wall boundary; and as producers of shared metabolites and of individual components of molecular complexes. Development of a nodule is a true cooperation between two genomes.

An understanding of the *Rhizobium*–legume symbiosis requires examining interactions between the two organisms from several points of view, and using various techniques, including studies such as physiological translocation of nutrients, microscopic examination, and biochemical and genetic analysis. We will touch on topics within this wide range, particularly trying to indicate ways in which these organisms influence and regulate each other.

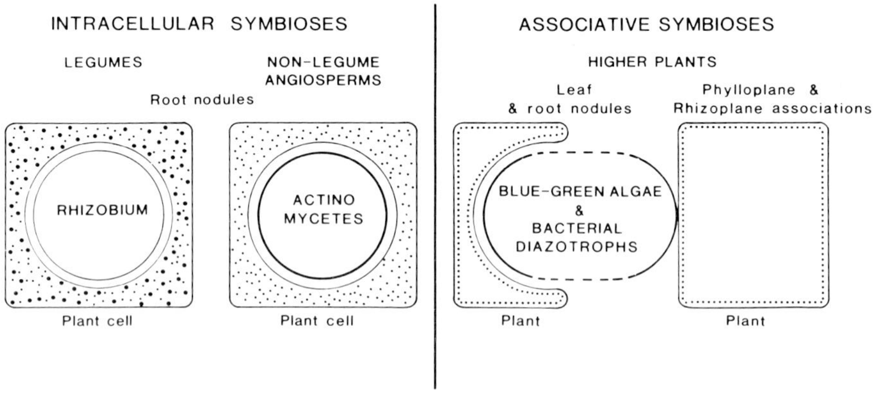

FIG. 1. Various associations between plants and diazotrophs leading to the symbiotic fixation of atmospheric nitrogen. Note the increase in complexity as association proceeds from extracellular to an intracellular state.

Understanding these interactions can provide ways to approach some of the major questions and problems relating to the *Rhizobium*–legume symbiosis. One such question is, how did this symbiosis evolve? One theory is that nitrogen-fixing soil bacteria acquired the ability to invade a host plant (see Dilworth and Parker, 1969). The reverse idea, that a plant pathogen acquired the gene complex for nitrogen fixation, has also been proposed. In both cases, the bacteria would need to acquire the ability to suppress host defenses (see Vance and Johnson, 1981; Verma, 1982). The usual consequence of microbial contact with a plant is a host reaction which combats the invader by various means including localized self-destruction (hypersensitive reaction). The challenge to *Rhizobium* and host is to recognize each other without triggering defense mechanisms.

Another important question is that of specificity. *Rhizobium* bacteria interact primarily with host plants in the family Leguminosae, although it is not clear that all legumes are capable of nodulation (see Vincent, 1974; Allen and Allen, 1981). Within this broad grouping, individual species of *Rhizobium* have been defined based on their ability to nodulate certain groups of legume plants. These groups may be small (a few genera only) or very broad. Characteristically, *Rhizobium* which will nodulate one host plant within a group will cross-inoculate other members of the group. However, there are exceptions to this: in some cases, different varieties of a host species have different symbiotic behavior in association with a particular strain of *Rhizobium*. The converse is also true. Promiscuity in *Rhizobium* has often been observed and the validity of cross-inoculation concept has been questioned (Wilson, 1944).

Thus, *Rhizobium*–host associations appear to show different, nested levels of specificity: discrimination within a given species, or between different cross-inoculation groups, or an entire family. This implies that specificity in this association is not due to a single molecular event, but to a series of interactions. Some events may be common to all *Rhizobium*–legume systems, while others distinguish only one cross-inoculation group. These differences, which are due to naturally occurring variation, and/or mutation(s), may not only reveal the genetic basis for specificity and effectiveness, but also help to dissect out the components of this complex interaction (Vincent, 1980).

II. Cellular Interactions

A. Events at the Cell Surface

The interaction of the two partners begins at their surfaces. The events occurring near or at the root surface—chemotaxis and attachment—have received a great deal of experimental and conceptual attention, both because of their importance to the symbiosis and because of their accessibility to experimental manipulation and biochemical analysis.

1. *Chemotaxis*

Chemotactic behavior, a familiar phenomenon in enteric and other bacteria, has been found in some *Rhizobium* strains (Vincent, 1974). Currier and Storbel (1974) found that *Rhizobium* cells are positively chemotactic for a plant host, and particularly for a glycoprotein isolated from that host. Swimming and chemotactic behavior of *R. meliloti* (Ames *et al.*, 1980) were described as being basically similar to that of *E. coli:* cells without stimulation move randomly by short "runs" (straight-line motion) interrupted by "tumbling" which causes a change in the direction of motion. Nonmotile mutants, such as Fla^- (lacking flagellae), Mot^- (nonmotile, with flagellae), and Che^- (nonchemotactic; abnormal motility with flagellae), are still able to form nodules when individually inoculated onto plants (Ames and Bergman, 1981). This indicates that neither motility itself, nor flagellar structures, are required for nodulation. However, when motile and nonmotile strains of *R. meliloti* are coinoculated onto alfalfa plants, the motile strains have a competitive advantage.

2. *Biochemistry of Rhizobium Cell Surface*

The mechanisms of attachment, recognition, and invasion are affected by the change in the biochemistry of the cell surfaces of both partners. At the surface, rhizobia and their hosts may show more diversity than universality. The *Rhizobium* surface, like that of many other Gram-negative bacteria, consists of a cell membrane and wall, with extracellular material often forming a capsule, and sometimes bearing flagellae. The surface carbohydrates of several *Rhizobium* species have been characterized; however, little is known about the membranes and/or proteins associated with various *Rhizobium* surface layers. For a detailed review of chemistry of *Rhizobium* cell surfaces, see Carlson (1982) and also parts of reviews by Bauer (1981) and Dazzo and Hubbell (1982).

The surface polysaccharides of *Rhizobium* fall into two general classes, lipopolysaccharides (LPS) and extracellular polysaccharide (ECPS). Both LPS and ECPS are complex carbohydrates, composed of several types of sugars. LPS in *Rhizobium* appear to contain lipid A, linked to a core polysaccharide by 2-keto-3-deoxyoctanoic acid (Carlson *et al.*, 1978; Zevenhuizen *et al.*, 1980). A repeating oligosaccharide, the O-antigen, is linked to the core polysaccharide. The fatty acid content of lipid A of LPS from four different *Rhizobium* species is reported to be identical (Zevenhuizen *et al.*, 1980). However, it appeared to differ from the lipid A found in Enterobacteriaceae. The polysaccharide composition of LPS varies in different strains of the same species (Carlson *et al.*, 1978; Zevenhuizen *et al.*, 1980).

Antibody raised against purified LPS of one *R. leguminosarum* agglutinated some but not all other *R. leguminosarum* strains, and agglutinated some *R. phaseoli* strains also (Zevenhuizen *et al.*, 1980). Thus, there does not appear to be a straightforward correlation of LPS composition and host specificity. The role of LPS may become more clear when their complete structures are elucidated (Carlson, 1982). There are experimental indications that LPS reacts with host plant lectins (Wolpert and Albersheim, 1976; Dazzo and Brill, 1979; Hrabak *et al.*, 1981) and in several systems, ECPS structure correlates with host specificity. However, there are many exceptions to this rule. Several fast-growing species, e.g., *R. trifolii, R. phaseoli,* and *R. leguminosarum* have identical ECPS structures (Carlson, 1982), which weakens the case for ECPS being a specific recognition molecule.

3. *Lectin Binding*

Several plant proteins are capable of binding bacterial cells and/or purified polysaccharide components from the cells; many of these bind to specific carbohydrates and fall under the generic name of "lectin." In many cases, there is a striking correlation between the infectivity of a *Rhizobium* strain on a particular host plant and its ability to bind to lectin from that plant (Bohlool and Schmidt, 1974). There are some exceptions to this pattern, however, and models of recognition are continually evolving to take into account more complex data.

Two systems which have been extensively studied are the clover—*R. trifolii,* and the soybean—*R. japonicum* symbioses. A glycoprotein isolated from clover (*Trifolium repens*) seeds specifically binds cells of *R. trifolii* but does not bind cells of other *Rhizobium* species. This protein, trifoliin-A, is also present on roots, as shown by immunolocalization (Dazzo *et al.*, 1978). The hapten sugar for this lectin, 2-deoxyglucose, prevents attachment of *R. trifolii* cells to clover roots (Dazzo *et al.*, 1976; Dazzo and Brill, 1977). However, 2-deoxyglucose was not found in *R. trifolii* LPS (Hrabak *et al.*, 1981) and its presence in ECPS has been debated. One possible reason for this is that the time at which *Rhizobium* cells show maximal attachment to roots and produce *trifoliin*-binding material (Dazzo *et al.*, 1979) is not the same as that at which the extracellular polysaccharide was isolated for structural studies. Chemical analysis performed on bacterial cultures at several phases of growth period may help resolve this present question. These inconsistencies point out that some of the phenomenon involved in early nodulation may be transient, in keeping with the possible multistep nature of the infection process (Bauer, 1981).

A temporal change in infectibility of *Rhizobium,* accompanied by differences in lectin-binding and in the chemistry of the bacterial cell surface, has also been reported in the soybean—*R. japonicum* system. Soybean lectin binds to *R. japonicum* cells (Bhuvaneswari *et al.*, 1977; Mort and Bauer, 1980; Tsien and Schmidt, 1981). The chemical composition of *R. japonicum* changes with the growth phase of the culture: as culture reaches late exponential phase, the amount of galactose decreases, and 4-*O*-methyl-galactose increases concomitantly (Mort and Bauer, 1980). Since soybean lectin recognizes galactose as a hapten, this modification may explain the change in infectibility which occurs in *R. japonicum* cultures with culture age.

Weighing against these interesting correlations is the discovery that a soybean line which completely lacks lectin is nonetheless able to form nodules in association with *R. japonicum* (Pull *et al.*, 1978; Stacey *et al.*, 1980). Furthermore, there are some infective strains of *R. japonicum* which are not bound by lectin (Bohlool and Schmidt, 1974). An agglutinating protein from alfalfa (*Medicago sativa*) binds all *R. meliloti* strains so far tested, and no other *Rhizobium* species with one exception (Paau *et al.*, 1981). Mutants of *R. meliloti* which were unable to nodulate fell into two categories; some did not bind agglutinin, and also did not adhere to alfalfa roots. The other nonnodulating mutants, however, were agglutinated normally by the alfalfa protein.

These complexities suggest, first, that recognition and infection occur by a multistep process, rather than a single-step mechanism (Bauer, 1981; Dazzo and Hubbell, 1982); and second, that various *Rhizobium*–plant pairs may differ in the repertory of recognition mechanisms that they use.

B. Development of Nodule Structure

Successful interaction of a rhizobial strain with its host results in the development of an organized structure, the root nodule, in which *Rhizobium* lives endosymbiotically within a group of infected host cells. There is a continuous interaction between the two organisms during the functional life of a nodule. Figure 2 shows some major temporal sequences of events that occur following inoculation of a legume plant by its compatible rhizobial strain.

1. *Invasion of Root Hair Cells*

One of the most crucial events in establishing a legume–*Rhizobium* symbiosis is the penetration of rhizobia into the plant cell. Numerous factors control and influence this step, including genetic specificity of the two partners, and environmental conditions such as mineral nutrition, pH, number of rhizobia, competition with other microbes, and time of inoculation.

a. *Root Hair Curling.* *Rhizobium* invades the plant principally by two methods: either via a host root hair or through openings which arise due to lateral root emergence as in peanut (*Arachis hypogea*) (Allen and Allen, 1940) and in lupine species (Chandler and Dart, 1973). Plants infected by the first method are often quite specific in symbiotic pairing. Dart (1977) has pointed out that the second mode of invasion may confer upon the host the ability to form nodules with a wider range of *Rhizobium* strains. *Arachis hypogea,* for example, nodulates with a variety of *Rhizobium* strains.

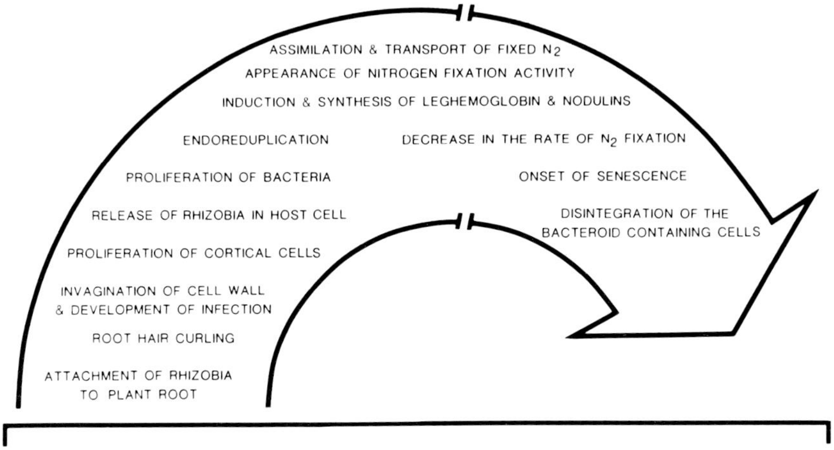

FIG. 2. Temporal sequence of events following inoculation of a legume plant with *Rhizobium* leading to symbiotic nitrogen fixation.

Root hair invasion is characterized by curling, contorted growth, and/or branching of the root hairs (Fahraeus, 1957; Nutman, 1959; Dart, 1974). The degree of deformation depends upon the plant species, the infecting rhizobial strain, and the position of infection on the root (Sahlman and Fahraeus, 1962; Yao and Vincent, 1969; Li and Hubbell, 1969). Markedly curled root hairs, or "shepherd's crooks," generally appear only in response to live bacteria of infective (homologous) strains (Yao and Vincent, 1969). Homologous strains of rhizobia produce a higher extent of curling than heterologous strains. Infected root hairs develop a bright spot on the root hair cell wall, often in the "pocket" of a "shepherd's crook" (Callaham and Torrey, 1981). This bright spot corresponds to the point where rhizobia enter the root (see Dart, 1977; Broughton, 1978; Newcomb, 1981). The mechanism of root hair curling is not completely understood. It appears to occur due to uneven rate of growth in the root hair wall.

b. *Infection Thread Formation.* *Rhizobium* invades the root hair by partial dissolution of the host cell wall with concomitant deposition of new cell wall-like material to form an "infection thread" along the path of invasion (Nutman, 1956). Callaham and Torrey (1981) observed that a new wall layer is deposited by the host cell internal to the existing wall as the old wall weakens during *Rhizobium* invasion. The infection thread is continuous with the newly deposited cell wall layer.

Root hair invasion and infection thread formation are postulated to involve pectinase and cellulase enzyme activities (Ljunggren, 1969; Verma *et al.*, 1978a; Dazzo and Hubbell, 1981). Action of these enzymes might produce changes in rhizobial and plant cell surfaces (Sahlman and Fahraeus, 1963; Napoli and Hubbell, 1975). A low level of pectolytic enzymes is detectable in free-living rhizobia (Hubbell *et al.*, 1978; Verma *et al.*, 1978a), and cellulase and pectinase activities increase in infected roots (Ljunggren and Fahraeus, 1961; Verma *et al.*, 1978a). Cellulase activity has been localized histochemically in the wall of the infection thread (Verma *et al.*, 1978a,b). However, involvement of cell wall hydrolyzing enzymes in the process of rhizobial infection remains an unresolved question (see Dart, 1974, 1977; Verma, 1982). Since infection is a localized, transient phenomenon, associated enzyme activity may not always be easily detectable.

Growth of the thread within the root hair cell is presumed to be due to bacterial divisions and increase in biomass. The thread appears to be thin-walled or even unwalled at the tip (Callaham, 1979; see also Newcomb, 1981). A balance between the rates of thread elongation and of cell wall deposition is important for successful thread growth to continue, but how this regulation works is not known. As it penetrates the root cortex, the infection thread crosses already-formed cell walls. Infection threads can grow intercellularly (within cell walls), or penetrate new cortical cells by branching, and the infection spreads laterally. A localized wall degradation followed by wall deposition appears to be the mechanism of thread invasion.

Histochemically, the thread wall is shown to contain pectic substances, hemicelluloses and cellulose, similar to the primary wall of the plant cells (Sahlman and Frahraeus, 1963; Higashi, 1966; Dart and Mercer, 1964). The bacteria are surrounded by a matrix which is finely granular and moderately electron dense (see Dart, 1974; Newcomb, 1981). On the basis of histochemical staining, it has been suggested that the matrix is made up of mucopolysaccharides synthesized and secreted by the bacteria (Mosse, 1964; see also Dart, 1974).

2. *Infection of Cortical Cells*

a. *Release of Bacteria.* Successful infection results from "release" of bacteria from the infection thread into the host cell. The release event is influenced by both host and bacterial genes. In the infected cell, "unwalled droplets" are found at the tip or side of the infection thread. These "unwalled droplets" contain bacteria and matrix materials surrounded by host membrane (Newcomb, 1976). Emergence of unwalled droplets from infection threads may be due to dissolution or lack of synthesis of the infection thread wall. Verma *et al.* (1978a) observed histochemically that infected host cells produce cellulase that is localized in the infection thread wall. This, along with the pectinase enzyme produced by rhizobia, appears to be responsible for degradation of the infection thread wall. Residual cell wall material appears to be sequestered into membrane vesicles through endocytosis (Basset *et al.*, 1977).

From the unwalled region of the infection threads, rhizobia move out into the cell cytoplasm where they remain bounded by a membrane of host origin (see below). In some, but not all species, bacteria continue to divide within these membrane envelopes. The extent of division is influenced by the host. The number of bacteroids in the membrane enclosing them is variable, from 1 or 2 in pea, up to as many as 16 bacteroids enclosed within a membrane envelope in mung bean. This variability in the number of rhizobia may in some cases be due to the fusion of membranes (Newcomb and McIntyre, 1981), or to different rates of membrane biosynthesis and bacteroid cell division (see Verma, 1982).

b. *Origin and Biogenesis of the Membrane Enclosing the Bacteroids.* The membrane enclosing rhizobia upon their release into the host cell was first observed by Bergersen and Briggs (1958) and is called "membrane envelope" or "peribacteroid membrane." Each of these terms has advantages and disadvantages (see Newcomb, 1981; Verma, 1982). In line with usage in a previous volume of this series (Newcomb, 1981), we have used in this article the term "peribacteroid membrane."

Regarding the origin of the peribacteroid membrane, there are three hypotheses:

1. The *Rhizobium* cells are enclosed by the plasma membrane of the host cell due to endocytosis at the time of their release (Bergersen and Briggs, 1958; Goodchild and Bergersen, 1966; Dixon, 1967; Tu, 1974; Newcomb, 1976; Verma *et al.*, 1978b).
2. It is derived from the host endomembranes, e.g., endoplasmic reticulum (ER) (Jordan *et al.*, 1963) or nuclear envelope (Prasad and De, 1971).
3. It is synthesized *de novo* (Dart and Mercer, 1963, 1964, 1966; Generozova and Yagodin, 1972; Jordan *et al.*, 1963; Jordan and Grinyer, 1965).

At present, evidence strongly favors the first hypothesis. Derivation of the peribacteroid membrane from the host plasma membrane has been demonstrated in several species, on the basis of membrane thickness (Dart and Mercer, 1963; Newcomb, 1976), staining characteristics (Dixon, 1967; Robertson *et al.*, 1974), particulate densities on faces of freeze fracture replicas of the two membranes (Tu, 1975a; Robertson *et al.*, 1978), their sensitivity to nonionic detergents, and enzymatic and peptide composition (Verma *et al.*, 1978b; Bal *et al.*, 1980).

The eventual behavior and function of this membrane is quite distinct from that of plasma membrane, however. For example, no cell wall is formed "outside" the membrane, and the transport functions it controls relate to the bacteria enclosed, not to adjoining plant cells. A key feature of this symbiosis is compartmentalization segregating eukaryotic and prokaryotic metabolisms, which is likely to be controlled by the peribacteroidal membrane. This membrane may contain specific transport system(s) for the passage of metabolites to and from bacteroids (Verma, 1982). Membrane formation and function must be controlled by both partners; mutations in *Rhizobium* which result in ineffective nodules change the pattern of development of this membrane (Mackenzie and Jordan, 1974, Basset *et al.*, 1977b; Werner *et al.*, 1980; Verma and Bal, unpublished). It is of great interest to know how further biogenesis and differentiation of the peribacteroidal membrane is controlled. The membrane synthesis must be intensive, since an infected cell with a large number of bacteroids has about 20–40 times more membrane area than an uninfected cell (Robertson *et al.*, 1974).

Cytoplasmic vesicles apparently produced by Golgi bodies have been reported to fuse with the peribacteroid membrane (Robertson *et al.*, 1978). It is not known how the cytoplasmic vesicles are directed to and recognize the target membrane(s). Once bacterial divisions cease, there appears to be little turnover of this membrane as evidenced by low level of radioactive amino acid incorporation in this membrane fraction (Zogbi, 1982).

c. *Intracellular Organization in the Infected Cells.* Cortical cells in which bacteria have been released show a number of morphological, cytological, and physiological changes. Infected cells undergo a proliferation of the endomembrane systems, including RER, Golgi bodies, and vesicles (Dart and Mercer, 1964, 1966; Goodchild and Bergersen, 1966). Free ribosomes and polyribosomes are abundant (Newcomb, 1976, 1981). Proplastids, mitochondria, amyloplasts, etc., also increase in number. In cells infected by fast-growing species of *Rhizobium,* a large central vacuole may develop due to the coalescence of small vacuoles. The nuclei increase several fold in volume and are often polyploid. At all times bacteria remain enclosed within the peribacteroid membranes failure of which results in senescence and death of the infected cell. An infected root nodule cell (Fig. 3) shows a high degree of cellular organization in spite of a large number of bacteroids being present in these cells. There is no central vacuole in soybean (cf. infection by fast-growing *Rhizobium* species, Newcomb, 1981) and mitochondria are present toward the periphery of the cell. Uninfected cells usually are smaller in size and often accumulate carbohydrate reserves.

d. *Bacteroid Differentiation.* The endosymbiont or bacteroid form of *Rhizobium* differs significantly from the free-living form. Bacteroids often exhibit distinct morphology. The type of differentiation varies from one host–microbe system to another, and is highly influenced by the host (Kidby and Goodchild, 1966; see also Dart, 1977; Sutton *et al.*, 1981). The transformation from free-living *Rhizobium* to bacteroid is often accompanied by reduction in bacterial wall thickness and rigidity, and changes in the outer membrane (Mackenzie *et al.*, 1973; Robertson *et al.*, 1978; Bal *et al.*, 1980). In *Vigna unguilata,* however, the wall of the bacteroids is very similar to free-living rhizobia (van Brussel *et al.*, 1979). Mesosome-like structures due to the invagination of bacteroidal cytoplasmic membrane are observed in nodules of several species (Dart and

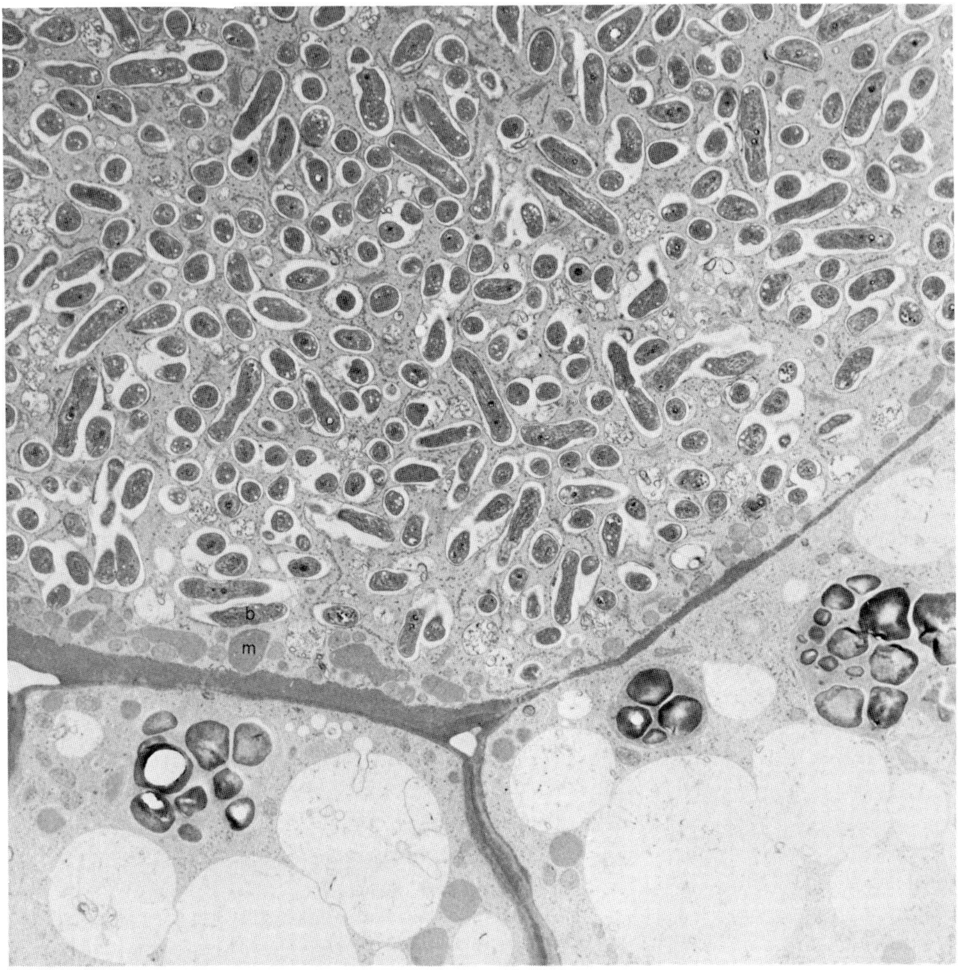

FIG. 3. Ultrastructure of soybean root nodule cell showing a part of an infected cell packed with bacteroids enclosed in peribacteroid membranes. Note the integrity of the membrane system and intracellular organization in infected vs uninfected cells. m, Mitochondria; b, bacteroids, ×6250. (From Verma *et al.*, 1978.)

Mercer, 1963b; Mosse, 1964; Hornez *et al.*, 1974; Tu, 1977; Gourret and Fernandez-Arias, 1974). A rapid accumulation of β-polyhydroxybutyrate (PHB) (Basset *et al.*, 1977a) occurs at both the poles in some species of bacteria, while in other species PHB which accumulates in bacteria during infection thread growth (Marks and Sprent, 1974; Werner and Morschel, 1978) disappears during bacteroid differentiation (Vance *et al.*, 1980).

The DNA content per bacteroidal nucleoid may be greater than that of free-living *Rhizobium,* by 1.6- up to 8-fold (Sutton *et al.,* 1978; Bisseling *et al.,* 1977). Some changes in bacteroid DNA buoyant density, denaturation temperatures, molecular weight, and kinetic complexity have also been observed (Sutton, 1974). On the other hand, RNA content per cell, RNA/DNA ratio, and even RNA/protein ratio decline with maturity of bacteroids (Dilworth and Williams, 1967; Sutton and Robertson, 1974; Bisseling *et al.,* 1979).

Bacteroids isolated from nodule cells are difficult to culture on a normal media and low frequency (0.5%) colony formation is usually observed (Sutton *et al.,* 1977; Sutton and Patterson, 1980; see also Jordan, 1962; Bergersen, 1974). Those colonies which do form may come from immature rather than fully differentiated bacteroids in the nodules. Supplementation of the growth media and osmotic protection are reported to increase the number of colonies (Gresshoff *et al.,* 1977; Gresshoff and Rolfe, 1978; Rolfe *et al.,* 1980).

3. *Organ Level Structural Interactions*

Rhizobium induced root nodules are completely unique organs, not homologous with other plant structures. As they develop, the tissues within nodules become organized according to patterns characteristic of each plant species (Allen and Allen, 1954; Libbenga *et al.,* 1973).

Nodules fall into two main morphological categories: spherical or oval (e.g., in *Vigna radiata, Glycine max,* and *Arachis hypogea*), and cylindrical, elongate, or club-shaped (e.g., in *Pisum sativum, Vicia faba, Trifolium,* and *Medicago*). The detailed morphogenesis of these nodules has been discussed and reviewed by Dart (1977) and Newcomb (1981). During ontogeny of pea root nodules, mitosis in the inner root can be observed soon after infection of root hairs. These cell divisions give rise to a nodule meristem. In most species, the cells that initiate nodule development are cortical, but in some plants, such as *Arachis hypogea,* the nodule may be derived from cells of the pericycle layer of the vascular tissue. The characteristic nodule shape is determined by orientation and duration of mitosis, and is controlled by the host (Dart, 1977). In cylindrical nodules, cells continually produced at the meristem base are infected by *Rhizobium;* by contrast, spherical nodules are more determinate, and infection occurs in a few localized zones of cells. In most plant species, such as pea, the cells stop dividing after rhizobial infection. In others, such as peanut, however, cell division continues for a short time (Syono *et al.,* 1976; Dart, 1977; Truchet, 1978; Newcomb and McIntyre, 1981).

For reasons not well understood, 20–50% of nodule cells remain uninfected. The DNA amount in these cells remains low (2 to 4C) (Truchet, 1978) as compared to the infected cells. Infected cells often have an increased ploidy level of 4C, 8C, or even 16C levels of DNA (Libbenga and Bogers, 1974). Evidence suggests that the DNA synthesis and increase in ploidy level occur after infection of host cells (see below). It has been proposed that ploidy increase results from endoreduplication caused by *Rhizobium*-produced cytokinin and auxin (Libbenga *et al.*, 1973; Libbenga and Torrey, 1973).

Ultrastructural studies indicate that some uninfected cells may specialize in metabolic pathways related to the assimilation of nitrogen (Gunning *et al.*, 1974). These include some cells with "transfer cell" morphology. Transfer cells have characteristic in-foldings of secondary cell wall, resulting in an increased plasma membrane area (Gunning and Pate, 1974), and are thought to function in short distance movement of solutes. Presence of similar cells in pericycle of nodular bundles (Pate *et al.*, 1969) or in roots (Gunning and Pate, 1974; Newcomb and Peterson, 1979) indicates a potential role of these cells in facilitating the movement of photosynthates from sieve elements toward infected cells, and/or export of ureides/amides from infected cells to the xylem elements of the nodule vascular tissue.

III. Metabolic Interactions

A continuous availability of metabolic energy and the protection of the oxygen labile nitrogenase system are two main requirements for the reduction of nitrogen. In the symbiotic state both of these requirements are met by the host; however, an extensive metabolic cooperation between the two partners is required for the success of this process. The enzyme, nitrogenase, is produced and located in the bacteroid cell. The substrate, some cofactors, and energy supply needed for its function must be supplied by, or at least move through, the host cytoplasm and the peribacteroid membrane. In addition, the host environment must adequately control oxygen tension in order to avoid inactivation of nitrogenase. The host transport system, cytoplasm, and membrane systems provide levels of control. We will deal with four main areas in this section: oxygen supply and the role of leghemoglobin, production and assimilation of ammonium, carbohydrate supply and utilization, and energy relations.

A. Oxygen

Nitrogenase is oxygen-sensitive, and all nitrogen-fixing organisms must have a mechanism for controlling oxygen. The microenvironment present in the cytoplasm of infected cells is well suited for bacteria in that there appears to be a limitation in oxygen availability (Tjepkema, 1971). Very low oxygen tensions occur at the center of mature nodules as shown by spectroscopy (Appleby, 1962, 1969) and by direct measurement with microelectrodes (Tjepkema and Yocum, 1974) of leghemoglobin oxygenation. The leghemoglobin present in the host cell cytoplasm (Verma and Bal, 9176; Verma *et al.*, 1978b, 1979) plays a role in transport of O_2 (Yocum, 1964; Bergersen, 1971; Tjekema, 1971; Appleby, 1974; Wittenberg *et al.*, 1974; Bergerson and Turner, 1975) by maintaining a sufficiently high pO_2 in the host cell cytoplasm for oxidative phosphorylation (see Verma, 1981), while providing a sustained low rate of O_2 to the bacteroids. Leghemoglobin has been shown to enhance acetylene reduction activity of purified soybean bacteroids *in vitro* at concentration of bacteroid and leghemoglobin similar to *in vivo* levels (Bergersen *et al.*, 1973).

There has been a controversy regarding the intracellular location of leghemoglobin in the infected nodule cell. While earlier studies (Dilworth and Kidby, 1968; Bergersen and Goodchild, 1973) have indicated that leghemoglobin is present inside the peribacteroid membrane vesicles, several lines of evidence strongly support its location to be inside the host cell cytoplasm. These are direct localization using antibodies against apo-leghemoglobin (Verma and Bal, 1976; Verma *et al.*, 1978b); lack of immunoreactive material inside the isolated peribacteroid membrane vesicles (Verma *et al.*, 1978b; Robertson *et al.*, 1978); and synthesis of apo-leghemoglobin on free polysomes in the host cell cytoplasm and the lack of a signal sequence on the newly synthesized peptides (Verma *et al.*, 1979). The latter evidence suggests that this protein may not be able to cross the membrane enclosing the bacteroids. The haem moiety of leghemoglobin made by *Rhizobium* must cross the peribacteroid membrane to be assembled with apo-leghemoglobin. Minor quantities of leghemoglobin observed inside the peribacteroid membrane appears to be due to some damage in the membrane and artifacts of detection techniques (cf. Bergersen and Appleby, 1981). If the host cell cytoplasm is the primary site for leghemoglobin, another oxygen carrier may be present in the peribacteroid membrane for its transport to the bacteroid surface.

B. Ammonia Assimilation

Ammonia is the primary product of the nitrogen fixation reaction. This is carried out in the bacteroids, as shown by the fact that nitrogen fixation is continued by bacteroids isolated from nodule cells (Bergersen and Turner, 1967; Dixon, 1968). Once formed, ammonia can be assimilated by various pathways. One way is by glutamate dehydrogenase (GDH) which forms glutamate by reaction of ammonia with α-ketoglutarate, a TCA cycle intermediate. Alternatively, net synthesis of glutamate can occur by a series of two reactions, glutamine synthetase (GS) and glutamate synthase (glutamine oxoglutarate amino transferase, GOGAT).

$$\alpha\text{-ketoglutarate} + NH_3 \xrightarrow{\text{GDH}} \text{glutamate} \quad (1)$$

$$\text{glutamate} + NH_3 \xrightarrow{\text{GS}} \text{glutamine} \quad (2)$$

$$\text{glutamine} + \alpha\text{-ketoglutarate} \xrightarrow{\text{GOGAT}} 2\ \text{glutamate}$$

Earlier theories on symbiotic N_2 fixation held that both fixation and assimilation occurred in the bacteroids. However, it appears that these functions are divided. Nitrogen fixation occurs in the bacteroids, and ammonia is transported into the plant cytoplasm, where it is assimilated (Miflin and Lea, 1976). When nitrogen-fixing *R. japonicum* bacteroids are isolated from soybean nodules and allowed to fix nitrogen *in vitro,* ammonia appears in the surrounding solution (see Bergerson, 1971). Consistent with the hypothesis, when *R. japonicum* are grown *in vitro* in conditions which permit nitrogenase induction, the bacterial cells excrete ammonia into the medium (O'Gara and Shanmugam, 1976a).

Miflin and Lea (1976) suggested that assimilation of ammonium occurs in the plant cytosol by the GS/GOGAT pathway, rather than by GDH. When soybean nodules were labeled with $^{13}N[N_2]$, Meeks *et al.* (1978) observed that the initial organic product labeled with ^{13}N was glutamine. Within 1 minute, however, glutamate was found to be more heavily labeled than glutamine, which explains why in some previous studies GDH had been indicated as the primary route of assimilation. Further evidence that GS plus GOGAT is the primary assimilation pathway comes from developmental studies of nodule enzyme content. Robertson *et al.* (1975a,b) reported that both GS and GOGAT rise during lupin nodule development in parallel with nitrogenase activity and leghemoglobin content. In alfalfa, both GS and GOGAT activities decline along with acetylene reduction when nodulating roots are deprived of their shoot system. These enzymes increase in activity again as acetylene reduction increases during nodule rebuilding (Groat and Vance, 1981). Alternative pathways for ammonia assimilation, acting on other amino acids such as asparagine, or on ketoacids, may also be important in nodules (Rawsthorne *et al.,*1978). The assimilation of ammonia in the plant cytoplasm has importance not only for plant nutrition, but because continued nitrogenase function in bacteroids can be affected by the nitrogen content of the plant cell environment. It has been pointed out that high GS activity would be important in keeping cellular NH_4^+ low, to maintain a gradient favoring continued NH_4^+ export by bacteroids (Rawsthorne *et al.,* 1978).

In addition, NH_4^+ may control nitrogenase production in *Rhizobium,* as in other nitrogen-fixing bacteria. This regulation has been postulated to occur with mediation of bacterial glutamine synthetase (GS). As bacterial glutamine levels rise and fall (in parallel with bacterial NH_4^+), bacterial GS is modified by adenylylation (Brenchley *et al.,* 1973). Adenylylated GS is inactive; removal of adenyl groups activates the enzyme, which is postulated to have a double role as an enzyme and as a positive regulator of nitrogenase expression (e.g., in *Klebsiella pneumoniae;* see Ausubel *et al.,* 1978). In *R. japonicum* and "cowpea" rhizobia, there is evidence that GS modification occurs in response to NH_4^+ and that such changes in GS accompany changes in nitrogen fixation (Bishop *et al.,* 1976; Bergersen and Turner, 1978; Ludwig, 1980; Ludwig and Signer, 1977). The situation is complicated by the finding that several *Rhizobium* systems have two forms of glutamine synthetase (Darrow and Knotts, 1977; Ludwig, 1980).

Only GS-I appears to undergo adenylylation, although activity of both GS-I and GS-II decreases in presence of NH_4^+. The assimilation enzymes of the plant cell are also subject to regulation, but the mechanisms for this are not well understood.

Beyond the infected cell, nitrogen is required for plant growth and development. Fixed, assimilated nitrogen must be carried to other parts of the plant, and this is regulated at the metabolic, cellular, and organismal levels. The ureides, allantoin and allantoic acid, are primary nitrogen-containing compounds exported from soybean, cowpea, and bean nodules (Matsumoto *et al.*, 1977; Herridge *et al.*, 1978). A physiological advantage of ureide export is that more nitrogen atoms can be carried per carbon atom in the basic molecular skeleton (Herridge *et al.*, 1978). This is important given the competition for fixed carbon by the double needs for energy generation and as skeletons for ammonium assimilation (Rawsthorne *et al.*, 1978). In contrast to ureide export by soybean and cowpea, other legumes export fixed nitrogen as amides, such as asparagine and glutamine (Pate, 1977). The overall pattern of carbohydrate metabolism is different in these plants, since asparagine and glutamine are derived from TCA intermediates while ureides derive from purine metabolic pathways.

C. Carbohydrate Metabolism

Carbohydrate and nitrogen metabolisms must be balanced so as to provide carbon skeletons for nitrogen transport (above), and also to provide ATP for energy and reducing power. Lack of photosynthate, experimentally imposed by darkening plants or by harvesting, results in decreased nitrogen fixation (Wong and Evans, 1971; Gerson *et al.*, 1978). There are several possible ways in which this might occur: through the effect of overall energy charge on nitrogenase activity or production (Ching *et al.*, 1975) or by the effect of increased NH_4^+ resulting from lack of carbon skeletons usable in assimilation.

Use of fixed carbon by nodules is intensive; in *Pisum*, it was found that around 30% of the net carbon fixed over a 9-day period was consumed by the root nodules (Minchin and Pate, 1973). In lupin, approximately 10% of photosynthate is consumed by nodules (Pate, 1980). The form in which carbon is translocated to the root nodules is most likely sucrose, as shown by labeling studies (Burley, 1961; Pate, 1962). However, Lawrie and Wheeler (1975) reported that when $^{14}CO_2$ was fed to *Vicia* shoots, radioactive sucrose was highest in total root tissue but glucose and fructose were the most labeled compounds in nodules, indicating rapid conversion. In a similar experiment with *Phaseolus*, sucrose was found to be very highly labeled in nodules (Antoniw and Sprent, 1978). Since *Rhizobium* species vary, especially between the fast-growing and the slow-growing groups, in their ability to metabolize various sugars (Vincent, 1977), the nature and metabolism of the carbohydrate provided by the plant may also vary.

D. Energy Relations and Hydrogenase

Significant energy inputs are required for symbiotic nitrogen fixation, and the level of N_2 fixation appears to be limited by energy supply (Hardy and Havelka, 1975). Long term CO_2 enrichment in the environment of growing plants to augment fixed carbon supply partly overcomes this limitation. Several recent reviews have concentrated on the energy relationships of legume symbiosis (Phillips, 1980; Schubert and Ryle, 1980; Pate *et al.*, 1981; Mahon, 1981). *In vitro* experiments suggest a value of 2 moles of ATP hydrolyzed per electron passing through the nitrogenase complex (see Orme-Johnson *et al.*, 1977; Mortenson and Upchurch, 1981). In addition, ATP generating cost and cost of ammonia assimilation as well as ancillary carbon metabolism must be provided by the plant host. In terms of carbohydrate consumption at least 0.55 mole of glucose (i.e., 21 moles of ATP) would be required for reduction of 1 mole of N_2 (Pate *et al.*, 1981).

Nitrogenase has the property that it reduces protons (H^+) to hydrogen gas (H_2), which uses ATP and wastes electrons. This adds to the ATP cost of nitrogen fixation. Some *Rhizobium* strains have an enzyme, uptake hydrogenase, which recycles hydrogen gas to protons, regenerating energy and electrons. While the capacity to recycle H_2 in nodules is determined by the *Rhizobium* (Carter *et al.*, 1978), its expression is influenced by the host genotype. The hydrogenase enzyme activity can be expressed in free-living cultures under the conditions of limited carbon supply and decreased oxygen tension (Maier *et al.*, 1978) and mutant strains that have lost the ability to utilize H_2 have been isolated. Study of these strains has confirmed that the presence of uptake hydrogenase system in *Rhizobium* increases nodule efficiency in symbiotic nitrogen fixation (Evans *et al.*, 1981). Inoculation of soybeans with either Hup^+ or Hup^- *R. japonicum* strains showed (Albrecht, 1979) that the plants with Hup^+ symbionts produced more dry matter and had a higher nitrogen content. Because of varying uptake hydrogenase activity and varying partitioning of electrons for proton and nitrogen reduction calculated symbiotic nitrogen fixation costs range from 22 to 70 moles of ATP per mole of N_2 (Pate *et al.*, 1981).

IV. Gene Level Interactions

A. Molecular Differentiation

1. *Bacteroid*

Rate of protein synthesis increases during differentiation of bacteroids (Sutton *et al.*, 1977; Planque *et al.*, 1978). The spectrum of proteins synthesized is dependent on the bacterial strain and the stage of differentiation (Coventry and Dilworth 1975; Sutton *et al.*, 1977). Bacteroids of the same strain in different hosts synthesize the same kind of polypeptides (Shaw and Sutton, 1979). It is not known which of these changes occur due to the microaerophyllic conditions and which due to the specific interaction with host.

a. *Heme Synthesis.* While the globin chains of leghemoglobin are encoded by plant DNA (Dilworth, 1969; Baulcombe and Verma, 1978; Sullivan *et al.*, 1981) and synthesized on plant cytoplasmic ribosomes (Verma *et al.*, 1974; Verma and Bal, 1976), it has been proposed that the heme moiety is primarily provided by the bacteroids. This suggestion is based on (1) the capacity of bacteroids to convert δ-aminolevulinic acid into heme (Cutting and Schulman 1969; Dilworth, 1969; Godfrey, 1972); and (2) increased rate of heme secretion *in vitro* under low partial pressure of oxygen (Avissar and Nadler, 1978). The transfer of the precursors of heme in between the symbionts may occur since heme biosynthetic enzymes are also present in the host cytoplasm. Recent results on heme auxotrophic mutants which form ineffective nodules lacking functional leghemoglobin (Nadler, personal communication) provide further support that heme moiety of leghemoglobin in nodules is primarily made by *Rhizobium.*

b. *Induction and Regulation of Nitrogenase Enzyme.* Although a few *Rhizobium* species, such as *R. japonicum* and "cowpea" rhizobia, can fix nitrogen *in vitro* under specific conditions, nitrogenase generally does not function in free-living *Rhizobium*. Control of *Rhizobium* nitrogenase appears to be at the level of transcription (Corbin *et al.*, 1982). Nitrogenase activity on a per cell basis increases 1000-fold or more within 3–4 days following induction of this enzyme (Sutton and Jepsen 1975). Synthesis of the two major nitrogenase components commences at different times during induction (Bisseling *et al.*, 1979, 1980). It has been suggested that glutamine synthetase may play a role in controlling relative level of nitrogenase (see Section III,B) but the mechanism for control is not well understood.

Nitrogenase constitutes about 10–12% of the total protein of the bacteroidal cell (Whiting and Dilworth, 1974). The enzyme is composed of two metalloproteins. The larger one, MoFe protein (component I) binds the reducible substrate; the smaller one, Fe protein (component II), transfers electrons to the MoFe protein. The enzyme requires ATP and Mg^{2+} for activity. Both the MoFe and Fe proteins are irreversibly inactivated by O_2. In pea nodules (*R. leguminosarum*), it appears that synthesis of MoFe protein commences first, and the ratio of Fe to MoFe component increases gradually as the nodule matures (Bisseling *et al.*, 1979, 1980). A marked homology exists among nitrogenase from various nitrogen-fixing organisms at both the protein and DNA levels (Ruvkun and Ausubel, 1980; Scott *et al.*, 1981) and it has been suggested that perhaps this enzyme emerged only once in evolution (Burris *et al.*, 1981).

2. *Leghemoglobin and Other Host Gene Products*

Although in most cases, host genes controlling symbiosis are unknown, or code for unknown functions, there are molecular and biochemical approaches which may lead to an understanding of these genes. Leghemoglobin, whose function as an oxygen carrier was described above, is one of the major plant gene products that is unique to legume nodules and has been related to the effectiveness of this symbiosis. This protein represents 25–30% of the total soluble protein in the infected cell. Biochemical studies (Appleby, 1974) and analysis of amino acid sequences (Elfolk, 1972; Hunt, 1972) indicate that its structure is closely related to that of myoglobin and animal hemoglobins. This has recently been extended to the level of genomic DNA sequences (Brisson and Verma, 1982; Hyldig-Nielsen *et al.*, 1982).

In soybean nodules, four major leghemoglobins have been resolved: Lba, c_1, c_2, and c_3 (Fuchman and Appleby, 1979). Differences in their primary sequence (Ellfolk and Sievers, 1971; Sievers *et al.*, 1978), and the fact that several translation products of polysomes that correspond to major leghemoglobins can be observed, suggest that they are different gene products. Other minor leghemoglobins present in soybean nodules seem to be posttranslational modifications of the major components (Whittaker *et al.*, 1981). However, the reason for this modification and their possible role in symbiosis is not clear. Substantial evidence indicates that leghemoglobins are encoded by a closely related family of plant genes (Sullivan *et al.*, 1981; Brisson *et al.*, 1982; Brisson and Verma, 1982). According to saturation hybridization data (Baulcombe and Verma, 1978), there may be as many as 40 copies of leghemoglobin genes in soybean. All legumes studied to date have more than one leghemoglobin in their nodules.

As uninfected plant cells do not synthesize detectable amounts of leghemoglobin, derepression of the genes appears to be under the control of *Rhizobium*. Leghemoglobin apoprotein can be detected 3 to 4 days prior to the appearance of nitrogenase (Verma *et al.*, 1979; Bisseling *et al.*, 1979). Leghemoglobin is found in nodules produced by nitrogenase-deficient mutant strains of *Rhizobium* (Verma *et al.*, 1981), suggesting that induction of leghemoglobin is independent of the nitrogen fixation ability of the nodules. The level of leghemoglobin production, however, appears to be correlated with levels of *Rhizobium* nitrogenase. The stability of the leghemoglobin is related to the effectiveness of the nodules in fixing nitrogen. There is a temporal difference in the induction of different leghemoglobins in soybean (Verma *et al.*, 1979), which indicates that they may perform different functions in root nodule symbiosis.

Host–*Rhizobium* interactions presumably involve other building blocks and functional molecules apart from leghemoglobin. Analysis of the soluble cytoplasmic proteins from uninfected soybean roots and from nodules formed by different strains of *R. japonicum* indicated the presence of several polypeptides specific to the nodules, called "nodulins" (Legocki and Verma, 1979, 1980). These constitute a minor fraction of the total cytoplasmic proteins and are mostly in the molecular weight range of 12,000–20,000. A polypeptide of 35,000 MW (nodulin-35) present in all nodules [developed by effective, as well as ineffective strains of rhizobia (Legocki and Verma, 1979)] has been purified. The nodulins are encoded by the plant genes and appear to be obligatory for the development of symbiosis. However, the function of these proteins is unknown.

In search for plant genes for development of symbiosis with *Rhizobium*, Auger *et al.* (1979) determined the total complexity of mRNA from uninfected and mature nodule tissues by RNA–cDNA hybridization and concluded that a similar number of structural genes is active in root and nodule tissues. Using differential hybridizations to nodule–cDNA, a nodule specific fraction has been isolated (Auger and Verma, 1981). These sequences were shown to be host specific and were found on host polysomes only after infection. It is not known if some of these sequences are encoding for nodulins. Molecular cloning of nodule–cDNA showed the presence in this library of several nodule-specific sequences (F. Fuller and D. P. S. Verma, unpublished data).

Further characterization of these genes and their products is essential in order to understand the molecular bases of interaction between the plant and *Rhizobium*. These cloned sequences will make it possible to test effects of mutations in *Rhizobium* and also of other stimuli, such as phytohormones, on the nodulation process (Verma *et al.*, 1982). Figure 4 summarizes our current understanding of various host genes which may be directly or indirectly involved in sym-

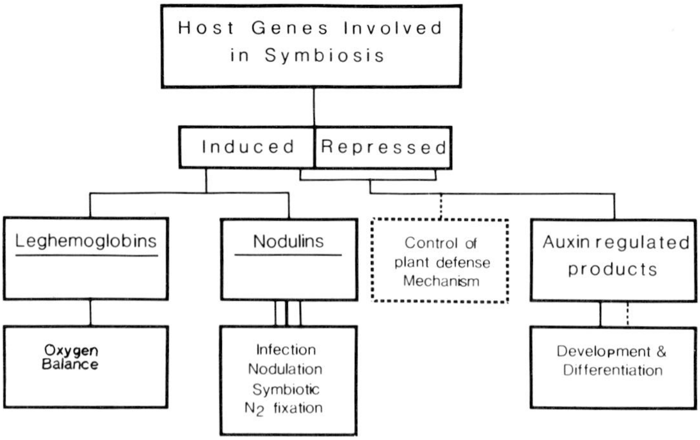

FIG. 4. Current understanding of the involvement of various host genes in symbiosis with *Rhizobium*.

biosis with *Rhizobium*. A subset of these genes appears to be regulated by phytohormones, e.g., indole acetic acid known to be produced by *Rhizobium*.

B. Genetic Analysis of the Symbiosis

The ultimate control of the interaction lies in the genes of host and bacterium. Genetic analysis provides an entry to the study of the bacterium–host signal system, and of the mechanism of differentiation. Symbiotic mutants are very useful in this regard. Mutants can reveal, through phenotype, the inner workings of a bacterial–plant interaction, as well as allowing the genes for those loci to be mapped and manipulated. A more complete treatment can be found in recent papers on bacterial genetics by Beringer *et al.* (1980), Vincent (1980), and Schwinghamer (1977).

1. *Bacterial Genes*

a. *Rhizobium Symbiotic Genetics.* Using transformation, transduction, and conjugation, and more recently plasmid studies and recombinant DNA analysis of individual regions and genes, circular linkage maps have been derived for several species: *R. meliloti, R. leguminosarum,* and *R. trifolii* (Johnston and Beringer, 1976; Meade and Signer, 1977; Kondorosi *et al.*, 1977; Casadesus and Olivares, 1979). This type of genetic analysis has been relatively undeveloped in the slow-growing species such as *R. japonicum* and the "cowpea" group.

Mutants which are unable to form symbiosis or specifically affect the symbiotic process have been isolated. These are classified as nonnodulating (Nod$^-$) or non-nitrogen-fixing (Fix$^-$). Mutant phenotypes are described in the context of nodule development, using structure to show which stage of development is the last to be completed. This is analogous to classifying auxotrophic mutants with respect to the metabolic pathway where it is blocked, by finding which pathway intermediates accumulate. Vincent (1980) has presented an overview of these studies in his reviews and has suggested a phenotypic code for mutants. Figure 5 shows a framework of various symbiotic sequence of events and phenotypic codes derived from mutations in *R. trifolii* (Rolfe *et al.*, 1981). Although descriptive analysis of *Rhizobium* mutants are available, few stage-specific biochemical markers have been established.

In some *Rhizobium* species, drug resistances are associated with loss of symbiotic effectiveness such as rifampicin resistance (Pain, 1979; Pankhurst, 1977), neomycin (Schwinghamer, 1964), and viomycin (Hendry and Jordan, 1969) resistance. It appears that changed membrane composition and permeability (MacKenzie and Jordan, 1970), possibly due to surface change, can cause or contribute to antibiotic resistance and the associated symbiotic defects. Resistance to bacteriophage, which may be mediated through changes in cell surface receptors, was found to be associated often with altered surface and defec-

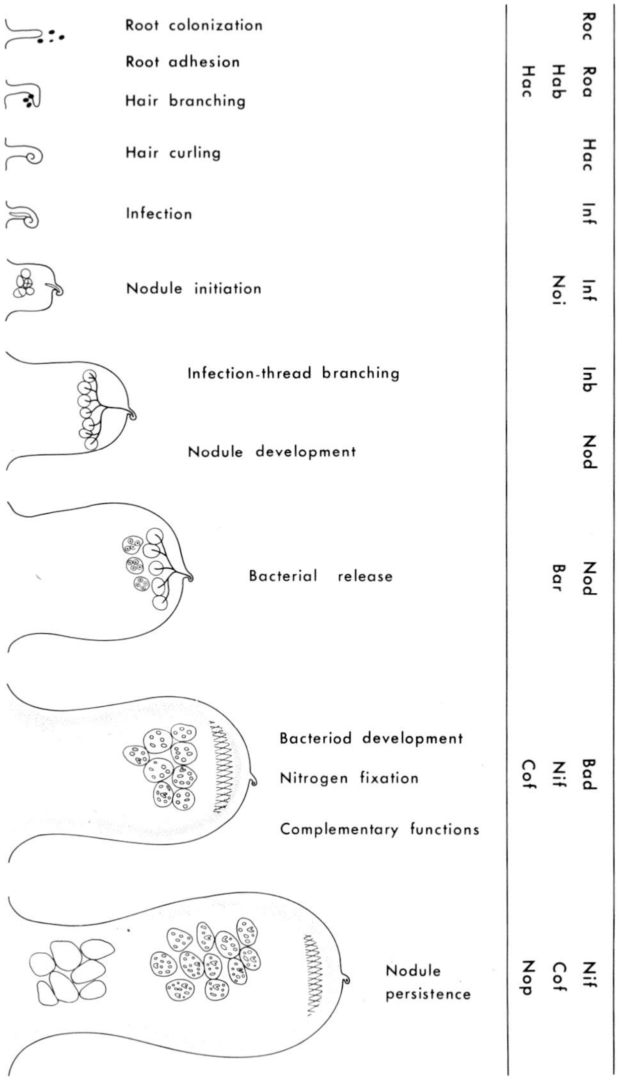

FIG. 5. Steps leading to a nitrogen-fixing nodule in white clover and phenotypic code of *R. trifolii* mutants affecting a particular stage of development. (From Rolfe *et al.*, 1981, reprinted with permission.)

tive nodulation in *R. phaseoli* (Raleigh, 1981). A mutant of *R. leguminosarum* isolated for its reduced capsule, Exo-1 (lack of exopolysaccharide), was found to be incapable of nodulation (Sanders *et al.*, 1978). However, when other exopolysaccharide mutants were isolated and examined, it showed that there is no consistent correlation with nodulation properties (Sanders *et al.*, 1981).

In certain *Rhizobium*–plant combinations, auxotrophy in the bacterial partner is associated with a specific block in the symbiosis. One metabolic group which is apparently important is the purines, particularly adenine. In *R. leguminosarum,* purine auxotrophs lost the ability to nodulate host pea plants, and all their prototrophic revertants were Nod^+. An auxotroph with a double growth requirement (adenine and thiamine) was found to be defective in nodule development (Schwinghamer, 1969; Pankhurst and Schwinghamer, 1974), causing the formation only of small bumps which never developed further. Addition of adenine to infected roots resulted in larger nodules with many more cells, but still no release of bacteria. Adenine metabolism in *Rhizobium* is of great interest since it plays a role not only in nucleic acid synthesis, but in energy metabolism, polysaccharide biosynthesis, and possibly in production of plant growth regulator compounds such as cytokinins.

Amino acid mutants are generally not affected in their symbiotic properties. Glycine-requiring mutants, for example, are more effective than their parental wild-type strain (Scherrer and Denarie, 1971). One notable exception to this is leucine, at least in *R. meliloti.* Leu^- mutants form abnormal nodules, within which the bacteria are not released from infection threads into the "target" cell cytoplasm. Leu^+ revertants are always normal, and Leu^- mutants form normal nodules if the infected roots are fed with leucine or a leucine precursor. It has been found that the initiation of the nodule meristem, and nodule organogenesis, are normal even without the release of infecting bacteria into the cytoplasm. However, polyploidization of the target cell nucleus does seem to depend on physical release of bacteria from the infection thread (Truchet *et al.,* 1980).

Riboflavin, which is important in energy metabolism, has been implicated in control of symbiotic development as indicated by studies of a riboflavin-requiring strain of *R. trifolii* [originally isolated for resistance to D-histidine (Schwinghamer, 1970)].

It is possible that auxotrophic mutations have a nonspecific effect on the symbiosis, due to general effect on vigor, for example. On the other hand, need at particular stage for these certain metabolites may occur because the bacterium produces a controlling substance at that point, such as a growth regulator.

Genes specific to the symbiosis have been identified by mutations. Several mutagenesis techniques have been used to generate symbiotic-defective mutants, including nitrosoguanidine and other point mutagens (Maier and Brill, 1976; Beringer and Johnston, 1976), heat treatments to cause deletions or curing of plasmids (Banfalvi *et al.,* 1982), and transposons (Beringer *et al.,* 1976).

Transposons are discrete DNA sequences which are capable of transposing or inserting, into nonhomologous DNA sequences elsewhere in the cell, thus creating mutations when they insert into or near a coding sequence. Ruvkun and Ausubel (1981) developed a method for creating site-specific mutations in cloned DNA sequences and "engineering" that mutant region back into a bacterium. By this means, a series of mutations were produced along a defined DNA sequence, and the effect of each determined. Using this approach they showed that the required regions for symbiotic function are located in the vicinity of the *nif* genes in *R. meliloti*.

b. *Role of Plasmids*. In addition to the chromosomal linkage group, *Rhizobium* typically harbor large molecular weight plasmids. Increasingly, it is being found that the information for symbiotic interaction with host plants is encoded on these plasmids particularly in fast-growing strains of *Rhizobium*.

Several lines of evidence have led to this conclusion. First was the observation that symbiotic properties are unstable, in that some strains spontaneously lost their symbiotic effectiveness at high frequency. This loss in some cases was accelerated by treatments known in other systems to promote plasmid loss, such as heat treatment (Zurkowski and Lorkiewicz, 1978). Also genetic studies indicated that host-range genes were not carried on the chromosome.

Transfer of a *R. leguminosarum* plasmid to a Nod$^-$ *R. leguminosarum* mutant allowed the recipient cell to nodulate peas (Johnston *et al.*, 1978; Brewin *et al.*, 1980). Plasmids can also restore Fix$^+$ phenotype to mutant recipients. This shows that nodulation and fixation genes are carried on this plasmid in these strains.

Heat-shock or long-term storage of strains often results in Nod$^-$ mutants, and these can be seen to lack a plasmid which is present in wild-type strains of *R. trifoli;* this association of plasmid loss and Nod$^-$ phenotype is also found in other strains. Moving a symbiotic *R. trifolii* plasmid into other *Rhizobium* species alters host range (Beynon *et al.*, 1980); Hooykaas, 1981). In *R. meliloti*, in addition to various medium range isolatable plasmids (Casse *et al.*, 1980), a very large "megaplasmid" is detected by direct lysis gels (Eckhardt, 1978) and is estimated to be greater than 300–450 megadaltons (Denarie *et al.*, 1981; Rosenberg *et al.*, 1981; Banfalvi *et al.*, 1981).

The basic biology of *Rhizobium* plasmids is little understood. Plasmid number and size ("plasmid profile") are widely variable from one strain of a species to another, and the relationships of these plasmids to each other is not clear. In some strains, *nif* genes are borne on a small plasmid, while in other strains, they are carried on a large one (Hirsch *et al.*, 1980; Johnston and Brewin, 1980). Localization of the nitrogenase (*nif*) genes in most cases has depended to a great extent on the identification of the *nif* sequences by sequence homology to previously cloned *nif* genes from *Klebsiella* pneumoniae (Cannon *et al.*, 1979). Using these cloned *nif* genes, Nuti *et al.* (1979) found cross-hybridization on a plasmid band in *R. leguminosarum*. In *R. meliloti nif* genes have been localized on the megaplasmid (Rosenberg *et al.*, 1982; Banfalvi *et al.*, 1982). Slow-growing strains of *Rhizobium*, including *R. japonicum*, have not been shown to carry *nif* sequences on plasmids (Haugland and Verma, 1981). It is possible that *nif* loci are integrated into the chromosome in these cases.

The demonstration that a single plasmid can both restore Fix and change Nod function in a recipient (Brewin and Johnston, 1980) and the finding that what appears to be a single deletion can eliminate nodulation function and the physically detectable *nif* sequences implied that nod and fix genes are linked (Rosenberg *et al.*, 1981; Banfalvi *et al.*, 1981). Physical and genetic data have confirmed that *nod* genes are located within 30 kilobase pairs of *nif* genes on the *R. meliloti* megaplasmid (Long *et al.*, 1982).

The existence of plasmids such as the *R. meliloti* megaplasmid raises the question of what else they code for, since the amount of DNA exceeds that likely to be needed for the few identifiable symbiotic genes. Krol *et al.* (1980) found that a symbiotic plasmid in *R. leguminosarum* was not expressed in free-living cells; however, RNA complementary to plasmid sequences was found in developing nodules. Scott (1980) found that a Nod$^-$ *R. trifolii* mutant, deleted for an entire plasmid, showed no difference in the two-dimensional protein pattern compared to wild-type cells, implying that this plasmid is not expressed to a major extent in free-living cells.

2. *Host Genetic Factors*

The involvement of host genetics in root nodule symbiosis has long been recognized. As with bacterial genes, plant symbiotic genes can be identified and analyzed by obtaining mutants and/or naturally occurring variants which are defective in the symbiosis. Such mutants and variants have been found in many host species (Caldwell and Vest, 1977; Vincent, 1980). In addition to changes of a qualitative nature, there are also many examples of quantitative effects of host genes, such as level of nitrogen fixation or different response to application of fertilizers (Gibson, 1968; Bethlenalvey, 1978). These responses have recently been reviewed by Caldwell and Vest (1977), Gibson (1980), Vincent (1980), and Nutman (1981).

Host genes control the host's selectivity among *Rhizobium* species and strains. The host plant also may control *Rhizobium*'s mode of infection. A particular "cowpea" *Rhizobium* strain infects *Ornithopus* through root hairs, but infects lupine through epidermal openings near lateral root emergence positions (Dart, 1977).

Among the other host genetic factors that are known to affect the symbiotic nodule development are time of nodule appearance (Conolly *et al.*, 1969) and nodule abundance and size (see Vincent, 1974; Nutman, 1981; Gibson, 1980). The host also has a determining effect on events later in symbiosis. Intracellular organization, including the number of bacteroids per envelope, is determined genetically by the host (Kidby and Goodchild, 1966). Bacteroid differentiation and morphology are greatly influenced by the host plant: when a single strain of *Rhizobium* was inoculated onto *Phaseolus angularis* and *Arachis hypogea*, the bacteroids in the former were long and electron dense, while those on the latter plant were round, with dispersed looking contents (Sutton *et al.*, 1982, Figs. 1 and 2; see also Dart, 1977). Also the number of bacteroids per peribacteroid membrane was different in two species.

Single plant genes, transmitted as Mendelian recessives, cause Nod^- phenotype in clover (Nutman 1954), soybean (Weber 1966), and peas (Lie, 1971; Holl, 1975; Degenhardt *et al.*, 1976). The recessive allele for Nod^- in soybean, rj_1, causes a conditional Nod^- response: the rj_1/rj_1 plants will nodulate with some *Rhizobium japonicum* strains but not others. The *Rhizobium* strains that are found to nodulate rj_1 soybean have different biochemical properties (Hubbel and Elkan, 1967), which may be a useful system to study the mechanism of host–bacterium compatibility.

Several genes for ineffectiveness can be reversed by modifying plant genes or by bacterial mutations. Examples have been identified in *Trifolium subterraneum* (Gibson, 1964), *Medicago* (Gibson, 1962; Viands *et al.*, 1979), *Pisum* (Holl, 1973, 1975; Lie and Timmermans, 1979), and in *Glycine max* (Vest, 1970; Vest and Caldwell, 1973).

Three dominant genes (Rj_2–Rj_4) conferring ineffectiveness have been identified in soybean: Rj_2 controls ineffective response to most strains of serogroups 1 and 122; Rj_3 is specific to a high strain, implying that a nonserological factor may be involved in the Rj_3 ineffectiveness (Caldwell and Vest, 1977). In pea a recessive gene (sym3) has been described; in this mutant, normal development of nodules occurs including induction of nitrogenase, but the enzyme does not function (Holl, 1973).

Four independent recessive genes have so far been identified in clover that influence nodule development at different stages. These stages are (1) release of bacteria from infection thread (genes n and d), (2) bacterial multiplication (gene ie), and (3) bacteroidal maturation [allele i_1] (Bergerson and Nutman, 1957). Another recessive factor, t, has been hypothesized to inhibit N_2 fixation after formation of normal nitrogen-fixing tissue (see Nutman, 1981).

A clear understanding of the physiological interactions of these various lesions is essential not only for understanding the molecular basis of this symbiosis, but also for a judicial selection of breeding systems for the improved performance in symbiotic nitrogen fixation of agriculturally important crops.

V. Perspectives

Rhizobium: Pathogen to Organelle

Most plant-associated bacteria are pathogens. Some, such as *Erwinia* and *Pseudomonas* colonize host tissues, developing lesions and proliferating extensively on the digested host material from killed cells. A contrasting pathogen,

Agrobacterium tumefaciens, one of *Rhizobium*'s closest relatives has evolved a unique mechanism to manipulate its host for the nutritional requirements. The bacterium inserts part of its DNA into the host plant genome which redirects host cell activity, inducing tumor (gall) formation and causing synthesis of specific compounds (opines). These compounds are utilized by *Agrobacterium* cells outside the host tissue. *Agrobacterium* and *Rhizobium* have several features in common; they share the ability to redirect plant morphogenesis via change in hormone balance and they both harbor very large, indigenous plasmids which bear genes for infection.

The way in which the various pathogens grow—uncontrolled bacterial growth at host expense or stimulation of uncontrolled host proliferation by *Agrobacterium*—contrasts with the unique symbiotic nature of the *Rhizobium*–legume interaction. Each partner influences and/or controls the other; there is division of labor for synthesis and for metabolism; and the two organisms coexist as cellular level partners: an intact bacterial cell penetrates a host cell without destroying it, and lives in the host cytoplasm as an endosymbiont.

It is apparent that the two organisms have coevolved to achieve a symbiotic state and any perturbation in their "metabolism" results in the abortion of this process, rendering it pathogenic. One may ask how a pathogenic bacterium acquired the ability to control the host defense mechanisms and why it can primarily be done with legume plants. The answer may come from understanding the molecular basis of interactions between these two organisms: How does *Rhizobium,* a facultative endosymbiont, compare with the "constitutive endosymbionts," chloroplasts and mitochondria? Is it evolving into an organelle? In the nodule, *Rhizobium* cell division rate and cell differentiation are highly influenced by the host plant. *Rhizobium's* functional role is to carry out an enzymatic process requiring compartmentation and special conditions, which is also the case with chloroplasts and mitochondria. *Rhizobium* differentiate into a form, bacteroids, which are bounded by a double set of membranes; they share and/or exchange metabolites and structures with the surrounding cytoplasm. On the other hand, *Rhizobium* maintains the ability to multiply outside its host, which true organelles cannot do. Furthermore, while rhizobial divison in the host cytoplasm is influenced by the host, it is still proliferative in nature: *Rhizobium* divides out of synchrony with the host cells, and dominates the cell in terms of function and physical space. Colonization of a host cell by *Rhizobium* is usually terminal for the host cell, if not for the *Rhizobium* (see above). It is interesting to speculate whether symbioses in which the host cells continue to divide for a short time after release of bacteria are more advanced, or whether this is an accidental arrangement.

Because bacteroid-containing cells usually do not divide, and because the invaded organ is removed from the germ line, it is somewhat moot to debate whether *Rhizobium* is evolving into an organelle. However, differentiated *Rhi-*

zobium (bacteroid) is undeniably organelle-like: and perhaps its ambiguous position provides us with insight into the early stages of endosymbiotic evolution. What would be the likely requirements for *Rhizobium* to become organellar? They might include loss of free-living growth; synchronization of bacteroidal divison with host cell division; loss of some genetic coding capacity to the host; and segregation of the essential ''organellar'' (symbiotic) genes onto a linkage group distinct from the chromosome. There is much speculation about the plasmids found in *Rhizobium*. Their large size is particularly intriguing since it is not known what all that DNA is coding for. Perhaps one angle to pursue is the ''organelle genome'' idea: Are megaplasmids *Rhizobium*'s way of making the equivalent of a chloroplast or mitochondrial genome? If so, this suggests specific genes, e.g., tRNA, ribosomal, ATPase components, and other membrane protein genes, etc., maybe localized on these plasmids. If some of these loci are found on *Rhizobium* plasmids, it may provide a glimpse of how organelles—true endosymbionts—behaved on their evolutionary path from free-living prokaryotes.

There are a number of rhizobial functions that are controlled by the host including its differentiation and metabolism. It is interesting to think why only the legume host has the ability to control these processes in *Rhizobium* in a coordinated fashion. This is analogous to some of the nuclear controls on organelles.

Acknowledgments

This study was supported by research grants from the Natural Sciences and Engineering Research Council of Canada. We would like to acknowledge the contribution of Dr. V. Kumar in preparation of some sections of this manuscript and wish to thank Dr. B. Sutton for reading and Miss Yvette Mark for typing this manuscript.

References

Albrecht, S. L., Maier, R. J., Hanus, F. J., Russell, S. A., Emerich, D. W., and Evans, H. J. (1979). Science **203,** 1255.

Allen, O. N., and Allen, E. K. (1954). *In* ''Abnormal and Pathological Plant Growth,'' Vol. 6, p. 209. Brookhaven Symp. Biol.

Allen, O. N., and Allen, E. K. (1940). *Bot. Gaz.* **102,** 121.

Allen, O. N., and Allen, E. K. (1981). ''The Leguminosae.'' Univ. of Wisconsin Press, Madison.

Ames, P., and Bergman, K. (1981). *J. Bacteriol.* **148,** 728.

Ames, P., Schluederberg, S., and Bergman, K. (1980). *J. Bacteriol.* **141,** 722.

Antoniw, L. D., and Sprent, J. I. (1978). *Phytochemistry* **17,** 675.

Appleby, C. A. (1962). *Biochim. Biophys. Acta* **60,** 226.

Appleby, C. A. (1969). *Biochim. Biophys. Acta* **188,** 222.

Appleby, C. A. (1974). *In* "The Biology of Nitrogen Fixation (A. Quispel, ed), pp. 521–554. North-Holland, Amsterdam.
Appleby, C. A., and Dilworth, M. J. (1974). *In* "Dinitrogen Fixation" (R. W. F. Hardy, ed.), Ch. 8, Vol. 1. Wiley-Interscience, New York.
Appleby, C. A., Nicola, N. A., Hurrell, J. G. R. and Leach, S. J. (1975). *Biochemistry* **14,** 4444.
Auger, S., and Verma, D. P. S. (1981). *Biochemistry* **20,** 1300.
Auger, S., Baulcombe, D., and Verma, D. P. S. (1979). *Biochim. Biophys. Acta* **563,** 496.
Ausubel, F. M., Bird, S. C., Durbin, K. J., Janssen, K. A., Margulskee, R. F., and Peskin, A. A. (1979). *J. Bacteriol.* **140,** 597.
Avissar, Y. J., and Nadler, K. D. (1978). *J. Bacteriol.* **135,** 782.
Bal, A. K., Shantharam, S., and Verma, D. P. S. (1980). *Can. J. Microbiol.* **26,** 1096.
Banfalvi, Z. V., Sakanyan, V., Konza, C., Kiss, A., Dusha, I., and Kondorosi, A. (1981). *Mol. Gen. Genet.* **184,** 318.
Basset, D., Goodman, R. N., and Novacky, A. (1977a). *Can. J. Microbiol.* **23,** 573.
Basset, D., Goodman, R. N., and Novacky, A. (1977b). *Can. J. Microbiol.* **23,** 873.
Bauer, W. D. (1981). *Annu. Rev. Plant Physiol.* **32,** 407.
Baulcombe, D., and Verma, D. P. S. (1978). *Nucl. Acids. Res.* **5,** 4141.
Benthenfalvay, G. J., and Phillips, D. A. (1977). *Plant Physiol.* **63,** 816.
Bergersen, F. J. (1971). *Annu. Rev. Plant Physiol.* **22,** 121.
Bergersen, F. J. (1974). *In* "The Biology of Nitrogen Fixation" (A. Quispel, ed.), pp. 473–498. North-Holland, Amsterdam.
Bergersen, F. J., and Appleby, C. A. (1981). *In* "Current Perspectives in Nitrogen Fixation" (A. H. Gibson and W. E. Newton, eds.). Aust. Acad. Sci., Canberra.
Bergersen, F. J., and Briggs, M. J. (1958). *J. Gen. Microbiol.* **19,** 482.
Bergersen, F. J., and Goodchild, D. J. (1973). *Aust. J. Biol. Sci.* **26,** 741.
Bergersen, F. J., and Nutman, P. S. (1957). *Heredity* **11,** 175.
Bergersen, F. J., and Turner, G. L. (1967). *Biochim. Biophys. Acta* **141,** 507.
Bergersen, F. J., and Turner, G. L. (1975). *J. Gen. Microbiol.* **91,** 345.
Bergersen, F. J., Turner, G. L., and Appleby, C. A. (1973). *Biochim. Biophys. Acta* **292,** 271.
Beringer, J. E., Hoggan, S. A., and Johnston, A. W. B. (1978). *J. Gen. Microbiol.* **104,** 201.
Beringer, J. E. Brewin, N. J., and Johnston, A. W. B. (1980). *Heredity* **45,** 161.
Beynon, J. L., Beringer, J. E., and Johnston, A. W. B. (1980). *J. Gen. Microbiol.* **120**/421.
Bhuvaneswari, T., Pueppke, S., and Bauer, W. (1977). *Plant Physiol.* **60,** 486.
Bishop, P. E. Quevara, J. G., Engelke, J. A., and Evans, H. J. (1976). *Plant Physiol.* **57,** 542.
Bisseling, T., Van den Bos, R. C. Van kammer, A., Van der Plueg, M., Van Duijn, P., and Houwers, A. (1977). *J. Gen. Microbiol.* **101,** 79.
Bisseling, T., Van den Bos, R. C., Weststrate, M. W., Hakkaart, M. J. J., and van Kammen, A. (1979). *Biochim. Biophys. Acta* **562,** 515.
Bisseling, T., Moen, A. A., Van Den Boss, R. C., and Van Kammen, A. (1980). *J. Gen. Microbiol.* **118,** 377.
Bohlool, B. B., and Schmidt, E. L. (1974). *Science* **185,** 269.
Brenchley, J. E., Prival, M. J., and Magasanik, B. (1973). *J. Biol. Chem.* **258,** 6122.
Brewin, N. J., Beringer, J. E., Buchanan-Wollaston, A. V., and Hirsch, P. R. (1980). *J. Gen. Microbiol.* **116,** 261.
Brisson, N., and Verma, D. P. S. (1982). *Proc. Natl. Acad. Sci. U.S.A.*, **79,** 4055–4059.
Brisson, N., Pombo-Gentile, A., and Verma, D. P. S. (1982). *Can. J. Biochem.* **60,** 272.
Broughton, W. J. (1978). *J. Appl. Bacteriol.* **45,** 165.
Burris, R. H., Arp. D. J., Hageman, R. V., Houchins, J. P., Sweet, W. J., and Tso, M. (1981). *In* "Current Perspectives in Nitrogen Fixation" (A. H. Gibson and W. E. Newton, eds.), pp. 56–66. Aust. Acad. Sci., Canberra.

Caldwell, B. E., and Vest, H. G. (1977). *In* "A Treatise on Dinitrogen Fixation" (R. W. F. Hardy and W. S. Silver, eds.), pp. 557–576. Wiley-Interscience, New York.
Callaham, D. A. (1979). M.S. Thesis, Univ. Mass., Amherst.
Callaham, D. A., and Torrey, J. G. (1981). *Can. J. Bot.* **59,** 1647.
Cannon, F. C., Reidel, G. E., and Ausubel, F. M. (1979). *Mol. Gen. Genet.* **174,** 59.
Carlson, R. (1982). *In* "Ecology of Nitrogen Fixation" (W. Broughton, ed.), Vol. 2, pp. 199–234. Oxford University Press, London.
Carlson, R., Sanders, R., Napoli, C., and Albesheim, P. (1978). *Plant Physiol.* **62,** 912.
Carter, K. R., Jennings, N. T., Hanus, F. J., and Evans, H. J. (1978). *Can. J. Gen. Microbiol.* **24,** 307.
Casadesus, J., and Olivares, J. (1979). *Mol. Gen. Genet.* **174,** 203.
Casse, F., Boucher, C., Julliot, J. S., Michel, M., and Dénarié, J. (1979). *J. Microbiol.* **113,** 229.
Chandler, M. R., and Dart, P. J. (1973). Rep. Rothansted Exp. Station Part I: P 5.
Ching, T. M., Hedtke, S., Russell, S. A., and Evans, H. J. (1975). *Plant Physiol.* **55,** 796.
Conolly, V., Masterson, C. L., and Conniffe, D. (1969). *Theor. Appl. Genet.* **39,** 206.
Corbin, D., Ditta, G., and Helinski, D. R. (1982). *J. Bacteriol.* **149,** 221.
Coventry, D. R., and Dilworth, M. J. (1975). *J. Gen. Microbiol.* **90,** 69.
Currier, W. M., and Strobel, G. A. (1974). *Science* **196,** 434.
Cutting, J. A., and Schulman, H. M. (1969). *Biochim. Biophys. Acta* **192,** 486.
Darrow, R. A., and Knotts, R. R. (1977). *Biochem. Biophys. Res. Commun.* **78,** 554.
Dart, P. J. (1974). *In* "The Biology of Nitrogen Fixation" (A. Quispel, ed), pp. 381–429. North-Holland, Amsterdam.
Dart, P. (1977). *In* "A Treatise on Dinitrogen Fixation" (R. W. F. Hardy and W. S. Silver, eds), pp. 367–472. Wiley-Interscience, New York.
Dart, P. J., and Mercer, F. V. (1963a). *J. Bacteriol.* **85,** 951.
Dart, P. J., and Mercer, F. V. (1963b). *Arch. Mikrobiol.* **46,** 382.
Dart, P. J., and Mercer, F. V. (1963c). *Arch. Mikrobiol.* **47,** 1.
Dart, P. J., and Mercer, F. V. (1964). *Arch. Mikrobiol.* **49,** 209.
Dart, P. J., andMercer, F. V. (1966). *J. Bacteriol.* **91,** 1314.
Dazzo, F. B., and Brill, W. J. (1977). *Appl. Environ. Microbiol.* **33,** 132.
Dazzo, F. B., and Brill, W. J. (1979). *J. Bacteriol.* **137,** 1362.
Dazzo, F., and Hubbell, D. (1982). *In* "Nitrogen Fixation in Rhizobium" (W. Broughton, ed.), Vol. 2, pp. 275–309. Oxford University Press, London.
Dazzo, F., Napoli, C., and Hubbell, D. (1976). *Appl. Environ. Microbiol.* **32,** 166.
Dazzo, F. B., Yanke, W. E., and Brill, W. J. (1978). *Biochim. Biophys. Acta* **539,** 276.
Dazzo, F., Urbano, M., and Brill, W. (1979). *Curr. Microbiol.* **2,** 15.
Degenhardt, T. L., Larue, T. A., and Paul, E. A. (1976). *Can. J. Bot.* **54,** 1633.
Dénarié, J., Rosenberg, C., Boistard, P., Truchet, G., and Casse-Delbart, F. (1981). *In* "Current Perspectives in Nitrogen Fixation" (A. H. Gibson and W. E. Newton, eds.), pp. 137–141. Aust. Acad. Sci., Canberra.
Dilworth, M. J. (1969). *Biochim. Biophys. Acta* **184,** 432.
Dilworth, M. J., and Kidby, D. K. (1968). *Exp. Cell. Res.* **49,** 148.
Dilworth, M. J., and Parker, C. A. (1969). *J. Theor. Biol.* **25,** 208.
Dilworth, M. J., and Williams, D. C. (1967). *J. Gen. Microbiol.* **48,** 31–36.
Dixon, R. O. D. (1967). *Arch. Microbiol.* **56,** 156.
Dixon, R. O. D. (1968). *Arch. Mikrobiol.* **62,** 272.
Dunican, L. K., and Cannon, F. C. (1971). *Plant Soil* **7,** 73.
Eckhardt, T. (1978). *Plasmid* **1,** 584.
Ellfolk, N. (1972). *Endeavour* **31,** 139.

Ellfolk, N., and Sievers, G. (1971). *Acta Chem. Scand.* **25,** 3532.
Fahraeus, G. (1957). *J. Gen. Microbiol.* **16,** 1374.
Fuchsman, W., and Appleby, C. A. (1979). *Biochim. Biophys. Acta* **579,** 314.
Generozova, I. P., and Yagodin, B. A. (1972). *Fiziol. Rast. (Moscow)* **19,** 348.
Gerson, T., Patel, J. J., and Wong, M. M. (1978). *Physiol. Plant* **42,** 420.
Gibson, A. H. (1962). *Aust. J. Agric. Res.* **13,** 388.
Gibson, A. H. (1968). *Aust. J. Agric. Res.* **19,** 907.
Gibson, A. H. (1980). *In* "Advances in Legume Science" (R. J. Summerfield and A. H. Bunting, eds.), pp. 69–75. Royal Botanic Gardens Kew.
Godfrey, C. A. (1972). Ph.D. thesis, University of Western Australia, Perth.
Goodchild, D. J., and Bergersen, F. J. (1966). *J. Bacteriol.* **92,** 204.
Gourret, J. P., and Fernandez-Arias, H. (1974). *Can. J. Microbiol.* **20,** 1169.
Groat, R. G., and Vance, C. P. (1981). *Plant Physiol.* **67,** 1198.
Gresshoff, P. M., and Rolfe, B. G. (1978). *Planta* **142,** 329.
Gresshoff, P. M., Skotnicki, M., Eadie, J., and Rolfe, B. G. (1977). *Plant Sci. Lett.* **10,** 299.
Gunning, B. E. S., and Pate, J. S. (1974). *In* "Dynamic Aspects of Plant Structure" (A. W. Robards, ed.), p. 441. McGraw-Hill, New York.
Gunning, B. E. S., Pate, J. S., Minchin, F. R., and Marks, I. (1974). *Symp. Soc. Exp. Biol.* **28,** 87.
Hardy, R. W. F., and Havelka, M. D. (1975). *Science* **188,** 633.
Haugland, R., and Verma, D. P. S. (1981). *J. Mol. Appl. Genet.* **1,** 205.
Hendry, G. S., and Jordan, D. C. (1969). *Can. J. Microbiol.* **15,** 671.
Hennecke, H. (1981). *Nature (London)* **291,** 354.
Herridge, D. F., Atkins, C. A., Pate, J. S., and Rainbrid, R. M. (1978). *Plant Physiol.* **62,** 495.
Higashi, S. (1966). *J. Gen. Appl. Microbiol.* **12,** 147.
Hirsch, P. R., Van Montagu, M., Johnston, A. W. B., Brewin, N. J., and Schell, J. (1980). *J. Gen. Microbiol.* **120,** 403.
Holl, F. B. (1973). *Can. J. Genet. Cytol.* **15,** 659.
Holl, F. B. (1975). *Euphytica* **24,** 767.
Hooykaas, P. J. J., Van Brussel, A. A. N., den Dulk-Ras, H., Van Slogteren, G. M. S., and Schilperoort, R. A. (1981). *Nature (London)* **291,** 351.
Hornez, J. P., Courtois, B., Defives, C., and Derieux, J. C. (1974). *C. R. Acad. Sci. (Paris) Ser. D* **278,** 157.
Hrabak, E., Urbano, M., and Dazzo, F. (1981). *J. Bacteriol.* **148,** 697.
Hubbell, D. H., and Elkan, G. H. (1967). *Phytochemistry* **6,** 321.
Hubbell, D. H., Morales, V. M., and Umali-Garcia, M. (1978). *Appl. Environ. Microbiol.* **35,** 210.
Hunt, L. T. (1972). *In* "Atlas of Protein Sequence and Structure" (M. O. Dayhoff, ed), pp. D369–D370. Nat. Biomedical Res. Found., Washington.
Hyldig-Nielsen, J. O., Jensen, E. P., Wilborg, O., Garrett, R., Jorgenesen, P., and Marcker, K. A. (1982). *Nucleic Acid Res.* **10,** 689.
Johnson, H. W., and Means, U. M. (1960). *Agron. J.* **52:** 651.
Johnston, A. W. B., and Beringer, J. E. (1977). *Nature (London)* **267,** 611.
Johnson, H. W., and Means, U. M. (1960). *Agron. J.* **52:** 651.
Johnston, A. W. B., and Brewin, N. J. (1981). *In* "Current Perspectives in Nitrogen Fixation" (A. H. Gibson and W. E. Newton, eds.), p. 404. Aust. Acad. Sci., Canberra.
Jordan, D. C. (1962). *Bacteriol. Rev.* **26,** 119.
Jordan, D. C. (1981). *In* "Current Perspectives in Nitrogen Fixation" (A. M. Gibson and W. E. Newton, eds.), pp. 317–320. Aust. Acad. Sci., Canberra.
Jordan, D. C., and Grinyer, I. (1965). *Can. J. Microbiol.* **11,** 721.
Jordan, D. C., Grinyer, I., and Coulter, W. H. (1963). *J. Bacteriol.* **86,** 125.

Kidby, D. K., and Goodchild, D. J. (1966). *J. Gen. Microbiol.* **45,** 147.
Kondorosi, A., Kiss, G. B., Forrai, T., Vincze, E., and Banfalvi, Z. (1977). *Nature (London)* **268,** 525.
Krol, A. J. M., Hontelez, J. G. J., Van den Bos, R. C., and Van Kammen, A. (1980). *Nuclic Acid Res.* **8,** 4337.
Lawrie, A. C., and Wheeler, C. T. (1975). *New Phytol.* **74,** 429.
Legocki, R. P., and Verma, D. P. S. (1979). *Science* **205,** 190.
Legocki, R. P., and Verma, D. P. S. (1980). *Cell* **20,** 153.
Li, D., and Hubbell, D. H. (1969). *Can. J. Microbiol.* **15,** 1133.
Libbenga, K. R., and Bogers, R. J. (1974). *In* "The Biology of Nitrogen Fixation" (A. Quispel, ed), pp. 430–472. North-Holland, Amsterdam.
Libbenga, K. R., and Torry, J. G. (1973). *Am. J. Bot.* **60,** 293.
Libbenga, K. R., Iren, Van, F., Bogers, R. J., and Schraag-Lamers, M. F. (1973). *Planta* **114,** 29.
Lie, T. A. (1971). *Plant Soil (Special volume)*, pp. 117–127.
Ljunggren, H. (1969). *In* "Mechanism and Pattern of Rhizobium Invasion into Leguminous Root Hairs" (C. Bloms and A. B. Baktryckeri, eds.), p. 18. Lund.
Ljunggren, H., and Fahraeus, G. (1961). *J. Gen. Microbiol.* **26,** 521.
Long, S. R., Buikema, W., and Ausubel, F. M. (1982). *Nature (London)*, in press.
Ludwig, R. A., and Singer, E. R. (1977). *Nature (London)* **267,** 245.
MacKenzie, C. R., and Jordan, D. C. (1970). *Biochem. Biophys. Res. Commun.* **40,** 1008.
Mackenzie, C. R., and Jordan, D. C. (1974). *Can. J. Microbiol.* **20,** 755.
Mackenzie, C. R., Vail, W. J., and Jordan, D. C. (1973). *J. Bacterial.* **113,** 387.
Mahon, J. D. (1981). *In* "Ecology of Nitrogen Fixation," Vol. 3.
Maier, R. J., and Brill, W. J. (1978). *J. Bacteriol.* **133,** 1295.
Maier, R. J., Postgate, J. R., and Evans, H. J. (1978). *Nature (London)* **276,** 494.
Marks, I., and Sprent, J. I. (1974). *J. Cell. Sci.* **16,** 623.
Matsumoto, T., Yatazawa, M., and Yamamota, Y. (1977). *Plant Cell Phys.* **18,** 673.
Meade, H. M., and Signer, E. R. (1977). *Proc. Natl. Acad. Sci. U.S.A.* **74,** 2076.
Meeks, J. C., Wolk, C. P., Schilling, N., Shaffer, P. W., Avissar, Y., and Chien, W-S. (1978). *Plant Physiol.* **61,** 980.
Miflin, B. J., and Lea, P. J. (1976). *Phytochemistry* **15,** 873.
Minchin, F. R., and Pate, J. S. (1973). *J. Exp. Bot.* **24,** 259.
Mort, A., and Bauer, W. (1980). *Plant Physiol.* **66,** 158.
Mosse, B. (1964). *J. Gen. Microbiol.* **36,** 49.
Napoli, C. A., and Hubbell, D. H. (1975). *Appl. Microbiol.* **30,** 1003.
Newcomb, W. (1976). *Can. J. Bot.* **54,** 2163.
Newcomb, W. (1981). *Int. Rev. Cytol. Suppl.* **13,** 247.
Newcomb, W., and Mc Intyre, L. (1981). *Can. J. Bot.* **59,** 2478.
Newcomb, W., and Peterson, R. L. (1979). *Can. J. Bot.* **57,** 2583.
Nuti, M. P., Lepidi, A. A., Prakash, R. K., Schilperoort, R. A., and Cannon, F. C. (1979). *Nature (London)* **282,** 533.
Nutman, P. S. (1954). *Heredity* **8,** 47.
Nutman, P. S. (1956). *Biol. Rev. Camb. Phil. Soc.* **31,** 109.
Nutman, P. S. (1959). *Symp. Soc. Exp. Biol.* **13,** 42.
Nutman, P. S. (1981). *In* "Current Perspectives in Nitrogen Fixation" (A. H. Gibson and W. E. Newton, eds.), pp. 194–204. Aust. Acad. Sci., Canberra.
O'Gara, F., and Shanmugam, K. T. (1976). *Biochim. Biophys. Acta* **437,** 313.
Orme-Johnson, W. H., Davis, L. C., Henzl, M. T., Averill, B. A., Orme-Johnson, N. R., Munck, E., and Zimmerman, R. (1977). *In* "Recent Developments in Nitrogen Fixation" (W. Newton, J. R. Postage, and C. Rodriguez-Barrueco, eds.), pp. 271–285. Academic Press, New York.

Paau, A. S., Jing, Y., and Brill, W. J. (1981). *Plant Physiol.* **67** (Suppl.), 518.
Pain, A. N. (1979). *J. Appl. Bacteriol.* **47,** 53.
Pankhurst, C. E. (1977). *Can. J. Microbiol.* **23,** 1026.
Pankhurst, C. E., and Schwinghamer, E. A. (1974). *Arch. Microbiol.* **100,** 219.
Pate, J. S. (1962). *Plant Soil* **17,** 333.
Pate, J. S. (1977). *In* "A Treatise on Dinitrogen Fixation (R. W. F. Hardy and W. S. Silver, eds.), pp. 473–517. Wiley, New York.
Pate, J. S. (1980). *Annu. Rev. Plant Phys.* **31,** 313.
Pate, J. S., Gunning, B. E. S., and Briarty, L. G. (1969). *Planta* **85,** 11.
Pate, J. S., Atkins, C. A., and Rainbird, R. M. (1981). *In* "Current Perspectives in Nitrogen Fixation" (A. H. Gibson and W. E. Newton, eds.), pp. 105–116. Aust. Acad. Sci., Canberra.
Peters, G. A., Mayne, B. C., Ray, T. B., and Toia, R. E. (1980). *In* "Nitrogen Fixation" (W. E. Newton and W. H. Orme-Johnson, eds.), Vol. 2, pp. 295–309.
Phillips, D. A. (1980). *Annu. Rev. Plant Physiol.* **31,** 29.
Planque, K., DeVries, G. E., and Kijne, J. W. (1978). *J. Gen. Microbiol.* **106,** 173.
Prakash, R. K., Hooykaas, P. J. J., Ledeboer, A. M., Schilperoort, R. A., Nuti, M. P., Lepidi, A. A., Casse, F., Boucher, C., Julliot, J. S., and Dénarié, J. (1980). *In* "Proc. III. Intern. Symp. of Nitrogen Fixation" (W. E. Newton and W. H. Orme-Johnson, eds.), Vol. II, pp. 139–163. Madison University Park Press, Baltimore.
Prasad, D. N., and De, D. N. (1971). *Microbios.* **4,** 13.
Pull, S. P., Pueppke, S. L., Hymowitz, T., and Orf, J. H. (1978). *Science* **200,** 1277.
Raleigh, E. (1981). Ph.D. Thesis, Massachusetts Institute of Technology, Boston.
Rawsthorne, S., Minchin, F. R., Summerfield, R. J., Cookson, C., and Coombs, J. (1980). *Phytochemistry* **19,** 341.
Reijnders, L., Visser, L., Aalbers, A. M. J., Van Kammen, A., and Houwers, A. (1975). *Biochim. Biophys. Acta* **414,** 206.
Roberston, J. G., Taylor, M. P., Craig, A., and Hopcroft, D. H. (1974). *In* "Mechanisms of Regulation of Plant Growth" (R. L. Bieleski *et al.*, eds.), Vol. 12, p. 31. Bull. R. Soc. N. Z.
Roberston, J. G., Farnden, K. J. F., Warbuton, M. P., and Banks, J-A.M. (1975a). *Aust. J. Plant Physiol.* **2,** 265.
Robertson, J. G., Warburton, M. P., and Farnden, W. J. F. (1975b). *FEBS Lett.* **55,** 22.
Robertson, J. G., Warburton, M. P., Lyllbeto, P., Fordyce, A. M., and Bullivant, S. (1978). *J. Cell Sci.* **30,** 151.
Rolfe, B. G., Gresshoff, P. M., and Shine, J. (1980). *Plant Science Lett.* **19,** 277.
Rosenberg, C., Boistard, P., Dénarié, J., and Casse-Delbart, F. (1981). *Mol. Gen. Genet.* **184,** 326.
Ruvkun, G. B., and Ausubel, F. M. (1980). *Proc. Natl. Acad. Sci. U.S.A.* **77,** 191.
Ruvkun, G. B., and Ausubel, F. M. (1981). *Nature (London)* **289,** 85.
Sahlman, K., and Fahraeus, G. (1962). *K. Lamtbr. Hogsk. Annlr.* **28,** 261.
Sahlman, K., and Fahraeus, G. (1963). *J. Gen. Microbiol.* **33,** 425.
Sanders, R. E., Carlson, R. W., and Albersheim, P. (1978). *Nature (London)* **271,** 240.
Sanders, R. E., Raleigh, E., and Signer, E. (1981). *Nature (London)* **292,** 148.
Scherrer, A., and Dénarié, J. (1971). *Plant Soil (special volume)*, pp. 39–45.
Schubert, K. R., and Evans, H. J. (1976). *Proc. Natl. Acad. Sci. U.S.A.* **73,** 1207.
Schubert, K. R., and Rayle, G. J. A. (1979). *In* "Proceedings of the Second International Legume Conference" (H. Bunting and R. J. Summerfield, eds.).
Schwinghamer, E. A. (1964). *Can. J. Microbiol.* **10,** 221.
Schwinghamer, E. A. (1970). *Aust. J. Biol. Sci.* **23,** 1187.
Schwinghamer, E. A. (1977). *In* "A Treatise on Dinitrogen Fixation" (R. W. F. Hardy and W. S. Silver, eds), pp. 577–622. Wiley-Interscience, New York.
Scott, K. F., Rolfe, B. G., Shine, J., Sundaresan, V., and Ausubel, F. M. (1981). *In* "Current

Perspectives in Nitrogen Fixation" (A. H. Gibson and W. E. Newton, eds.), pp. 393–395. Aust. Acad. of Sci. Canberra.
Shaw, B. D., and Sutton, W. D. (1979). *Biochim. Biophys. Acta* **563,** 216.
Sievers, G., Huhtala, M. L., and Ellfork, N. (1978). *Acta. Chem. Scand. B* **31,** 723.
Stacey, G., Paau, A., and Brill, W. (1980). *Plant Physiol.* **66,** 609.
Sullivan, D., Brisson, N., Goodchild, B., Verma, D. P. S., and Thomas, D. Y. (1981). *Nature (London)* **289,** 516.
Sutton, W. D. (1974). *Biochim. Biophys. Acta* **336,** 1.
Sutton, W. D., and Jepsen, N. N. (1975). *Plant Physiol.* **56,** 665.
Sutton, W. D., and Patterson, A. D. (1980). *Planta* **148,** 207.
Sutton, W. D., and Robertson, J. G. (1974). *In* "Mechanism of Regulation of Plant Growth" (R. L. Bieleski, H. R. Ferguson, and M. M. Cresswell, eds.), pp. 23–30. Royal Soc. N.Z. Bull. 12, Wellington.
Sutton, W. D., Jepsen, N. M., and Shaw, B. D. (1977). *Plant Physiol.* **59,** 741.
Sutton, W. D., Pankhurst, C. E., and Craig, A. S. (1981). *Int. Rev. Cytol. Suppl.* **13,** 149.
Syono, K., Newcomb, W., and Torrey, J. G. (1976). *Can. J. Bot.* **54,** 2155.
Tjepkema, J. D. (1971). Ph.D. Thesis, University of Michigan, Ann Arbor.
Tjepkema, J. D., and Yocum, C. S. (1974). *Planta* **119,** 351.
Truchet, G. (1978). *Ann. Sc. Nat. Bot. Biol. Vég.* **19,** 3–38.
Truchet, G., Michel, M., and Dénarié, J. (1980). *Differentiation* **16,** 163.
Tsien, H., and Schmidt, E. (1981). *J. Bacteriol.* **145,** 1063.
Tu, J. C. (1974). *J. Bacteriol.* **119,** 986.
Tu, J. C. (1977). *Can. J. Bot.* **55,** 35.
Van Brussell, A. A. N., Costerton, J. W., and Child, J. J. (1979). *Can. J. Microbiol.* **25,** 352.
Vance, C. P., and Johnson, L. E. B. (1981). *Plant Disease* **65,** 118.
Vance, C. P., Johnston, L. E. B., and Hardarson, G. (1980). *Physiol. Plant Pathol.* **17,** 167.
Verma, D. P. S. (1982). *In* "Molecular Biology of Plant Development" (H. Smith and D. Grierson, eds.), pp. 437–466. Blackwell Scientific Publ., Oxford.
Verma, D. P. S., and Bal, A. K. (1976). *Proc. Natl. Acad. Sci. U.S.A.* **73,** 3843.
Verma, D. P. S., Zogbi, V., and Bal, A. K. (1978a). *Plant Sci. Lett.* **13,** 137.
Verma, D. P. S., Kazazian, V., Zogbi, V., and Bal, A. K. (1978b). *J. Cell Biol.* **78,** 919.
Verma, D. P. S., Hunter, N., and Bal, A. K. (1978c). *Planta* **138,** 107.
Verma, D. P. S., Ball, S., Guerin, C. W., and Wanamaker, L. (1979). *Biochemistry* **18,** 476.
Verma, D. P. S., Haugland, R., Brisson, N., Legocki, R. P., and Lacroix, L. (1981). *Biochim. Biophys. Acta* **653,** 98.
Viands, D. R., Vance, C. R., Heidel, G. H., and Barnes, D. K. (1979). *Crop Sci.* **19,** 905.
Vincent, J. M. (1974). *In* "The Biology of Nitrogen Fixation" (A. Quispel, ed.), pp. 265–341. North-Holland, Amsterdam.
Vincent, J. M. (1977). *In* "A Treatise on Dinitrogen Fixation" (R. W. F. Hardy and W. S. Silver, eds.), pp. 277–366. Wiley-Interscience, New York.
Vincent, J. M. (1980). *In* "Nitrogen Fixation" (W. E. Newton and W. H. Orme-Johnson, eds.), Vol. 2, pp. 103–129. Univ. Park Press, Baltimore.
Vose, P. B., and Ruschel, A. P. (1981). "Associative Nitrogen Fixation." CRC Press, Cleveland.
Weber, C. R. (1966). *Agron. J.* **58,** 43.
Werner, D., and Mörschel, E. (1978). *Planta* **141,** 169.
Werner, D., Mörschel, E., Stript, R., and Winchenbach, B. (1980). *Planta* **147,** 320.
Whiting, M. J., and Dilworth, M. J. (1974). *Biochim. Biophys. Acta* **371:** 337.
Whittaker, R. G., Lennox, S., and Appleby, C. A. (1981). *Biochem. Int.* **3,** 117.
Wilson, J. K. (1944). *Soil Science* **58,** 61.

Wittenberg, J. B., Bergersen, F. J., Appleby, C. A., and Turner, G. L. (1974). *J. Biol. Chem.* **249,** 4057.
Wolpert, J. S., and Albersheim, P. (1976). *Biochem. Biophys. Res. Commun.* **70,** 729.
Wong, P. P., and Evans, H. J. (1971). *Plant Physiol.* **47,** 750.
Yao, P. Y., and Vincent, J. M. (1969). *Aust. J. Biol. Sci.* **22,** 413.
Yocum, C. S. (1964). *Science* **146,** 432.
Zevenhuizen, L., Shoten-Koerselman, I., and Posthumus, M. (1980). *Arch. Microbiol.* **125,** 1.
Zogbi, V. (1982). M.Sc. Thesis, McGill University.
Zurkowski, W., and Lorkiewicz, Z. (1979a). *Genet. Res.* **32,** 311.
Zurkowski, W., and Lorkiewicz, Z. (1978b). *Acta Microbiol. Pol.* **27,** 309.

Analysis of Possible Gene Transfer between an Insect Host and Its Bacteria-like Endocytobionts

W. SCHWEMMLER

Institut für Pflanzenphysiologie und Zellbiologie, Freie Universität Berlin, Berlin, Federal Republic of Germany

I. Definition of Endocytobiosis

Symbiosis is defined as a partnership between different species of organisms which is useful for both partners. The larger partner is called the host, the smaller is the symbiont. A special form is intracellular symbiosis or endocytobiosis in which the symbionts—referred to here as endocytobionts—live in special cells of their host.

Endocytobionts are to be found in almost all plant and animal phyla (Buchner, 1965). Nitrogen-assimilating bacteria which are located in special cells, the nodules, of the rootstock allow certain legumes such as beans and peas to live on nitrogen-poor soils. Insects which are nutritional specialists, in so far as they consume only mammalian blood, phloem, or hard-to-digest plant substances such as cellulose and wood, often require endocytobionts. In these cases, the host and symbiont show the highest degree of mutual dependence. A typical example is the endocytobiosis of plant-phloem-sucking cicadas, a sister group of the lice. This is one of the few endocytobioses for which important aspects of the structural, functional, and genetic interdependence between the eukaryotic host cell and the prokaryotic endocytobiont are known (Schwemmler, 1978, 1981a). It serves here as an example which is better than a review-like presentation.

In the following, the developmental cycle of the cicada host and its symbiont will be described. On the basis of this information, the possibility of gene transfer between the bacterial endocytobiont and the host cell, also called the

ISBN 0-12-364375-9

bacteriocyte, will be discussed, as well as the functional consequences of a possible gene transfer.

gene transfer?
bacteriocyte ⟷ endocytobiont
(eukaryote) (prokaryote)

II. Endocytobiosis Cycle

The small cicada *Euscelis incisus* K.[1] is found throughout Europe. Its developmental cycle consists of egg, larval, and adult phases. In the inner circle of the developmental diagram (Fig. 1), the host stages are illustrated, and the corresponding symbiont organs are seen in the outer circle. The time values in the center of the diagram refer to standard cultivation conditions.

Euscelis harbors two essential types of bacteria-like endocytobionts which are considered to belong to the as such newly described, intracellular prokaryote group known as Protoplastoides (Schwemmler, 1971; Schwemmler *et al.*, 1975). Type *a*, which is also found in lice and in primitive forms of lice and cicadas, can occur alone as the primary symbiont; the type *t* auxiliary symbiont appears, when present, only together with a type *a* symbiont (M̈uller, 1949). A third, rickettsia-like endosymbiont is not essential for the cicada. It can be referred to as the accessory symbiont KR_E. The essential primary and auxiliary endocytobiont types are transferred to the next egg generation by way of egg infection. This occurs with the help of a special infectious stage enclosed in a double membrane. The infectious stage is inserted, in the form of a symbiont ball consisting of approximately 200 individuals, into the posterior pole area between the egg cell and the egg coat. The fate of the endocytobionts then varies (Körner, 1969; reviewed by Schwemmler, 1973, 1978, 1981a). The *a*-endocytobionts are soon taken up by polyploid host cells of the egg plasma, the so-called a_1-bacteriocytes. These, however, disintegrate during a later stage. The released *a*-endocytobionts then enter the final or a_2-bacteriocytes which are cells of gonadal origin with two cell nuclei. Only during this stage are the *t*-endocytobionts phagocytosed polyploidic *t*-bacteriocytes which have previously wandered down from certain relatively far-removed abdominal segments of the embryo. Within these bacteriocytes the *a*- and *t*-endocytobionts then differentiate into their reproductive forms, the so-called vegetative stages, which are surrounded by a triple membrane (Fig. 2).

In the further course of development, the *a*- and *t*-bacteriocytes merge in a

[1]The earlier name *Euscelis plebejus* is no longer accepted. According to Metcalf's general catalog (1967) the species must now be called *Euscelis incisus* Kbm. 1858. The name *E. plebejus*, introduced in 1806 by Fallén, was based on the rules of priority, but must be dropped because of primary homonomes (see Strübing and Hasse, 1974).

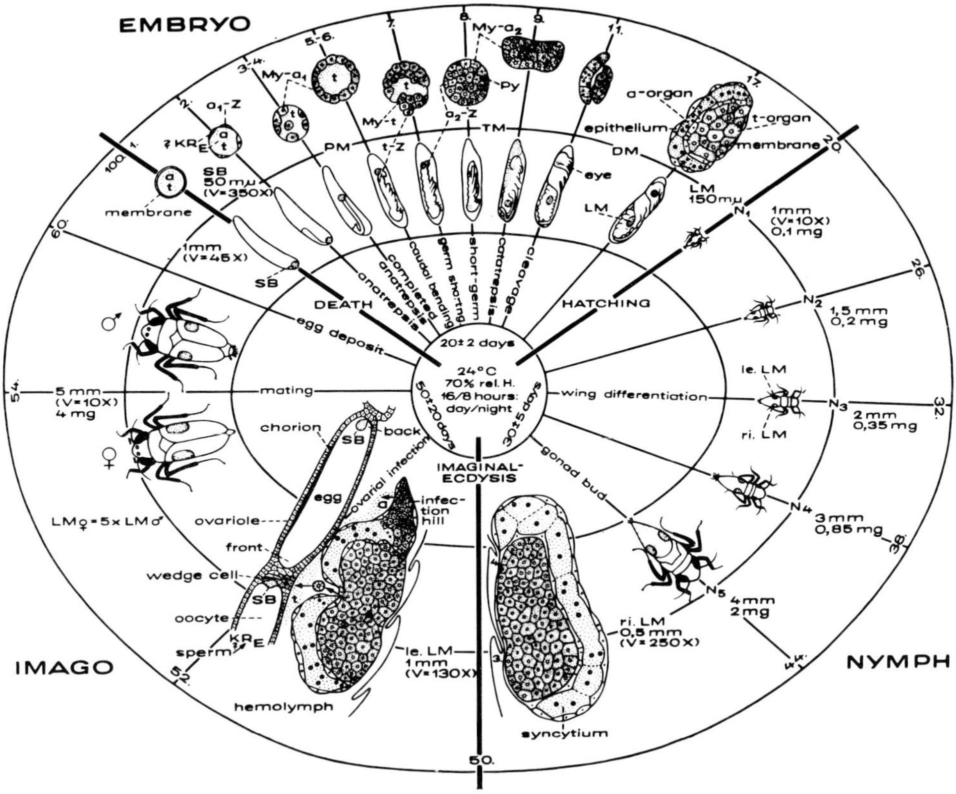

FIG. 1. Development cycle of the host and endocytobionts of the cicada *Euscelis incisus* (embryonic stage according to Körner, 1969). The time span refers to the given standardized breeding conditions. *a,* Primary symbiont *a;* $a_{1,2}$-Z, $a_{1,2}$-cell; DM, double bacteriome; I, imaginal stage; KR_E, rickettsia-like accessory symbiont; LM, lateral bacteriome (ri, right; le, left); My-$a_{1,2}$, $a_{1,2}$-bacteriocyte; My-*t*, *t*-bacteriocyte; N, nymphal stage; PM, primary bacteriome; Py, pyknosis; SB, symbiont ball; TM, transitory bacteriome; *t,* auxiliary symbiont *t;* *t*-Z, *t*-cell; V, enlargement. (Modified from Schwemmler, 1973; see text for details.)

strictly predetermined fashion to form a common symbiont organ, the bacteriome. This organ then divides to build two lateral bacteriomes which lie on the left and right sides of the host abdomen. Each consists of a central *t*-bacteriome surrounded by an *a*-bacteriome. The lateral bacteriomes are still in this state when the embryo hatches and undergo little change in the course of larval development. Further differentiation occurs only in the bacteriomes of adult females. Ovary cells migrate into the *a*-bacteriome and form the so-called infectious cell mound ("Infektionshügel"; Buchner, 1965), in which the *a*-infectious stages mature and from which they are released into the hemolymph. The *t*-

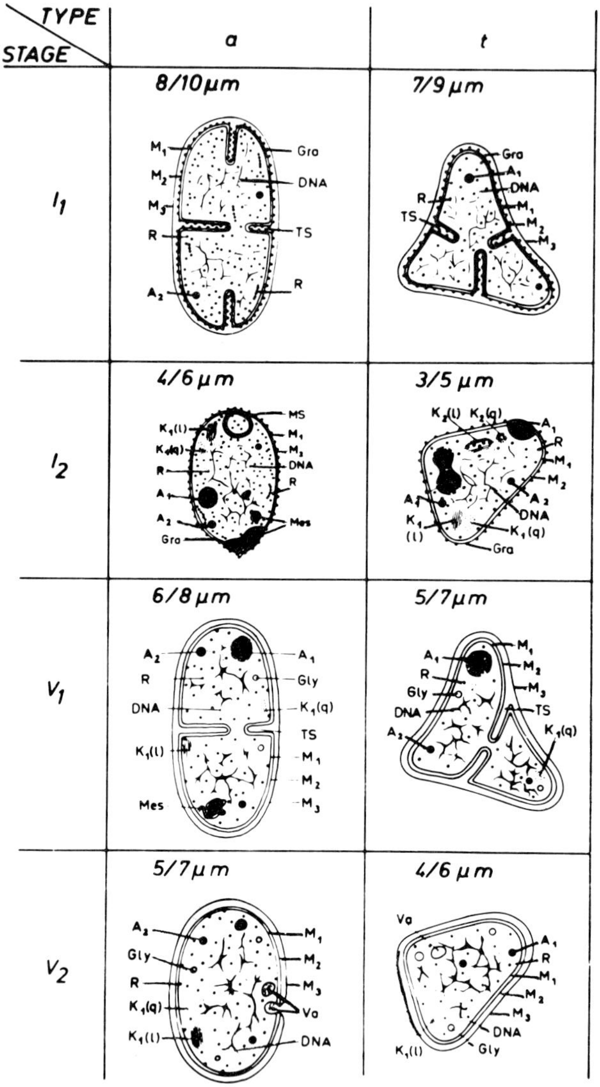

FIG. 2. Substructure scheme of infectious and vegetative stages of *a* and *t* leafhopper endocytobionts. A_1, Large, amorphous inclusion body; A_2, small, amorphous inclusion body; Gly, glucogen inclusion; Gra, granula; I_1, infectious division stage; I_2, infectious stage; K_1, spherical, paracrystal inclusion body (q, lateral; l, longitudinal); K_2, hexagonal, paracrystal inclusion body; M_1, symbiotic cytoplasm membrane; M_2, symbiotic membrane with cell wall rudiments; M_3, bacteriocytic cytoplasm membrane of symbiont; Mes, mesosome; MS, membrane layer body; R, ribosome; TS, division septum; V_1, vegetative division stage; V_2, vegetative stage; Va, vacuole (b, arising; r, peripheral). (Modified from Schwemmler, 1980; see text for details.)

infectious stages are formed at places of contact between the ovarioles and the *t*-organ, in the motile *t*-bacteriocytes. These migrate into the hemolymph and release the symbionts after bursting. Both infectious stages reach the ovarioles by way of the hemolymph. Here they infect the egg, before it is enclosed in the chorion by entering special wedge cells. The infectious stages of the KR_E-symbiont are probably transferred into the egg by sperm impregnation. The developmental cycle is thus completed.

The cicada host seems to control the cellular specificity of incorporation of its endocytobionts by *a*- and/or *t*-specific systems of membrane receptor proteins and the lysis specificity of the endocytobionts by the amount of antibacterial lysozyme which is produced differently in its cells; furthermore, the host regulates multiplication of the endocytobiotic infectious form by direct coupling to its sexual hormone system (Table I, Fig. 3; Schwemmler, 1974a,b).

The question which arises at this point is whether this extensive structural interdependence between the cicada host and its symbiont is based on a reciprocal genetic integration of the two systems. As a first step toward answering this question, the DNA molecular weight of the endocytobionts was determined.

TABLE I

THE CONTROL MECHANISMS OF LEAFHOPPER ENDOCYTOBIOSIS[a,b]

No incorporation of *a*, *t*	Incorporation of *a*, *t*
Epithelial cells (ovariole)	
Epithelial cells (bacteriome)	a-$_{1,2}$-, *t*-Bacteriocytes
Other cells	Wedge cells (ovarioles)
Cause: membrane receptor proteins?	
Lysis of *a*, *t*	No lysis of *a*, *t*
Wedge cells	$a_{1,2}$-, *t*-Bacteriocytes
Cause: lysozymes?	
	No production of infectious *a*, *t*;
Production of infectious *a*, *t*	production of vegetative *a*, *t*
a_1-Bacteriocytes, mature	a_2-Bacteriocytes, syncytial (*a*-organ)
Bacteriocytes of infectious mount (*a*-organ)	
Migratory bacteriocytes (*t*-organ)	*t*-Bacteriocytes (*t*-organ)
Cause: hormones?	

[a]Modified from Schwemmler (1974).

[b]Classification of the different cells in the endocytobiotic system, corresponding to the incorporation, lysis, and production of endocytobionts and the possible distribution of lysosomes, hormones, and of recognition-specific receptor proteins of the glycocalix.

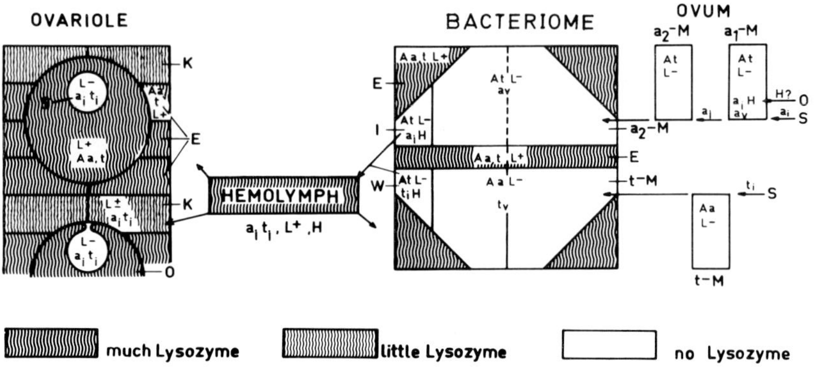

FIG. 3. Preliminary model of the control mechanisms of leafhopper endocytobiosis. The cellular specificity of incorporation of the endocytobionts is due to *a*- and /or *t*-specific systems of glycocalix recognition proteins, while lysis specificity results from the different action of antimicrobial lysozymes, and the specificity of infectious endocytobionts production is determined by the host's sexual hormones. The incorporation specificity was tested by antibiotic inhibition; the lysozyme activity was determined by immunbiological plaque tests, measuring the lysis of membranes from *Micrococcus lysodeikticus;* the endocytobiotic production specificity was studied by application of leafhopper gonad extract to bacteriocyte cultures. $a_{1,2}$-M, $a_{1,2}$-Bacteriocytes; E, epithelial cell; I, infectious mount cell; K, wedge cell, O, oocyte, S, symbiont ball; *t*-M, *t*-bacteriocyte; W, migratory cell; $A_{a,t}$, membrane proteins (prohibiting incorporation of *a* and *t*); $a_{i,v}$, *a*-symbiont: infectious, vegetative stage; H, hormones (inducing production of a_i and t_i); L+, ±, −, lysozymes: much, little, none (effecting lysis of a_i, t_i); $t_{i,v}$, *t*-symbiont: infectious, vegetative stage. (Modified from Schwemmler, 1974.)

III. Endocytobiont Genome

Since the symbiont ball of a single egg contains only approximately 0.2% of the symbiont content of a single bacteriome, it was not possible to prepare enough symbiont ball material to perform density and sedimentaiton analysis. For density and sedimentation analyses, approximately 400 bacteriomes per test were used. These bacteriomes had previously been marked with [6-^{3}H]thymidine and prepared from adult cicada females (Schwemmler *et al.*, 1975).

The density analysis of such bacteriome preparations in CsCl gradients produces two bands: one large band, and a smaller, less dense one (Fig. 4, I,A). Since the smaller band is totally absent in control preparations of absolutely symbiont-free host cells (Fig. 4, I,B), it was assumed that it represents the endocytobiont DNA. By comparison with a marker a buoyant density of ρ = 1.707 for the host cell DNA and of ρ = 1.700 for the endocytobiont DNA was determined.

The DNA fractions of the host and symbiont were then purified several times in CsCl gradients until visual fluorescence tests at 360 nm showed no further

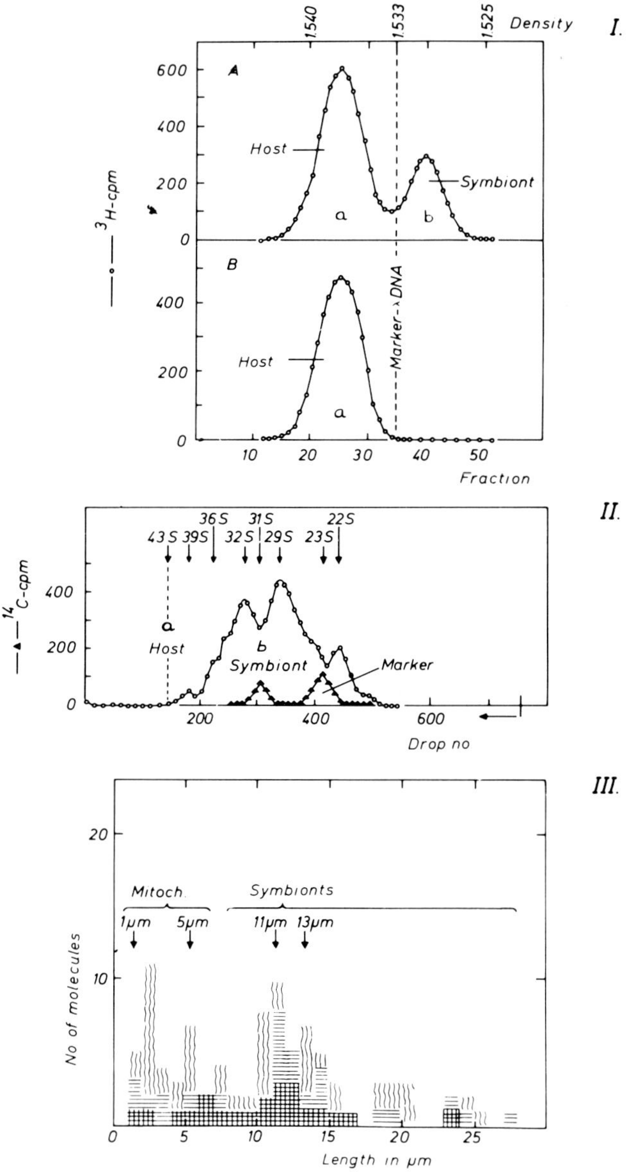

FIG. 4. Host and symbiont DNA of the leafhopper. (I) Density of thoracic bacteriomal *Euscelis* DNA. [^{14}C]Thymidine-labeled bacteriophage λ DNA was added as a marker. (A) Endocytobiont and host DNA from the bacteriome, (B) symbiont-free host DNA from thorax cells. (II) Sedimentation of *Euscelis* endocytobiont DNA. The marker used here was [^{14}C]thymidine-labeled circular DNA. (a) Host DNA (density $\rho = 1.707$); (b) endocytobiont DNA (density $\rho = 1.700$). (III) Conformation of *Euscelis* endocytobiont DNA molecules. Hatched lines, endocytobiont DNA from the symbiont ball; vertical lines, endocytobiont DNA from the male bacteriomes (density $\rho = 1.711$); wavy lines, endocytobiont DNA from female bacteriomes (density $\rho = 1.700$). (Modified from Schwemmler *et al.*, 1975; see text for details.)

noticeable impurities. The fractions were then subjected to sedimentation analysis in neutral sucrose gradients. The host cell DNA was shown to be a homogeneous population of fragments of about 6×10^7 MW. The fraction of endocytobiont DNA, on the other hand, was determined to consist of two dominating, likewise linear molecular populations, i.e., one of 22×10^6 and one of 26×10^6 MW, as well as a number of small components: 11×10^6, 43×10^6, 50×10^6, and 32×10^6 MW (Fig. 4, II).

Finally, electron microscopic or conformational analyses were carried out with the purified bacteriome DNA, as well as with the unpurified, but also radioactively marked symbiont ball DNA (Fig. 4, III). The preparations of host DNA contain randomly broken DNA fragments, which are however all significantly larger than the individual species of endocytobiont DNA molecules. The DNA histrogram of the endocytobiont fragments shows, in contrast, two series of oligomers: 11, 22, 32, and 43×10^6 and 26 and 50×10^6 MW, whereby the species of 22×10^6 and 26×10^6 dominate. These two series are in good agreement with the sedimentation analyses, since DNA molecules of double size

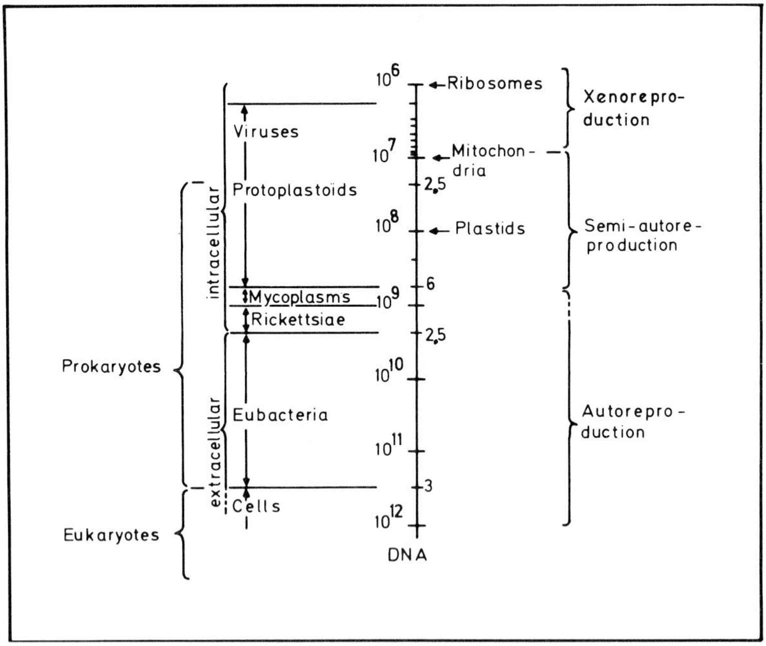

FIG. 5. Classification of eukaryotes, prokaryotes (including protoplastoid endocytobionts, mycoplasma, rickettsiae, and eubacteria) and viruses according to their DNA quantity in comparison to the DNA molecular weight of the DNA-containing or RNA-containing cell organelles (ribosomes, mitochondria, plastids) and the degree of their ability for self-reproduction. (Modified from Schwemmler, 1971; see also Schwemmler *et al.*, 1975; see text for details.)

have double the radioactivity. Control preparations with mitochondrial DNA ($\rho = 1.707$) contain significantly shorter fragments and a totally different frequency distribution. The most attractive interpretation of these data is that the two dominating types of molecules represent the *a*- and *t*-endocytobiont DNA. Accordingly, the cicada endocytobionts have a DNA molecular weight of maximally 10^8. This corresponds to approximately 150,000 bases with an informational capacity of coding for 150 proteins with a mean molecular weight of approximately 40,000 (cf. Fig. 5).

In comparison, mycoplasma and rickettsiae, which are among the prokaryotes with the smallest known genome, have a DNA molecular weight which is 10 times greater at about 10^9, corresponding to a coding capacity sufficient for approximately 2000 proteins. It would thus appear to be interesting to study how great the autonomy of the endocytobionts is from their hosts, in view of their low DNA molecular weight. To this end, *in vitro* culture experiments with the endocytobionts outside of their host cell were undertaken.

IV. Endocytobiont Autonomy

Extracellular, *in vitro* culture experiments with the endocytobionts were performed in a tissue culture medium which was otherwise successfully employed for the cultivation of host cells (Fig. 6, Schwemmler, 1973a). However, neither reproduction nor growth of the endocytobionts occurs in such cell-free cultures. Only the addition of certain cell metabolites, such as ATP, into the medium, or the presence of host cells in the cultivation vessel leads to a few solitary divisons. However, these division stages also show no growth.

If, however, *in vitro* cultures of symbiont-free host cells, so-called pseudobacteriocytes, are infected with the endocytobionts formerly held *in vitro*, the latter are phagocytosed by the pseudobacteriocytes and immediately begin to divide and grow intracellularly. This corresponds to *in vivo* conditions, where proliferation of the endocytobionts also has been observed only intracellularly, never extracellularly. Apparently, the endocytobionts are capable of reproduction only within the host cell. The endocytobionts are only semiautonomous, a feature which they share with the DNA-containing cell organelles, for example, mitochondria and plastids. Thus, the *in vitro* behavior of plastids is comparable to that of the endocytobionts (Fig. 7; Schwemmler and Luuring, 1981). The supposed symbiont's DNA molecular weight of 2.2 to 2.6×10^7 also lies exactly between that of mitochondria, with approximately 10^7 and that of the plastids with about 10^8 (cf. Fig. 6; Schwemmler, 1978, 1981, 1982).

In the further course of work, we attempted to investigate whether the host is dependent in turn on the endocytobiont for its reproduction.

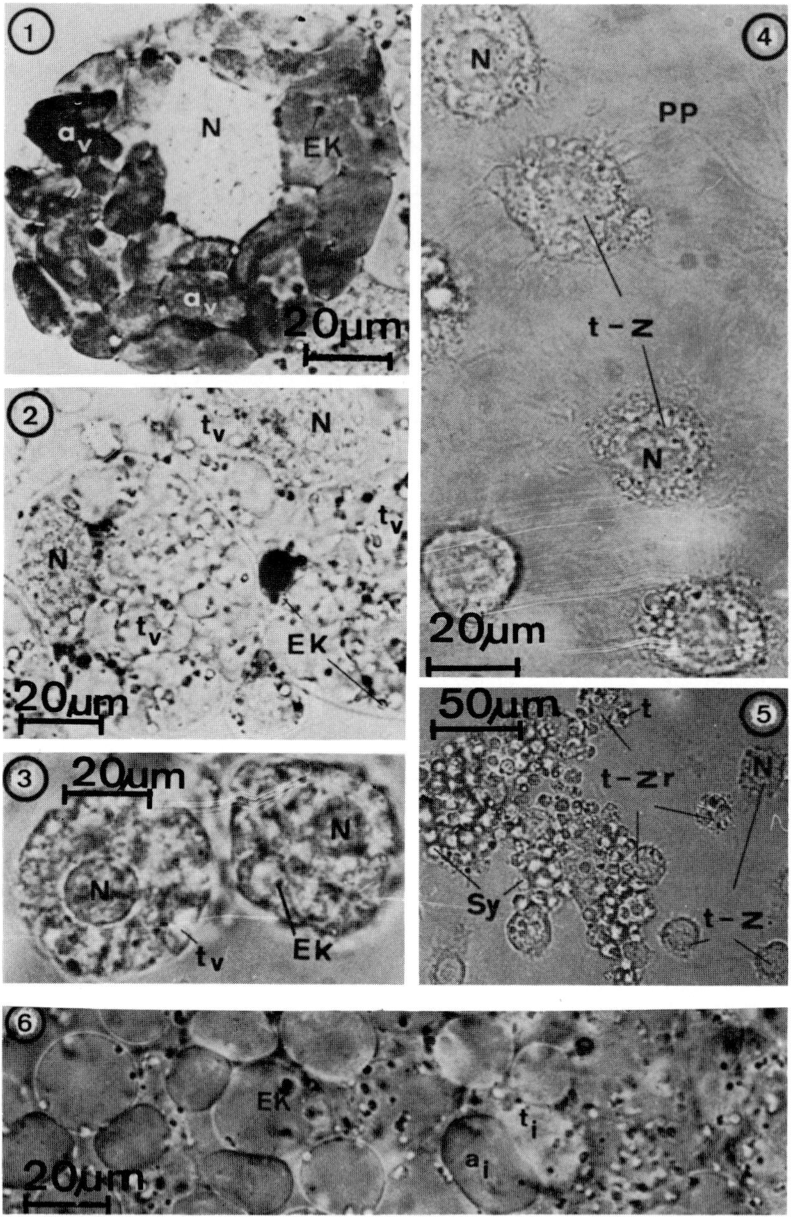

FIG. 6. *In vitro* reinfection tests with cicada endocytobionts. (1) Vitally stained *a*-bacteriocyte with *a*-endocytobionts *in vitro* (×893). (2) Vitally stained *t*-bacteriocyte with *t*-endocytobionts *in vitro* (×893). (3) Two-week-old *in vitro* culture of *t*-bacteriocytes carrying symbionts (×893). (4) Two-week-old culture of symbiont-free *t*-cells (prospective *t*-bacteriocytes) (×850). (5) Reinfected *t*-cell cultures with *in vitro* conserved *t*-endocytobionts (×298). (6) *In vitro* conservation of infectious *a*- and *t*-endocytobionts of the symbiont ball for 1 week (×893). $a_{i,v}$, Primary symbiont *a* (infectious, vegetative); EK, symbiotic inclusion body; HS, heterochromatin; N, nucleus; PP, pseudopodia; Sy, symbiont; $t_{i,v}$, auxiliary symbiont *t* (infectious, vegetative); *t*-Z (r), *t*-cell (reinfected). (Modified from Schwemmler, 1973; see also Schwemmler and Luuring, 1981.)

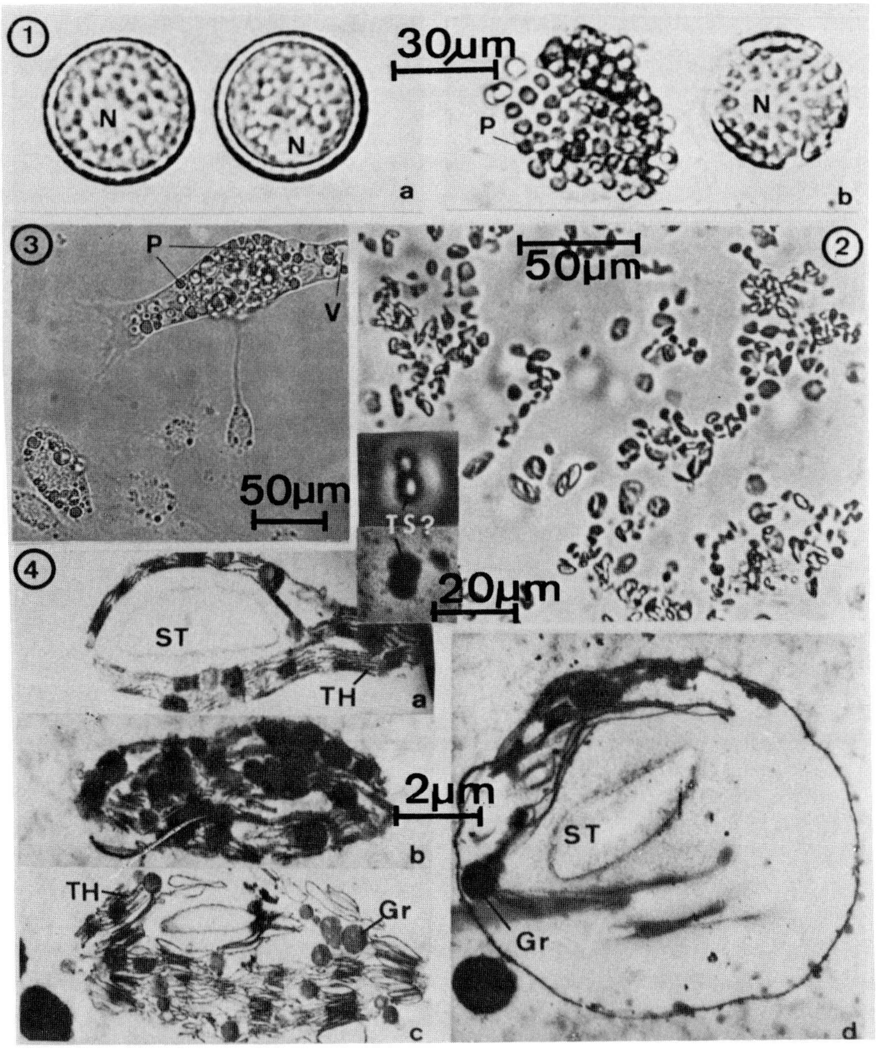

FIG. 7. *In vitro* reinfection tests with plastids. (1) Isolation of sterile plastids. (a) *In vitro* fraction of free, etiolized protoplasts from wheat plants kept in the dark (×694). (b) *In vitro* fraction of free plastids (protochloroplasts) after rupture of the protoplast cell membrane (×694). (2) *In vitro* conservation of wheat plastids (chloroplasts) in a synthetic tissue culture medium for 1 week with possible division stages (×515; inset: ×893). (3) "Infection" of animal cultures with the plastids kept *in vitro* (×340). (4) Analysis of the fine structure plastids kept *in vitro* for different long periods (chloroplasts): all transitions from intact (a) and poor conservation (b) to total degeneration (c, d) are shown (×8500). Gr, Granula; N, nucleus; TH, thylakoid; TS, division stages; ST, starch particle; P, plastid. (Modified from Schwemmler and Luuring, 1981).

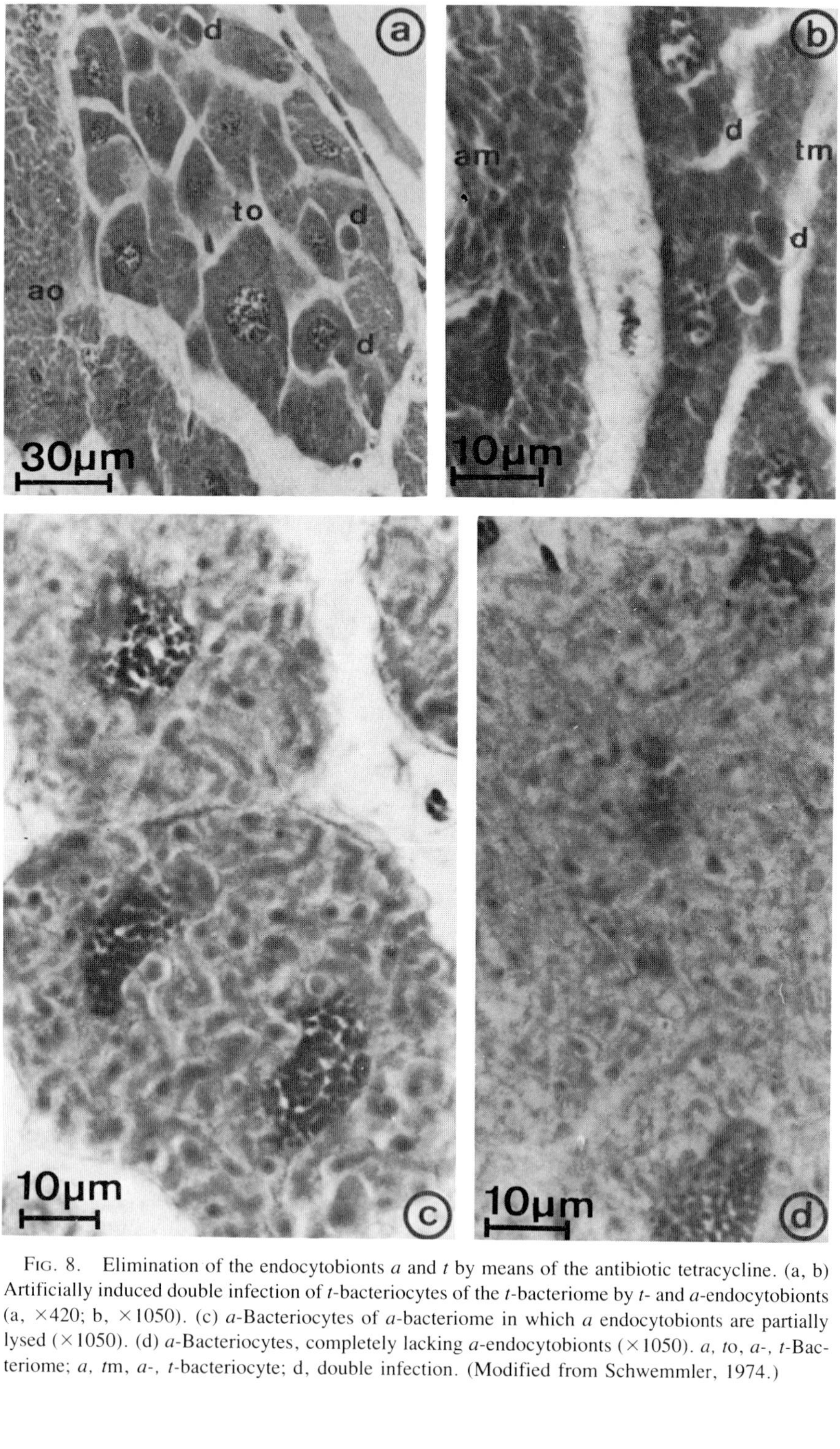

FIG. 8. Elimination of the endocytobionts *a* and *t* by means of the antibiotic tetracycline. (a, b) Artificially induced double infection of *t*-bacteriocytes of the *t*-bacteriome by *t*- and *a*-endocytobionts (a, ×420; b, ×1050). (c) *a*-Bacteriocytes of *a*-bacteriome in which *a* endocytobionts are partially lysed (×1050). (d) *a*-Bacteriocytes, completely lacking *a*-endocytobionts (×1050). *a, to, a-, t*-Bacteriome; *a, tm, a-, t*-bacteriocyte; d, double infection. (Modified from Schwemmler, 1974.)

V. Host Autonomy

By feeding antibiotics to larval or adult hosts, one can totally eliminate their endocytobionts (Fig. 8; Schwemmler, 1973b, 1974a,b; Schwemmler *et al.*, 1973). Hosts which have lost their symbionts in this manner die within a short period of time. The elimination of the symbionts during egg development also has devastating consequences. If egg infection is prevented by applying lysozyme to the cicada female, symbiont-free eggs are laid. These develop into cephalothorax embryos without an abdomen (Fig. 9; Schwemmler, 1974b, 1980a). Spontaneously symbiont-free eggs also develop into cephalothorax embryos (Sander, 1959). If, however, the symbiotic infectious mass is separated from the egg after egg laying, normal embryos with symbiont-free bacteriomes, so-called pseudobacteriomes, develop (Sander, 1968) (the symbiont-free host

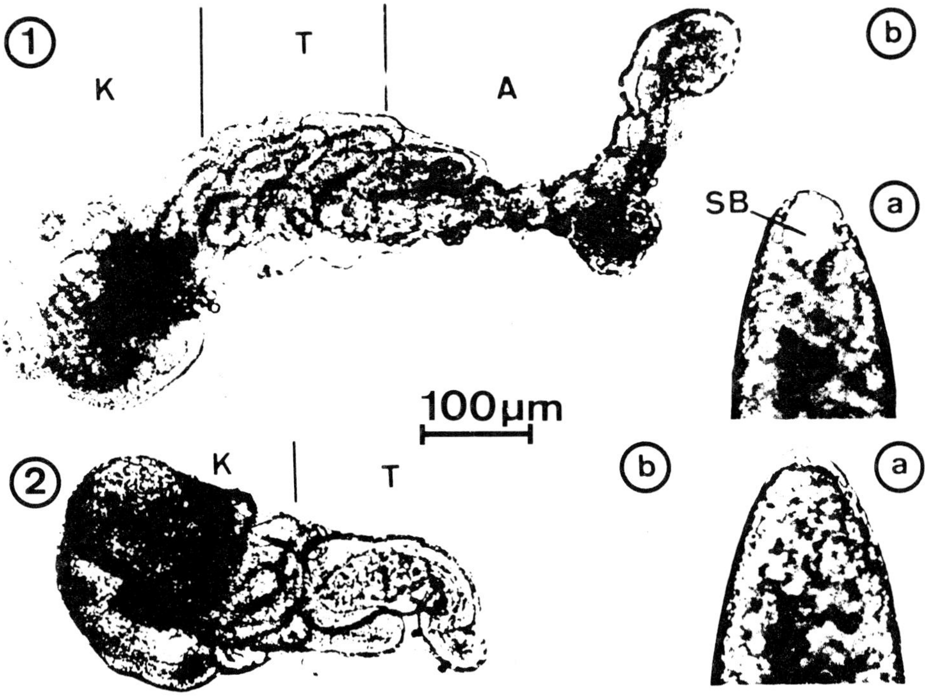

FIG. 9. Prevention of the infection of *Euscelis incisus* eggs. (1a) Posterior pole of a normal egg with normal SB (~ 90,000 μ^3). ×100. (1b) Six-day-old normal embryo with normal primary bacteriome from the above egg. ×100. (2a) Posterior pole of an egg without a symbiont ball, laid by a tetracycline-treated female. The results are similar after lysozyme treatment. ×100. (2b) Six-day-old cephalothorax embryo without an abdomen from an X-ray-treated egg, similar to embryos from symbiont-free eggs, from females treated with tetracycline or lysozyme. ×100. A, abdomen; K, head; SB, symbiont ball; T, thorax. (Modified from Schwemmler, 1973, 1974.)

cell cultures arise from these pseudobacteriomes). The provisions for endocytobiosis are thus preprogrammed before the egg is laid; the endocytobionts have no more influence on this process. Before the egg is laid, however, the endocytobionts seem to directly or indirectly influence the activities of the factors which control the formation of the abdomen (Schwemmler, 1980a). The endocytobionts thus represent a kind of complement to the host genome which must be passed on from generation to generation, thereby enabling the host to reproduce identically. Thus the host is, from a genetic standpoint, also only a semiautonomous system. What are the functional consequences of this extensive reciprocal genetic interdependence?

VI. Endocytobiont Function

A comparison of the minimal diet (completely synthetic food meeting minimal requirements in quality and quantity) of normal cicadas with that of symbiont-free or semisymbiont-free cicadas shows that the endocytobionts synthesize, in normal cases, choline, almost all essential amino acids, and approximately half of the necessary vitamins for their host (Schwemmler, 1978; Schwemmler *et al.*, 1973). Feeding experiments with radioactively marked substances show furthermore that the endocytobionts are capable of producing cholesterol, as well as those amino acids which contain sulfur, using only sulfate as a sulfur source (Ehrhardt, 1968, 1969; Houk, 1974). The endocytobionts are therefore essential factors for the host's anabolism. The endocytobionts probably use host catabolites to synthesize metabolites for themselves and their hosts. This is inferred from the observation that animals freed of symbionts suffer an abnormal accumulation of wastes such as uric acid, particularly in their fat bodies (Schwemmler *et al.*, 1973).

During the developmental cycle, the host and symbiont furthermore undergo characteristically correlated changes in their respective pH values. The host goes from alkaline during the early embryonal phase to acid during the late embryonal and the entire larval phase and finally back to alkaline in the adult phase, whereas the pH changes in the endocytobiont run parallel in the other direction, changing from acid to alkaline and back to acid (Fig. 10; Schwemmler, 1978, 1981). The osmotic values of host and symbiont also fluctuate several times during the developmental cycle in a similar, antagonistic manner. Finally, the pH and osmotic pressure values in the hemolymph and in the egg plasma of normal animals are lower than the corresponding values of symbiont-free cicadas. This evidence appears to indicate a regulation of pH and osmotic pressure between host and symbiont; however, such a regulation has yet to be proved with absolute certainty. The energy metabolism and endogenous rhythms of host and symbiont

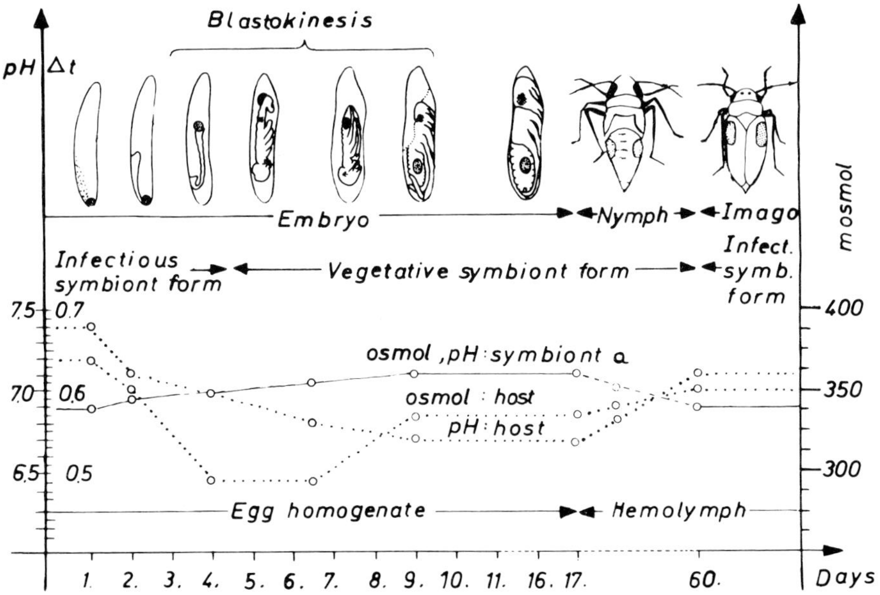

FIG. 10. pH and osmole values of the host cicada (eggs, larvae, imagos) and its symbionts (infectious and vegetative forms). The data points represent the averages of three measurements. The scatter for the pH and osmole values were at most 0.2 units. Host pH values are always on the acid side when the symbiont values are alkaline, and vice versa. The osmole values of the host and symbiont are also antagonistic. (Modified from Schwemmler, 1971, 1974; sketches of the embryonic stages from Körner, 1969.)

also appear to be correlated. Whenever the host cell exhibits a high energy level, that of the symbiont is low, and vice versa (Fig. 11; Schwemmler, 1977; Schwemmler and Herrmann, 1979, 1980).

In conclusion, it can be determined that the endocytobionts are so firmly integrated into their cicada hosts, not only structurally and functionally but genetically that they cannot be removed without causing extensive disturbance in the system.

VII. Endocytobiont Gene Transfer

The low DNA molecular weight and the genetic semiautonomy of the endocytobionts indicate that, in the course of an assumed 200 million year common evolution with their hosts, a transfer of genes from the symbiont genome into the cicada genome probably took place. Accordingly, the part of metabolites which

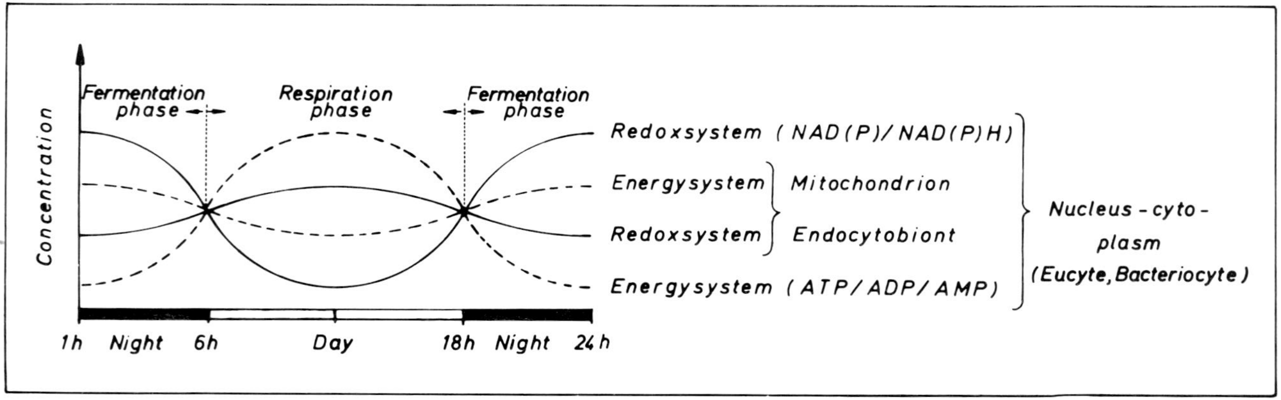

FIG. 11. Hypothetical, idealized diagram, according to which the energy metabolism and the reduction–oxidation system between the host and the symbiont of cicada reacts antagonistically, comparable to the possible relationship between the nucleocytoplasm system and mitochondria. (Modified from Schwemmler and Herrmann, 1979, 1980.)

the endocytobionts need for reproduction are coded by the nucleocytoplasm system of the host and are also synthesized there. The semiautonomy of the host also indicates a possible gene transfer in the opposite direction, that is, from the host genome into the symbiont genome, resulting in the coding and synthesizing of essential host metabolites by the endocytobiont DNA (Schwemmler, 1981).

In this context it is interesting to note that the endocytobiont of the single cell organism *Euplotes aediculatus*, called *omikron*, has a DNA molecular weight of only 0.5×10^9, corresponding to about one-fourth of nonparasitic bacteria (Heckmann, 1980). This endocytobiont is structurally and functionally so extensively integrated into the host system that it has been considered by many researchers to be a DNA-containing cell organelle. Observations of metabolic products and gene products of the unicellular organism *Paramecium aurelia* give indirect evidence of gene transfer between the host and symbiont genome (Preer *et al.*, 1974).

The only direct evidence to date of gene transfer has been provided by Schell and his co-workers (1980) for the endocytobiont DNA of nitrogen-fixing bacteria and the genome of their plant cells. The DNA information of the endocytobionts is integrated into the host genome in the form of episomes. This allows the host cells to harbor and reproduce the endocytobionts in their cytoplasm.

VIII. Summary and Conclusions

The mutual structural, functional, and genetic integration of such endocytobionts in the nucleocytoplasm system calls to mind an analogous situation in the

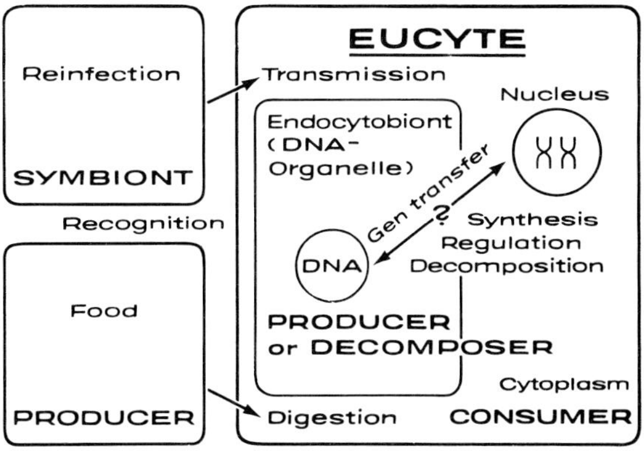

FIG. 12. The bacteriocyte formulated as an intracellular ecosystem. The endocytobionts function as decomposers of waste products of the nucleocytoplasm system (consumer) or of the food (producer), as synthesizers of metabolites for the nucleocytoplasm system and for themselves and as regulators of the pH, osmotic pressure, and probably of certain endogenous rhythms of the nucleocytoplasm system. (Modified from Schwemmler, 1978, 1981, 1982, 1983; see text for details.)

case of the mitochondria. Both endocytobionts and mitochondria synthesize anabolites for their system by using its catabolites (Fig. 12; Schwemmler, 1978, 1981). In doing so, both apparently regulate the pH and osmotic pressure values of their nucleocytoplasm system. The various inclusion bodies found in the endocytobiont and mitochondria, among them Ca^{2+}, HPO_4^{2-}, amino acids, and/or proteins, are the most visible manifestation of this regulatory activity. Furthermore, both appear to have a regulatory influence on certain endogenic rhythms of their particular system. Thus, the endocytobionts and mitochondria mediate between the consuming nucleocytoplasm system and its nutrition-producing environment, on the chemical, as well as on the physiological level.

According to a working hypothesis, the so-called organelle endocytobiosis hypothesis, such endocytobionts are comparable to DNA-containing organelles, whereas the mitochondria, on the other hand, have evolved from formerly free-

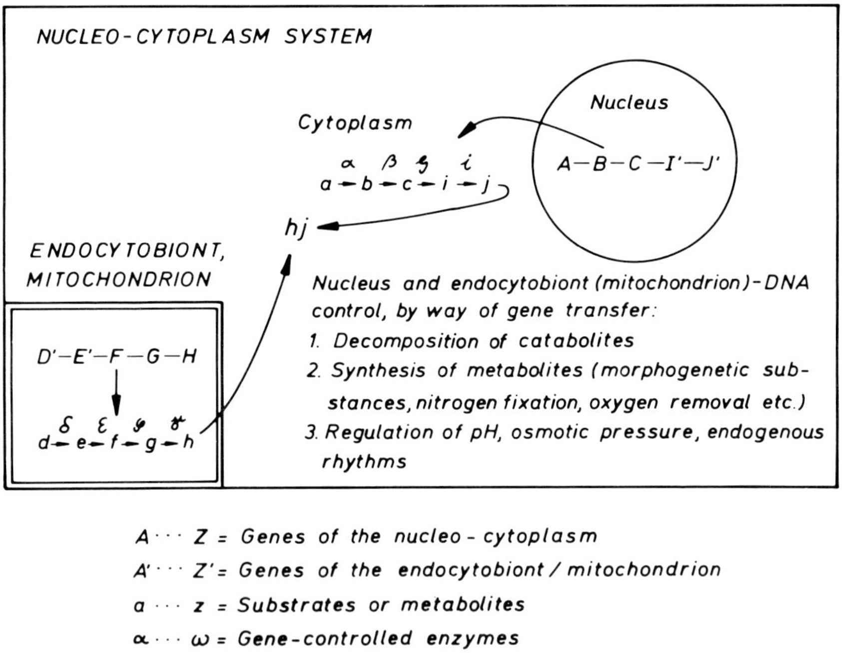

FIG. 13. According to the "organelle endocytobiosis hypothesis," both endocytobionts and mitochondria produce metabolites for themselves and their nucleocytoplasm system using the waste products and regulate the pH, osmotic pressure, and, probably certain endogenous rhythms, which is directly controlled by the genes of both system, as a result of gene transfer between nuclear DNA and endocytobiont or mitochondrion DNA during the million years of their common evolution. (Modified from Schwemmler, 1978, 1980, 1981.)

living endocytobionts (Fig. 13; Schwemmler, 1980b). This working hypothesis has gained greater significance since the endocytobiont nature of plastids, including chloroplasts, has come to be generally accepted today. One possibility for proving or disproving this hypothesis for mitochondria would be to analyze the genetic system of the cicada endocytobionts. In order to determine possible homologies to the mitochondrial system, endocytobiont DNA or RNA could be cloned and subjected to sequencing or hybridization. Cicada endocytobionts should prove to be particularly suited for such studies since, as a very old phylogenetic group, they possibly represent a kind of living fossil in the form of a missing link that has survived for millions of years under the protection of its cicada host.

For such investigations, a combination of symbiosis and cell biological methods is necessary. Thus, a new synthetic field of research is being established between symbiosis research and cell biology—endocytobiology (e.g., Schwemmler and Schenk, 1980).

References

Buchner, P. (1965). "Endosymbiosis of Animals with Plant Microorganisms." Wiley, New York.

Ehrhardt, P. (1968). *Experientia* **24,** 82–83.

Ehrhardt, P. (1969). *Biol. Zentralbl.* **88,** 335–348.

Heckmann, K. (1980). *Int. Colloq. Proc.* **I,** 393–400.

Houk, E. J. (1974). *J. Insect Physiol.* **20,** 471–478.

Körner, H. (1969). *Oecologia (Berlin)* **2,** 319–346.

Müller, H. J. (1949). *Biol. Zentralbl.* **68,** 343–368.

Preer, J. R., Preer, L. B., and Jurand, A. (1974). *Bacteriol. Rev.* **38,** 113–163.

Sander, K. (1959). *Roux Arch.* **151,** 430–497.

Sander, K. (1968). *Dev. Biol.* **17,** 16–38.

Schell, J., Van Montagu, M., Holsters, M., Hernalsteens, J. P., Willmitzer, L., O'Farrell, P. Z., De Picker, A., Engler, G., Lemmers, M., and Schröder, J. (1980). *Int. Colloq. Proc.* **I,** 491–509.

Schwemmler, W. (1971). *Cytobiologie* **3,** 427–429.

Schwemmler, W. (1973a). *Cytobios* **8,** 63–73.

Schwemmler, W. (1973b). *Biol. Zentralbl.* **92,** 749–772.

Schwemmler, W. (1974a). *In* "Contemporary Topics in Immunobiology" (E. L. Cooper, ed.), Vol. 4, pp. 179–187. Plenum, New York.

Schwemmler, W. (1974b). *J. Insect Physiol.* **20,** 1467–1474.

Schwemmler, W. (1977). *In* "Advances in Invertebrate Reproduction" (K. G. Adiyodi and R. G. Adiyodi, eds.), Vol. I, pp. 482–489. Peralam-Kenoth, Karivellur, Kerala.

Schwemmler, W. (1978). "Mechanismen der Zellevolution. Grundriβ einer modernen Zelltheorie." De Gruyter, Berlin.

Schwemmler, W. (1980a). *Naturwiss. Rdschau.* **33** (2), 52–59.

Schwemmler, W. (1980b). *BioSystems* **12,** 111–122.

Schwemmler, W. (1981). *Ber. Dsch. Bot. Ges.,* **94,** 591–598.

Schwemmler, W. (1982). *Acta Biotheoret.* **31,** 45–68.

Schwemmler, W. (1983). *Int. Colloq. Proc.* **II** (in press).

Schwemmler, W., and Herrmann, M. (1979). *Cytobios* **25,** 45–62.

Schwemmler, W., and Luuring, B. (1981). *Ann. N.Y. Acad. Sci.* **361,** 284–297.
Schwemmler, W., and Schenk, H. E. A. (1980). *Int. Colloq. Proc.* **I,** 2.
Schwemmler, W., Duthoit, J. L., Kuhl, G., and Vago, C. (1973). *Z. Morphol. Tiere* **74,** 297–322.
Schwemmler, W., Hobom, G. R., and Egel-Mitani, M. (1975). *Cytobiologie* **10** (2), 249–259.
Strübing, H., and Hasse, A. (1974). *Zool. Beiträge (Berlin)* **20** (3).

INTERNATIONAL REVIEW OF CYTOLOGY, SUPPLEMENT 14

Cellular and Molecular Mechanisms of Intracellular Symbiosis in Leishmaniasis

K.-P. CHANG

The Rockefeller University, New York, New York

I. Introduction

A. TRYPANOSOMATID PROTOZOA AND SYMBIOSIS

Most trypanosomatid protozoa (Order: Kinetoplastidae; Class: Mastigophora) live in one way or another with multicellular organisms (Vickerman, 1976). Best known are infectious agents of medical and veterinary importance mainly in the tropics. Exotic diseases, such as the African sleeping sickness of man, Nagana of cattle, Chagas disease in South America, and leishmaniases in many parts of the world, are caused by a number of species in the genus, *Trypanosoma* or *Leishmania*. Nonpathogenic species of leishmanias also exist in reptiles; those of trypanosomes in fish, amphibians, reptiles, birds, bats, and other mammals; and those of *Phytomonas* in lactiferous plants. The transmission of these trypanosomatids from one host to the other is accomplished frequently by blood-sucking or plant-sucking insects. In these vectors, the protozoa may undergo considerable development and thus have a digenetic life cycle. Far more common are the monogenetic species of the genera *Crithidia, Blastocrithidia, Leptomonas,* and others, which live in arthropods and are transmitted by environmental contamination (McGhee and Cosgrove, 1980).

The microecological niche of these trypanosomatids in their hosts is mostly

ISBN 0-12-364375-9

extracellular, ranging from insect gut to cloaca of chameleons to mammalian blood vessel (Vickerman, 1976). Some species have, however, become adapted to live an intracellular life. Examples of such intracellular symbiosis include *Leptomonas karyophilus* in the macronucleus of *Paramecium trichium, Endotrypanum* in the red cells of sloth, *T. cruzi* in the heart muscle and leishmanias in mononuclear phagocytes of man. Sometimes, trypanosomatid flagellates serve as hosts for other microorganisms or agents. Virus-like particles have been observed by electron microscopy in *L. hertigi*—a species originally isolated from porcupines (Croft and Molyneux, 1979). More frequently encountered are, however, intracellular bacteria in monogenetic species of insect origin (McGhee and Cosgrove, 1980).

Trypanosomatid flagellates are unique cells in several ways (Vickerman and Preston, 1976). One outstanding feature of all trypanosomatids is the ubiquitous kinetoplast, a segment of the highly branched mitochondrion located just beneath the basal body of the flagellum. The mitochondrial DNA in the form of maxicircles and minicircles is condensed into regularly arranged fine fibrils in the kinetoplast. Investigations are underway to determine their function and the mode of replication (Simpson, 1972; Englund, 1981). Another unique feature of all trypanosomatids is their microtubular system. There are four functional types, i.e., flagellar, subpellicular, cytopharyngeal, and mitotic microtubules, the first two being more prominant (Vickerman and Preston, 1976).

During their transition from one kind of host to another, the digenetic species undergo cell differentiation, manifested by both morphological and metabolic changes. Presumably, these changes are crucial for the adaptation of these organisms to live in a new environment.

This article deals with a group of digenetic trypanosomatids, leishmanias. A part of their life cycle involves a stage, which lives within mammalian host cells, thereby providing an excellent system for studying the intracellular association of one eukaryote with the other.

B. Leishmanias and Leishmaniases

Leishmaniases are chronic diseases, existing on all continents, except Australia (Zuckerman and Lainson, 1977). Of the many species in the genus *Leishmania,* only *L. braziliensis, L. donovani, L. mexicana, L. tropica,* and their subspecies/strains are pathogenic to man. In the life cycle of leishmanias, there are two different stages: extracellular promastigotes in the digestive tract of sandfly vectors and intracellular amastigotes in the mononuclear phagocytes of the reticuloendothelial system of the mammalian hosts (Bray, 1974). Promastigotes are motile cells of variable dimension (about 2 μm by 2–20 μm), each equipped with an anterior flagellum as long as or longer than the body length. The promastiogote stage of leishmanias can be cultured *in vitro* in a variety of undefined, semidefined, or defined media (Chang and Fish, 1982). Amastigotes are nonmotile small entities of about 2–4 μm in diameter; their flagella are nonfunctional and are confined within the flagellar pocket of the cell. Multiplication of the amastigotes in macrophages results in all the pathology and clinical symptoms associated with leishmaniases. There are several types of leishmaniases, depending on the species of leishmanias involved (Hommel, 1978). Species like *L. tropica* cause simple cutaneous leishmaniasis; the lesion is self-limiting and self-healing in the course of several months. Species such as *L. braziliensis* and sometimes *L. mexicana* cause mucocutaneous leishmaniasis; facial disfigurement in such infection is not uncommon. *Leishmania donovani* and related species cause kala azar or visceral leishmaniasis and fatal consequence may ensue in untreated cases. The pathogenic mechanism in leishmaniases is unknown. It is likely that the clinical pathology and symptoms observed are related to the immune response of the hosts to the infection rather than the intracellular residence of leishmanias in the macrophages (Turk and Bryceson, 1971).

II. Leishmania–Macrophage Association *in Vitro*

The most intriguing aspect of intracellular symbiosis in leishmaniasis is the ability of amastigotes to live exclusively in macrophages—mammalian cells specialized in defense against invaders. Macrophages are well equipped to ingest and digest invading microorganisms by virtue of their possession of surface receptors, high phagocytic activity, and a battery of enzyme systems for degradation (Silverstein *et al.*, 1977; Elsbach, 1980). How leishmanias manage to counter these extremely hostile activities of macrophages bears not only on the pathogenic mechanism of leishmanias but also on the evolution of intracellular symbiosis (see Section III for further discussion). These questions can be clarified by our understanding the mechanisms of how leishmanial infection of macrophages is established and maintained. Essential for the establishment of intracellular symbiosis are four sequential events: recognition, intracellular entry, survival, and multiplication (Trager, 1974; Smith, 1979; Moulder, 1979b). Leishmania–macrophage interactions have been studied *in vitro* with the aim of understanding the cellular and molecular mechanisms governing these sequential events. Available information is summarized below in the order of the sequential events for the establishment of intracellular symbiosis in leishmaniasis.

A. Leishmania–Macrophage Binding—Recognition

Leishmanias, like any other intracellular agent, must first come close and then attach themselves to the host cells before entry. The importance of these events can be envisioned readily in both the primary infection by promastigotes delivered into susceptible hosts by the sandfly vector as well as in the secondary infection by amastigotes released from heavily parasitized cells during the course of leishmaniasis. These extracellular leishmanias must be confronted with multifarious cell types in the hosts. It is uncertain whether there may be signal molecules in the form of chemotactic factors that leishmanias may release to attract macrophages or vice versa. It is known, however, that there is a predilection of leishmanias for binding to their normal host cells. Thus, promastigotes bind avidly to phagocytes, less well or poorly to lymphocytes, and not at all to erythrocytes (Benoliel *et al.*, 1980; Chang, unpublished). The binding of leishmanias to macrophages has been studied *in vitro* by light microscopy under phase contrast or bright field after Giemsa staining as well as by scanning and transmission electron microscopy. Quantitation of the binding has been made by counting macrophage-associated leishmanias (Zenian, 1981; Benoliel *et al.*, 1980) or by a radioisotopic assay (Chang, 1981a). Unfortunately, the paucity of information and the disparity of methodology applied to different systems defy, at present, general discussion in a unified fashion. Findings obtained by different authors are summarized below according to the leishmania–macrophage systems used.

1. *Leishmania donovani–Hamster Macrophage System*

It has been observed *in vitro* by light microscopy that a large number of promastigotes becomes attached to the cells in a relatively short time period (Figs. 1 and 2). The binding is apparently strong enough to withstand the shearing force produced by vigorous rinsing of the culture with buffer or medium (Chang, 1981a). Since there is no specific orientation for the binding (Chang, 1979), the binding ligand(s) of leishmanias must be distributed in both their flagellar and somatic sections of the plasma membrane. There are multiple points of attachment instead of a continuous surface abuttment at the leishmania–macrophage interface (Chang, 1979), suggesting perhaps that the binding is mediated by discrete membrane molecules, but not by nonspecific surface interactions.

A radioisotopic assay was developed to quantitate and to characterize the binding between *L. donovani* promastigotes and hamster peritoneal macrophages *in vitro* (Chang, 1981a). The binding was found to follow saturation kinetics and to be temperature-dependent (Fig. 1). The molecular property of the adhesion in leishmania–macrophage binding was probed by this assay coupled with binding competition experiments and those after modification of the surface of the interacting cells. Figures 3–5 represent some examples of these types of experiment. Sialic acid, glucose, mannose, and their related monosaccharides compete for leishmania–macrophage binding more effectively than others (Fig. 3), consistent with the effect of pretreating promastigotes with glycosidases, which

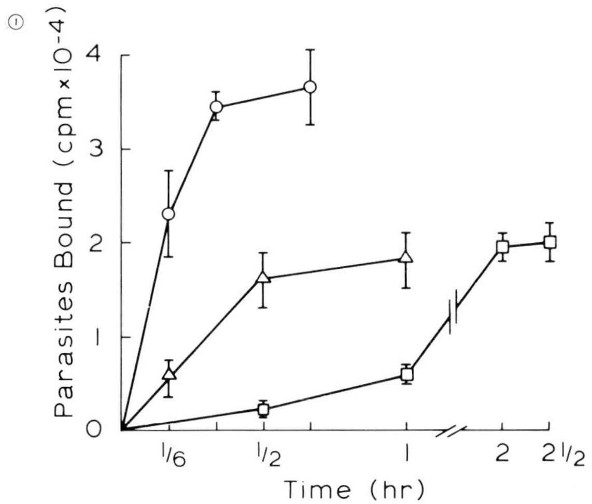

FIG. 1. Kinetics of *Leishmania donovani* promastigotes–hamster macrophage binding *in vitro* at different temperatures. ○——○, 37°C; △——△ 25°C; □——□ 4°C. See Chang (1981a) for the radioisotopic assay. Note the binding reaches saturation in 20–30 minutes at 37°C, 30–60 minutes at 25°C, and 2 hours at 4°C. The level of binding at lower temperatures only reaches about 50% of that at 37°C.

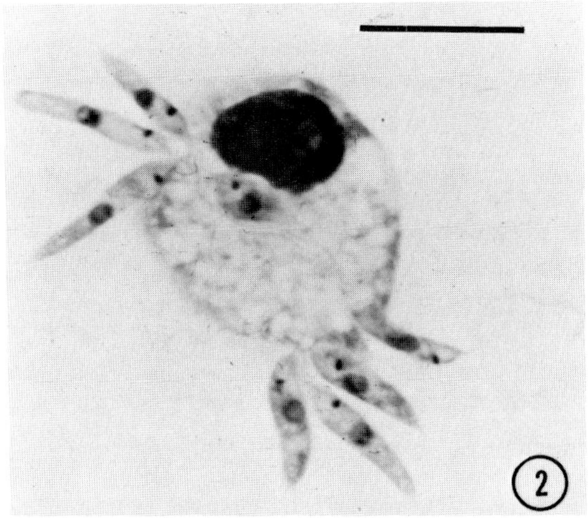

FIG. 2. A representative hamster macrophage in coverslip culture infected with promastigotes of *L. donovani* for 20 minutes at 37°C, 5% CO_2 in air, sowing one intracellular parasite and several others attached to the macrophage cell surface with either end of the organisms. Bar = 10 μm.

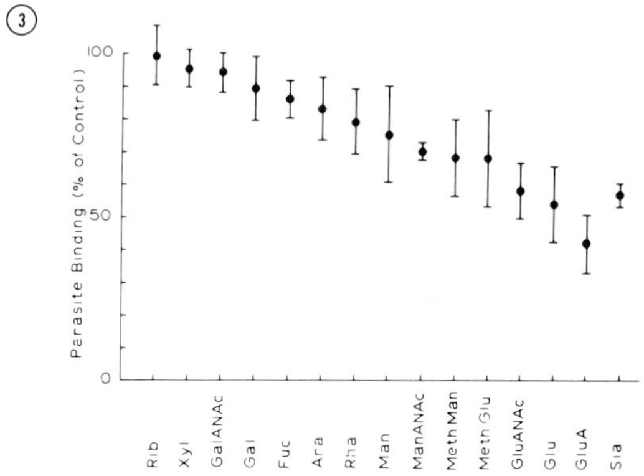

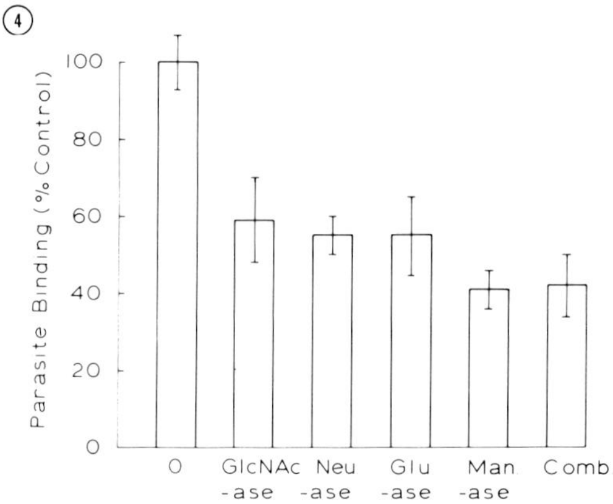

FIG. 3. Inhibition of hamster macrophage–*Leishmania donovani* binding by various monosaccharides. Rib., D-ribose; Xyl., D-xylose; GalANAc, *N*-acetyl-D-galactosamine; Gal., D-galactose; Fuc., L-fucose; Ara., D-arabinose; Rha., D-Rhamnose; Man., D-mannose; ManANAc, *N*-acetyl-D-mannosamine; Meth. Man., α-methyl-D-mannoside; Meth. Glu., α-methyl-D-glucoside; GluAnAc., *N*-acetyl-D-glucosamine; Glu., D-glucose; GluA, D-glueosamine; Sia., sialic acid. Binding assays were carried out at 37°C for 20 minutes with the final concentration of all sugar at 100 m*M*.

FIG. 4. Effect of prior treatment of *Leishmania donovani* promastigotes with various glycosidases on their subsequent binding to hamster macrophages. O, Control; GlcNAc.-ase, α-*N*-acetyl glucosaminidase; Neu.-ase; neuraminidase; Glu.-ase, β-glucosidase; Man.-ase, α-mannosidase; Comb., a combination of the last three enzymes. See Chang (1981a) for conditions of enzyme treatment. Binding assays were carried out at 37°C for 20 minutes.

specifically cleave these sugars of their surface (Fig. 4). Prior treatment of macrophages with either rabbit antimacrophages antiserum (Fig. 5) or trypsin significantly reduces their capacity of binding promastigotes. Results from these and other similar experiments lead to the following conclusions (Chang, 1981a): (1) *L. donovani* promastigote–hamster peritoneal macrophage binding appears to

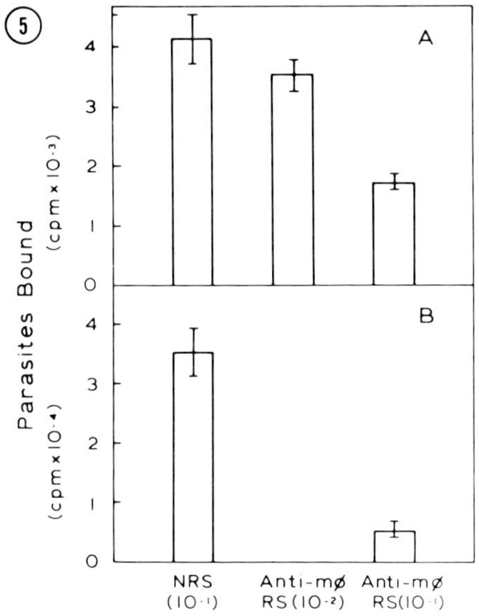

FIG. 5. Effect of prior opsonization of hamster macrophages on their subsequent binding to *Leishmania donovani* promastigotes at 4°C (A) or 37°C (B). NRS, Normal rabbit serum; Anti-MϕRS (10^{-2}), 1/100 dilution of antimacrophage rabbit antiserum; AntiϕRS (10^{-1}), 1/10 dilution of antimacrophage antiserum. Opsonization of macrophages was carried out at 4°C for 1 hour. Binding assays were done in 1 hour at 4°C (A) and in 20 minutes at 37°C (B).

be a type of ligand–receptor interaction, being Ca-dependent, but serum-independent; (2) the putative receptor(s) of macrophages for leishmanias are antigenic membrane proteins; and (3) the binding ligands of leishmanias probably involve their surface antigenic glycoproteins, with sialic acid, glucose, mannose, mannosamine, and *N*-acetyl-glucosamine as possible binding determinants. It is suggested that the promastigote–macrophage binding may be akin to receptor-mediated endocytosis of zymosan granules (yeast cell wall) or lysosomal glycosidases by macrophages (Chang, 1981a); this mannose/*N*-acetyl-glucosamine-specific recognition system presumably serves the important function of nonspecific cellular defense (Weir, 1980) and/or the retrieval of secreted lysosomal enzymes (Stahl and Schlesinger, 1980).

2. *Leishmania tropica–Mouse Macrophage System*

In this system, promastigotes have been observed to bind to macrophages primarily through their flagellar end (Zenian *et al.*, 1979) and the binding was quantitated by microscopic counting of leishmanias in macrophages (Zenian, 1981). The stated aim of the work was to determine promastigote–macrophage binding in terms of biochemical reactions and metabolic requirements of the interacting cells. Results obtained were interpreted to indicate the following: (1)

the binding requires divalent cations, with magnesium being more crucial than calcium; (2) it is temperature- and serum factor-dependent; (3) the binding promoting factor(s) in the fetal bovine serum are nondializable, heat-sensitive, and precipitable by ammonium sulfate; (4) binding is inhibited by 2-deoxy-D-glucose and various metabolic inhibitors, i.e., sodium azide, sodium fluoride, and iodoacetate; and (5) it is enhanced in the presence of glucose or mannose, but not xylose, galactose, or fucose at 5.5 m*M*. Whether the compounds used acted on the macrophages or the leishmanias, and whether the effects observed were an inhibition of binding or uptake, are, however, not very clear. The author concluded that the promastigote–macrophage binding is a complex phenomenon involving biochemical reactions at the cellular level.

3. *Leishmania mexicana–Rat Macrophage System*

The binding of *L. mexicana* and related species to macrophages from animals other than the rat is limited to morphological examination using the promastigote stage (Alexander, 1975; Lewis, 1974; Merino *et al.*, 1977). As has been observed with *L. tropica,* promastigotes of these species attach to macrophages via their flagellar end predominantly.

Binding of *L. mexicana* to rat macrophages was assayed at room temperature using amastigotes isolated from infected hamsters and peritoneal exudate cells of the rat (Benoliel *et al.*, 1980). Binding was quantitated by counting macrophage-adherent leishmanias in a hemacytometer. In this work, leishmanial amastigotes were used simply as one of the five different test particles for studying the "nonspecific" binding by macrophages *in vitro.* Transmission electron microscopy showed close apposition of the surface membrane for a considerable length between amastigotes and macrophages. Their binding under the experimental conditions was found to require neither serum factors nor divalent cations. Metabolic inhibitors, e.g., sodium azide and iodoacetate, and physicochemical inhibitors, e.g., ethyleneglycol, urea, and dimethyl sulfoxide, had no effect on the binding, indicating that it is independent of cell energy metabolism and hydrophobic or hydrogen bonding. From the results obtained with leishmanias and other particles, it is suggested that the macrophage membrane has several binding structures specific for different particles and that the binding mechanism for leishmanias is different from that of latex particles and immune complexes.

Experimental data summarized above do not allow any general conclusion with regard to the mechanism of leishmania–macrophage binding. Whether the inconsistent results observed are due to differences intrinsic to various systems or experimental conditions used must await further investigations.

4. *Complexity and Complications*

The dynamic fluidity as well as the molecular complexity and heterogeneity are well-known surface properties of eukaryotes, to which both leishmania and macrophage belong. Thus, their binding is more comparable to lymphocyte–macrophage adhesion than to that between mammalian cells and prokaryotes whose surface molecular structure is more homogeneous. The leishmania–macrophage binding is probably far more complicated than that of receptor-mediated endocytosis of toxin, viral, and hormonal molecules (Neville and Chang, 1978). Irrelevant to these systems concerning homogeneous binding ligands are several complications, which must be considered in eukaryote–eukaryote binding, such as that between macrophages and leishmanias. First, it is possible that multiple ligand–receptor sites may be involved. Second, such binding may be affected by the surface activities not only of macrophages but also of leishmanias, e.g., clustering, capping, shedding, and/or recycling phenomena of their membrane molecules. These surface membrane activities have been demonstrated in leishmanias (Dwyer, 1976), and capping and shedding of the amastigote surface antigens are considered to play a role in antibody-mediated inhibition of phagocytosis in *L. donovani*–human phagocyte interactions *in vitro* (Chang, 1981c). Finally, the similarity of the metabolic pathways between macrophages and leishmanias precludes the use of many chemical and pharmacological reagents, which have been effectively used in other systems to study the binding mechanisms.

Receptor theory has been evoked or implied to explain the leishmania–macrophage binding mechanism in two of the three studies summarized above (Chang, 1981a; Benoliel *et al.*, 1980). Conclusive evidence for this must await further study of binding under stringent conditions using purified ligand(s) and/or receptor(s). We favor the hypothesis of molecular mimicry whereby leishmania surface component(s) may fit preexisting receptor(s) of macrophages for other important biological functions, e.g., the retrieval of secreted lysosomal enzymes (Stahl and Schlesinger, 1980). Whether the acid phosphatase found recently on the surface of *L. donovani* (Gottlieb and Dwyer, 1981) may represent one of the ligands for their binding to macrophages deserves consideration.

It should be noted that additional factors may mediate leishmania–macrophage binding in natural infection. Leishmanias are exposed to humoral factors of the mammalian hosts, e.g., complement and natural antibodies, and those of insect vectors, e.g., their saliva and gut regurgitant (which may be introduced into the mammalian hosts together with promastigotes). The possibilities that these factors may play a role in leishmania–macrophage interactions have not been studied.

B. Entry of Leishmanias into Macrophages

Leishmanias attached to macrophages ultimately gain entry into these cells under physiological conditions. It appears that leishmanias depend on the phagocytic activity of the macrophages for intracellular entry, aided perhaps by their surface compatibility for adhesion and the motility of the organisms for an increased chance encounter in the case of promastigotes. This is further supported by the hypothetical argument that, when the host cell is already phagocytic, there is probably no evolutionary pressure on the part of leishmanias to develop an additional mechanism for cell invasion.

Special interest has been focused on promastigotes, as they each possess an anterior flagellum which bestows on them not only motility but also easily recognized cellular polarity, and both may bear on the mode of their entry into macrophages. Motility may conceivably increase their encounter with macrophages, thereby helping the process of entry indirectly. Cellular polarity raises the question of random versus directed mode of entry into macrophages. The latter would imply active participation on the part of promastigotes, involving perhaps polar organelles for the invasion of host cells. Flagella of leishmanias have been found to lead the direction of entry into nonphagocytic (Chang, 1978a) or low-phagocytic (Lewis, 1974) cells. How leishmanias can induce phagocytic activity of normally nonphagocytic cells is unknown. It is known, however, that not all species can effect such "induced phagocytosis" (Chang, 1978a). When the phagocytic activity of macrophages is inhibited by cytochalasin B, the interactions between promastigotes and macrophages become limited to the flagellar end (Chang, 1979), although their entry into cells is inhibited. Whatever the mechanism, the flagellar end of promastigotes here plays a definitive role in their interactions with the mammalian cells.

The orientation of promastigotes during their entry into macrophages under normal conditions has been a matter of some controversy. The mode of entry was described as predominantly flagellar end first in the following promastigote–macrophage systems: *L. donovani*–hamster (Miller and Twohy, 1967), *L. tropica*–mouse (Zenian *et al.*, 1979), *L. brasiliensis*–hamster (Merino *et al.*, 1977), *L. mexicana*–mouse (Alexander, 1975), and *L. enriettii*–guinea pig (Rahman and Sethi, 1978). Contradictory to this is the finding that promastigotes of *L. donovani* enter hamster macrophages predominantly posterior end first (Akiyama and Haight, 1971). It now appears certain that there is really no specific orientation during the entry of *L. donovani* promastigotes into mononuclear phagocytes. Quantitative scanning electron microscopy provides evidence indicating that head-first and tail-first modes of entry are approximately the same in frequency of occurrence (Chang, 1979). This finding is supported by a similar study using human monocyte-derived macrophages (Pearson *et al.*, 1981). Whether this is also true in other promastigote–macrophage combinations remains to be determined. It is likely that the flagella of promastigotes play a minor role in their entry into macrophages.

The entry of promastigotes into macrophages is a cellular event akin to phagocytosis of other ingestible particles by these cells, as seen by electron microscopy (Figs. 6 and 7) (Chang, 1979). There is, however, exaggerated response of the macrophage surface membrane around the promastigotes, manifested by the formation of extensive ruffles and whorled pseudopodia (Fig. 8) (Chang, 1979). The absence of these structures during phagocytosis of amastigotes or killed promastigotes by macrophages suggests that their formation may be attributable to the mechanical agitation caused by the motility of the organisms. It is thus likely that the role of leishmania flagella in the intracellular entry of promastigotes is limited to their physical function rather than their chemical property. Certainly, they cannot be equated to the polar organelles of coccidian protozoa, which may release substances to help their entry into host cells (Kilejian, 1976).

Conditions required for the entry of leishmanias into macrophages have not been defined. Available information indicates that macrophages require the same conditions for ingesting leishmanias as they do for phagocytosis of other particles. Thus, the entry of leishmanias into macrophages is inhibited by cytochalasin B (Alexander, 1975; Chang, 1979; Ardehali *et al.*, 1979; Zenian *et al.*, 1979) and iodoacetate (Benoliel *et al.*, 1980), suggesting the participation of actin filaments and the requirement of energy generated by glycolysis of the host cells. The entry of leishmanias into phagocytic cells including macrophages is accompanied by chemiluminescence (Meshnick and Eaton, 1981) and the release of H_2O_2 (Murray, 1981a)—phenomena usually triggered by phagocytosis. Stimulating conditions for the entry of leishmanias into macrophages include the presence of sera and the normal temperature of mammalian cells. Whether these conditions favor the process of entry directly or indirectly by promoting the adhesion between leishmanias and macrophages or their *in vitro* survival is not clear.

C. Intracellular Survival Mechanisms of Leishmanias

Once inside the macrophages, leishmanias must deal with an array of hostile activities, with which these cells are equipped to kill ingested microbes. Here, leishmanias must overcome a formidable challenge in order to establish intracellular symbiosis. To be discussed in this section are three areas: (1) the intracellular fate of leishmanias in relation to lysosome–phagosome fusion; (2) their susceptibility to cell-mediated microbicidal activities; and (3) differentiation of leishmanias, which prepares them to live in an intracellular environment.

1. *Lysosome–Phagosome Fusion*

The most significant event of leishmania–macrophage association is the fact that leishmanias are processed by macrophages as usual for degradation, but do not suffer the consequences. Supporting evidence for this is obtained by following the intracellular fate of leishmanias in relation to prelabeled secondary lysosomes of macrophages by electron microscopy. In *L. mexicana*–mouse mac-

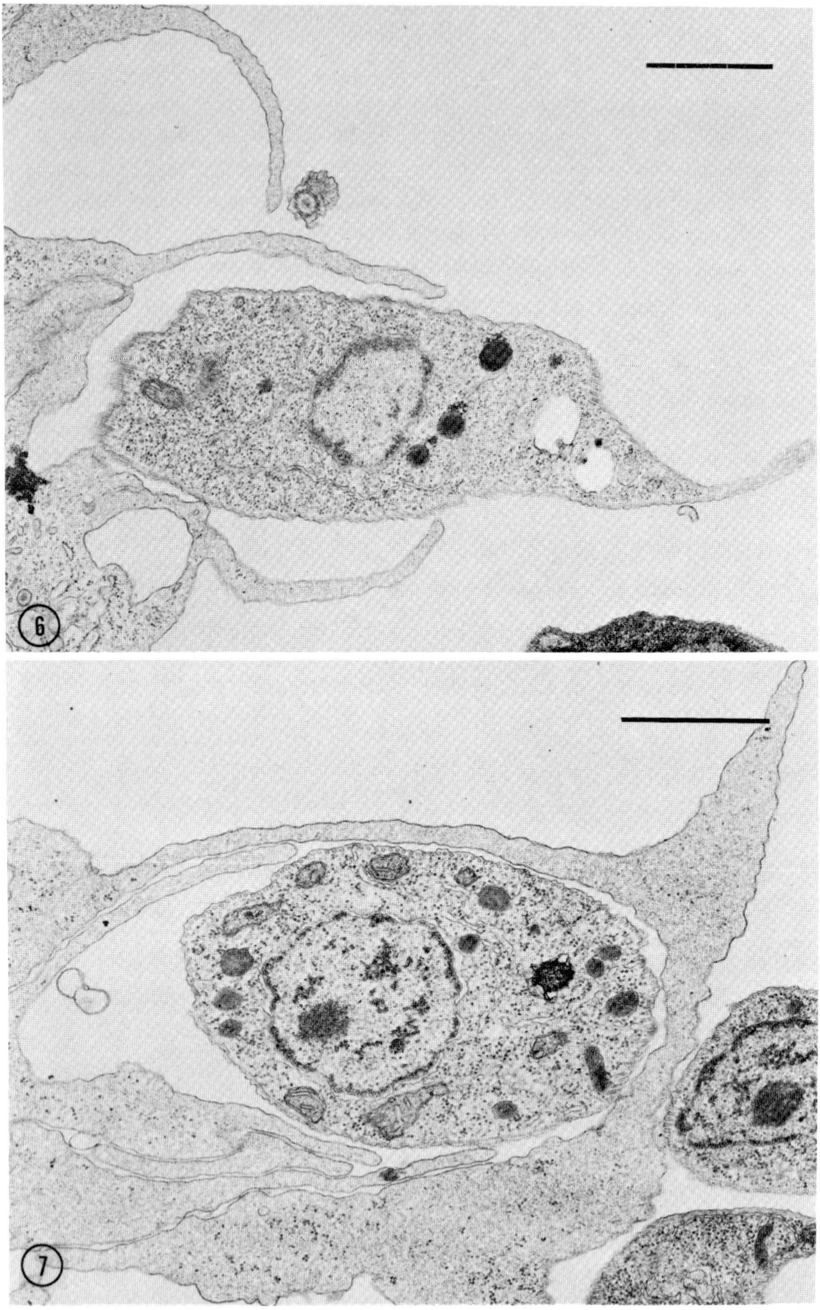

FIGS. 6 AND 7. Transmission electron micrographs showing promastigotes in oblique or tangential sections in early interactions with pseudopodia of macrophages for 10 min. Bar = 0.1 μm. Figure 6. A promastigote flanked by slender cytoplasmic extensions originating from macrophage. Note that the promastigote–macrophage adhesion is limited to few points of contact interrupted by wide irregular spaces. Figure 7. A promastigote completely encircled by multilayered pseudopodia of macrophage. Note the absence of continuous membrane abuttment at the host–parasite interface.

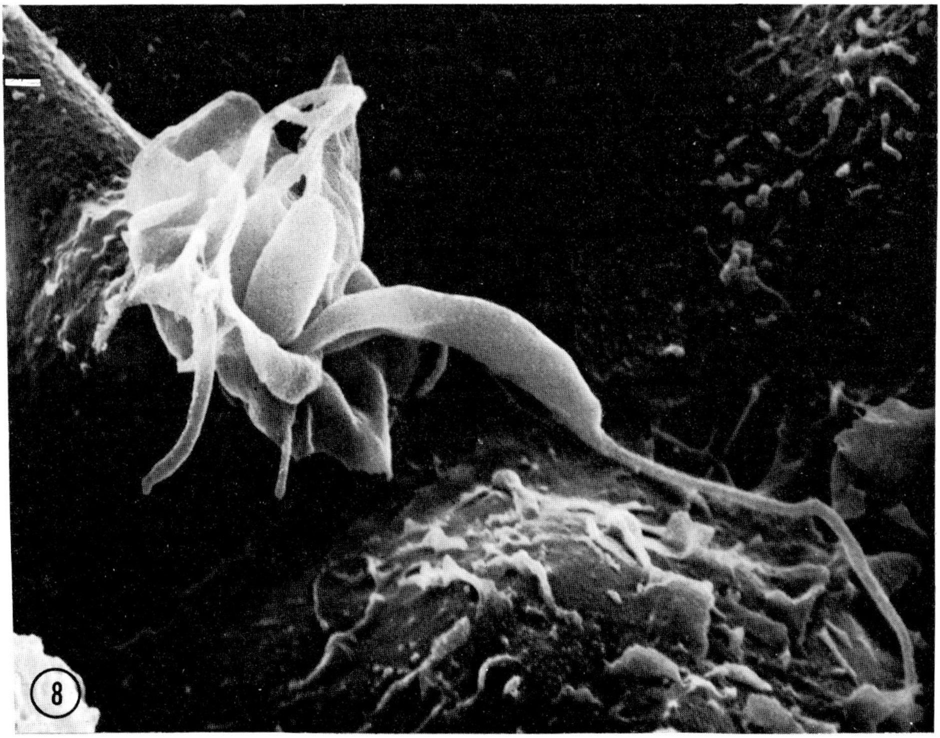

FIG. 8. Scanning electron micrograph of a macrophage ingesting several promastigotes simultaneously through an opening with membranous veils arranged like flower petals after infection for 30 minutes. Bar = 1 μm.

rophage (Alexander and Vickerman, 1975) and *L. donovani*–hamster macrophage (Chang and Dwyer, 1976) *in vitro* systems, it has been seen that amastigotes are first lodged in phagosomes, which are subsequently fused with the secondary lysosomes (Fig. 9). In phagolysosomes, they not only survive but continue to multiply. The fusion event has been subsequently confirmed in other leishmania–macrophage *in vitro* systems (Lewis and Peters, 1977; Ebert *et al.*, 1976; Berman *et al.*, 1979). Fusion of leishmania-containing phagosomes with primary lysosomes has been demonstrated in human phagocytes (Chang, 1981b). These cells have abundant primary lysosomes, which contain peroxidase—an enzyme readily demonstrable by ultrastructural cytochemistry and absent from the surface of leishmanias. Diaminobenzidine reaction products for peroxidase have been localized in the leishmania-containing vacuoles of human monocytes infected with *L. donovani* (Chang, 1981b). From the above observations, it has been proposed that intracellular survival mechanisms of leishmanias in mac-

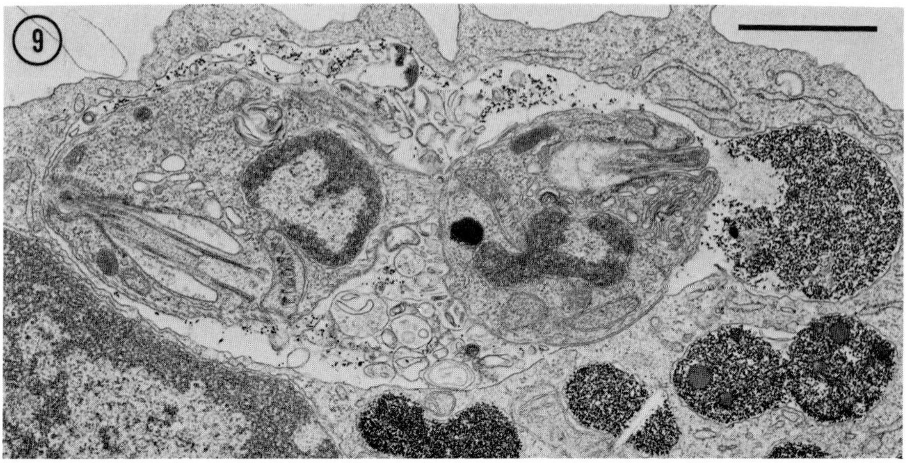

FIG. 9. Portion of a thorotrast-prelabeled hamster macrophage infected subsequently with *Leishmania donovani* for 1 hour showing the fusion of a secondary lysosome and a parasitophorous vacuole containing two amastigotes. Note the membrane debris lodged with the parasites in the same vacuole. Bar = 1 μm.

rophages must be based on their resistance to the enzymatic degradation in the lysosome and to other microbicidal factors present in this intracellular microenvironment.

2. *Resistance of Leishmanias to Lysosomal Enzyme Degradation*

How leishmanias resist the degradation by the hydrolytic enzymes in the lysosomal compartment of macrophages is a question that remains to be an-

---

FIG. 10. Autoradiogram after SDS–polyacrylamide gel electrophoresis of radioiodinated *Leishmania mexicana amazonesis*. Promastigotes were grown in HEPES-buffered medium 199 plus 10% heat-inactivated fetal bovine serum at 27°C, washed three times in Hanks' balanced salt solution and surface iodinated at 4°C at 100 μCi $K^{125}I/10^8$ cells for 20 minutes. See Fong and Chang (1981) for sodium dodecyl sulfate–polyacrylamide slab gel electrophoresis and autoradiography. A, Iodination of living cells in the presence of IODO-GEN; B, iodination of living cells in the absence of IODO-GEN; C, affinity binding of NP-40 solubilized sample to concanavalin A Sepharose; D, C in the presence of 0.1 *M* α-methyl-D-mannoside and 0.1 *M* α-methyl-D-glucoside; E, elution of concanavalin A bound sample from C with the sugar solution in D; F, control for E without sugars; G, immunoprecipitation of NP-40 solubilized sample followed by affinity binding to protein A Sepharose; H, control for G using normal mouse serum.

FIG. 11. Autoradiogram after SDS–polyacrylamide gel electrophoresis of [^{35}S]methionine-labeled *Leishmania mexicana amazonensis*. A, Detergent solubilized (0.1% SDS and 1% NP-40) sample bound to concanavalin A Sepharose; B, immunoprecipitation of NP-40-solubilized sample followed by affinity binding to protein A Sepharose; C, immunoprecipitation of cell surface antigens by the method of Nogueira *et al*. (1981); D, control for C using normal mouse serum. See Fong and Chang (1981) for other experimental conditions.

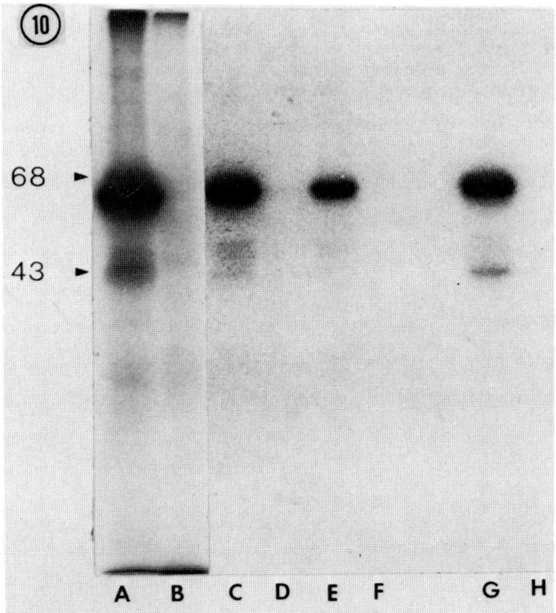
10
68
43
A
B
C
D
E
F
G
H

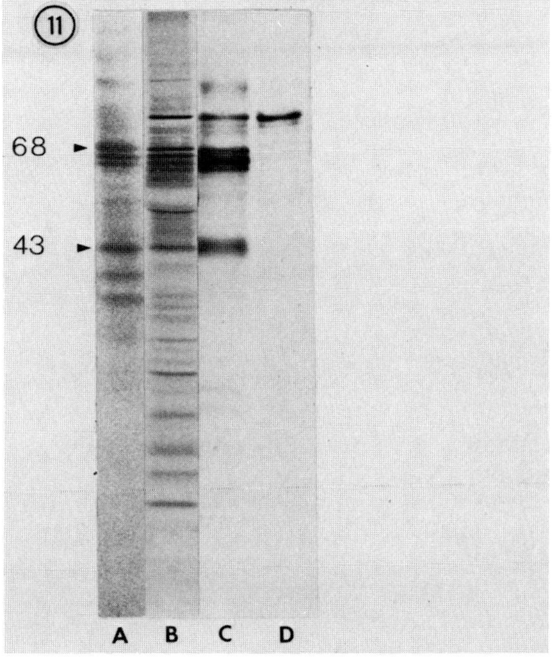
11
68
43
A
B
C
D

swered. There are at least two possibilities (Alexander and Vickerman, 1975): (1) leishmanias may have a surface, which is naturally resistant to lysosomal enzyme degradation (Chang and Dwyer, 1976); and (2) leishmanias may release factors, which inactivate lysosomal enzymes directly by acting as enzyme inhibitors (El-On *et al.*, 1980) or indirectly by changing the intralysosomal pH (Coombs, 1982). It should be noted that both possibilities may be working in concert, but conclusive evidence to support either has yet to be obtained.

That leishmanias may have a lysosomal enzyme-resistant surface stems from the observation that extraneous particles lodged in the leishmania-containing vacuoles disappear, whereas amastigotes do not (Chang and Dwyer, 1976). To provide direct evidence for this possibility, we have recently begun to identify leishmania surface antigens/glycoproteins and subjected them to treatment with lysosomal enzymes (Chang and Fong, 1983). Leishmanial membrane proteins and surface antigens have been studied (Dwyer, 1980; Handman *et al.*, 1981; Hernandez *et al.*, 1981). We have selected *L. mexicana amazonensis* promastigotes for investigation, as they can be grown for prolonged periods *in vitro* without losing their ability to survive in macrophages (Chang, 1980a). Promastigotes surface-iodinated or labeled biosynthetically with [^{35}S]methionine were lysed with nonionic detergent; with or without immunoprecipitation, samples were subjected to affinity binding using protein A or concanavalin A Sepharose 4B. Leishmania glycoproteins and antigens collected were subsequently identified as protein bands after sodium dodecyl sulfate–polyacrylamide slab gel electrophoresis followed by autoradiography. Radioiodination of living leishmanias yielded two major surface proteins: a heavy band at 68 kilodaltons and a lighter one at 43 kilodaltons (Fig. 10). Both species are apparently antigenic glycoproteins, as they can be immunoprecipitated and bound to concanavalin A beads (Fig. 10). Identification of surface antigens (Nogeira *et al.*, 1981) using [^{35}S]methionine-labeled materials gave essentially the same results, except that the heavy band of 68 kilodaltons in the radioiodinated sample now appeared to be a triplet species (Lane C, Fig. 11). The triplet bands may represent either a single antigen or closely related molecules sharing the same antigenic determinants, as they can be identified or immunoprecipitated with three different monoclonal antibodies (Fig. 12) (Fong and Chang, 1982). This antigen is different in molecular weight from that identified on the surface of the same leishmania species using monoclonal antibodies prepared by Pratt and David (1981). All four bands in our preparations are present in [^{35}S]methionine-labeled samples after concanavalin A sepharose affinity binding (Lane A, Fig. 11) or after immunoprecipitation for total cellular antigens (Lane B, Fig. 11). Surface antigenic glycoproteins collected were treated at pH 5, 37°C with Triton X-100-solubilized lysosomes purified by isopycnic sedimentation from the rat liver. At least one surface glycoprotein at 68 kilodaltons and the other at 43 kilodaltons persisted throughout the entire period of incubation, while other irrelevant protein mole-

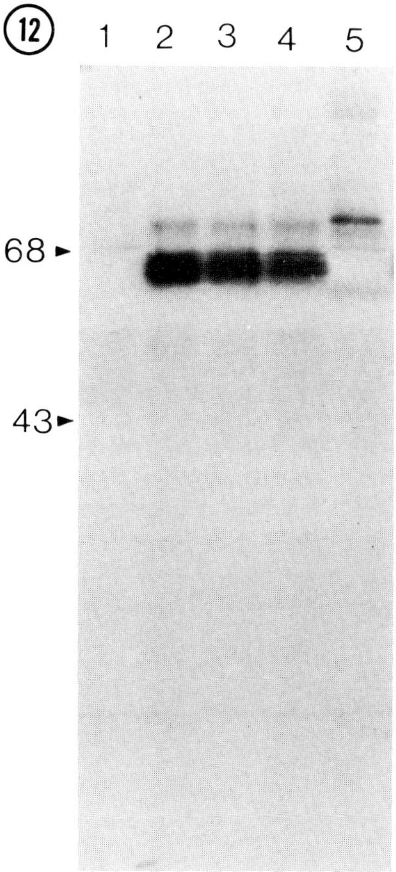

FIG. 12. Surface antigens of *L. mexicana amazonensis* promastigotes identified by immunoprecipitation using four different monoclonal antibodies prepared by hybridoma technique. BALB/c mice were injected intravenously with living promastigotes at 10^8 cells/mouse three times. Splenic cells were prepared and fused with SP2/0 cells. Standard procedures were followed for cell fusion, clone selection, and expansion. Culture supernatants of growth-positive clones were assayed for antibodies specific to cell surface antigens by immunofluorescent microscopy and ELISA. Culture supernatants were collected and used for the immunoprecipitation experiments. Lanes from left to right represent (1) control using a growth-positive, but surface antigen-reactivity-negative clone; (2–4) three different clones which produce IgG_1 reactive apparently to same surface antigens of the organisms; (5) a fourth clone, which produces Ig_{2a} reactive apparently to different surface antigens. See Fong and Chang (1981) as well as Chang and Fong (1982) for [^{35}S]methionine-labeling conditions and SDS–PAGE.

cules disappeared or faded, resulting apparently from the hydrolytic action of the lysosomal enzymes (Fig. 13). When radioiodinated living promastigotes were fed to macrophages, the labeled surface molecules also persisted for up to 36 hours without being degraded. These results offer direct evidence for the refractory property of leishmanial surface antigenic glycoproteins to lysosomal degradation. It is therefore possible that they may provide a protective "shell" for leishmanial promastigotes to resist the action of hydrolytic enzymes, thereby surviving in the lysosomal compartment of macrophages.

The molecular basis of resistance of leishmania membrane glycoproteins to lysosomal degradation remains to be determined. The functional and structural

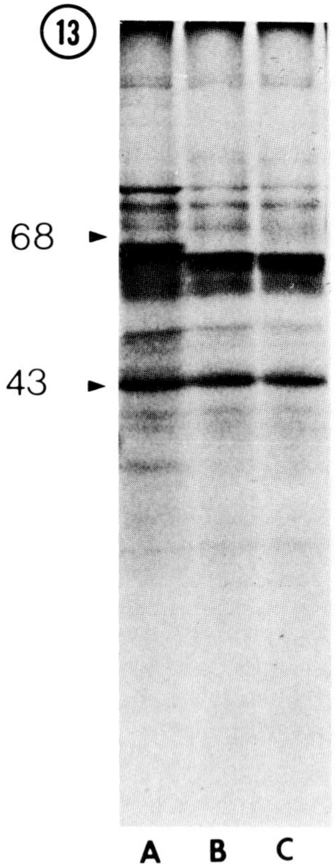

FIG. 13. Autoradiogram after SDS–polyacrylamide gel electrophoresis of [^{35}S]methionine-labeled and detergent-solubilized parasite antigens after treatment with rat liver lysosomal enzymes. A, 0 hour; B, 8 hour; C, 24 hour. Controls at 37°C in the absence of lysosomal enzymes are identical to A. Rat liver lysosomes were purified by isopycnic sedimentation and their purity was verified by standard assays for lysosomal enzymes. Lysosomal degradation experiments were carried out at pH 5, 37°C in 1% Triton X-100.

integrity of lysosomes depends on the resistance of their limiting membrane to the activities of their constituent enzymes and on the prevention of their mutual degradation. The carbohydrate moieties of the lysosomal enzymes and membrane have been proposed as the stabilizing factors (Avila and Convit, 1976). Whether the leishmanial surface glycoproteins may bear structural resemblance to these lysosomal constituents and thus escape from the attack by the hydrolytic enzymes is worthy of consideration. The recent finding of acid phosphatase on the surface of *L. donovani* promastigores (Gottlieb and Dwyer, 1981) lends additional credence to this hypothesis.

The proposal that leishmanias may release lysosomal enzyme inhibitor(s) is based on the observation of macrophages infected with *L. mexicana* (Alexander and Vickerman, 1975). The formation of extremely large leishmania-containing vacuoles (Fig. 14) in this infection does raise the possibility that leishmania-released substances may accumulate in them. It is hypothesized that amastigotes may produce ammonia, which may accumulate in the vacuoles, thereby raising the intralysosomal pH to inactivate the hydrolytic enzymes (Coombs, 1982). However, this possibility has yet to be substantiated with experimental evidence. More work is accomplished with "excretory factors," which are released by a number of leishmania species and are highly negatively charged, carbohydrate-rich antigenic molecules of variable molecular weights (El-On *et al.*, 1980). Excretory factors derived from *L. donovani* appear to inhibit lysosomal β-galatosidase of macrophages, but not their acid phosphatase, β-glycuronidase and β-glucosaminidase; electrostatic interactions between the excretory factors and the enzymes are proposed to be the mechanism of inhibition (El-On *et al.*, 1980). Similar exometabolites have been produced under more stringent conditions, making possible their characterization as a glycopeptidophosphingolipid, which forms micellar structures of a high molecular weight in the aqueous solution (Semprevivo and McLeod, 1981). How such supramolecules can act as enzyme inhibitors is not immediately apparent.

3. *Resistance of Leishmanias to Microbicidal Oxygen Metabolites*

In phagolysosomes, leishmanias must defend themselves against not only hydrolytic enzymes but also other microbicidal factors. Most powerful are perhaps the anionic free radicals, e.g., H_2O_2, O_2,·OH, and $^1O_2$, which are formed by enzymatic reduction of molecular oxygen during the phagocytosis-induced respiratory burst (Babior, 1978). Presumably, these products are fluxed into phagolysosomes to effect oxidative killing of microorganisms therein, although the mechanism of killing is not clearly understood and probably varies with the organisms concerned. Human polymorphonuclear leukocytes, e.g., neutrophils and eosinophils, represent the archetype of the phagocytes, which are endowed with elaborate mechanisms for this kind of microbicidal activity. In these cells, both amastigotes and promastigotes of *L. donovani* are degraded in phagolyso-

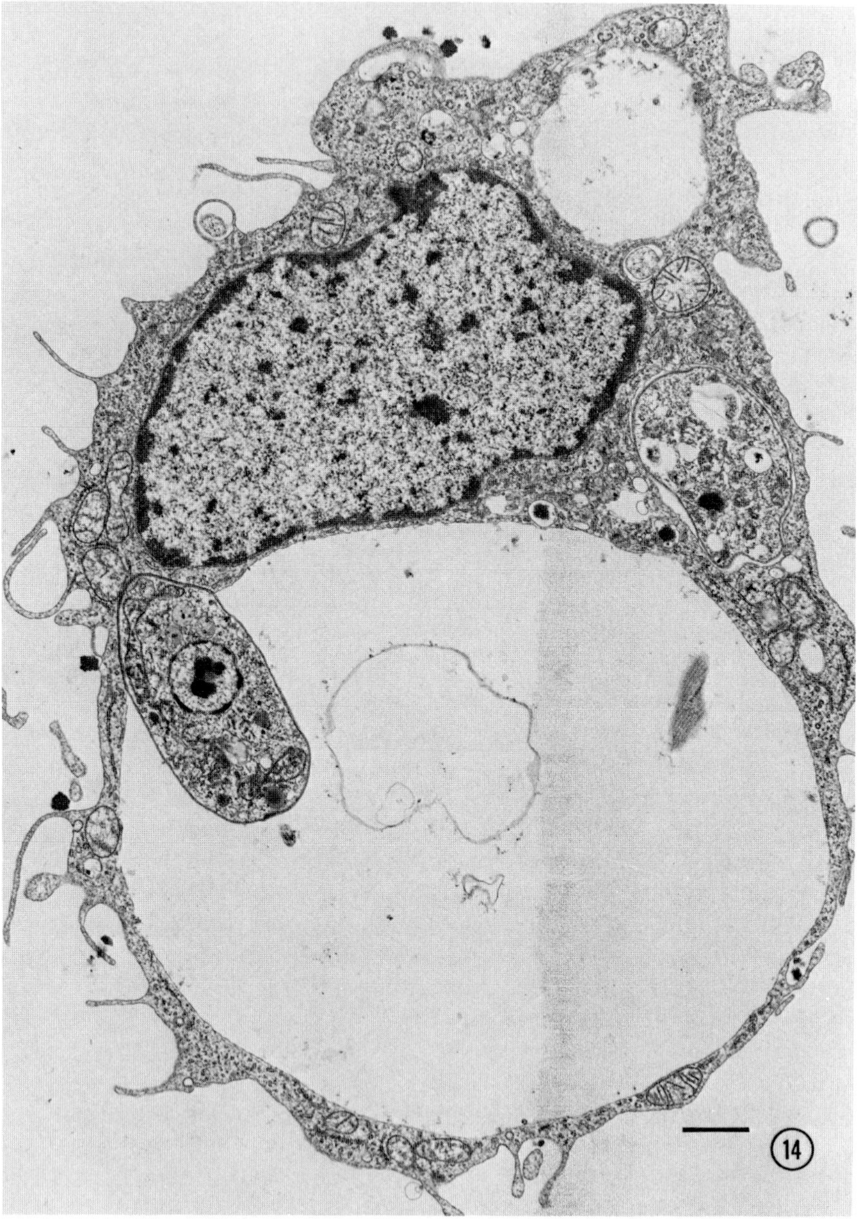

FIG. 14. A hamster peritoneal macrophage infected with *L. mexicana amazonensis* for 7 days. Note the large vacuole containing an amastigote which adheres to the inner lining of the vacular membrane, most of the vacular space being empty. Bar = 1 μm.

somes (Chang, 1981b; Pearson and Steigbigel, 1981). The leishmanicidal activity of these cells is apparently mediated by oxidative mechanisms, particularly the myeloperoxidase–H_2O_2–halide microbicidal system (Klebanoff and Hamon, 1972). Additional support for this is the survival of leishmanias in cells from patients with chronic granulomatous disease (Pearson and Steigbigel, 1981; Chang, 1981b), which is marked by an inability of oxidative killing due to genetic defect for the production of the oxygen reducing enzyme. It is suggested that a lower level of this oxygen-dependent microbicidal activity associated with monocytes or macrophages may account for the ability of amastigotes to survive in these cells (Chang, 1981b).

How leishmanias detoxify the oxidative metabolites of monouclear phagocytes is unknown. In a preliminary report, regular peroxidases, superoxide dismutase, and other related enzymes of *L. mexicana mexicana* were considered not to play a significant role in its intracellular survival mechanism, as the activities of these enzymes were either undetectable or unimpressive as compared to normal aerobic cells (Ghafoor and Coombs, 1980). Promastigotes of *L. tropica* were, however, found to have a relatively high activity of superoxide dismutase, but very little or no catalase and glutathione peroxidase (Meshnick and Eaton, 1981). Thus, under cell-free conditions, they are relatively sensitive to H_2O_2 (10^{-4} *M*) and are resistant to O_2^- (also $\cdot OH$ and $^1O_2$) (Murray, 1981a; Pearson *et al.*, 1981). Hydrogen peroxide has been localized to the leishmania-containing vacuoles of human monocytes infected with *L. donovani* amastigotes (Fig. 15) (Chang, 1981b), indicating that intracellular leishmanias are exposed to H_2O_2. It appears, then, that the sensitivity of leishmanias to this oxygen metabolite may be dose-dependent. The crucial question of exactly how much H_2O_2, to which leishmanias are exposed in macrophages has yet to be determined. Additional questions worthy of consideration are how they dispose of H_2O_2 independently of the usual enzyme systems and whether they may exploit the activity of macrophages to perform the task for them. Until these questions are resolved, the role of H_2O_2 in the intracellular survival mechanisms of leishmanias and in the leishmanicidal activity of lymphokine-activated macrophages cannot be certain (Murray, 1981b; Buchmüller and Mauel, 1981).

4. *Intracellular Differentiation of Leishmanias*

In macrophages, promastigotes differentiate into amastigotes, presumably a necessary event for the transition of leishmanias from an extracellular to an intracellular life. Intracellular cytodifferentiation of leishmanias precedes their multiplication and is thus a crucial step for the establishment of intracellular symbiosis. Previous work on leishmanial differentiation has largely been limited to the transformation from amastigotes into promastigotes (see Fong and Chang, 1981, for recent review). Intracelluair differentiation in the reverse direction has not been studied, except for morphological and ultrastructural observations (Akiyama and McQuillen, 1972; Chang, 1978a; Ebert *et al.*, 1979). It has been

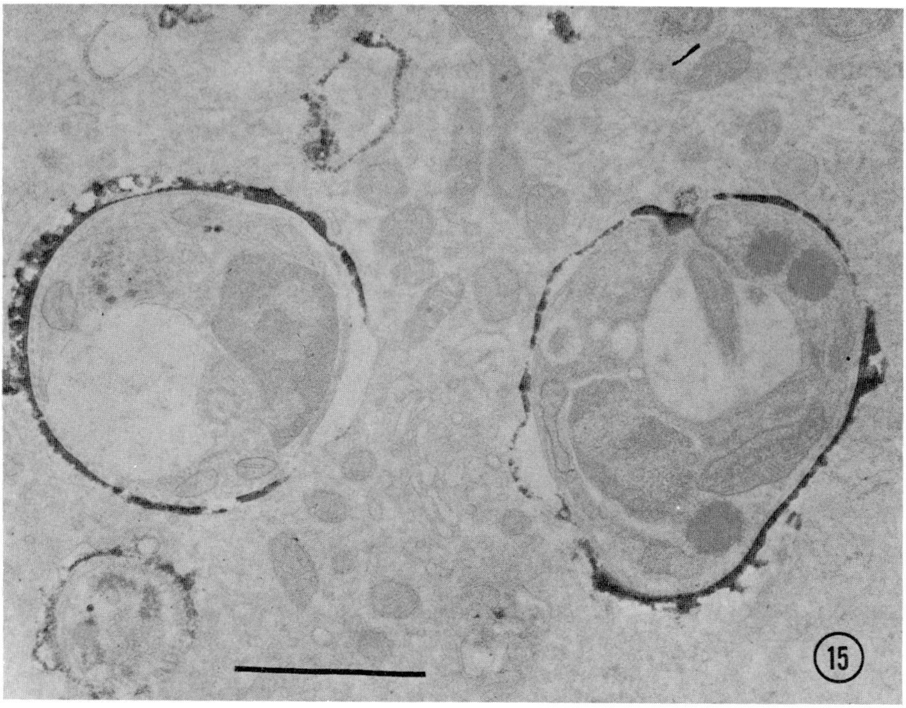

FIG. 15. Portion of a human monocyte infected with *L. donovani* amastigotes showing diaminobenzidine reaction products for H_2O_2 in the vacular space of the leishmania-containing vacuole. See Chang (1981b) for detail. Bar = 1 μm.

observed that promastigotes in the host cells become shorter and lose their flagella, but the details of these changes cannot be observed because of the asynchronism of the event. The intracellular microenvironment and body temperature of the mammalian hosts are both important for promastigote-to-amastigote differentiation in most cases (Lamy *et al.*, 1967; Frothingham and Lehtimaki, 1969; Akiyma and Taylor, 1970). Lysosomal pH and lysosomal constituents seem to play no role in the induction of leishmanial cytodifferentiation (Chang, 1978a). A temperature shift alone may be sufficient to trigger the morphological changes of some South American species, e.g., *L. mexicana amazonensis* (Hendricks *et al.*, 1978). Further investigations are needed to determine the control mechanisms for intracellular differentiation of leishmanias.

Recently, the antigenic and molecular changes accompanying the intracellular differentiation of leishmanias have been studied in an *in vitro* system, consisting of *L. mexicana amazonensis* and a macrophage line, J774G8 (Chang, 1980a; Chang and Fong, 1982; Fong and Chang, 1981). This *in vitro* system is particularly amenable to this and other investigations because of the following advantages (1) *L. mexicana amazonensis* and J774G8 line can be continuously grown separately as promastigotes and tissue culture cells without losing their respective traits; (2) cocultivation of the two under appropriate conditions yields intracellular amastigotes in perpetual association with the macrophages; (3) intracellular amastigotes in such culture can be purified with a high rate of recovery by a simple procedure, i.e., cell disruption followed by centrifugation in a single stepped gradient of Percoll; and (4) purified amastigotes in cell-free medium readily differentiate into promastigotes, which can be cultured continuously as such or used to infect J774G8 cells. Thus, a complete developmental cycle of *L. mexicana amazonensis* in its natural infection is duplicated in this *in vitro* system without resorting to the use of vector sandflies and laboratory animals. The potential application of this system for studying all aspects of intracellular symbiosis is obvious. We have begun to exploit the convenience of this system for studying leishmanial differentiation because of the paucity of information in this area and the lack of other systems.

The *L. mexicana amazonensis*–J774G8 macrophage system proves to be an excellent model for studying the molecular mechanisms of intracellular leishmanial differentiation with particular reference to tubulin biosynthesis. Intracellular differentiation occurs at 35°C and takes about a week for completion (Fong and Chang, 1981). Here again, the change in cell morphology is highly asynchronous and is marked by the occurrence of a morphological form intermediate between amastigotes and promastigotes. There are changes in the protein species during this transformation, as determined by slab gel electrophoresis of leishmanias labeled with [^{35}S]methionine. Particularly evident is a doublet protein species (apparent M_r = 55,000), which decreases in intensity during leishmanial differentiaon (Fig. 16). These two protein bands are identified as α- and β-tubulin molecules on the basis of their comigration with these molecules of *Chlamydomonas* axonemal proteins in two-dimensional gel and their immunoprecipitation with rabbit antiserum raised against β-tubulin of *Chlamydomonas* (Fig. 17). This change in tubulin biosynthesis is well-correlated with differentiation-related changes of leishmanial morphology, i.e., shortening of flagella and subpellicular microtubules. Obviously, a decrease in demand for this structural protein calls for a reduction of biosynthesis for the tubulin subunits. It is known that the tubulin biosynthesis in other systems of eukaryotic differentiation is mostly under transcriptional control, i.e., a production of mRNAs for tubulin molecules before their biosynthesis. Because of this and the fact that leishmanial differentiation from amastigotes to promastigotes is inhibited by actinomycin D (Simpson, 1968), it is proposed that a similar mechanism may also control the change in tubulin biosynthesis observed in our system (Fong and Chang, 1981). This possibility has now been studied in an *in vitro* cell-free translation system using rabbit reticulocyte lysates and poly(A)+ mRNAs iso-

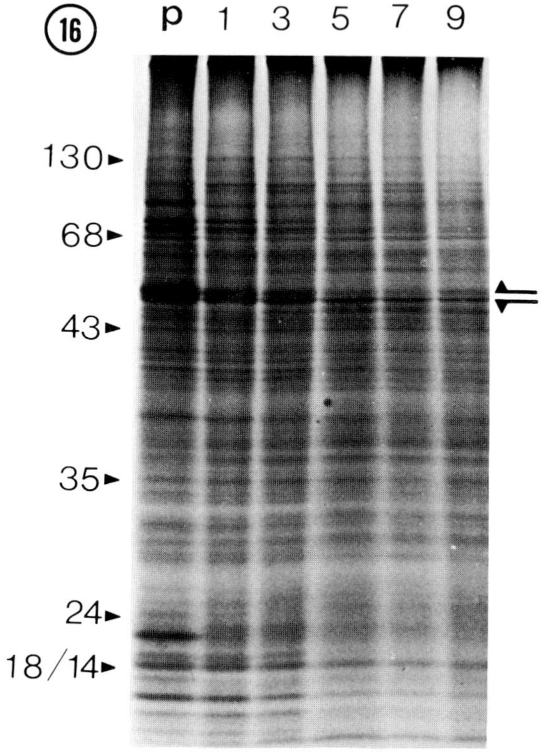

FIG. 16. SDS–polyacrylamide gel electrophoresis of [^{35}S]methionine-labeled proteins of *L mexicana*. Lanes: p, promastigotes labeled at 27°C; 1–9, leishmanias isolated from macrophages during the course of promastigote-to-amastigote transformation at different time points (days) as indicated. The molecular weight markers were β-galactosidase (130,000), bovine serum albumin (68,000), ovalbumin (43,000), pepsin (35,000), trypsinogen (24,000), β-lactoglobulin (18,000), and lysozyme (14,000). Arrow indicates tubulin bands.

lated from both stages of leishmanias (Wallach *et al.*, 1982). Cell-free translation products directed by poly(A)+ mRNAs from promastigotes and amastigotes were compared to each other and to the *in vivo* metabolically labeled proteins (Fig. 18). Unexpectedly, the results of these experiments indicate a post-transcriptional control for the tubulin biosynthesis during leishmanial differentiation. Tubulin bands are, as expected, the predominant species in the metabolically labeled promastigotes and in the *in vitro* translation products directed by the promastigote mRNAs, but not in metabolically labeled amastigotes (Fig. 18). Unexpected is the finding that the mRNAs from amastigotes translate just as much tubulin molecules as the promastigote samples (Fig. 18). To ascertain the identity and, to some extent, the quantity of tubulin translated *in vitro*, immunoprecipitation of samples from both amastigotes and promastigotes was carried

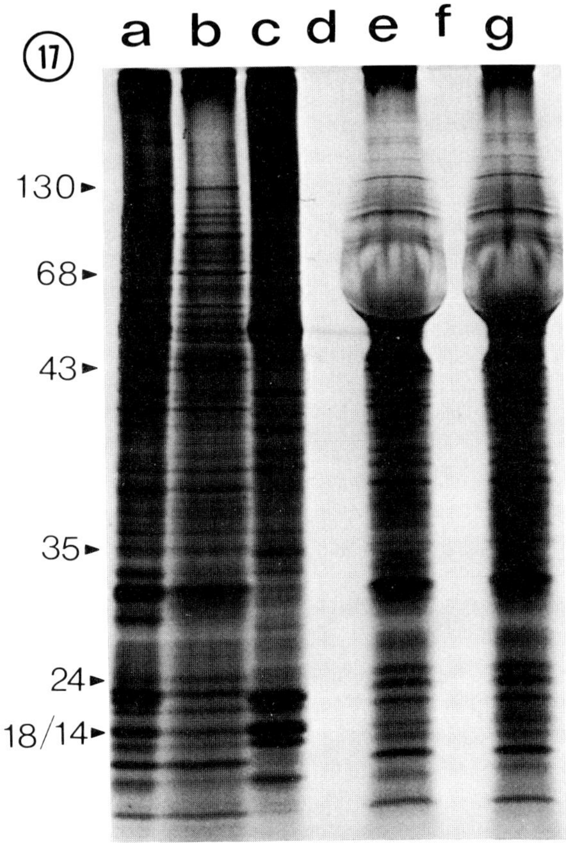

FIG. 17. Identification of β-tubulin of *L. mexicana* by monospecific rabbit serum against β-tubulin of *Chlamydomonas*. See Fong and Chang (1981) for details. Lanes: a, whole cell lysate; b, NP-40 soluble fraction; c, NP-40 insoluble fraction; d, immunoprecipitate of lane b using rabbit anti-β-tubulin of *Chlamydomonas;* e, soluble fraction of lane d; f, control for lane d using normal rabbit serum; g, soluble fraction of lane f. See Fig. 16 for molecular weight markers.

out using monospecific antiserum raised against the β-tubulin of *Chlamydomonas*. β-Tubulin bands were of equal intensity after specific immunoprecipitation from the translation products of total cellular poly(A)+ mRNAs from amastigotes and promastigotes (Fig. 19). Thus, similar amounts of translatable tubulin mRNAs must be present in both stages of leishmanias. Possibly, the tubulin mRNAs of amastigotes may be sequestered in some way, thus unavailable for protein biosynthesis in intact cells. The retention of tubulin mRNAs by amastigotes may be a manifestation of their incomplete adaptation in the mammalian hosts. It may certainly confer the advantage on them for a rapid

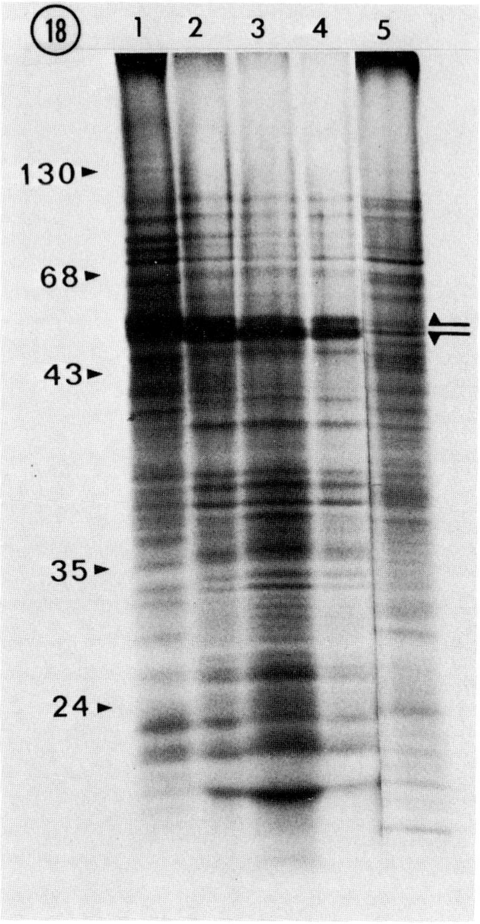

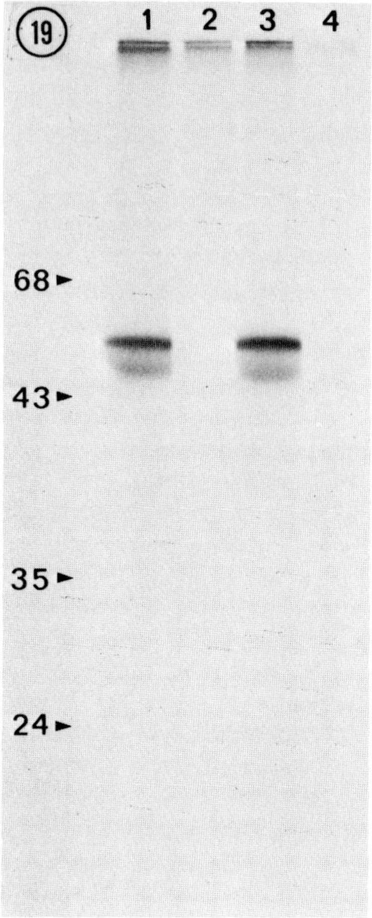

FIG. 18. Comparison of proteins synthesized *in vitro* by promastigote and amastigote mRNAs to those metabolically labeled *in vivo*. Promastigotes of *L. mexicana* were obtained as described in the legend to Fig. 10. Amastigotes were cultured in cells of the macrophage line J774 and purified by Percoll gradient centrifugation; isolated amastigotes remained viable and free from the contamination of macrophage materials. Total poly(A)+ RNA was extracted by standard procedures of SDS–phenol extraction followed by oligo(dT)-cellulose column purification. Proteins synthesizes in the rabbit reticulocyte cell-free system using poly(A)+ RNA (1 μg) and metabolically labeled proteins of intact organisms in the presence of [^{35}S]methionine were analyzed by SDS–PAGE and autoradiography. Lane 1 shows proteins synthesized by promastigotes *in vivo*. Cell-free products directed by promastigote mRNA (lane 2), mRNA from 9-day-old cultures of amastigotes (lane 3), and mRNA from 3-week-old cultures of amastigotes (lane 4) are shown. Lane 5 shows the proteins synthesized by amastigotes *in vivo*. Molecular weight markers are indicated as in Fig. 10. Arrow = tubulin bands.

FIG. 19. Immunoprecipitation for β-tubulin of *in vitro* translation products of amastigotes and promastigote mRNAs. Protein samples translated in the reticulocyte system were diluted to about 1.5 $\times 10^6$ cpm in 50 μl PBS containing 1% NP-40 and incubated for 30 minutes at 4°C. Each sample was divided into two equal portions; to each was added 10 μl normal rabbit serum as control or rabbit

tubulin biosynthesis during their differentiation into promastigotes when occasion calls for such an activity, for example, in the sandfly gut.

Even more changes in protein biosynthesis during intracellular leishmanial differentiation can be demonstrated by immunoprecipitation using hyperimmune sera (Fig. 20) (Chang and Fong, 1982). These antigenic changes were noted mostly among protein bands between 24 and 68 kilodaltons. Within this region, six were identified as common antigens of both leishmanial stages, six to seven were promastigote-specific, and three to five were amastigote-specific. At the higher molecular weight region (> 68,000) were two bands one being predominant in amastigotes and the other in promastigotes. There may be a differentiation-specific protein band (apparent $M_r = 20{,}000$). The transition of these stage-specific antigens varies widely with different protein species and may play important roles in the leishmanial differentiation in macrophages. While the origin of all these antigenic bands awaits further determination, it is likely that the triplet species at 68 kilodaltons (region D) and the one at 43 kilodaltons (band H) may correspond to the surface antigenic glycoproteins identified earlier by a different method of immunoprecipitation using polyclonal (Lane C, Fig. 11) and monoclonal antibodies (Fig. 12). These protein bands appear to be promastigote-specific surface antigens, as they become inapparent when the differentiation is completed (Fig. 20). This possibility is further supported by the finding that the monoclonal antibodies specific to these antigens do not react with amastigotes, as determined by immunofluorescent microscopy (Fong and Chang, 1982). It would be of interest to determine the biological functions of the surface antigens in relation to leishmanial cytodifferentiation.

D. Intracellular Multiplication of Leishmanias

Growth of leishmanias in macrophages represents the last, but not the least, important prerequisite for the establishment of intracellular symbiosis. *In vitro* systems for growing amastigotes in different leishmania–macrophage combinations and related methodology have been reviewed (Chang and Fish, 1982). Intracellular multiplication of leishmanias is usually assessed by microscopic

---

antiserum against β-tubulin of *Chlamydomonas* flagellar axonemal proteins (Fong and Chang, 1981). After 18 hours at 4°C, each sample received a 50 μl 33% Protein A-Sepharose 4B in PBS and incubated at 4°C, for 2 hours under constant agitation. Immunoprecipitates collected in the Protein A beads were extensively washed. All samples were subjected to SDS–PAGE followed by autoradiography. Lanes 1 and 3 are immunoprecipitated β-tubulin from amastigote and promastigote samples, respectively. Lanes 2 and 4 are their respective controls using normal rabbit serum. Note that the 55,000-dalton β-tubulin bands immunoprecipitated from amastigotes and promastigote samples are of equal intensity and that there are no protein bands in the controls. The light band below tubulin are artifacts due to the presence of abundant heavy chains of IgG, which is banded at about the same location and brings down some of radiolabeled tubulins.

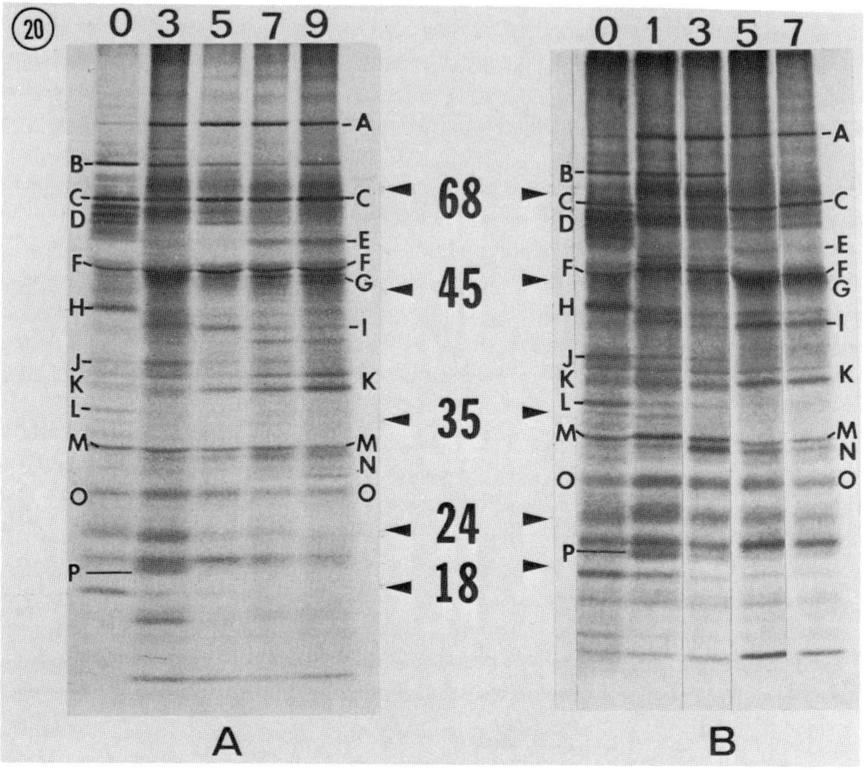

FIG. 20. Antigenic changes during intracellular differentiation of *Leishmania mexicana amazonensis* from promastigotes to amastigotes in J774G8 macrophages *in vitro*. Results from two separate experiments (A and B) are presented. Promastigotes cultured at 27°C and those allowed to transform at 35°C in macrophages for different days (numbers 0–9 at the top) were isolated, labeled with [^{35}S]methionine, and lysed in NP-40. Detergent-soluble fractions after centrifugation at 100,000 *g* were immunoprecipitated with hyperimmune sera raised in mice. Immune complexes formed were removed by using Protein A Sepharose 4B beads and solubilized in electrophoresis buffer. Antigens were identified by sodium dodecyl sulfate–polyacrylamide gel (10–15%) electrophoresis followed by autoradiography as bands (A–O). Numbers in the center represent molecular weight standards (× 10^3).

observations and counting of amastigotes in Giemsa-stained culture preparations. The general conclusion from these studies is that amastigotes are fastidious, slow-growing organisms. Their continuous culture *in vitro* has not been possible with the exception of *L. mexicana amazonensis* (Chang, 1980a; see Section II,A,4) and perhaps one strain of *L. tropica* (Handman and Spira, 1977). Most species, however, undergo multiplication in macrophages on a short-term basis. The rate of their multiplication depends on the culture conditions, sources of

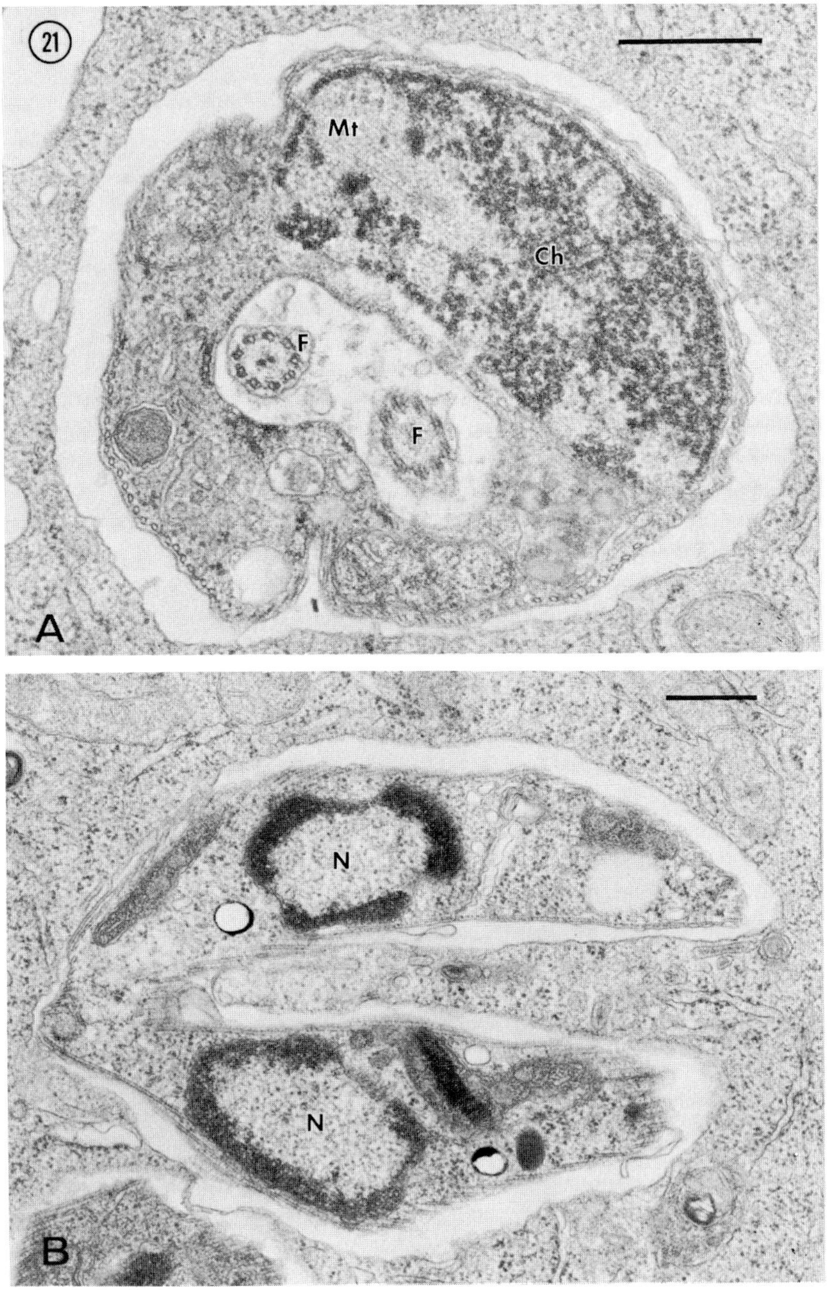

FIG. 21. Dividing amastigotes of *Leishmania donovani* in J774G8 cells in cross section (A) and longitudinal section (B) profiles. The vacuolar space around *L. donovani* is generally smaller than that produced by *L. mexicana amazonensis* in these cells. Replication of amastigotes is indicated by the presence of condensed chromatin granules (Ch), nuclear spindle microtubules (Mt), and biflagellation (F) in A, and by the presence of binucleation (N) and the division furrows in 7. Bar = 0.5 μm.

macrophages, and species and stages of the organisms. The biological and biochemical bases of these differences have just begun to receive attention and their relevance to the natural infection is uncertain. Several aspects of phenomenology and their interpretations are given below.

1. *General Requirement*

Virtually nothing is known about the physical conditions and the nutritional factors conducive to the multiplication of amastigotes, except for their requirement for an intracellular environment and the body temperature of their mammalian hosts. A temperature of 34–35°C favors the growth of cutaneous leishmanias, e.g., *L. tropica* (Berman and Neva, 1981), *L. mexicana* (Alexander and Vickerman, 1975; Chang, 1980a), and *L. enriettii* (De Castro and Pinto, 1960). This temperature is presumably more comparable to that of their natural site of infection. Other strains of *L. tropica* can, however, grow at 37°C (Handman and Spira, 1977). It is uncertain whether the thermotolerance is an inherent property of different species/strains or related to the differences in the culture conditions. The growth of one strain of *L. tropica* can apparently be improved substantially by using macrophages cultured in polypropylen tubes (Nacy and Diggs, 1981). The underlying mechanism for this is uncertain.

2. *The Intracellular Site*

Leishmanial amastigotes grow in membrane-lined vacuoles of infected macrophages (Fig. 21). Three types of such leishmania-containing vacuoles have been described in hamster macrophages infected with *L. donovani,* i.e., amastigotes in tight fitting vacuoles, those in loose fitting vacuoles, and the latter with amastigotes partially abutting the inner lining of the vacuolar membrane (Chang and Dwyer, 1978). The last vacuolar type is most striking in *L. mexicana*-infected macrophages (Mattock and Peters, 1975; Scorza *et al.*, 1979). The vacuolar space is so enormous that it becomes the most prominent structure of the infected cells (Fig. 14). When infected cultures are treated with antileishmanial drugs, the large vacuoles disappear together with amastigotes (Rabinovitch *et al.*, 1982).

The leishmania-containing vacuoles appear to be a continuum of the lysosome–phagosome vacuolar system of macrophages, as they behave like the secondary lysosome. Into this cellular compartment of macrophages exogenous macromolecules or particles are usually conveyed. Indeed, fluorescein isocyanate-conjugated dextran and thorotrast colloidal granules are shuttled by macrophages, apparently via their phagosome–lysosome vacuolar system, into *L. mexicana*-containing (Chang, 1980b) and *L. donovani*-containing (Chang and Dwyer, 1978) vacuoles, respectively. Moreover, there is no appreciable difference in the intralysosomal pH between *L. mexicana*-infected and noninfected J774G8 cells, even though the leishmania-containing vacuoles in the former are simultaneously determined (Chang, 1980b). Intralysosomal residence of leishmanias is further indicated by the localization in their vacuoles of acid phosphatase and peroxidase (Lewis and Peters, 1977; Ebert *et al.*, 1976; Chang, 1981b)—known constituent enzymes in the lysosomes of macrophages and human monocytes, respectively. The evidence with the acid phosphatase becomes somewhat equivocal in view of the finding of this enzyme on the surface of *L. donovani* promastigotes (Gottlieb and Dwyer, 1981). It is uncertain, however, as to whether this enzyme is also present on the surface of amastigotes and of other leishmania species. Whether the leishmania-containing vacuole can be equated to a secondary lysosome must await biochemical analysis of these cell organelles for the determination of their constituents, e.g., macrophage acid hydrolases. For such investigation, it is crucial to develop a cell fractionation technique for the isolation of leishmania-containing vacuoles and to obtain a sound knowledge of acid hydrolases of leishmanias. It is predicted from the available information that the biochemical approach mentioned above would verify leishmanias as a true intralysosomal organism.

The implications of intralysosomal residence of leishmanias have been discussed previously in relation to their intracellular life (Chang and Dwyer, 1978). These and additional points can now be further elaborated in the light of recent work. Clearly, the intralysosomal residence of leishmanias means that they can interact not only with intracellular constiuents of macrophages but also with extracellular milieu via the pinosome–lysosome and phagosome–lysosome vacuolar system of the host cells. Interactions of leishmanias with macrophages may involve the exchange of molecules produced by their respective biosynthetic machinery, as demonstrated by autoradiographic evidence (Bhattacharya and Janovy, 1975). Whether this represents nutritional flow between the interacting cells is uncertain. There is an intimate membrane association between amastiogotes and the inner lining of the host vacuoles, manifested by the formation of a tight junction-like structure (de Souza, 1981). It again remains to be determined whether molecular exchanges may occur at this junction. The interaction of leishmanias with lysosomal constituent enzymes has been studied and discussed in the context of their intracellular survival mechanism (see Section II,C,2). Another possible interaction is for leishmanias to exploit the digestive capacity of these enzymes for their own benefits. Conceivably, macromolecules broken down in this cellular compartment may be utilized to satisfy their nutritional requirement (Alexander and Vickerman, 1975; Chang and Dwyer, 1978). Alexander argues against this possibility, as he found that a blockade of lysosome–phagosome fusion after the entry of *L. mexicana* into macrophages resulted in an increase in the rate of amastigote growth (Alexander, 1981a). He reasons justifiably that the opposite would be expected, should the intralysosomal microenvironment be beneficial to amastigotes in any way. This notion would also seem incompatible with the supposition that leishmanias may release lysosomal enzyme inhibitors (see Section II,C,2) and with the finding that *L. donovani* infection of macrophages lowers the digestive capacity of these cells toward other related, but nonpathogenic organisms (Kutish and Janovy, 1981). It should be noted, however, that no biochemical data for the lysosomal enzyme activity were presented in these studies, and that none was concerned directly with the utilization of extracellular substrates by intracellular leishmanias in macrophages. Thus, whether the lysosomal enzymes of infected macrophages may play a role in the nutrition of leishmanias is still an open question.

3. *Variation among Different Species and Stages*

It is well known that the ability of leishmanias to grow in macrophages varies with species and stages of the organisms. It is generally true in this regard that cutaneous species are less fastidious than the visceral species. Thus, *L. mexicana* and *L. tropica* grow readily and continuously in macrophages (Handman and Spira, 1977; Chang, 1980a), whereas *L. donovani* does not (Chang and Dwyer, 1978). It is also generally accepted that amastigotes can establish infection in macrophages much more readily than promastigotes. There are, however, considerable variations among different species. Irrespective of the stages used, *L. mexicana amazonensis* grows in the macrophage line, J744G8 (Chang, 1980a). On the other hand, the intracellular growth of *L. donovani* often depends on the stages used for inoculation. Consistent proliferation of this species in hamster macrophages ensues only when amastigotes are used for infection (Chang, 1979). Infection of these cells with cultured promastigotes of *L. donovani* often results in the disappearance of the latter in a matter of several days. This variability has been accounted for by the following possibilities: (1) the inherent property of different leishmania species; (2) *in vitro* selection of nonpermissive clones of promastigotes with respect to either their intracellular survivability (Pearson *et al.*, 1981) or their ability of intracellular promastigote to amastiogte differentiation (Lewis and Peters, 1977); and (3) inadequate culture conditions and/or host cells (see below). Distinction between possibilities (1) and (2) can be accomplished by studying the infectivity of sandfly-derived promastigotes to *in vitro* cultured macrophages. Preliminary attempts to determine this are frustrated by the difficulty of obtaining promastigotes (*L. mexicana amazonensis*) from the proboscis of the infected flies; however, those from the gut are identical to cultured promastigotes in their infectivity to macrophages *in vitro* (Chang *et al.*, 1981).

4. *Heterogeneity of Host Cells*

With the exception of *L. mexicana amazonensis,* which appears to grow equally well in mononuclear phagocytes from man, mice, hamster, and rat (Chang, unpublished), all leishmania species have some preference for particular sources of host cells. For example, *L. donovani* grows in hamster peritoneal macrophages (Chang and Dwyer, 1978), in a mouse macrophage line (Berens and Marr, 1979), and in human peripheral blood monocyte-derived macrophages (Berman *et al.*, 1979), but not in mouse peritoneal macrophages (Herman, 1980; Chang and Chiao, 1981; Haidaris and Bonventre, 1981). Also, the peritoneal macrophages from susceptible strains of mice support the growth of *L. tropica*, while those from resistant strains do not (Handman *et al.*, 1979). Intracellular multiplication of leishmanias also varies with the macrophage subpopulations from the same animal. Thus, a strain of *L. tropica* grows better in skin-derived adherent cells than in peritoneal macrophages (Gorczynski *et al.*, 1981). Similarly, the growth of *L. enriettii* differs with the subpopulations of guinea pig peritoneal macrophages separated by velocity sedimentation (Gorzynski and McRae, 1982). Most dramatic is perhaps the finding that *L. mexicana mexicana* grows in peritoneal macrophages, but is killed in bone marrow macrophages of mice (Alexander, 1981b). The author suggested that the cell cycle-dependent maturation or differentiation of mononuclear phagocytes might account for their heterogeneity in supporting the intracellular multiplication of leishmanias. This proposal is consistent with the finding of stimualtion by granulocyte–macrophage colony-stimulating factor of *L. tropica* killing by macrophages (Handman and Burgess, 1979). Another possibility is the change of macrophages in their "physiological state" after prolonged *in vitro* culture as adherent cells, making them somehow unfavorable as host cells for the growth of amastigotes. This is suggested by the fact that *L. donovani* amastigotes fail to grow in hamster peritoneal macrophages beyond the first 1–2 weeks; however, their continuous growth can be made possible by weekly subcultures in freshly isolated cells (Chang, 1978b).

5. *Perturbation of Leishmanial Replication and Changes of Infected Macrophages*

Judging from the chronicity of leishmaniasis, the leishmanial amastigote and the host cell in an infected culture should be able to form a stable association whereby both grow together for a prolonged time period. This has been indeed the case in *L. mexicana–J774G8 cell in vitro* system (Chang, 1980a). In other leishmania–macrophage combinations, amastigotes grow only for a short time and often disappear completely from the culture for reasons that have yet to be elucidated (see Section II,D,3). In all these culture systems, the growth of amastigotes can be selectively interrupted by chemotherapeutic agents (Mattock and Peters, 1975; Rabinovitch *et al.*, 1982; Carson and Chang, 1981; Berman and Wyler, 1980; Alexander, 1981a; Berens *et al.*, 1980; Bachrach *et al.*, 1980) or by immunoactive lymphokine preparations (Chang and Chiao, 1981; Nacy *et al.*, 1981; Haidaris and Bonventre, 1981; Mauel and Behin, 1980). Inhibition or destruction of intracellular amastigotes observed in these studies is often interpreted in immunological or chemotherapeutic terms. The possible effects of the treatment on the leishmania–macrophage interactions have rarely been taken into account.

Leishmanial infection of macrophages may affect their activities, but most of these studies are concerned with subjects unrelated to the intracellular replication of amastigotes. It is known that infected macrophages have a higher rate of pinocytosis (Chang, 1980b), and that they incorporate more [^{3}H]thymidine or [^{3}H]uridine (Nacy *et al.*, 1981; Handman and Spira, 1977) than uninfected cells. Whether the increase may represent simply cumulative activities of the leishmania and the macrophage or their mutual stimulation leading to a heightened activity of one participant remains to be determined. Leishmanial infection of macrophages has also been shown to result in expression of leishmania-specific antigens on their surface (Berman and Dwyer, 1981) and to render them immunosupressive in the lymphocyte function assays (Scott and Farrell, 1981). It would be of interest to determine if multiplication or other activities of amastigotes may be invovled in these phenomena of obvious importance to the animal immunity in leishmaniasis.

III. Intralysosomal Symbiosis in Leishmaniasis—Model for a New Evolutionary Pathway?

The restriction of leishmanial infection in mammalian hosts to macrophages is of particular interest, as they are the very cell type responsible for the disposal of invading microorganisms. What may be envisioned from this for the evolution of leishmaniasis is a conventional cellular defense of hosts against intruding promastigotes, but the latter somehow manage to survive. At the cellular level, leishmanias are recognized, ingested, and delivered into the lysosomal compartment of macrophages for degradation just as any foreign objects. What makes leishmanias unique is their ability not only to survive but also to multiply in this extreme environment. With the exception of *Mycobacterium lepraemurium* (Hart *et al.*, 1972) and *Coxiella burnetii* (Hackstadt and Williams, 1981), other intracellular organisms circumvent the problem of intracellular digestion by residing in the nonlysosomal compartment of host cells (Jones, 1974; Goren, 1977), e.g., *Toxoplasma gondii* (Jones and Hirsch, 1972) and *Trypanosoma cruzi* (Kress *et al.*, 1977; Nogueira and Cohn, 1976) in macrophages, rickettsiae in tissue culture cells (Weiss and Dash, 1981; Moulder, 1979a), algal symbionts in *Paramecium* (Karakashian and Karakashian, 1973; Karakashian and Rudzinska, 1981), bacterial symbionts in amoeba (Jeon, 1980; Jeon and Jeon, 1976). While neither the mechanism adopted by leishmanias nor that by other organisms for avoiding lysosomal digestion has been determined, these two mechanisms may be viewed as two divergent pathways for the survival of intracellular organisms during the evolution of symbiosis.

That leishmanial infection of macrophages may form a stable association has been proposed (Vickerman, 1980) and is supported by observations of the *L. mexicana amazonensis*–J774G8 *in vitro* system (Chang, 1980a). In this culture, a cytopathic effect leading to a rapid cytolysis never occurs as it does in those infected by other pathogenic organisms (Jones, 1974; Moulder, 1974, 1977; Weiss, 1973). It may be speculated that a rapid killing of macrophages does not benefit leishmanias, but only deprives them of their normal habitat. Several lines of evidence suggest that leishmanial amastigotes have actually become obligate intralysosomal organisms of macrophages. It has been so far impossible to grow amastigotes extracellularly at the temperature of their mammalian hosts (Chang and Fish, 1982). In them, it is highly unusual to find leishmanias in cells other than mononuclear phagocytes. Consistent with this is the finding *in vitro* that nonphagocytic cells are generally nonsusceptible to leishmanial infection or, when infected, do not support the growth of amastigotes (Chang, 1978a).

In the extant intracellular symbiosis, there is a spectrum of associations with graded symbiote–host relationships, ranging from parasitism to mutualism. The position of leishmania–macrophage associations in this spectrum can be considered at two different levels. At the cellular level, this association is close to a harmonious one: amastigotes are destined to live in the lysosomes of macrophages, although they may not be fully integrated into the physiology of the host cells (and might never be, given the extreme conditions of the ecological niche). At the organismal level, the association is shifted to the extreme of parasitism manifested as leishmaniases. Perhaps the lack of a total integration in the association at the cellular level leads to the recognition of leishmanias by the sophisticated mechanisms of immune surveillance in the mammalian hosts. The subsequent immune response, be it effective as in cutaneous leishmaniasis or ineffective as in visceral leishmaniasis, may be frustrated by the intracellular residence of the organisms, resulting in the chronic disease state. Thus, elucidation of the cellular and molecular mechanisms of leishmania–macrophage interactions *in vitro* will prove fruitful not only for the formualtion of a strategy to combat leishmaniasis, but also for understanding the evolutionary pathway of intracellular symbiosis.

IV. Concluding Remarks

Continuing effort is clearly needed to elucidate leishmania–macrophage molecular interactions essential for our full understanding of the mechanisms of intracellular symbiosis in leishmaniasis. Emphasis should be given to further identification and purification of both amastigote and promastigote surface molecules. Some of these molecules undoubtedly play pivotal roles for the successful intracellular symbiosis in all aspects, i.e., leishmania–macrophage binding, entry of leishmania into macrophages, resistance of leishmanias to lysosomal enzymes for their intracellular survival, and uptake of nutrients for their intracellular multiplication. We and others have begun to identify externally exposed surface molecules of leishmanias by biochemical and immunological means. Monoclonal antibodies produced by the hybridoma technique should prove particularly useful for characterization and purification of surface antigens to facilitate the evaluation of their biological functions in leishmania–macrophage interactions. Also of interest is leishmanial transformation in macrophages which is not only crucial for the intracellular survival of this organism but also a potentially useful system for studying regulatory mechanisms of cytodifferentiation. Initial attempts have been made to underlie gene expression during the differentiation of *L. mexicana amazonensis* by determining its molecular and antigenic changes. We found reduced tubulin biosynthesis and the disappearance of a triplet polypeptide surface antigen, accompanying the morphological changes. Experimental evidence indicates posttranscriptional control for the tubulin biosynthesis during leishmanial differentiation. It would be of interest to determine whether changes of other molecules, e.g., the triplet surface polypeptide antigen, may be similarly regulated and whether the control mechanism for tubulin biosynthesis lies at the level of mRNA sequestration/repression or at the level of protein biosynthesis machinery. One last point to be mentioned is the possible significance of intralysosomal residence of amastigotes in leishmaniasis that has been discussed in terms of the intracellular survival mechanism of leishmanias (Section II,C,2), their utilization of nutrients in macrophages (Section II,D,2), and the evolutionary hypothesis of intracellular symbiosis (Section III). We will gain much insight into these areas by further study of leishmania–macrophage associations *in vitro* at the cellular and molecular level.

Acknowledgments

I thank Drs. W. Trager, D. Fong, and M. Wallach for their critical reviews of this manuscript; Mrs. C. Feirman and Mr. J. Stanorski for their excellent technical help; and the financial support from NIAID–NIH Grant No. AI-11916,19183, UNDP/WHO Special Programme for Research and Training in Tropical Diseases Contract No. 780040, and Irma T. Hirschl Career Scientist Award.

References

Akiyama, H. J., and Haight, R. D. (1971). *Am. J. Trop. Med. Hyg.* **20,** 539–545.

Akiyama, H. J., and McQuillen, N. K. (1972). *Am. J. Trop. Med. Hyg.* **21,** 873–879.

Akiyama, H. J., and Taylor, J. C. (1970). *Am. J. Trop. Med. Hyg.* **19,** 249–254.

Alexander, J. (1975). *J. Protozool.* **22,** 237–240.

Alexander, J. (1981a). *Exp. Parasitol.* **52,** 261–270.

Alexander, J. (1981b). *In* "Heterogeneity of Mononuclear Phagocytes" (O. Forster and M. Landy, eds.), pp. 447–454. Academic Press, New York.

Alexander, J., and Vickerman, K. (1975). *J. Protozool.* **22,** 502–508.
Ardehali, S. M., Khoubyar, K., and Rezai, H. (1979). *Acta Trop.* **36,** 15–21.
Avila, J. R., and Convit, J. (1976). *Biochem. J.* **160,** 129-
Babior, B. M. (1978). *N. Engl. J. Med.* **298,** 721–725.
Bachrach, U., Schnur, L. F., El-On, J., Greenblatt, E. L., Pearlman, E., Robert-Gero, M., and Lederer, E. (1980). *Febs. Lett.* **121,** 287–291.
Benoliel, A. M., Capo, C., Bongrand, P., Ryter, A., and Depieds, R. (1980). *Immunology* **41,** 547–560.
Berens, R. L., and Marr, J. J. (1979). *J. Protozool.* **26,** 453–456.
Berens, R. L., Marr, J. J., Nelson, D. J., and LaFon, S. W. (1980). *Biochem. Pharmacol.* **29,** 2397–2398.
Berman, J. D., and Dwyer, D. M. (1981). *Clin. Exp. Immunol.* **44,** 342–348.
Berman, J. D., and Neva, F. A. (1981). *Am. J. Trop. Med. Hyg.* **30,** 318–321.
Berman, J. D., and Wyler, D. J. (1980). *J. Infect. Dis.* **142,** 83–86.
Berman, J. D., Dwyer, D. M., and Wyler, D. J. (1979). *Infect. Immun.* **26,** 375–379.
Bhattacharya, A., and Janovy, J., Jr. (1975). *Exp. Parasitol.* **37,** 353–360.
Bray, R. S. (1974). *Annu. Rev. Microbiol.* **28,** 189–217.
Buchmüller, Y., and Mauel, J. (1981). *J. Reticuloendothel. Soc.* **29,** 181–192.
Carson, D. A., and Chang, K.-P. (1981). *Biochem. Biophys. Res. Commun.* **100,** 1377–1383.
Chang, K.-P. (1978a). *Am. J. Trop. Med. Hyg.* **27,** 1084–1096.
Chang, K.-P. (1978b). *J. Parasitol.* **64,** 931–933.
Chang, K.-P. (1979). *Exp. Parasitol.* **48,** 175–189.
Chang, K.-P. (1980a). *Science* **209,** 1240–1242.
Chang, K.-P. (1980b). *In* "The Host-Invader Interplay" (H. Van den Bossche, ed.), pp. 231–234. Elsvier, Amsterdam.
Chang, K.-P. (1981a). *Mol. Biochem. Parasitol.* **4,** 67–76.
Chang, K.-P. (1981b). *Am. J. Trop. Med. Hyg.* **30,** 322–333.
Chang, K.-P. (1981c). *Am. J. Trop. Med. Hyg.* **30,** 334–339.
Chang, K.-P., and Chiao, J. W. (1981). *Proc. Natl. Acad. Sci. U.S.A.* **78,** 7083–7087.
Chang, K.-P., and Dwyer, D. M. (1976). *Science* **193,** 678–680.
Chang, K.-P., and Dwyer, D. M. (1978). *J. Exp. Med.* **147,** 515–529.
Chang, K.-P., and Fish, W. R. (1982). *In* "*In Vitro* Cultivation of Protozoa Pathogenic to Man and Domestic Animals" (J. B. Jensen, ed.), CRC Press, Boca Raton, Florida (in press).
Chang, K.-P., and Fong, D. (1982). *Infect. Immun.* **36,** 430–431.
Chang, K.-P., and Fong, D. (1983). *In* "Cytopathology of Parasitic Diseases." Ciba Foundation Symposium 99 (in press).
Chang, K.-P., Bray, R. S., and Leaney, A. J. (1981). *Trans. R. Soc. Trop. Med. Hyg.* **75,** 475–476.
Coombs, G. H. (1982). *Parasitology* **84,** 149–185.
Croft, S. L., and Molyneux, D. H. (1979). *Ann. Trop. Med. Parasitol.* **73,** 213–226.
De Castro, M. P., and Pinto, S. C. (1960). *J. Protozool.* 7 (Suppl.), 7.
De Souza, W. (1981). *Z. Parasitenkd.* **66,** 25–29.
Dwyer, D. M. (1976). *J. Immunol.* **117,** 2081–2091.
Dwyer, D. M. (1980). *J. Protozool.* **27,** 176–182.
Ebert, F., Enriguez, G. L., and Mühlpfordt, H. (1976). *Behring Inst. Mitt.* **60,** 65–74.
Ebert, F., Buse, E., and Mühlpfordt, H. (1979). *Z. Parasitenkd.* **59,** 31–41.
El-On, J., Bradley, D. J., and Freeman, J. C. (1980). *Exp. Parasitol.* **49,** 167–174.
Elsbach, P. (1980). *Rev. Infect. Dis.* **2,** 106–128.
Englund, P. T. (1981). *In* "Biochemistry and Physiology of Protozoa" (M. Levandowsky and S. H. Hutner, eds.), pp. 334–384. Academic Press, New York.
Fong, D., and Chang, K.-P. (1981). *Proc. Natl. Acad. Sci. U.S.A.* **78,** 7624–7628.
Fong, D., and Chang, K.-P. (1982). (In preparation).

Frothingham, T. E., and Lehtimaki, E. (1969). *J. Parasitol.* **55,** 196–199.
Ghafoor, M. Y., and Coombs, G. H. (1980). *Parasitology.* **81,** xxviii.
Goren, M. B. (1977). *Annu. Rev. Microbiol.* **31,** 507–533.
Gorczynski, R. M., and MacRae, S. (1982). *Cell. Immunol.* **67,** 74–89.
Gorczynski, R. M., MacRae, S., Kuba, R., and Price, G. B. (1981). *Cell. Immunol.* **60,** 367–375.
Gottlieb, M., and Dwyer, D. M. (1981). *Science* **212,** 939–941.
Hackstadt, T., and Williams, J. C. (1981). *Proc. Natl. Acad. Sci. U.S.A.* **78,** 3240–3244.
Haidaris, C. G., and Bonventre, P. F. (1981). *Infect. Immun.* **33,** 918–926.
Handman, E., and Burgess, A. W. (1979). *J. Immunol.* **122,** 1134–1137.
Handman, E., and Spira, D. T. (1977). *Z. Parasitenkd.* **53,** 75–81.
Handman, E., Ceredig, R., and Mitchell, G. F. (1979). *Aust. J. Exp. Biol. Med. Sci.* **57,** 9–29.
Handman, E., Mitchell, G. F., and Goding, J. W. (1981). *J. Immunol.* **126,** 508–512.
Hart, D. D., Armstrong, J. A., Brown, C. A., and Draper, P. (1972). *Infect. Immun.* **5,** 803–807.
Hendricks, L. D., Wood, D. E., and Hajduk, M. E. (1978). *Parasitology* **76,** 309–316.
Herman, R. (1980). *Infect. Immun.* **28,** 585–593.
Hernandez, A. G., Arguello, C., Dagger, F., Infante, R. B., Stojanovich, D., Dawidowicz, K., Riggione, F., and La Riva, G. (1981). *In* "The Biochemistry of Parasites" (G. M. Slutzky, ed.), pp. 47–65. Pergamon, Oxford.
Hommel, M. (1978). *Bull. Inst. Pasteur* **75,** 5–102.
Jeon, K. W. (1980). *In* "Cellular Interactions in Symbiotic and Parasitic Relationships" (C. B. Cook, P. Pappas, and G. Rudolph, eds.). Ohio State Univ. Press, Columbus.
Jeon, K. W., and Jeon, M. S. (1976). *J. Cell. Physiol.* **89,** 337–347.
Jones, T. C. (1974). *J. Reticuloendothel. Soc.* **15,** 439-
Jones, T. C., and Hirsch, J. G. (1972). *J. Exp. Med.* **136,** 1173–1194.
Karakashian, M. W., and Karakashian, S. J. (1973). *Exp. Cell Res.* **81,** 111–119.
Karakashian, S. J., and Rudzinska, M. A. (1981). *Exp. Cell Res.* **131,** 387–393.
Kilejian, A. (1976). *J. Protozool.* **23,** 272–277.
Klebanoff, S. J., and Hamon, C. B. (1972). *J. Reticuloendothel. Soc.* **12,** 170–196.
Kress, Y., Tanowitz, H., Bloom, B., and Wittner, M. (1977). *Exp. Parasitol.* **41,** 385–396.
Kutish, G. F., and Janovy, J. Jr. (1981). *J. Parasitol.* **67,** 457–462.
Lamy, L., Wonde, T., and Lamy-Roux, L. (1967). *C.R. Acad. Sci. (D)* **264,** 1889–1891.
Lewis, D. H. (1974). *Ann. Trop. Med. Parasitol.* **68,** 327–336.
Lewis, D. H., and Peters, W. (1977). *Ann. Trop. Med. Parasitol.* **71,** 295–312.
McGhee, R. B., and Cosgrove, W. B. (1980). *Microbiol. Rev.* **44,** 140–173.
Mattock, N. M., and Peters, N. (1975). *Ann. Trop. Med. Parasitol.* **69,** 359–392.
Mauel, J., and Behin, R. (1980). *In* "Biochemistry and Physiology of Protozoa" (M. Levandowsky and S. H. Hutner, eds.), 2nd Ed., Vol. 4, pp. 385–430. Academic Press, New York.
Merino, F., Ajjam, E., Hernandez, A., Dawidowicz, K., and Merino, G. J. (1977). *Int. Arch. Allergy Appl. Immunol.* **55,** 487–495.
Meshnick, S. R., and Eaton, J. W. (1981). *Biochem. Biophys. Res. Commun.* **102,** 970–976.
Miller, H. C., and Twohy, D. W. (1967). *J. Protozool.* **14,** 781–789.
Moulder, J. W. (1974). *J. Infect. Dis.* **130,** 300–306.
Moulder, J. W. (1977). *Trans. Kans. Acad. Sci.* **78,** 105–118.
Moulder, J. W. (1979a). *Proc. R. Soc. London Ser. B* **204,** 199–210.
Moulder, J. W. (1979b). *In* "Host-Parasite Interfaces" (B. B. Nickol, ed.), pp. 127–138. Academic Press, New York.
Murray, H. W. (1981a). *J. Exp. Med.* **153,** 1302–1315.
Murray, H. W. (1981b). *J. Exp. Med.* **153,** 1690–1695.
Nacy, C. A., and Diggs, C. L. (1981). *Infect. Immun.* **34,** 310–313.
Nacy, C. A., Meltzer, M. S., Leonard, E. J., and Wyler, D. J. (1981). *J. Immunol.* **127,** 2381–2386.

Neville, D. M., and Chang, T.-M. (1978). *Curr. Top. Membr. Transp.* **11,** 66–137.
Nogueira, N., and Cohn, Z. (1976). *J. Exp. Med.* **143,** 1402–1420.
Nogueira, N., Chaplan, S., and Tydings, J. D. (1981). *J. Exp. Med.* **153,** 629–636.
Pearson, R. D., and Steigbigel, R. T. (1981). *J. Immunol.* **127,** 1438–1443.
Pearson, R. D., Romito, R., Symes, P. H., and Harcus, J. R. (1981). *Infect. Immun.* **32,** 1249–1253.
Pratt, D. M., and David, J. R. (1981). *Nature (London)* **291,** 581–583.
Rabinovitch, M., Dedet, J.-P., Ryter, A., Robineaux, R., Topper, A., and Brunet, E. (1982). *J. Exp. Med.* **155,** 415–431.
Rahman, A. A., and Sethi, K. K. (1978). *Experientia* **34,** 598-
Scorza, J. V., Valera, M., Scorza, C. D., Carnevali, M., Merino, E., and Lugo-Hernandez, A. (1979). *Trans. R. Soc. Trop. Med. Hyg.* **73,** 293–298.
Scott, P. A., and Farrell, J. P. (1981). *J. Immunol.* **127,** 2395–2400.
Semprevivo, L. H., and McLeod, M. E. (1981). *Biochem. Biophys. Res. Commun.* **103,** 1179–1185.
Silverstein, S. C., Steinman, R. A., and Cohn, Z. A. (1977) *Annu. Rev. Biochem.* **46,** 669–722.
Simpson, L. (1968). *J. Protozool.* **13,** 286–290.
Simpson, L. (1972). *Int. Rev. Cytol.* **32,** 139–207.
Smith, D. C. (1979). *Proc. R. Soc. London Ser. B* **204,** 115–130.
Stahl, P. D., and Schlesinger, P. H. (1980). *Trends Biochem. Sci.* **5,** 194–196.
Trager, W. (1974). *Science* **183,** 269–273.
Turk, J. L., and Bryceson, A. D. M. (1971). *Adv. Immunol.* **13,** 209–266.
Vickerman, K. (1976). *In* "Biology of the Kinetoplastida" (W. H. R. Lumsden and D. A. Evans, eds.), Vol. I, pp. 1–34. Academic Press, New York.
Vickerman, K. (1980). *In* "Molecular Basis of Microbial Pathogenicity" (H. Smith, J. J. Skehel, and M. J. Turner, eds.), pp. 247–260. Verlag Chemie, Weinheim.
Vickerman, K., and Preston, T. M. (1976). *In* "Biology of the Kinetoplastidae" (W. H. R. Lumsden, and D. A. Evans, eds.), Vol. I, pp. 35–130. Academic Press, New York.
Wallach, M., Fong, D., and Chang, K.-P. (1982). *Nature (London),* in press.
Weir, D. M. (1980). *Immunol. Today* **1,** 45–51.
Weiss, E. (1973). *Bacteriol. Rev.* **37,** 259–283.
Weiss, E., and Dasch, G. A. (1981). *In* "The Prokaryotes—A Handbook on Habitats, Isolation, and Identification of Bacteria" (M. P. Starr, J. Stolp, H. G. Trüper, A. Balows, and H. G. Schlegel, eds.), Springer Verlag, Berlin and New York.
Zenian, A. (1981). *Exp. Parasitol.* **51,** 175–188.
Zenian, A., Rowles, P., and Gingell, D. (1979). *J. Cell Sci.* **39,** 187–199.
Zuckerman, A., and Lainson, R. (1977). *In* "Parasitic Protozoa" (J. P. Kreier, ed.), Vol. I, pp. 58–134. Academic Press, New York.

INTERNATIONAL REVIEW OF CYTOLOGY, SUPPLEMENT 14

Symbiotic Interactions between *Legionella pneumophila* and Human Leukocytes

MARCUS A. HORWITZ

Laboratory of Cellular Physiology and Immunology, The Rockefeller University, New York, New York

I. Introduction

Legionnaires' disease, a severe respiratory illness, was first recognized in July 1976 when it struck 221 people, mostly American Legionnaires, and claimed 34 lives in Philadelphia (Fraser *et al.*, 1977). About 6 months later, *Legionella pneumophila,* a gram-negative bacterium, was identified as the etiologic agent of Legionnaires' disease (McDade *et al.*, 1977). Subsequently, it became apparent that Legionnaires' disease was newly recognized but not new; examination of tissue and serum samples stored away from previously unsolved outbreaks of respiratory illness revealed that the illnesses were caused by *L. pneumophila* (McDade *et al.*, 1977; Eickhoff, 1979). The earliest known isolate is from a patient who had had a febrile respiratory illness in 1947 (McDade *et al.*, 1979).

II. Epidemiology

It is now appreciated that Legionnaires' disease occurs worldwide in epidemic and endemic form (Eickhoff, 1979; Gregory *et al.*, 1979; Storch *et al.*, 1979;

ISBN 0-12-364375-9

Bartlett, 1979; Sanford, 1979). The disease is an important cause of pneumonia in the United States (Renner *et al.*, 1971; Fraser and McDade, 1979). The incidence has been variously estimated to be 25–100,000 cases per year. The disease is apparently responsible for a substantial proportion of fatal hospital-acquired pneumonia (Haley *et al.*, 1979; Cohen *et al.*, 1979). Legionnaires' disease affects all age groups, but particularly people over 30 years old, and men are affected more than twice as often as women (Sanford, 1979; Baine, 1979). It has a predilection for immunocompromised patients, particularly those with cancer and those who have received renal transplants (Saravolatz *et al.*, 1979; Owens, 1978; Bock *et al.*, 1978). It also has a predilection for people who smoke cigarettes or abuse alcohol, travellers, and construction workers (Storch *et al.*, 1979; Fraser and McDade, 1979). The disease occurs most frequently in the summer.

In addition to Legionnaires' disease, *L. pneumophila* has been linked to another syndrome, dubbed Pontiac Fever after an outbreak that occurred in Pontiac, Michigan in July 1968 (Glick *et al.*, 1978). Pontiac Fever differs from Legionnaires' disease both clinically and epidemiologically. Whereas Legionnaires' disease is a severe form of pneumonia that is frequently fatal (15% case-fatality rate), Pontiac Fever is a mild nonpneumonic self-limited febrile illness that has not been fatal (Glick *et al.*, 1978; Fraser *et al.*, 1979). Legionnaires' disease occurs after a 2- to 10-day incubation period and affects only a small proportion, less than 5%, of exposed individuals. In two recognized outbreaks of Pontiac Fever, the incubation period was much shorter, usually less than 2 days, and 95–100% of exposed individuals were affected (Glick *et al.*, 1978; Fraser *et al.*, 1979). Why *L. pneumophila* should cause two distinct syndromes is one of its most intriguing mysteries.

Legionnaires' disease primarily affects the lung. At autopsy, the most consistent lesion is an acute fibrinopurulent bronchopneumonia; the lung alveoli are filled with many neutrophils and macrophages and abundant fibrin (Chandler *et al.*, 1977; Carrington, 1979; Blackmon *et al.*, 1978; Winn *et al.*, 1978). Bacteria are found extracellularly and intracellularly in both macrophages and neutrophils; the intracellular organisms are found in membrane-bound vacuoles.

III. Transmission and Ecology

Legionnaires' disease is transmitted by the airborne route; person-to-person transmission has not been documented. In several outbreaks, the vehicle of transmission appears to have been aerosols containinated with *L. pneumophila* and generated in cooling towers or evaporative condensers (Fraser and McDade, 1979); *L. pneumophila* has been isolated from water in such air-conditioning systems (Morris *et al.*, 1979). More recently, shower heads have been implicated as generators of contaminated aerosols in hospital outbreaks of Legionnaires' disease (Cordes *et al.*, 1981; Meenhorst *et al.*, 1981). *L. pneumophila* has been isolated from hospital water supplies (Meenhorst *et al.*, 1981; Tobin *et al.*, 1981; Stout *et al.*, 1982).

The reservoir for *L. pneumophila* in the environment appears to be water; the organism has been isolated from scores of lakes, rivers, and streams (Fliermans *et al.*, 1979, 1981). How this supposedly fastidious organism survives and multiplies in nature is not yet clear. The organism has been isolated from an algal-bacterial mat community growing at 45°C in a man-made thermal effluent and the bacterium reportedly multiplies when cocultured with blue-green algae (Tison *et al.*, 1980). It is possible that the organism utilizes extracellular products of blue-green algae for growth in natural aquatic environments (Tison *et al.*, 1980).

IV. Bacteriology

The causative agent of Legionnaires' disease was first isolated by McDade and co-workers from infected lung tissue of Legionnaires' disease victims by a technique previously used to isolate rickettsia (McDade *et al.*, 1977; Fraser and McDade, 1979). These investigators ground up the infected lung tissue and injected it intraperitoneally into guinea pigs. When the guinea pigs became ill, the investigators removed the spleen, ground it up, and injected it into the yolk sacs of embryonates hens' eggs. When the embryos died, the causative agent of Legionnaires' disease was found on microscopic smears of the yolk sac.

Subsequently, artificial media were developed for the culture of *L. pneumophila*. Currently, charcoal yeast extract agar is the medium of choice (Feeley *et al.*, 1979).

L. pneumophila is an aerobic gram-negative bacterium which grows best at 35–37°C at pH 6.9 (Weaver and Feeley, 1979). The bacterium measures 0.3 to 0.9 μm in diameter and ≥ 2 μm in length (Chandler *et al.*, 1979); the length varies according to culture conditions. Filamentous forms occur when the organism is grown on artificial medium (Pine *et al.*, 1979). Flagella have been demonstrated on the organism by use of the Leifson flagella stain (Thomason *et al.*, 1979). Ultrastructurally, the organism appears typically gram-negative; it is enclosed by a double unit membrane (Chandler *et al.*, 1979). In papain-treated cells, a layer which appears to be peptidoglycan-containing is observed (Flesher *et al.*, 1979); a peptidoglycan-like component has been isoalted from *L. pneumophila* which resembles the peptidoglycan of gram-negative bacteria (Keel *et al.*, 1979; Guerrant *et al.*, 1979). The organism frequently contains vacuoles; these stain for fat with Sudan black B (Weaver and Feeley, 1979). The organism divides by a pinching nonseptate process (Chandler *et al.*, 1979).

The bacterium has endotoxin-like activity which differs from the classic endotoxicity of gram-negative bacteria. The bacterium is highly reactive in the limulus amebocyte lysate assay but *in vivo* it has little activity; for example, it has low pyrogenicity in rabbits (Wong *et al.*, 1979). Analysis by gas–liquid chromatography has indicated that *L. pneumophila* does not contain hydroxy fatty acids usually associated with lipid A of endotoxin (Wong *et al.*, 1979).

Biochemically, the organism is catalase, gelatinase, and weakly oxidase positive (Weaver and Feeley, 1979), and produces a β-lactamase which acts primarily as a cephalosporinase (Thornsberry and Kirven, 1978; Fu and Neu, 1979). It does not reduce nitrates or degrade urea (Weaver and Feeley, 1979). When cultured on medium containing L-tyrosine or L-phenylalanine, *L. pneumophila* produces a brown soluble pigment (Baine *et al.*, 1978; Baine and Rasheed, 1979). When cultured in broth, it exhibits yellow fluorescence at 366 nm ultraviolet light (Weaver and Feeley, 1979). The bacterium elaborates a substance with hemolytic activity (Baine *et al.*, 1979).

On analysis by gas–liquid chromatography, *L. pneumophila* has a distinctive fatty acid profile with branched chain fatty acids comprising >77% of the total fatty acid; such a high content of branched chain fatty acids is unusual among gram-negative bacteria (Moss *et al.*, 1977, 1979).

L. pneumophila DNA has a guanine plus cytosine content of 39% (Brenner *et al.*, 1978). By reassociation kinetics, the genome size of *L. pneumophila* has been estimated at 2.5×10^9 daltons (Brenner *et al.*, 1978).

On analysis by the DNA hybridization technique, *L. pneumophila* was found unrelated to all other organisms studied; a new family, Legionellaceae, and a new genus and species was proposed for *L. pneumophila* (Brenner *et al.*, 1978, 1979). Six serogroups of *L. pneumophila* have been identified (McKinney *et al.*, 1979, 1980; England *et al.*, 1980); the great majority of clinical isolates are serogroup 1.

As a fallout of the discovery of *L. pneumophila,* several other pneumonia-causing organisms have been identified that are similar to *L. pneumophila* in many respects. These include the so-called Pittsburgh Pneumonia agent, now tentatively named *Legionella micdadei* (Myerowitz *et al.*, 1979; Rogers *et al.*, 1979; Pasculle *et al.*, 1979; Hebert *et al.*, 1980), and *Legionella bozemanii, Legionella dumoffii* (Brenner *et al.*, 1980), and *Legionella longbeacheae* (McKinney *et al.*, 1981). In addition, organisms named *Legionella gormanii* (Morris *et al.*, 1980) and *Legionella jordanis* (Cherry *et al.*, 1982) have been isolated from environmental samples; no human disease resulting from these organisms has been reported.

V. Intracellular Multiplication of *L. pneumophila*

My work has focused on the immunobiology of *L. pneumophila,* the interaction of *L. pneumophila* with the human immune system (Horwitz and Silverstein,

1980a,b,c, 1981; Horwitz, 1982). As a first step toward understanding this subject, I sought to categorize the bacterium as an intracellular or extracellular pathogen. I suspected the bacterium was an intracellular pathogen for two main reasons. First, in lung specimens from patients with Legionnaires' disease, clusters of apparently intact organisms were frequently observed inside alveolar macrophages. Second, the organism was isolated originally not from routine bacteriologic media but from guinea pigs and chicken embryos by a technique usually used to isolate rickettsia, which are obligate intracellular parasites.

In all the studies described below, I used virulent egg yolk grown *L. pneumophila*, Philadelphia 1 strain. I used egg yolk grown organisms because of the evidence that agar-adapted bacteria lose their virulence for chicken embryos and guinea pigs (McDade and Shepard, 1979). I isolated human leukocytes from blood by established techniques and cultured them under antibiotic-free conditions at 37°C in 5% CO_2–95% air in tissue culture medium containing 15% fresh normal human AB serum. This serum was from a person without a history of Legionnaires' disease and contained no antibody to *L. pneumophila* detectable by the indirect fluorescent antibody assay.

I first incubated *L. pneumophila* with mononuclear leukocytes, i.e., monocytes and lymphocytes. At various intervals thereafter, I sampled the medium and assayed it for colony-forming units of *L. pneumophila* on a modified charcoal yeast extract agar.

L. pneumophila multiplied several logs when incubated with the total mononuclear cell population or when incubated with the adherent monocyte-enriched fraction of that population (Fig. 1). *L. pneumophila* did not multiply when incubated with the nonadherent lymphocyte-enriched fraction of the mononuclear cell population. In other experiments, *L. pneumophila* also did not multiply when incubated with human polymorphonuclear leukocytes; nor did it multiply when incubated in tissue culture medium.

Thus *L. pneumophila* requires the presence of leukocytes to grow under tissue culture conditions, and it grows in the presence of monocytes but not lymphocytes or polymorphonuclear leukocytes.

In cultures of human monocytes, *L. pneumophila* multiplied until all the monocytes in the culture were destroyed. The level to which *L. pneumophila* multiplied was proportional to the number of mononuclear cells in the culture. The log phase doubling time of *L. pneumophila* in monocyte cultures was about 2 hours, a rate of growth faster than that observed in any of the specialized media used to grow the organism.

I followed the course of *L. pneumophila* infection of monocytes by fluorescence microscopy. I infected monocytes in monolayer culture with *L. pneumophila* and, at various intervals thereafter, I stained them with a fluorescent anti-*L. pneumophila* antibody. Immediately after a light infection, few (<5%) of the monocytes examined had bound or ingested *L. pneumophila*, and these

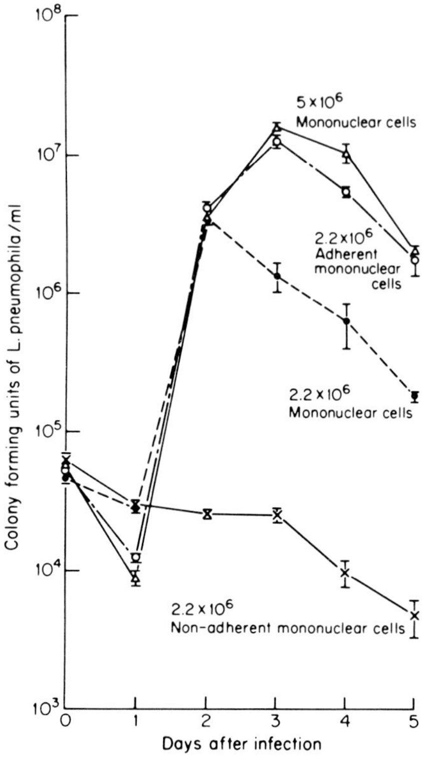

FIG. 1. *L. pneumophila* multiplies in the presence of monocytes (adherent cells) but not lymphocytes (nonadherent cells). *L. pneumophila* (10^5 CFU) were added to petri dishes containing 2.2 ml RPMI, 15% serum, and (a) 2.2 or 5×10^6 mononuclear cells; (b) 2.2×10^6 adherent mononuclear cells selected from 5×10^6 mononuclear cells; (c) 2.2×10^6 nonadherent mononuclear cells selected from the same mononuclear cell population as in (b). The cultures were incubated at 37°C in 5% CO_2–95% air on a gyratory shaker (100 rpm) for 3 hours and under stationary conditions thereafter. CFU of *L. pneumophila* in each petri dish were determined daily. Each point represents the average for three replicate petri dishes ±SE. (Reproduced from Horwitz and Silverstein, 1980a.)

monocytes usually had only one cell-associated bacterium (Fig. 2A). Over the next few days, as the number of colony-forming units in the culture increased 4–5 logs, the number of monocytes containing *L. pneumophila* increased in parallel with bacterial growth, and the monocytes contained many bacteria (Fig. 2B). At the peak of infection, monocytes appeared packed full with organisms.

Several lines of evidence indicated that *L. pneumophila* multiplied within the monocytes and not simply in association with them. First, in the absence of leukocytes, *L. pneumophila* did not grow in tissue culture medium with or without serum even if the medium was conditioned for 4 days by viable mono-

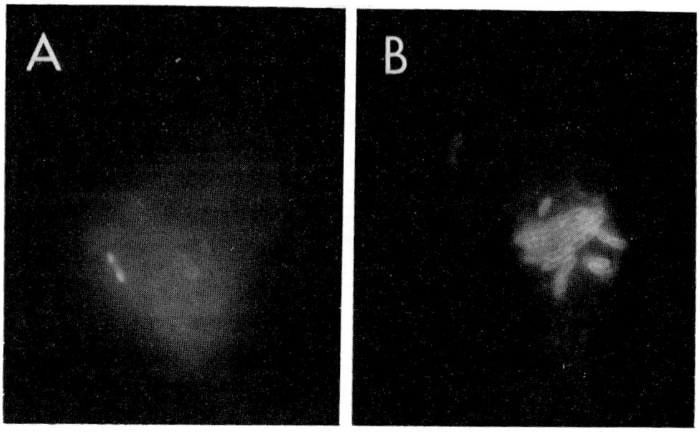

FIG. 2. Course of *L. pneumophila* infection of monocytes as visualized by fluorescence microscopy. *L. pneumophila* were incubated with 24-hour explanted monocytes on glass coverslips for 1 hour at 37°C in 5% CO_2–95% air at a 5:1 ratio of CFU:monocytes and nonadherent bacteria were washed away. The monocytes were fixed, incubated with fluorescein-conjugated rabbit anti-*L. pneumophila* antibody, and examined by fluorescence microscopy immediately after the infection (A) and 4 days later at the peak of the infection (B). ×1200. (Reproduced from Horwitz and Silverstein, 1980a.)

cytes. Second, *L. pneumophila* did not multiply in sonicated mononuclear cells. Moreover, *L. pneumophila* required intact monocytes for multiplication at every phase of its growth. Lysis of infected monocytes at various times during logarithmic growth of *L. pneumophila* was followed by cessation of bacterial multiplication. Growth resumed when intact mononuclear cells were added back to the cultures. Third, in experiments utilizing parabiotic chambers, *L. pneumophila* multiplied only when in direct contact with monocytes. In these experiments, I separated the two compartments of each chamber with a filter that contained pores large enough to allow ready passage of molecules but too small to allow passage of bacteria or mononuclear cells from one compartment to the other. I placed mononuclear cells on one side of the filter and *L. pneumophila* on the same side of the filter or on the opposite side of the filter from the mononuclear cells. The bacteria multiplied only when placed on the same side of the filter as mononuclear cells.

VI. The Unique Intracellular Site of *L. pneumophila* Replication

I infected monocyte monolayers with *L. pneumophila* and examined the monocytes by electron microscopy. Two days after infection, the monocytes were packed full with bacteria (Fig. 3A). The bacteria were inside the monocytes and not on the surface. All bacteria were in membrane-bound cytoplasmic vac-

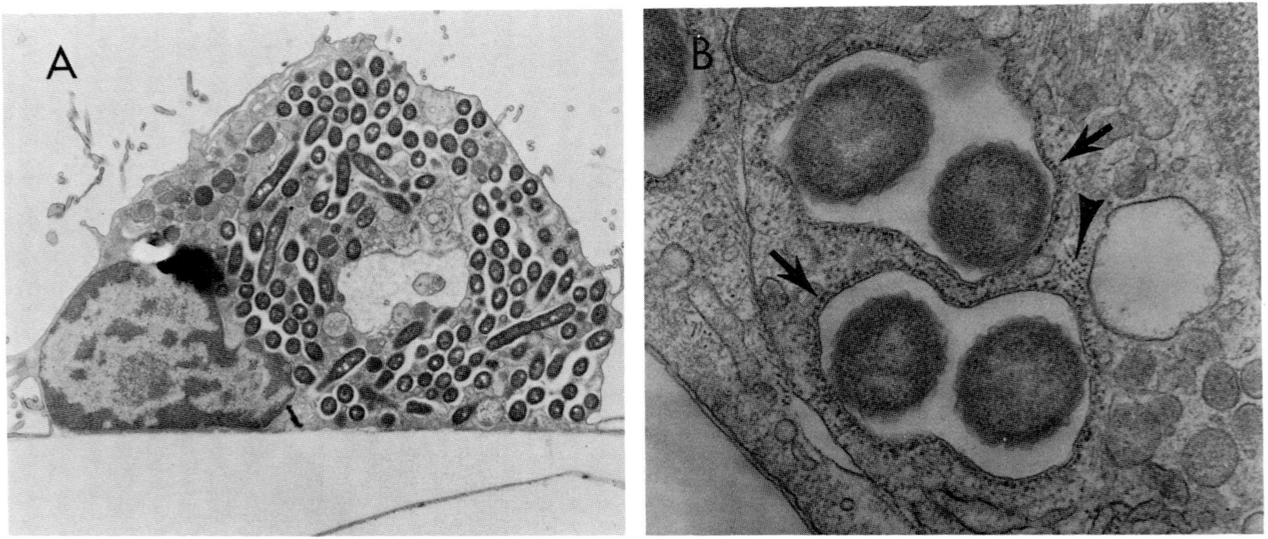

Fig. 3. Electron micrographs of monocytes infected with *L. pneumophila*. Monocytes in a monolayer culture in plastic petri dishes were infected with *L. pneumophila*, incubated in RPMI and 15% serum at 37°C for 2 days, fixed in glutaraldehyde and osmium, and processed for electron microscopy. (A) Monocyte heavily infected with *L. pneumophila*. ×5400. (B) *L. pneumophila* in vacuoles studded with ribosome-like structures. Each vacuole contains two bacilli. The ribosome-like structures (arrows) appear to be separated from cytoplasmic face of the vacuolar membrane by a gap of ~100 Å. Numerous 10-nm filaments in longitudinal and cross-section (arrowhead) are visible in the monocyte cytoplasm. Note that their diameter is smaller than that of the ribosome-like structures. ×32,400. (Reproduced from Horwitz and Silverstein, 1980a.)

uoles (Fig. 3B). Strikingly, the cytoplasmic sides of the membrane-bound vacuoles surrounding the *L. pneumophila* were studded with structures resembling monocytes ribosomes (Fig. 3B). This morphologic feature has also been observed in infected alveolar macrophages in human lung tissue specimens obtained from patients with Legionnaires' disease (Glavin *et al.*, 1979). *L. pneumophila* is evidently unique among bacterial pathogens in promoting the formation of these vacuoles, the role of which is unknown.

The studies show that *L. pneumophila* multiplies intracellularly in human monocytes, and that under tissue culture conditions, the bacterium multiplies only intracellularly. Since *L. pneumophila* can multiply extracellularly on complex media, this bacterium is a facultative intracellular parasite. *L. pneumophila* thus falls into a select category of bacterial pathogens that can evade host defenses by parasitizing monocytes. Other members of this category are the mycobacteria including *Mycobacterium tuberculosis* and *Mycobacterium leprae, Listeria monocytogenes, Brucella* sp., *Salmonella* sp., and *Francisella tularensis*.

The mechanism by which *L. pneumophila* resists monocyte microbicidal systems is unknown. Other intracellular parasites resist these systems by a variety of strategies. *M. tuberculosis* (Armstrong and Hart, 1971; Hart and Armstrong, 1974) and *Toxoplasma gondii* (Jones and Hirsch, 1972) interfere with host phagosome–lysosome fusion. *Mycobacterium lepraemurium* (Hart *et al.*, 1972), antibody-treated *M. tuberculosis* (Armstrong and Hart, 1975), *Leishmania mexicana* (Alexander and Vickerman, 1975), *Leishmania donovani* (Chang and Dwyer, 1976), and *Leishmania tropica* (Berman *et al.*, 1981) are able to thrive within the normally inhosptiable milieu of the phagolysosome. It remains to be determined if phagosomes containing *L. pneumophila* fuse with monocyte lysosomes.

VII. Influence of Antibody and Complement on the Interactions Between *L. pneumophila* and Human Leukocytes

Patients with Legionnaires' disease respond to the infection by producing antibody against *L. pneumophila*. To assess the role of humoral immunity in host defense against *L. pneumophila*, I have studied the influence of antibody and complement on the interactions of virulent *L. pneumophila* with human polymorphonuclear leukocytes (PMN) and monocytes *in vitro*, under antibiotic-free conditions (Horwitz and Silverstein, 1980b,c).

I first examined if antibody promotes killing of *L. pneumophila* by complement. I incubated the bacteria with fresh normal human serum at 37°C for 1 hour in the presence and absence of high-titer rabbit or human anti-*L. pneumophila* antibody. *L. pneumophila* in concentrations ranging from 10^3 to 10^6 CFU/ml were completely resistant to the bactericidal effects of 0–50% fresh normal human serum, even in the presence of high concentrations of rabbit or human specific antibody. The fresh normal human serum used in these experiments reduced CFU of a serum-sensitive strain of *Neisseria gonorrhea* by >4 logs. The resistance of *L. pneumophila* to complement was not due to a failure to fix complement. Fluorescence microscopy using rhodamine-conjugated anti-human complement C3 IgG showed that complement component C3 was fixed to the bacterial surface in the presence of antibody; however fixation of complement did not result in death of the bacterium.

To determine the influence of antibody and complement on phagocytosis of *L. pneumophila* by human PMN, I incubated PMN with *L. pneumophila* in the presence of serum that contained anti-*L. pneumophila* antibody and/or complement. I then centrifuged the PMN onto glass slides, stained the PMN-associated bacteria with fluorescein-conjugated rabbit anti-*L. pneumophila* antiserum, and enumerated the number of PMN-associated bacteria by fluorescence microscopy (Table I).

In the presence of normal serum (a source of complement alone) or heat-inactivated immune serum (a source of antibody alone), relatively few PMN (21 and 14% respectively in Table I) bound or ingested *L. pneumophila,* and those that did had an average of less than two bacteria/PMN (Table I). In the presence of fresh immune serum (a source of both antibody and complement), 93.5% of

TABLE I
CAPACITY OF PMN TO BIND OR TO INGEST *L. pneumophila* IN THE PRESENCE OF ANTIBODY AND/OR COMPLEMENT[a,b]

Serum preparation	Percentage of PMN binding or ingesting *L. pneumophila*	Average number of *L. pneumophila*/PMN with ≥1 associated bacteria	Index of PMN *L. pneumophila* association[c]
Normal serum	24	1.2	25
Heat-inactivated immune serum	14	1.8	25
Fresh immune serum	93.5	5.1	475

[a]Reproduced from Horwitz and Silverstein (1980b).

[b]PMN (5×10^6) and 5×10^5 CFU of *L. pneumophila* ($\sim 2.5 \times 10^7$ bacterial particles) were incubated in the presence of 10% of the indicated serum preparation for 30 minutes at 37°C on a gyratory shaker. At the end of the incubation, the cells were cytocentrifuged onto glass slides, fixed, stained with fluorescein-conjugated rabbit anti-*L. pneumophila* antiserum, examined by fluorescence microscopy, and 200 consecutive PMN counted.

[c]The index of PMN–*L. pneumophila* association is the percentage of PMN with bound or ingested *L. pneumophila* multiplied by the average number of *L. pneumophila*/PMN with ≥1 associated bacteria.

PMN bound or ingested *L. pneumophila;* these had an average of 5.1 *L. pneumophila*/PMN (Table I). In the presence of antibody and complement, nearly all the bacterial cells in the culture were bound to or ingested by PMN.

Electron microscopy revealed that the opsonized bacteria were internalized by the PMN. The bacteria were found in membrane-bound cytoplasmic vacuoles within the PMN.

Thus, antibody and complement markedly promote phagocytosis of *L. pneumophila* by PMN.

To determine if PMN can kill *L. pneumophila,* I incubated PMN with *L. pneumophila* in the presence of a source of antibody and/or complement and determined the total number of CFU in the cultures initially and at the end of the incubation. PMN killed *L. pneumophila* only when both a source of antibody and of complement were present. Even then, PMN reduced CFU of *L. pneumophila* by only 0.5 log.

The experiment depicted in Fig. 4 illustrates the remarkable difference in the capacity of PMN to kill *L. pneumophila* and an encapsulated serum-resistant

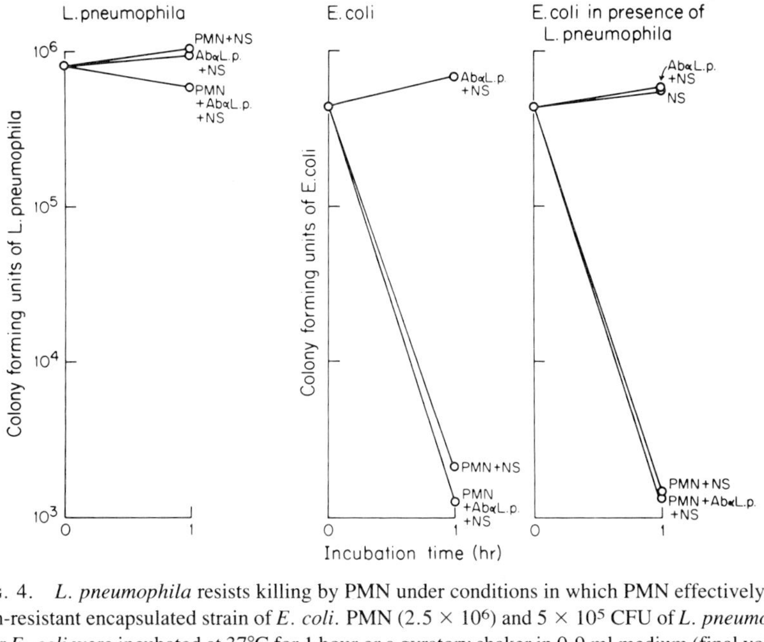

FIG. 4. *L. pneumophila* resists killing by PMN under conditions in which PMN effectively kill a serum-resistant encapsulated strain of *E. coli*. PMN (2.5×10^6) and 5×10^5 CFU of *L. pneumophila* and/or *E. coli* were incubated at 37°C for 1 hour or a gyratory shaker in 0.9 ml medium (final volume) containing 10% fresh normal human serum (NS) alone or 10% NS plus 8.5 agglutinating units/ml of dialyzed rabbit anti-*L. pneumophila* antiserum (Abα-L.p.). CFU of *E. coli* and *L. pneumophila* were determined initially and at the end of the incubation. (Reproduced from Horwitz and Silverstein, 1980b.)

strain of *Escherichia coli*. In this experiment, I incubated PMN with *L. pneumophila* alone, *E. coli* alone, or both types of bacteria at the same time. PMN killed few *L. pneumophila* even in the presence of high-titer anti-*L. pneumophila* antibody and fresh serum. In contrast, under these same conditions, PMN very effectively killed an encapsulated serum-resistant strain of *E. coli*, reducing the number of viable bacteria by 2½ logs. [PMN require both antibody and complement to kill encapsulated *E. coli* (Horwitz and Silverstein, 1980d); both are present in fresh normal human serum, which contains low levels of natural antibody against this strain of *E. coli*.] When the *E. coli* and *L. pneumophila* were mixed together in the same test tube, PMN killed the *E. coli* as effectively as when the *E. coli* were present alone, indicating that the killing capacity of PMN is not reduced in cultures containing *L. pneumophila*.

Other experiments demonstrated that when *L. pneumophila* are incubated with PMN in the presence of antibody and complement, the surviving bacteria are PMN-associated; the ineffective killing of opsonized *L. pneumophila* by PMN is a result of a failure of PMN to kill these bacteria after they become PMN-associated.

Thus antibody and complement promote phagocytosis of *L. pneumophila* by PMN but not effective killing.

Experiments with monocytes showed that antibody and complement markedly promote phagocytosis of *L. pneumophila* by monocytes. As with PMN, monocytes require both antibody and complement to kill any *L. pneumophila*, but even then, monocytes kill only a limited proportion (0.25 to 0.5 log) of an inoculum (Horwitz and Silverstein, 1980c).

VIII. Influence of Antibody and Complement on the Intracellular Multiplication of *L. pneumophila*

I next investigated if antibody and complement affect the capacity of *L. pneumophila* to multiply in monocytes. I preincubated *L. pneumophila* with fresh normal serum (a source of complement) or fresh immune serum (a source of both antibody and complement) and then incubated the bacteria with monocytes in the presence of the same type of serum in which they were preincubated (Fig. 5). In cultures containing antibody, a small proportion of the inoculum was killed during the first hour of incubation with monocytes; no killing occurred in cultures lacking antibody. Thereafter, CFU in cultures with and without specific antibody changed in parallel, first decreasing slightly and then increasing 3 logs. Thus antibody and complement fail to inhibit the rate of intracellular multiplication of *L. pneumophila* in monocytes.

I have thus examined three ways that antibody might function to promote host defense. I have found (1) that antibody fails to promote killing of *L. pneumophila*

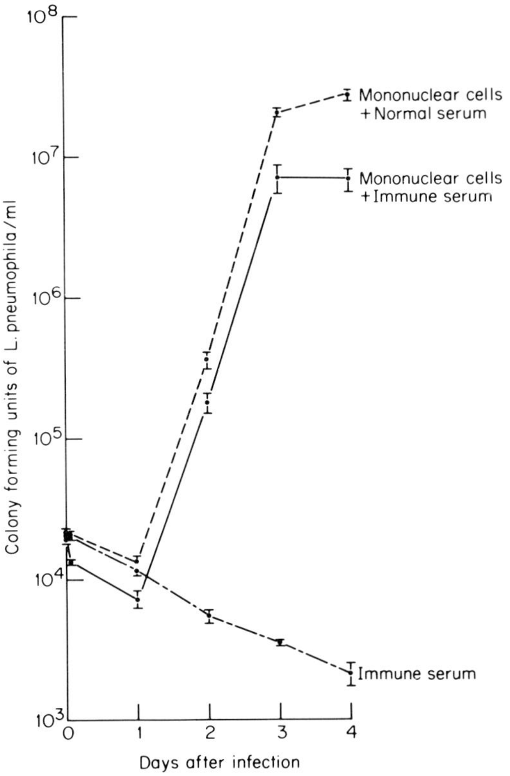

FIG. 5. Specific antibody fails to inhibit the multiplication of *L. pneumophila* in monocytes. *L. pneumophila* (2.5×10^4 CFU) were incubated at 37°C for 10 minutes in 33% fresh normal human serum or 33% fresh immune human serum. The bacteria were then incubated at 37°C in 5% CO_2–95% air with 5×10^6 mononuclear cells (or RPMI-1640 as control) in medium that contained a final concentration of 10% of the same type of serum to which the bacteria were initially exposed. The cultures were shaken for 1 hour and incubated under stationary conditions thereafter for 4 days. CFU were determined at 0, 1, 24, 48, 72, and 96 hours after infecting the monocytes. Each point represents the average for five replicate tubes ±SE. (Reproduced from Horwitz and Silverstein, 1980c.)

by complement; (2) that antibody and complement together fail to promote effective killing of *L. pneumophila* by PMN and monocytes; and (3) that antibody and complement fail to inhibit the rate of intracellular multiplication of *L. pneumophila* in monocytes. These findings suggest tht humoral immunity may not be an effective host defense against *L. pneumophila* and that a vaccine that resulted only in antibody production agianst *L. pneumophila* may not be efficacious.

IX. Interaction between *L. pneumophila* and Activated Human Monocytes

Cell-mediated immunity plays an important role in host defense against intracellular pathogens. Lymphokines produced by sensitized lymphocytes have been shown to activate animal or human mononuclear phagocytes such that they have an enhanced capacity to resist or kill *M. tuberculosis* (Patterson and Youmans, 1970; Cahall and Youmans, 1975), *L. monocytogenes* (Simon and Sheagren, 1972; Fowles *et al.*, 1973; Cole, 1975; Jones and Youmans, 1973), *Leishmania enrietti* (Mauel *et al.*, 1978), *Toxoplasma gondii* (Sethi *et al.*, 1975; Shirahata *et al.*, 1977; Jones *et al.*, 1975; Borges and Johnson, 1975; Anderson and Remington, 1974; Anderson *et al.*, 1976), *Trypanosoma cruzi* (Nogueira *et al.*, 1977; Nogueira and Cohn, 1978), *Rickettsia tsutsugamushi* (Nacy and Meltzer, 1979), and *Coxiella burnetti* (Hinrichs and Jerrells, 1976). To investigate the role of cell-mediated immunity in host defense against *L. pneumophila*, I have examined the interaction between *L. pneumophila* and *in vitro* activated human monocytes (Horwitz, 1982). To activate the monocytes, I incubated them with lymphocytes in the presence of the mitogen concanavalin A (Con A).

I incubated freshly explanted human blood monocytes for 24 hours with Con A only, peripheral blood lymphocytes only, neither Con A nor lymphocytes, or both Con A and lymphocytes (Fig. 7). I then added *L. pneumophila* to the cultures. *L. pneumophila* multiplied several logs in all monocyte cultures except those containing both Con A and lymphocytes, where multiplication was inhibited (Fig. 6). Other experiments showed that Con A exhibited a dose–response pattern in sensitizing mononuclear cell cultures to inhibit *L. pneumophila* multiplication. Con A was consistently maximally effective at ≥4 μg/ml.

These studies indicate that monocytes in Con A-sensitized mononuclear cell cultures acquire the capacity to inhibit the multiplication of *L. pneumophila* and that this phenomenon is lymphocyte-dependent. This suggested that a cytokine produced in Con A-sensitized mononuclear cell cultures might be involved in activating monocytes such that they inhibit *L. pneumophila* multiplication.

To investigate this, I preincubated monocytes in monolayer culture with the cell-free filtered supernatant of mononuclear cell cultures sensitized for 48 hours with 15 μg/ml Con A. After 24 hours I infected the monocytes with *L. pneumophila* (Fig. 7). Monocytes treated with supernatant from Con A-sensitized mononuclear cell cultures inhibited *L. pneumophila* multiplication, and the degree of inhibition was proportional to the concentration of supernatant added.

I conducted other experiments to define experimental conditions for generating supernatants with enhanced potency. These showed that the most potent supernatant is obtained from mononuclear cell cultures incubated with ≥ 15 μg/ml Con A for 48 hours. Other experiments defined optimal conditions for activating monocytes. These showed (1) that the degree of monocyte inhibition of *L.*

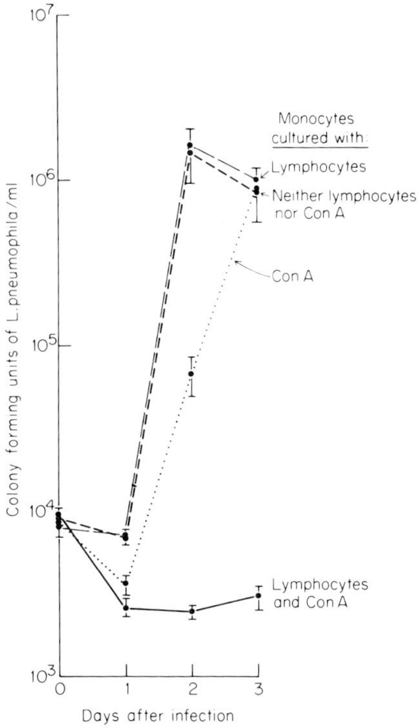

FIG. 6. Monocytes incubated with both Con A and lymphocytes inhibit *L. pneumophila* multiplication. Mononuclear cells (6×10^6) in 2 ml medium containing 25% fresh human serum were incubated in 35 mm plastic petri dishes for 1.5 hour at 37° C in 5% CO_2–95% air to allow monocytes to adhere. Some cultures were vigorously washed to remove nonadherent (predominantly lymphoid) cells. These and the other cultures were then further incubated for 24 hours with or without Con A (16 μg/ml). *L. pneumophila* (2×10^4 CFU) were then added to each culture and CFU in each culture determined daily. Each point represents the average for three replicate petri dishes ±SE. (Reproduced from Horwitz and Silverstein, 1981.)

pneumophila multiplication is proportional to the length of time monocytes are preincubated with supernatant (48 >24 >12 hours), (2) that monocytes treated with supernatant daily after infection are more inhibitory than monocytes treated initially only, and (3) that with time in culture, monocytes progressively lose their capacity to respond to supernatant with inhibition of *L. pneumophila* multiplication.

Activated monocytes differ morphologically from nonactivated monocytes. Con A supernatant-treated monocytes are larger, more spread, and more stretched out than nonactivated monocytes (Fig. 8). Activated monocytes cover 1.8 times the surface area of nonactivated monocytes on petri dishes.

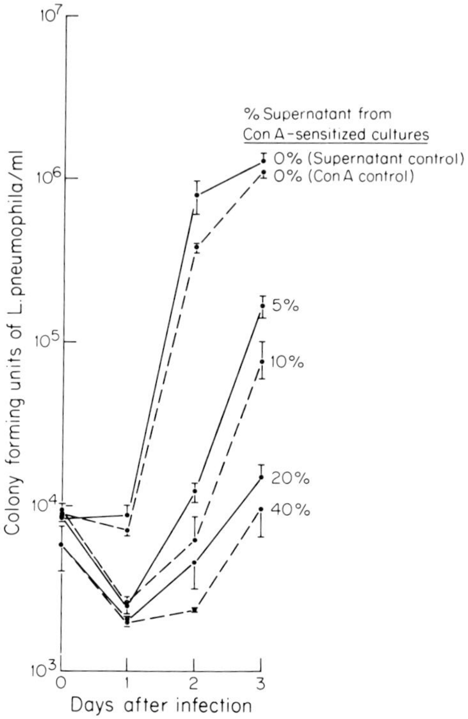

FIG. 7. Monocytes incubated with the supernatant of Con A-sensitized mononuclear cell cultures inhibit *L. pneumophila* multiplication. Monocytes in monolayer culture were incubated at 37°C in 5% CO_2–95% air in 2 ml RPMI medium containing 15% fresh normal human serum and 0–40% cell-free supernatant from Con A-sensitized mononuclear cell cultures. Control monolayers were incubated with supernatant from a mononuclear cell culture from which Con A was omitted (supernatant control) or with 15 μg/ml Con A (Con A control). After 24 hours, *L. pneumophila* (2×10^4 CFU) were added to the cultures and CFU in each culture determined daily. Each point represents the average for three replicate petri dishes ±SE. (Reproduced from Horwitz and Silverstein, 1981.)

Supernatant-activated monocytes inhibit *L. pneumophila* multiplication in two ways. First, they phagocytize approximately 50% fewer bacteria than nonactivated monocytes and thereby restrict access of these bacteria to the intracellular milieu they require for multiplication (Horwitz and Silverstein, 1981). Second, activated monocytes markedly slow the multiplication rate of bacteria that are internalized; the doubling time of *L. pneumophila* in activated monocytes during the first 22 hours after infection was 3.1 times longer than in nonactivated monocytes (Horwitz and Silverstein, 1981).

Activated monocytes do not have a greater capacity to kill *L. pneumophila* than nonactivated monocytes; in the presence of fresh normal serum, containing

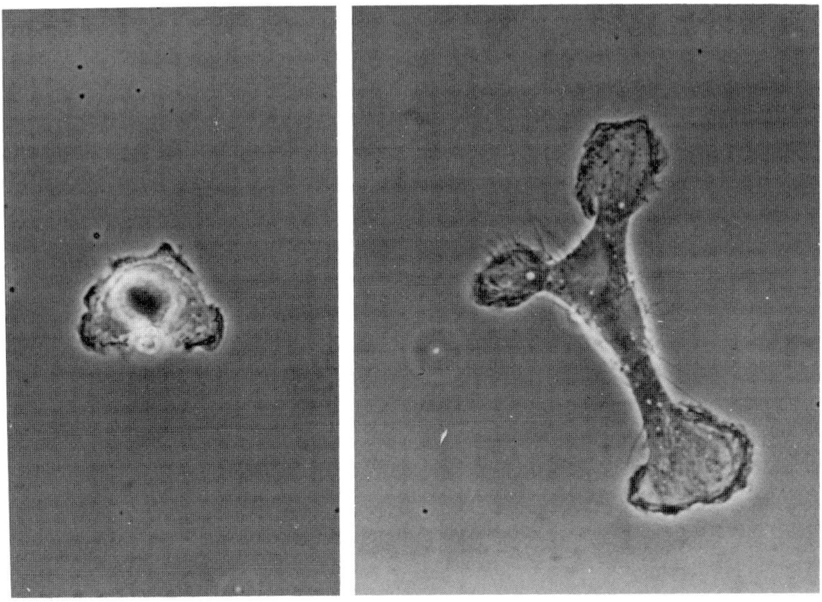

FIG. 8. The morphology of monocytes changes with activation. Freshly explanted monocytes in monolayer culture were incubated in RPMI medium containing 15% fresh normal serum and either 10% control medium (left) or Con A-induced supernatant (right). After 48 hours, both control and supernatant-activated monocytes were fixed and photographed at the same magnification, using a phase contrast microscope. ×850. (Reproduced from Horwitz and Silverstein, 1981.)

complement but not antibody, monocytes did not kill *L. pneumophila* whether or not the monocytes were activated (Horwitz and Silverstein, 1981). In the presence of fresh immune serum, containing both anti-*L. pneumophila* antibody and complement, both activated and nonactivated monocytes killed a small proportion of the original inoculum (<0.25 log). The activated monocytes did not kill more *L. pneumophila*; in fact they consistently killed fewer bacteria than the nonactivated monocytes, perhaps because they phagocytize less avidly.

These experiments show that human monocytes can be activated to inhibit the multiplication of *L. pneumophila*. Thus, inhibition of *L. pneumophila* multiplication is accomplished by activating the monocytes and not by coating the bacteria with antibody and complement.

More recently, I have examined peripheral blood leukocytes from patients recovered from Legionnaires' disease for evidence of specific cell-mediated immunity. I have found that when mononuclear cells from these patients are incubated with killed *L. pneumophila*, they produce supernatants capable of endowing normal monocytes with an enhanced capacity to inhibit *L. pneumophila* multiplication (Horwitz, 1982).

These studies indicate that cell-mediated immunity likely plays a major role in host defense against *L. pneumophila* as it does against other intracellular pathogens. Consistent with this view, Wong and co-workers have demonstrated that guinea pigs sensitized by injection of *L. pneumophila* antigens exhibit a cutaneous delayed-type hypersensitivity reaction in response to *L. pneumophila* antigens (Wong *et al.*, 1980).

X. Concluding Remarks

Since its discovery 5 years ago, *L. pneumophila* has provided its share of surprises to the scientific and medical communities. The first surprise was its very existence. Here was a major human pathogen that had somehow escaped detection until the mid-1970s. And the organism was newly recognized but not new—epidemics and sporadic cases had been occurring for years throughout the world. A second surprise was its pathogenetic versatility. It caused two distinct clinicoepidemiologic syndromes—Legionnaires' disease and Pontiac Fever. A third surprise was its ubiquitousness. The organism that was so hard to find was everywhere in aquatic environments. The organism that was so hard to grow in the laboratory was thriving in the great outdoors. Finally, a fourth surprise was its capacity to survive and multiply in human leukocytes—cells specially equipped to destroy human pathogens and so defend the body against infectious diseases.

At this point, we can only speculate as to how *L. pneumophila* acquired its remarkable capacity to survive and multiply in human monocytes. Are humans accidental hosts, unwitting victims of their own air conditioning and water dispersal technology, for an organism that normally parasitizes simpler beings such as amoebae? If so, Dr. Jeon's (see Jeon, this volume) illuminating work suggests one pathway which the bacterium may have followed in acquiring the capacity to live intracellularly. In any case, what is the role of the unique vacuole in which *L. pneumophila* resides within the human monocyte? Do the ribosome-like structures surrounding the vacuole serve an essential function for the bacterium? Are they a protective response of the host? Or are these structures an epiphenomenon incidental to the survival and multiplication of the bacterium in the monocyte? Further study is needed to answer these questions.

Much has been learned about *L. pneumophila* since its fateful debut in 1976 when it struck down 32 persons in Philadelphia. But this fascinating and multifaceted bacterium still holds many mysteries that continue to challenge students of a variety of disciplines including the medical practitioner, the immunologist, the microbiologist, the cell biologist, and the plant biologist.

ACKNOWLEDGMENTS

Dr. Horwitz is supported by the American Cancer Society, the John A. Hartford Foundation, National Institutes of Health grant AI 17254, and the Heiser program for Research in Leprosy.

REFERENCES

Alexander, J., and Vickerman, K. (1975). *J. Protozool.* **22,** 502–508.

Anderson, S. E., and Remington, J. S. (1974). *J. Exp. Med.* **139,** 1154–1174.

Anderson, S. E., Bautista, S., and Remington, J. S. (1976). *J. Immunol.* **117,** 381–387.

Armstrong, J. A., and Hart, P. D. (1971). *J. Exp. Med.* **134,** 713–740.

Armstrong, J. A., and Hart, P. D. (1975). *J. Exp. Med.* **142,** 1–15.

Baine, W. B. (1979). "Legionnaires': The Disease, the Bacterium and Methodology" (G. L. Jones and G. A. Hebert, eds.), pp. 13–16.HEW, Public Health Service, Center for Disease Control, Bureau of Laboratories, Atlanta, Ga. No. (CDC) 79-8375.

Baine, W. B., and Rasheed, J. K. (1979). *Ann. Int. Med.* **90,** 619–620.

Baine, W. B., Rasheed, J. K., Feeley, J. C., Gorman, G. W., and Casida, L. E., Jr. (1978). *Curr. Microbiol.* **1,** 93–94.

Baine, W. B., Rasheed, J. K., Mackel, D. C., Bopp, C. A., Wells, J. G., and Kaufmann, A. F. (1979). *J. Clin. Microbiol.* **9,** 453–456.

Bartlett, C. L. R. (1979). *Ann. Intern. Med.* **90,** 592–595.

Berman, J. D., Fioetti, T. B. and Dwyer, D. M. (1981). *J. Protozool.* **28,** 239–242.

Blackmon, J. A., Hicklin, M. D., Chandler, F. W. (1978). *Arch. Pathol. Lab. Med.* **102,** 337–343.

Bock, B. V., Edelstein, P. H., Snyder, K. M., Hatayama, C. M., Lewis, R. P., Kirby, B. D., George, W. L., Owens, M. L., Haley, C. E., Meyer, R. D., and Finegold, S. M. (1978). *Lancet* **I,** 410–413.

Borges, J. S., and Johnson, W. D., Jr. (1975). *J. Exp. Med.* **141,** 483–496.

Brenner, D. J., Steigerwalt, A. G., Weaver, R. E., McDade, J. E., Feeley, J. C., and Mandel, M. (1978). *Curr. Microbiol.* **1,** 71–76.

Brenner, D. J., Steigerwalt, A. G., and McDade, J. E. (1979). *Ann. Intern Med.* **90,** 656–658.

Brenner, D. J., Steigerwalt, A. G., Gorman, G. W., Weaver, R. E., Feeley, J. C., Cordes, L. G., Wilkinson, H. W., Patton, C., Thomason, B. M., and Lewallen-Sasseville, K. R. (1980). *Curr. Microbiol.* **4,** 111–116.

Cahall, D. L., and Youmans, C. P. (1975). *Infect. Immun.* **12,** 833–840.

Carrington, C. B. (1979). *Ann. Intern Med.* **90,** 496–499.

Chandler, F. W., Hicklin, M. D., and Blackmon, J. A. (1977). *N. Engl. J. Med.* **297,** 1218–1220.

Chandler, F. W., Cole, R. M., Hicklin, M. D., Blackmon, J. A., and Callaway, C. S. (1979). *Ann. Intern. Med.* **90,** 642–647.

Chang, K. P., and Dwyer, D. M. (1976). *Science* **193,** 678–680.

Cherry, W. B., Gorman, G. W., Orrison, L. H., Moss, C. W., Steigerwalt, A. G., Wilkinson, H. W., Johnson, S. E., McKinney, R. M., and Brenner, D. J. (1982). *J. Clin. Microbiol.* **15,** 290–297.

Cohen, M. L., Broome, C. V., Paris, A. L., Martin, W. T., and Allen, J. R. (1979) *Ann. Intern. Med.* **90,** 611–613.

Cole, P. (1975). *Infect. Immun.* **12,** 36–41.

Cordes, L. G., Wiesenthal, A. M., Gorman, G. W., Phair, J. P., Sommers, H. M., Brown, A., Yu, V. L., Magnussen, M. N., Meyer, R. D., Wolf, J. S., Shands, K. N., and Fraser, D. W. (1981). *Ann. Intern. Med.* **94,** 195–197.

Eickhoff, T. C. (1979). *Ann. Intern. Med.* **90,** 499–502.

England, A. C., III, McKinney, R. M., Skaliy, P., and Gorman, G. W. (1980). *Ann. Intern. Med.* **93,** 58–59.

Feeley, J. C., Gibson, R. J., Gorman, G. W., Langford, N. C., Rasheed, J. K., Mackel, D. C., and Baine, W. B. (1979). *J. Clin. Microbiol.* **10,** 437–441.

Flesher, A. R., Ito, S., Mansheim, B. J., and Kasper, D. L., (1979). *Ann. Intern. Med.* **90,** 628–630.

Fliermans, C. B., Cherry, W. B., Orrison, L. H., and Thacker, L. (1979). *Appl. Environ. Microbiol.* **37,** 1239–1242.

Fliermans, C. B., Cherry, W. B., Orrison, L. H., Smith, S. J., Tison, D. L., and Pope, D. H. (1981). *Appl. Environ. Microbiol.* **41,** 9–16.

Fowles, R. E., Fajardo, I. M., Leibowitch, J. L., and David, J. R. (1973). *J. Exp. Med.* **138,** 952–964.

Fraser, D. W., and McDade, J. E. (1979). *Sci. Am.* **241,** 82–99.

Fraser, D. W., Tsai, T. R., Orenstein, W., Parkin, W. E., Beecham, H. J., Sharrar, R. G., Harris, J., Mallison, G. F., Martin, S. M., McDade, J. E., Shepard, C. C., Brachman, P. S., and the Field Investigation Team. (1977). *N. Engl. J. Med.* **297,** 1189–1197

Fraser, D. W., Duebner, D. C., Hill, D. L., and Gilliam, D. K. (1979). *Science* **205,** 690–691.

Fu, K. P., and Neu, H. C. (1979). *Antimicr. Agents Chemother.* **16,** 561–564.

Glavin, F. L., Winn, W. C., Jr., and Craighead, J. E. (1979). *Ann. Intern. Med.* **90,** 555–558.

Glick, T. H., Gregg, M. B., Berman, B., Mallison, G. F., Rhodes, W. W., Jr., and Kassanoff, I. (1978). *Am. J. Epidemiol.* **107,** 149–160.

Gregory, D. W., Schaffner, W., Alford, R. H., Kaiser, A. B., and McGee, Z. A. (1979). *Ann. Intern. Med.* **90,** 518–521.

Guerrant, G. O., Lambert, M. S., and Moss, C. W. (1979). *J. Clin. Microbiol.* **10,** 815–818.

Haley, C. E., Cohen, M. L., Halter, J., and Meyer, R. D. (1979). *Ann. Intern. Med.* **90,** 583–586.

Hart, P. D., and Armstrong, J. A. (1974). *Infect. Immun.* **10,** 742–746.

Hart, P. D., Armstrong, J. A., Brown, C. A., and Draper, P. (1972). *Infect. Immun.* **5,** 803–807.

Hebert, G. A., Steigerwalt, A. G., and Brenner, D. J. (1980). *Curr. Microbiol.* **3,** 255–257.

Hinrichs, D. J., and Jerrells, T. R. (1976). *J. Immunol.* **117,** 996–1003.

Horwitz, M. A. (1982). *Clin. Res.* **30,** 369A.

Horwitz, M. A., and Silverstein, S. C. (1980a). *J. Clin. Invest.* **66,** 441–450.

Horwitz, M. A., and Silverstein, S. C. (1980b). *J. Exp. Med* **153,** 386–397.

Horwitz, M. A., and Silverstein, S. C. (1980c). *J. Exp. Med.* **153,** 398–406.

Horwitz, M. A., and Silverstein, S. C. (1980d). *J. Clin. Invest.* **65:** 82–94.

Horwitz, M. A., and Silverstein, S. C. (1981). *J. Exp. Med.* **154,** 1618–1635.

Jones, T. C., and Hirsch, J. G. (1972). *J. Exp. Med.* **136,** 1173–1194.

Jones, T., and Youmans, G. P. (1973). *Cell. Immunol.* **9,** 353–362.

Jones, T. C., Len, L., and Hirsch, J. G. (1975). *J. Exp. Med.* **141,** 466–483.

Keel, J. A., Finnerty, W. R., and Feeley, J. C. (1979). *Ann. Intern. Med.* **90,** 652–655.

McDade, J. E., and Shepard, C. C. (1979). *J. Infect. Dis.* **139,** 707–711.

McDade, J. E., Shepard, C. C., Fraser, D. W., Tsai, T. R., Redus, M. A., Dowdle, W. R., and the Laboratory Investigation Team. (1977). *N. Engl. J. Med.* **297,** 1197–1203.

McDade, J. E., Brenner, D. J., and Bozeman, F. M. (1979). *Ann. Intern. Med.* **90,** 659–661.

McKinney, R. M., Thacker, L., Harris, P. P., Lewallen, K. R., Hebert, G. A., Edelstein, P. H., and Thomason, B. M. (1979). *Ann. Intern. Med.* **90,** 621–624.

McKinney, R. M., Wilkinson, H. M., Sommers, H. M., Fikes, B. J. Sasseville, K. R., Yungbluth, M. M., and Wolf, J. S. (1980). *J. Clin. Microbiol.* **12,** 395–401.

McKinney, R. M., Porchen, R., Edelstein, P. N., Bisset, M. J., Harris, P. P., Bondell, S. P.,

Steigerwalt, A. G., Weaver, R., Ein, M., Lindquist, D. S., and Brenner, D. J. (1981). *Ann. Intern. Med.* **94,** 739–743.

Mauel, J., Buchmüller, Y., and Behin, R. (1978). *J. Exp. Med.* **148,** 393–407.

Meenhorst, P., Reingold, A., Gorman, G., Wilkinson, H., Feeley, J., Band, J., and van Furth, R. (1981). *Intersci. Conf. Antimicr. Agents Chemother., 21st, Nov. Abstract 298.*

Morris, G. K., Patton, C. M., Feeley, J. C., Johnson, S. E. Gorman, G., Martin, W. T., Skaliy, P., Mallison, G. F., Politi, B. D., and Mackel, D. C. (1979). *Ann. Intern. Med.* **90,** 664–666.

Morris, G. K., Steigerwalt, A. G., Feeley, J. C., Wong, E. S., Martin, W. T., Patton, C. M., and Brenner, D. J. (1980). *J. Clin. Microbiol.* **12,** 718–721.

Moss, C. W., Weaver, R. E., Dees, S. B., and Cherry, W. B. (1977). *J. Clin. Microbiol.* **6,** 140–143.

Moss, C. W., Weaver, R. E., Dees, S. B., and Cherry, W. B. (1979). *In* "Legionnaires, The Disease, the Bacterium and Methodology" (G. L. Jones and G. A. Hebert, eds.), pp. 48–54. U.S. Public Health Service, Centers for Disease Control (CDC), Bureau of Laboratories, Atlanta, Ga. Publ. No. (CDC) 79-8375.

Myerowitz, R. L., Pasculle, A. W., Dowling, J. N., Pazin, G. J., Puerzer, M., Yee, R. B., Rinaldo, C. R., Jr., and Hakala, T. R. (1979). *N. Engl. J. Med.* **301,** 953–958.

Nacy, C. A., and Meltzer, M. S. (1979). *J. Immunol.* **123,** 2544–2549.

Nogueira, N., and Cohn, Z. A. (1978). *J. Exp. Med.* **148,** 288–300.

Nogueira, N., Gordon, S., and Cohn, Z. (1977) *J. Exp. Med.* **146,** 157–171.

Owens, M. L. (1978). *N. Engl. J. Med.* **298,** 1089.

Pasculle, A. W., Myerowitz, R. L., and Rinaldo, C. L., Jr. (1979). *Lancet* **2,** 58–61.

Patterson, R. J., and Youmans, G. P. (1970). *Infect. Immun.* **1,** 600–603.

Pine, L., George, J. R., Reeves, M. W., and Harrell, W. K. (1979). *In* "Legionnaires, the Disease, the Bacterium and Methodology" (G. L. Jones and G. A. Hebert, eds.), pp. 27–40. U.S. Public Health Service, Centers for Disease Control, Bureau of Laboratiories, Atlanta, Ga. Publ. No. (CDC) 79-8275.

Renner, E. D., Helms, C. M., Hierholzer, W. J., Jr., Hall, N., Wong, Y. W. Viner, J. P., Johnson, W., and Hauster, W. J., Jr. (1971). *Ann. Intern. Med.* **90,** 603–606.

Rogers, B. H., Donowitz, G. R., Walker, G. K., Harding, S. A., and Sande, M. A. (1979). *N. Engl. J. Med.* **301,** 959–961.

Sanford, J. P. (1979). *N. Engl. J. Med.* **300,** 654–656.

Saravolatz, L. D., Burch, K. H., Fisher, E., Madhavan, T., Kiani, D., Neblett, T., and Quinn, E. L. (1979). *Ann. Intern. Med.* **90,** 533–537.

Sethi, K. K., Pelster, B., Suzuki, N., Piekarski, G., and Brandis, H. (1975). *J. Immunol.* **115,** 1151–1158.

Shirahata, T., Shimizu, K., Noda, S., and Suzuki, N. (1977). *Z. Parasitenkd.* **53,** 31–40.

Simon, H. B., and Sheagren, J. N. (1972). *Cell. Immunol.* **4,** 163–174.

Storch, G., Baine, W. B., Fraser, D. W., Broome, C. V., Clegg H. W., II, Cohen, M. L., Goings, S. A. J., Politi, B. D., Terranova, W. A., Tsai, T. F., Plikaytis, B. D., Shepard, C. C., and Bennett, J. V. (1979). *Ann. Intern. Med.* **90,** 596–600.

Stout, J., Yu, V. L., Vickers, R. M., Zuravleff, J., Best, M., Brown, A., Yee, R. B., and Wadowsky, R. (1982). *N. Engl. J. Med.* **306,** 466–468.

Thomason, B. M., Chandler, F. W., and Hollis, D. G. (1979). *Ann. Intern. Med.* **91,** 224–226.

Thornsberry, C., and Kirven, L. A. (1978). *Curr. Microbiol.* **1,** 51–54.

Tison, D. L., Pope, D. H., Cherry, W. B., and Fliermans, C. B. (1980). *Appl. Environ. Micr.* **39,** 456–459.

Tobin, J. O'H., Swann, R. A., and Bartlett, C. L. R. (1981). *Br. Med. J.* **282,** 515–517.

Weaver, R. E., and Feeley, J. C. (1979). *In* "Legionnaires, the Disease, the Bacterium and

Methodology'' (G. L. Jones and G. A. Hebert, eds.), pp. 19–25. U.S. Public Health Service, Centers for Disease Control, Bureau of Laboratories, Atlanta, Ga. Publ. No. (CDC) 79-8375.

Winn, W. C., Jr., Glavin, F. L., Perl, D. P., Keller, J. L., Andres, T. L., Brown, T. M., Coffin, C. M., Sensecqua, J. E. Roman, L. N., and Criaghead, J. E. (1978). *Arch. Pathol. Lab. Med.* **102,** 344–350.

Wong, K. H., Moss, C. W., Hochstein, D. H., Arko, R. J., and Schalla, W. O. (1979). *Ann. Intern. Med.* **90,** 624–627.

Wong, K. H., McMaster, P. R. B., Feeley, J. C., Arko, R. J., Schalla, W. O., and Chandler, F. W. (1980). *Curr. Microbiol.* **4,** 105–110.

INTERNATIONAL REVIEW OF CYTOLOGY, SUPPLEMENT 14

Plastids—Past, Present, and Future

JEAN M. WHATLEY

Botany School, University of Oxford, Oxford, England

I. Introduction

A. Background

About 100 years ago Schimper (1883) observed that chloroplasts never arose *de novo* within the cell, but were able to grow and divide by fission in a more or less autonomous manner. He suggested that this might be because plastids had evolved from blue-green algal symbionts. Since that time an endosymbiotic (bacterial) origin has been proposed for a number of other types of organelle,

ISBN 0-12-364375-9

including mitochondria (Altman, 1890; Mereschowsky, 1905; Ris and Plaut, 1962), flagella (Margulis *et al.*, 1979, 1981), hydrogenosomes (Müller, 1975), omikron in *Euplotes* (Heckmann, this volume), and kappa and other particles in the ciliate *Paramecium* (Sonneborn, 1959). The hypothesis has aroused widespread interest and has been the subject of several symposia (Fredrick, 1981; Richmond and Smith, 1979; Schwemmler and Schenk, 1980) and though it has been strongly criticized by some (e.g., Cavalier-Smith, 1975; Mahler, 1980; Raff and Mahler, 1972, 1975; Uzzell and Spolsky, 1974, 1981), that portion of it relating to chloroplasts in particular has recently gained ground (Dodge, 1979, 1980; Gibbs, 1978, 1981a,b; Taylor, 1979; Whatley *et al.*, 1979; Whatley and Whatley, 1981).

If some present day types of organelle in eukaryotic cells did indeed arise from endosymbionts, how, when, and in what sequence did this take place? How is the partnership controlled? What modifications have taken place in the metabolism and structure of the partners? Do similar endosymbiotic partnerships still become established today and are any of them likely to lead to the evolution of further new forms of organelle? Each of these questions has been the subject of much speculation but few definitive answers are possible. What is certain is that closely integrated relationships between host cells and both intra- and extracellular symbionts are much more widespread than was formerly realized. The intention in this article is to examine the suggested lines of evolution followed by chloroplasts, including some more recent modifications, methods of population control for both symbionts and organelles, and the possibilities for the future evolution of new forms of organelle.

B. How Organelles May Have Evolved from Endosymbionts

The endosymbiotic theory for the origin of organelles suggests that free-living organisms were taken up into vacuoles by host cells. Once inside the host, the invading cells managed to avoid digestion, a permanent endosymbiotic mode of existence became established, and eventually the formerly free-living organisms became transformed into organelles (Jeon, this volume; Lee, this volume). How was this accomplished?

A free-living organism is completely autonomous and carries out all its own synthetic, metabolic, and reproductive functions. An organelle derived from a symbiont has lost much of this autonomy. The functions retained at least in part by the organelle are essentially those concerned with reproduction and with that part of its former metabolic activity in which it has become a specialist in cooperation with the host cell, e.g., respiration in mitochondria and photosynthesis in chloroplasts.

The way in which mitochondria may have evolved from gram-negative bacteria with characteristics resembling those of the present-day *Paracoccus de-*

nitrificans or *Rhodopseudomonas sphaeroides* has been discussed many times (e.g., John and Whatley, 1975; but see Weitzman, 1981). During such a transformation several modifications must have taken place:

1. The ATPase and the constitutive features of the bacterial respiratory chain, which closely resemble those of mitochondria, were retained; adaptive features of the bacterial respiratory chain were lost.
2. The mode of operation of carriers was altered and an adenine nucleotide (ATP/ADP) carrier was acquired to facilitate controlled exchange of energy between the two organisms.
3. Genetic information which coded for bacterial proteins was taken over by the host nucleus, perhaps by a form of plasmid transfer. The host also became responsible for regulating the two populations.

Transformation of the proposed ancestors of chloroplasts would have followed a pattern of modification and loss of adaptive features parallel to that described for mitochondria, but biased toward the retention of photosynthetic rather than respiratory activity.

II. Chloroplast Evolution

Photosynthetic bacteria do not evolve oxygen and have only one light reaction for photosynthesis. By contrast, all prokaryotic algae and chloroplasts of eukaryotic algae and land plants have two light reactions and evolve oxygen. For this reason prokaryotic algae rather than photosynthetic bacteria are considered the endosymbiotic ancestors of chloroplasts. The photosynthetic apparatus in prokaryotic algae and chloroplasts contains at least two photosynthetic pigments. One of these, the primary pigment, is always chlorophyll a. The accessory pigments include phycobilins, chlorophyll b, and chlorophyll c. On the basis of differences in the chloroplast accessory pigments the eukaryotic algae are often divided into three main groups: (1) Rhodophyta (red algae) with accessory phycobilins; (2) Chlorophyta (green algae) and euglenoids with accessory chlorophyll b; (3) Chromophyta (so designated by Christensen, 1964) with accessory chlorophyll c.

It is largely because of these differences in pigmentation that a polyphyletic origin for chloroplasts has been proposed (e.g., Schiff and Epstein, 1965; Raven, 1970). Indeed a major criticism of the endosymbiotic theory in the past was that only one possible ancestral group (the blue-green algae) was known. However, the recent discovery of the prochlorons, prokaryotic algae with chlorophyll a and accessory chlorophyll b, means that modern relatives of possible ancestors of green algal chloroplasts do indeed exist (Lewin and Withers, 1975; Newcomb and Pugh, 1975). This leaves unsolved the ancestry of the chromophyte chloroplasts. However, the problem for the latter is not only one of pigmentation. The chloroplasts of red and green algae, like mitochondria, are always surrounded by two membranes and these are usually assumed to be homologous with the plasma membrane of the prokaryotic endosymbiont and the vacuolar membrane of the host. Chromophyte chloroplasts are surrounded by more than two membranes. For a number of reasons (to be discussed below) it has now been proposed that chromophyte chloroplasts were probably derived from eukaryotic rather than prokaryotic algal symbionts, i.e., they are the result of two successive acts of endosymbiosis rather than one. This wider concept of chloroplast origin is now becoming increasingly accepted (e.g., Cavalier-Smith, 1980, 1981; Dodge, 1979; Gibbs, 1981b; Whatley and Whatley, 1981). It certainly provides a reasonable explanation of some apparently anomalous features which are hitherto not easy to account for.

A. Surrounding Membranes

Mitochondria, hydrogenosomes, and the chloroplasts of red and green algae and land plants are surrounded by two membranes. The inner membrane of the pair is considered homologous with the plasma membrane of the prokaryote and the outer membrane homologous with the vacuolar membrane of the host. Certainly in mitochondria some structural components of the inner membrane are coded for by the mitochondrion itself and the outer membrane, which more closely resembles eukaryotic endomembrane, is coded for entirely by the "host" cell nucleus. This interpretation of membrane origin means that topologically the organelles, like endosymbionts, lie outside the host cell cytoplasm. Chloroplasts of all other algae are surrounded either by three (euglenoids and dinoflagellates) or by four membranes. Possible homologies between these membranes and those which surround various types of photosynthetic endosymbionts known today remain uncertain, but some possible clues as to their precise derivation are provided by the chloroplasts of one class of chromophyte alga, the cryptomonads.

The complex chloroplasts of the cryptomonads are surrounded by four membranes. The inner pair represent the chloroplast envelope proper; the outer pair are usually designated chloroplast endoplasmic reticulum. Between the two pairs lies the perichloroplastic or periplastidial space. This space contains large (80 S) ribosomes, starch, and a double membrane-bound structure which has been given the name nucleomorph (Greenwood, 1974). In the matrix of the nucleomorph lie osmiophilic globules and fibrilogranular material; the surrounding membranes are interrupted by pores or slits. It has been suggested that the nucleomorph may represent the reduced remains of a former nucleus. Though no

DNA has been positively identified, recent cytological studies have tentatively concluded that the nucleomorph contains nucleic acid (Gillot and Gibbs, 1980).

If the nucleomorph is indeed a reduced nucleus, then the perichloroplastic space with its large ribosomes may represent the cytoplasm of a formerly free-living eukaryotic alga. The presence of starch in the perichloroplastic space is comparable with the presence of starch in the cytoplasm of a red alga. The chloroplast proper with its double envelope would represent a prokaryotic alga taken up and subsequently modified inside a eukaryotic host, possibly a red alga. This modified red or "prechromophyte" alga (Whatley and Whatley, 1981) must, in its turn, have been taken up by a second eukaryotic host cell, within which it has become further modified into the complex cryptomonad chloroplast that we know today.

In other classes of algae with complex chloroplasts there is no nucleomorph and the perichloroplastic space is much reduced (in chromophyte algae with chloroplasts enclosed in four membranes) or is completely absent (in euglenoids and dinoflagellates, with chloroplasts enclosed in three membranes). It is probable that, in one way or another, all algae with complex chloroplasts represent the result of two successive acts of symbiosis involving, first, a prokaryotic and, later, a eukaryotic endosymbiont. The various possible endosymbiotic pathways which may have been associated with chloroplast evolution (Fig. 1) have recently been analyzed in greater detail elsewhere (Gibbs, 1981a; Whatley and Whatley, 1981).

B. Pigments and Thylakoids of Photosynthetically Functional Chloroplasts

The chloroplasts of different algal phyla vary considerably in their pigmentation and in the arrangement of their thylakoids. Nevertheless, if one accepts that chloroplasts had a polyphyletic origin, then along each separate evolutionary line there have been only minor modifications of the photosynthetic apparatus; this is most obvious in the line represented by blue green-algae and the red-algal chloroplast (Fig. 1).

The blue-green algae have chlorophyll a as their primary photosynthetic pigment, phycobilins as accessory pigments, and single thylakoids with protruding phycobilisomes (site of the phycobilins). The photosynthetic apparatus thus closely resembles that in a red-algal chloroplast. There are, however, minor differences. Phycocyanin is usually the major phycobilin in blue-green algae whereas phycoerythrin is preponderant in most red algae. The carotenoids also differ. Blue-green and red-algae both contain the ubiquitous β,β-carotene, but the former have monocyclic carotenoid glycosides whereas red algae have ϵ-ring carotenes and bicyclic xanthophylls (Liaaen-Jensen, 1977, 1979). In spite of

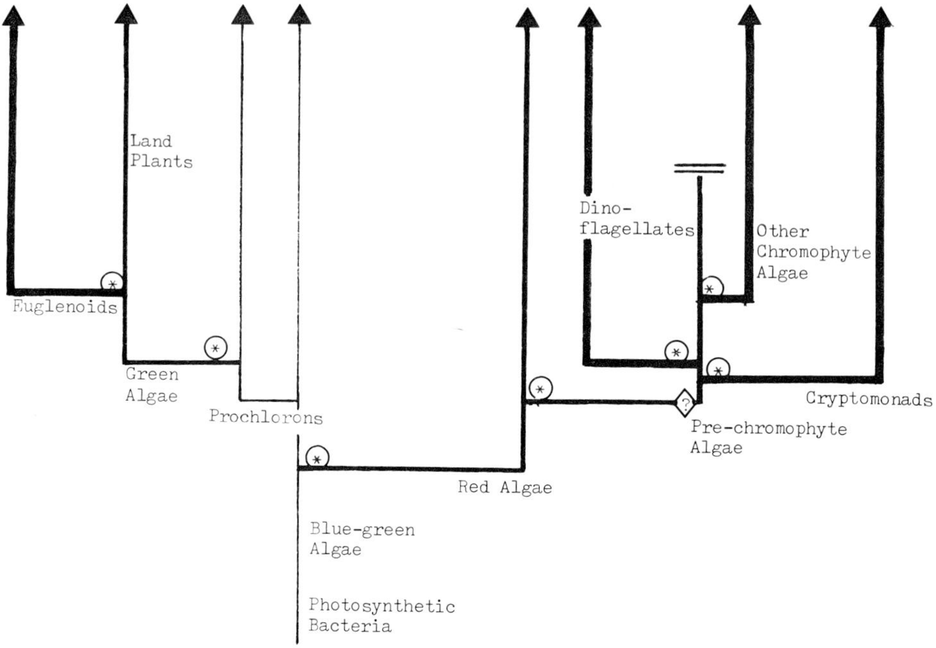

FIG. 1. The possible course of evolution of chloroplasts from free-living photosynthetic ancestors taken up as symbionts (circled asterisk) by a variety of different host cells. Thick lines, plastids = eukaryotic symbionts; medium lines, plastids = prokaryotic symbionts; thin lines, free-living photosynthetic prokaryotes; circled asterisk, uptake of symbionts.

these differences in pigmentation between blue-green algae and red-algal chloroplasts, it is nevertheless reasonable to conclude that the ancestors of red-algal chloroplasts resembled present day unicellular blue-green algae.

Some minor modifications in thylakoid arrangement appear to have taken place in the photosynthetically active chloroplasts of the more complex red algae. Phycobilisomes cannot always be distinguished, but this may be a problem of fixation (Dodge, 1973). Pyrenoids are absent from chloroplasts of most members of the more advanced Florideae and the main group of thylakoids within the chloroplasts of this subclass are often encircled by a distinctive peripheral thylakoid or inner limiting disc (Coombs and Greenwood, 1976).

The second main line of chloroplast evolution is represented today by the prochlorons and the chloroplasts of green algae and land plants. The prochlorons themselves probably evolved from blue-green algae, following the loss of phycobilins and the appearance of accessory chlorophyll b. The P700 chlorophyll a protein (Complex I) and the size of the photosynthetic unit in prochlorons are intermediate between those of blue-green algae and the chloroplasts of green algae and land plants (Withers *et al.*, 1978). However a recent report on partial sequencing of 16 S ribosomal RNA shows a high sequence homology between *Prochloron* and blue-green algae, particularly *Nostoc*, but a low similarity coefficient between *Prochloron* and "green" chloroplasts (Seewaldt and Stackebrandt, 1982).

The thylakoid arrangement in prochlorons is variable. Instead of lying singly in the cytoplasm, as they do in blue-green algae, the thylakoids lie closely appressed, mainly in irregularly interconnected pairs, but also in deeper but ill-defined stacks which extend only over short distances. In the chloroplasts of green algae, too, the thylakoids are closely appressed but in irregular bands of two to six, or in occasional larger stacks (Coombs and Greenwood, 1976; Gibbs, 1970; Whatley and Whatley, 1981). In most land plants and in those green algae which are thought to have preceded them (e.g., members of the Charales), thylakoid stacking in the chloroplasts is more highly regulated and the stacks are organized into discrete grana. A modified arrangement of the thylakoid system is also found in chloroplasts of some siphonaceous green algae, e.g., members of the Caulerpales. Their plastids are distinctive in having a terminal structure made up of concentric double-membrane rings. The plastids are enclosed in a multilayered peripheral integument which is resistant to digestion with pronase and lysozyme (e.g., Borowitzka and Larkum, 1974; Giles and Sarafis, 1974). Isolated chloroplasts of *Codium* and *Caulerpa*, which are resistant to osmotic shock, are taken up as symbionts by some marine molluscs (Trench *et al.*, 1973).

The carotenoids of prochlorons resemble those of blue-green algae but differ from those of green algae. β-Carotene and zeaxanthin make up almost 90% of prochloron carotenoids. Green-algal carotenoids, some of which are concentrated in eyespots (to be discussed below), are much more varied but none of those specific to green algae has been found in prochlorons (Withers *et al.*, 1978). Many siphonaceous green algae contain siphonaxanthin. These algae and (uniquely) the nonsiphonaceous *Ulva japonica* have been found to use siphonaxanthin as an accessory photosynthetic pigment when they are growing in deep water. As this pigment absorbs light at 540 nm, these species are exploiting for photosynthesis a part of the spectrum not normally used by green algae (Kageyama *et al.*, 1977; Yokohama and Kageyama, 1977). In land plants, particularly angiosperms, the range of carotenoids is greater than in green algae. Nuclear mutations are responsible for the additional carotenoids synthesized in angiosperm plastids (Kirk and Tilney-Bassett, 1978).

The differences in the types of carotenoid pigments found in blue-green algae and red-algal chloroplasts on the one hand and prochlorons and green-algal chloroplasts on the other, may, in part, result from their (presumed) endosymbiotic ancestry. Most nonphotosynthetic eukaryotes (i.e., host cells) are unable to synthesize carotenoids *de novo*, though they are able to modify carotenoids from organisms taken up as food. Some of the variation in carotenoids of eukaryotic algae may be the result of such modifications effected following endosymbiosis (Whatley and Whatley, 1981).

The chloroplasts of euglenoids resemble those of green algae in lacking phycobilins but having chlorophyll b as their accessory pigment. The thylakoids are closely appressed, often in irregularly connected bands of three. Euglenoid plastids are believed to have evolved following the symbiotic uptake of a eukaryotic green alga or its chloroplasts, i.e., following two successive acts of symbiosis (Gibbs, 1978; Whatley *et al.*, 1979). The euglenoids, like several classes of chromophyte algae, but unlike green algae, have acetylenic carotenoids; 4-keto-carotenoids, present in the euglenoid eyespots which are extraplastidial, are also found in blue-green algae and in chloroplasts of green algae suffering from nitrogen deficiency (Liaaen-Jensen, 1977, 1979).

The third main line of chloroplast evolution is represented by the chromophyte algae. Neither a prokaryotic alga with accessory chlorophyll c, nor the hypothetical eukaryotic prechromophyte alga is known today. The stage at which chlorophyll c first evolved (before or after uptake by a host cell) cannot therefore be determined. However, it has been reported that chloroplasts belonging to the dinoflagellate genus *Prorocentrum* may be enclosed by only two rather than by the three membranes characteristic of plastids in other dinoflagellate genera. This could perhaps indicate that these plastids developed from a prokaryotic rather than a eukaryotic algal symbiont, i.e., that these dinoflagellates may be direct rather than indirect descendants of the prechromophyte alga. Further investigation of these interesting dinoflagellates is currently being carried out by Dr. Maret Vesk of Sydney University.

Possibly the least advanced chlorophyll c-containing algal plastids are those of the cryptomonads; these chloroplasts are in some respects intermediate between those of red algae and those of other chromophyte algae. Unlike the chloroplasts of other algae, those of the cryptomonads contain, in addition to the primary pigment, chlorophyll a, two different types of accessory pigment, phycobilins and chlorophyll c_2. The phycobilins in red-algal chloroplasts are concentrated in the phycobilisomes which protrude into the stroma from single thylakoids. The cryptomonad phycobilins differ chemically from those of red algae and lie within the thylakoid sacs. As Greenwood (1974) pointed out, the absence of protruding phycobilisomes makes possible the closer apposition of adjacent thylakoids. The thylakoid arrangement in cryptomonad chloroplasts is somewhat variable; the thylakoids are either single (as in red algae) or apposed in pairs over long distances, though they are never firmly bonded as they are in prochlorons.

Dinoflagellate chloroplasts, too, have accessory chlorophyll c_2 but they lack phycobilins. The thylakoid arrangement is again somewhat variable. The chloroplasts of all other chromophyte algae also lack phycobilins, but they differ from dinoflagellates in having both chlorophylls c_1 and c_2. The thylakoids are usually associated, but not firmly bonded, in bands of three. These triple bands are extensive and the outermost thylakoid of one band may pass over to join the next band in a precisely regulated manner (Coombs and Greenwood, 1976). In some chromophyte algae (members of the Bacillariophyceae, Chrysophyceae, Phaeophyceae, Raphidophyceae, and Xanthophyceae—see Coombs and Greenwood, 1976) the main triple thylakoid bands are encircled by a distinctive girdle lamella which is itself a triple thylakoid band. When a girdle lamella is present, the chloroplast DNA lies in a distinctive ring which encircles the chloroplast immediately inside the girdle.

The carotenoids of cryptomonads are relatively simple. The major carotenes are β,β- and β,ϵ-; the xanthophylls are acetylenic rather than simple (as they are in red algae) or epoxides (as they are in other chromophyte algae) (Liaaen-Jensen, 1977, 1979; Hibberd, 1979). Dinoflagellates are distinctive in producing peridinin. On the basis of carotenoid biosynthetic pathways, particularly that of fucoxanthin—a carotenoid which participates in photosynthesis—Ragan and Chapman (1978) have concluded that the chromophyte algal classes Bacillariophyceae, Chrysophyceae, Phaeophyceae, and Prymnesiophyceae have a close phylogenetic relationship. On the other hand, members of the Eustigmatophyceae, Raphidophyceae, and Xanthophyceae all form vaucheriaxanthin.

C. Eyespots

Eyespots are small concentrations of carotenoid globules which are found in flagellated cells of many eukaryotic algae. (Other carotenoids are located in the plastid envelope and in the thylakoids.) The eyespots of most chromophyte algae and the euglenoids lie close to the flagellar bases. Those of green algae and cryptomonads show no such close flagellar association (Dodge, 1973). Most commonly eyespot globules are arranged in one or more layers inside but close to the periphery of one plastid in each cell. However, in some algae the eyespots are found elsewhere in the cell, free in the cytoplasm (in zoospores of members of the Eustigmatophyceae, Hibberd, 1979), delimited by membranes (in some dinoflagellates, e.g., Dodge and Crawford, 1969), or as a complex of discrete membrane-bound packets of globules (in some euglenoids, Walne and Arnott, 1967). Recent work on some of these extraplastidial eyespots has shed new light on chloroplast evolution and has suggested possible alternative lines of investigation.

Anomalous carotenoids (fucoxanthin instead of peridinin) have been found in three dinoflagellate species (Jeffrey, 1976). Two of these species have been examined in the electron microscope and have been found to contain eyespots which appear to be reduced dinoflagellate chloroplasts. In addition these algae have been found to contain chrysophyte algae as photosynthetic endosymbionts (having both chlorophyll c_1 and c_2 as well as chlorophyll a). It is these symbionts which appear to be the source of the fucoxanthin (a characteristic chrysophyte carotenoid) as well as the chlorophyll pigments (Jeffrey and Vesk, 1976; Tomas and Cox, 1973; Withers and Haxo, 1978; Withers *et al.*, 1977).

The extraplastidial "eyespots" of some members of the dinoflagellate family *Warnowiaceae* are particularly interesting. They have a membrane envelope which surrounds both the pigment globules and a paracrystalline structure (the retinoid), but they are always found in close and precisely organized association with other organelles (mitochondria, microtubules, microfibrils, and specialized vacuoles, one of which may act as a lens). The whole integrated complex is called an ocellus and in many respects it resembles a primitive eye but its precise function has not been determined (Dodge, 1973, 1979; Francis, 1967; Taylor, 1980). Greuet (1978) has shown that the membrane-enclosed "eyespot" portion of the ocellus is indeed derived from a chloroplast.

It is not unreasonable to conclude that in species with both chloroplasts and extraplastidial membrane-bound eyespots photosynthetic symbionts have been taken up at two different periods of evolution and perhaps even from two different taxonomic sources. Photosynthetic activity has been retained by one symbiont population; the other symbiont has provided the eyespot.

D. Storage Products

The photochemical segment (light reactions) of photosynthesis is carried out in the thylakoid system. The dark reactions of photosynthesis take place in the surrounding matrix, i.e., the cytoplasm of prokaryotic algae, or, its homolog, the stroma, of chloroplasts. Several types of storage product can accumulate in the cytoplasm of free-living prokaryotic algae. These include (1) carbohydrates (e.g., in glycogen granules), (2) proteins (in cyanophycean or structured granules and in carboxysomes), and (3) linear polyphosphates (in polyphosphate granules). Storage bodies which accumulate in the plastid stroma are much less common and less varied: (1) starch, in green algae and land plants only and (2) protein, including Fraction I (ribulose bisphosphate carboxylase/oxygenase). The protein is stored in the pyrenoids—perhaps the chloroplast equivalent of the carboxysomes—which are found in plastids of many simpler eukaryotic algae and a few of the more primitive land plants, e.g., the hornwort, *Anthoceros* (Dodge, 1973; Kirk and Tilney-Bassett, 1978). The plastids of more complex algae and most land plants lack pyrenoids. However, protein is also accumulated in membrane-bound bodies in the plastids of a few species of angiosperms. Nevertheless the capacity of chloroplasts for storage is very much less than that of their free-living counterparts.

Many metabolites (even though they may be produced in the plastids) are commonly stored in the cell cytoplasm—e.g., the storage carbohydrates of most algae and polyphosphate granules in green algae, for example (Dodge, 1973). The tannins which first accumulate in the plastids of brown algae are later extruded by way of ER into the cytoplasm where they fuse to form fucosan vesicles or physodes (Evans and Holligan, 1972; Pellegrini, 1980; Gibbs, 1981a). The capacity for storage of many materials thus seems to have been lost by the symbionts (plastids) but retained by the host cells.

The derivation of chloroplasts from both prokaryotic and eukaryotic symbionts fits well with what is known about the accumulation of storage carbohydrate (Whatley *et al.*, 1979; Whatley and Whatley, 1981). Initially both symbiont and host cell would, in the free-living state, have been capable of storing excess carbohydrate. In green algae and land plants the capacity to do so has uniquely been retained by the symbiont but lost by the "host" cell, i.e., starch accumulates in the plastid stroma. In the red algae the reverse is the case and floridean starch is accumulated in the cytoplasm.

In both the red and the green algae the storage product is an α-1,4-glucan, a product also accumulated by blue-green algae and prochlorons. The starch of blue-green algae such as *Nostoc* is a phytoglycogen, and it has recently been shown that a similar starch is present in *Prochloron* (Fredrick, 1980). The starch of eukaryotic green algae such as *Chlorella* contains two components, an amylopectin and an amylose, and Fredrick has shown that *Prochloron,* too, may contain an amylose component. This could suggest that the storage products of *Prochloron* are intermediate in character between those of blue-green algae (from which they probably evolved) and those of the plastids of green algae (into which their presumed endosymbiotic relatives became modified).

The storage product of cryptomonads is also an α-1,4-glucan, but it is accumulated in the perichloroplastic space, i.e., in what may be the reduced cytoplasm of a former modified eukaryotic red algal cell. In all other chromophyte algae and in the euglenoids the site of carbohydrate accumulation is the cytoplasm of the second host. The storage product of dinoflagellates is again an α-1,4-glucan, but the varied products accumulated by all other chromophyte algae (leucosin, laminarin, etc.) and the paramylon of the euglenoids are β-1,3-glucans. This would suggest that in the euglenoids and in these other chromophyte algae (1) the capacity to store carbohydrate has been lost by both the initial prokaryotic symbiont and by the later eukaryotic symbiont, but has been retained by the second eukaryotic host and (2) the second act of symbiosis took place later into final host cells which were already diverse—witness their modified and varied storage products (Whatley and Whatley, 1981).

III. The Evolution of New Forms of Plastid

The evolution of increasingly complex algae has been accompanied by only a limited increase in the diversity of their chloroplasts. Those in less advanced species tend to be cup-shaped, stellate, or to take the form of parietal plates.

Plastids of more complex species are generally small and more numerous. They tend to be discoid in shape but some are spherical or pleomorphic. Photosynthetically functional chloroplasts are highly conservative organelles. As indicated in Section II, along each main line of algal evolution major internal changes within the chloroplasts have been comparatively few and have led to the establishment of the three main lines of chloroplast evolution. Subsequently, grana and terminal bodies became established in some green algae, the peripheral thylakoid or inner limiting disc in some red algae, and the girdle lamella in some chromophyte algae. However, all these modifications, too, are of widespread occurrence and are therefore, presumably, of long standing. Since these early modifications, photosynthetically functional chloroplasts seem to have undergone little change. Indeed it is difficult to distinguish between a chloroplast of the green alga, *Chara,* for example, and one from the angiosperm, *Phaseolus vulgaris* (Fig. 2a and b), in spite of the very different phylogenetic positions and natural environments of these two species.

Most of the more recent changes which have taken place in plastids have been purely *quantitative,* involving either some loss of photosynthetic activity and a reduction in the extent of the thylakoid system or the accumulation of greater or smaller amounts of various ancillary products. Since the three main lines of plastid evolution were established, there appear to have been comparatively few *innovative* changes leading either to a major modification of the photosynthetic pathway and substantial rearrangement of the thylakoid system or to the introduction of completely new ancillary products. This is perhaps not surprising in an organelle which is restricted in function and which has lost much of its former autonomy.

A. The Evolution of Nonphotosynthetic Plastids

The basic function of a chloroplast is photosynthesis. Both prokaryotic algae and the chloroplasts of unicellular eukaryotic algae usually retain their thylakoid

---

FIG. 2. (a,b) Chloroplasts from the green alga, *Chara* sp. (a) and from the angiosperm, *Phaseolus vulgaris* (b) are very similar in structure. (a) ×26,000; (b) ×18,000. (c) Proplastid (Stage 1, eoplast) in the primary leaf of *Phaseolus,* 24 hours after the start of germination. ×62,500. (d) Proplastid (Stage 2, amyloplast) in the primary leaf of *Phaseolus,* 48 hours after the start of germination ×21,500. (e) Pregranal plastid (Stage 4) in the light-grown primary leaf of *Phaseolus,* 5 days after the start of germination: note the prolamellar body type structure (arrow) which lacks the precise organization of the true prolamellar body in (f). ×26,500. (f) Etioplast from the primary leaf of a seedling of *Phaseolus* grown for 5 days in darkness: note the regularly paracrystalline structure of the prolamellar body (arrow). ×21,500. (g) Eoplast from the hypocotyl of a *Phaseolus* seedling grown in darkness for 24 hours which contains a large deposit of phytoferritin (P). ×31,500. (h) Plastid in the hypocotyl of a *Phaseolus* seedling grown in darkness for 5 days. A prolamellar body (arrow) is just beginning to form whereas in a leaf of the same age (f) such a structure is already present. ×25,000.

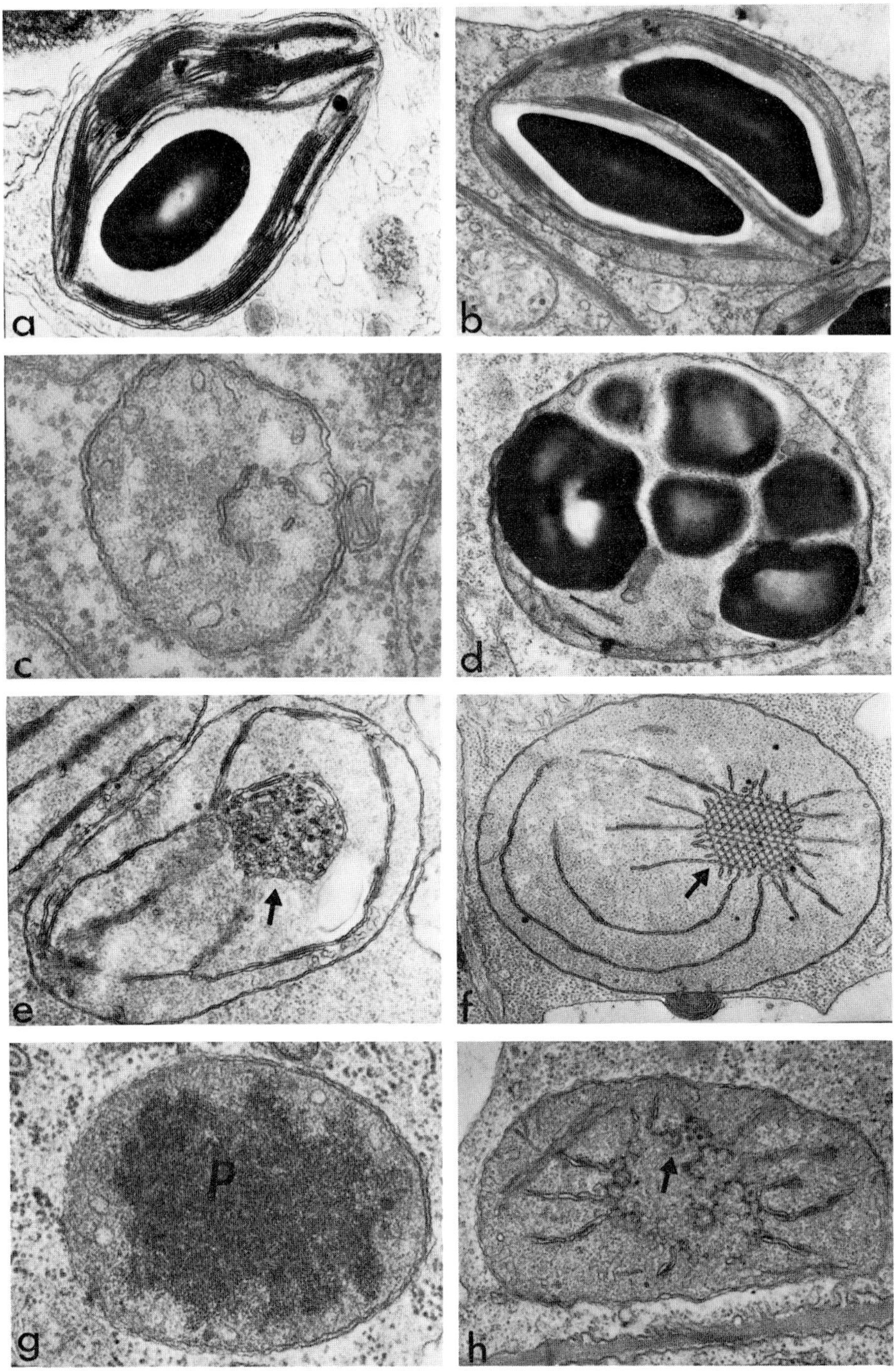
a
b
c
d
e
f
P
g
h

systems at all times. Indeed this is essential in an organism which is completely autotrophic. Nevertheless the evolution of most new forms of plastid seems to have depended, in the first instance, on some loss of photosynthetic activity. Among unicellular species only those which are preadapted to an alternative mode of nutrition can support the absence of a photosynthetic apparatus for any length of time. Such an absence may be temporary or permanent, complete or partial, regularly or spasmodically effected, and environmentally or genetically controlled. Schnepf (1980), for example, distinguishes between permanent genetic effects and temporary effects associated with the activation and repression of key nuclear genes.

The origin of an organelle from an endosymbiont implies that the host organism once had a phagocytic mode of nutrition. In spite of the frequent presence of a secondary wall, some photosynthetic unicells remain phagocytic today (dinoflagellates and chrysophyte algae have phagocytic vacuoles, for example, and dinoflagellates have ejectile organelles which closely resemble those in some invertebrates where they are used for capturing prey). Other unicellular eukaryotic algae too, retain features of a phagocytic life style (e.g., the gullet of euglenoids), or contain endosymbionts within their cytoplasm even though there is no obvious mode of uptake—e.g., bacterial symbionts in the green alga, *Bryopsis hypnoides* (Burr and West, 1970). In species such as these which remain phagocytic prolonged absence of photosynthesis can be tolerated.

There exist today unicellular organisms which lack photosynthetic pigments but which are obviously close relatives of normally pigmented algae. These "colorless" algae are rare but of widespread phylogenetic distribution (Pringsheim, 1941, 1963). Some of these so-called "colorless" species (which may nevertheless contain carotenoids) appear to completely lack plastids, e.g., species of the "euglenoid," *Astasia* (Kivic and Vesk, 1974); others contain highly dedifferentiated plastids, e.g., *Prototheca ciferrii,* related to *Chlorella* (Menke and Fricke, 1962); *Polytoma papillatum* (e.g., Gaffal, 1978), related to *Chlamydomonas;* and the cryptomonad, *Chilomonas paramecium* (Sepsenwohl, 1973), all of which accumulate starch in spite of the absence of photosynthesis, and the diatom, *Nitschia alba* (Schnepf, 1969). However, new plastid forms have not evolved from such "dead-end" species as these in which mutations have brought about a permanent absence of photosynthetic pigments. Rather it is from species in which loss of photosynthetic activity is only temporary, that new types of plastid have arisen.

Though most prokaryotic algae and chloroplasts of many unicellular algae normally retain their pigments and thylakoids at all times these are nevertheless subject to minor temporary modification in response to changes in environmental conditions: e.g., in wavelength or intensity of light, in supply of nutrients or water (e.g., Lang and Whitton, 1973; McLean, 1967; Wolf and Cox, 1981). Associated with this fine-tuning of the photosynthetic apparatus to the prevailing conditions there is often a change in the quantity and variety of storage products accumulated. Under particularly severe conditions specialized thick-walled cells (e.g., the akinetes of filamentous blue-green and of green algae and other forms of cyst or resting spore) may be produced. Because of difficulty in preparation, few such dormant spores of eukaryotic algae have been examined in the electron microscope (Dodge, 1973). However, in these, the thylakoid system can (but does not always) become less extensive, polysaccharides and carotenoids may accumulate, and the rate of metabolism is reduced. When conditions improve, active growth of the dormant cell begins again, the fully functional photosynthetic apparatus is restored, and the metabolic rate increases. It may well be this capacity for temporary response to environmental conditions which has provided the basis for the more recent evolution of new plastid forms but under internal rather than external control.

B. Sites of Nonphotosynthetic Plastids

The chloroplasts of some unicellular algae regularly undergo dedifferentiation and redifferentiation of their thylakoid system (i.e., behave as proplastids) during sexual reproduction and this process, like plastid dedifferentiation itself, can be induced by nutritional deficiency (e.g., Trainor, 1970). For example, in the zygote of *Chlamydomonas reinhardtii* (Cavalier-Smith, 1976), thylakoids in the fused chloroplast derived from the two gametes first dedifferentiate but later redifferentiate as the dormant zygote germinates. The origin of the proplastid proper, as a regular feature of the plastid "life cycle," is probably associated with the induction of sexual reproduction as a normal and internally controlled part of the algal life cycle rather than (as in resting spores) as a spasmodic event controlled by the external environment. Subsequent evolution of proplastids seems to be associated with the extension of the temporary nonphotosynthetic state both backward (first in the male and later in the female gamete) and forward (in the embryo) through the algal life cycle from its origin in the dormant zygote (Whatley, 1982).

Evolution is associated with a trend toward the production of smaller male and larger female gametes. In the developing male gamete of such oogamous species, the plastids commonly lose their thylakoids, and become smaller and less numerous (e.g., Bisalputra *et al.*, 1971; Burr and West, 1970; Caceres and Robinson, 1981; Pickett-Heaps, 1975). Subsequently the paternal plastids may disintegrate completely (examples are known from red and green algae and from land plants, e.g., Braten, 1973; Duckett and Peel, 1978; Peel and Duckett, 1975; Whatley, 1982), or may be debarred from entry to or discarded from the gamete (examples are known from red, chromophyte and green algae and from land plants, e.g., Moestrup, 1970; Myles, 1978; Peel and Duckett, 1975; Wheeler and Page, 1974). Dedifferentiation of plastids during development of male gametes might be thought merely to represent an early stage in disintegration but this is not necessarily so. There is unfortunately no direct information available about the redifferentiation potential of the dedifferentiated paternal plastids in algae, though gametes which fail to participate in fertilization are sometimes able to act as spores and later germinate. Among land plants, plastid inheritance can be maternal, biparental, or essentially paternal. The dedifferentiated paternal plastids of species showing paternal or biparental plastid inheritance are obviously capable of redifferentiating their thylakoid system, i.e., they are proplastids *sensu stricto*. Even when plastid inheritance is maternal the dedifferentiated paternal plastids of some species are known to be viable (Whatley, 1982).

As the life span of the male gamete is so brief, absence of photosynthetic activity is no problem. In any case the dedifferentiated paternal plastids in green algae and the cytoplasm in red algae often contain starch, though this is not always made use of (Peel and Duckett, 1975; Whatley, 1982). In zygotes also, temporary loss of photosynthetic activity during dormancy presents no serious difficulty as the rate of metabolism is low. However, the extension both forward and backward through the life cycle of a nonphotosynthetic mode into other types of cell both prior to and following reproduction only seems to have taken place in more complex multicellular species in which adjacent photosynthetic cells can provide nutrition, e.g., the egg cell of the green alga, *Chara* (Pickett-Heaps, 1975); embryos and meristems in some red, green, and brown algae and in land plants; rhizoids in *Chara* and roots of land plants; the various floral parts in angiosperms. The plastids in many but not all of these sites are capable of redifferentiating into chloroplasts, i.e., they are proplastids, *sensu stricto*.

In land plants the proportion of the total organism given over to nonphotosynthetic cells is much greater than in any alga. However, in the multicellular red, green, and brown algae and in land plants more or less thylakoid-free plastids have evolved in many of the same types of cell, though not by any means in all species. The parallel coevolution in red, green, and brown algae of meristems with proplastids is particularly striking. As the meristematic cells differentiate so, too, do their plastids. For each class of alga or land plant successive characteristic stages of development from proplastid to mature chloroplast can be identified at increasing distances from the meristematic apex (e.g., Aghajanian and Hommersand, 1980; Dawes and Barilotti, 1969; Craig and Gibor, 1971; Evans and Holligan, 1972; Whatley, 1978).

Most proplastids retain some fragment of their main thylakoid system; in addition those of red algae tend to retain their peripheral thylakoid (e.g., Borowitzka, 1978; Bouck, 9162; Wetherbee, 1980), those of brown algae their girdle lamella (e.g., Bisalputra *et al.*, 1971; Evans, 1968; Evans and Holligan, 1972), and those of some siphonaceous green algae a portion of their terminal body (Borowitzka and Larkum, 1974), all of these being sites from which new thylakoids may develop.

In addition to cells or organs associated with or arising as a result of sexual reproduction, undifferentiated algal plastids are sometimes found in sieve elements of brown algae (e.g., Schmitz and Srivastava, 1974, 1975), and in rhizoids such as those of the green algae, *Chara* (e.g., Bold and Wynne, 1978; Sievers, 1965) or *Caulerpa* (Dawes and Rhamstine, 1967). The development of hairs, like the development of dedifferentiated plastids, seems to be promoted by nutrient deficiency (e.g., Schonbeck and Norton, 1979). Some "colorless" hairs contain nonpigmented plastids but in others plastids are absent (e.g., Pickett-Heaps, 1974).

Because different types of cell in so few species have been systematically examined in the electron microscope, the list of sites given for undifferentiated plastids or proplastids in algae is highly tentative and certainly incomplete. Among land plants the variety of sites associated with nonphotosynthetic plastids is very much greater.

C. Stages in Plastid Development

In land plants chloroplast dedifferentiation does not often take place and it has not been followed in detail in the electron microscope. However, the reverse more rapid and synchronous process of development of a proplastid into a photosynthetically mature chloroplast has been investigated in several species. A sequence of five successive developmental stages has been described for *Phaseolus vulgaris,* for example, and these are shown in Figs. 2b–e and 3 (Whatley, 1977). Sometimes this sequence is blocked, temporarily (e.g., while in the dark), or permanently (e.g., in most roots), and the plastids remain at an intermediate stage. When this happens, precursor materials may accumulate. The forms which these accumulation bodies take depend on the nature of the blockage and the position within the developmental sequence where it occurs. At other times the plastid may be diverted from its normal developmental pathway. Then different proportions of normal plastid components or completely different components may be synthesized.

In addition to these variations in the basic pathway of plastid development, there are other differences related to the position of the plastids within a plant organ. Within the leaf of *Phaseolus* or the green root of the fern, *Azolla pinnata,* for example, the plastids in each cell file follow characteristic patterns of differential development. At maturity each file has plastids which differ quantitatively (e.g., in size, extent of the thylakoid system, or amount of starch) from those of adjacent files. These quantitative differences can be considerable (Whatley, 1979; Whatley and Gunning, 1981).

It is convenient to consider changes in chloroplast structure primarily in terms of the photosynthetic apparatus, with modifications in ancillary products playing a secondary role. The mechanisms which control the photosynthetic apparatus

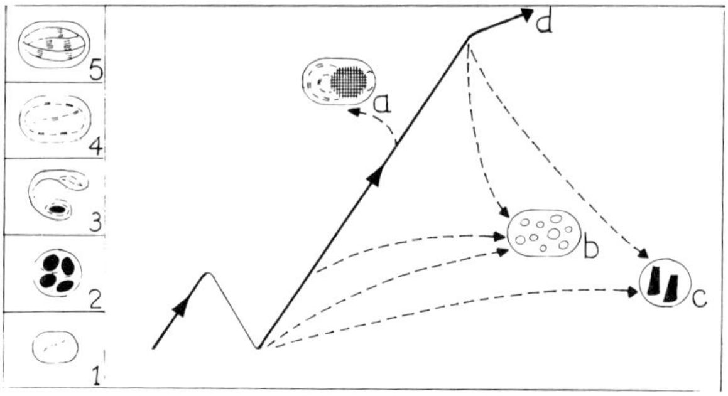

FIG. 3. The interrelationships between some different forms of plastid. Solid arrow and thick line, differentiation of the thylakoid system; thin solid line, dedifferentiation of the thylakoid system; dashed arrow, interconversion of one form of plastid to another. 1–5, Successive stages of plastid development from the simplest form of proplastid (1 = eoplast) to mature chloroplast (5). (a) Etioplast; (b) chromoplast; (c) sieve element plastid; (d) continuing development of the mature chloroplast.

and the production of ancillary materials often operate independently, i.e., similar amounts of the same ancillary material can be produced in a proplastid, a mature chloroplast or at any stage between. Many of the variations in plastid structure found in angiosperms, for example, represent the permutations and combinations which result from these two superimposed systems. The different types of plastid and the ways in which they develop have recently been reviewed by Schnepf (1980).

D. New Forms of Plastid

1. *Etioplasts (Fig. 2f)*

Most algae and many land plants are able to synthesize chlorophyll and form a thylakoid system in the dark. Exceptions among algae include *Euglena gracilis* (Klein *et al.*, 1972), *Ochromonas danica* (Gibbs, 1962), and mutants like Y-1 of *Chlamydomonas reinhardtii* (Ohad *et al.*, 1967). Among land plants, angiosperms appear to be the only group which is totally unable to synthesize chlorophyll in the dark, although in some gymnosperms this capacity has been retained only by seedlings (Kirk and Tilney-Bassett, 1978). When angiosperm seedlings are grown in darkness, the normal pathway of plastid development in leaves becomes blocked at Stage 4, i.e., before grana are formed (Whatley, 1977). Some thylakoid extension takes place, and incipient grana (bithylakoids) are formed, but the most conspicuous structural feature of these plastids (etioplasts) is a paracrystalline lattice, the prolamellar body. The prolamellar body has frequently been described in detail (summarized in Gunning and Steer, 1975; Kirk and Tilney-Bassett, 1978) and its structure will not be discussed further here. The prolamellar body is believed to represent the accumulation of precursor material which cannot be incorporated into the blocked thylakoid system. When dark-grown plants are placed in the light the blockage is removed and the etioplast is transformed into a normal chloroplast. Prolamellar bodies can also develop in the leaf plastids of many nonangiosperms.

The formation of prolamellar bodies is frequently equated with growth in darkness, but this is not strictly correct. In plants of *Phaseolus vulgaris* grown in darkness, plastid development in the hypocotyl is temporarily blocked at a stage during which phytoferritin is accumulated (Fig. 2g). A second blockage (Fig. 2h) also precedes prolamellar body formation and at this time an amorphous structure is formed (Whatley, 1977). Only after a further period of darkness do prolamellar bodies finally appear. Plastids in permanently darkened roots of *Phaseolus* never seem to form true prolamellar bodies, though a less precisely organized nonparacrystalline structure which may be related to a prolamellar body sometimes develops (Newcomb, 1967). Thus, in *Phaseolus,* there appears to be a decline in the capacity to form prolamellar bodies extending downward from the leaves toward the roots. For a few species (e.g., *Secale cereale*) an even greater divergence from the accepted pattern of prolamellar body formation has been described in the plastids of roots (Badenhuizen and Salema, 1967; Salema, 1971). In these plastids, prolamellar body-like structures develop, but only after prolonged exposure of the roots to *light;* subsequently these plastids (unlike those in roots of most other angiosperm species) develop into fully functional chloroplasts. Although we know that different forms of plastid (such as etioplasts) can be produced when normal development is blocked, as yet we know nothing about the different ways in which this is controlled or what causes the different patterns of behavior in different types of cell.

2. *Proteinoplasts (Fig. 4a)*

The plastids of most land plants and of some of the more complex algae lack pyrenoids. However, a new form, perhaps equivalent, of protein storage appears to be present in the plastids of some angiosperm species. These proteinoplasts have been observed in many different types of cell and in plastids at various stages of development, e.g., meristematic cells, differentiating cells, and epidermal and guard cells, and in mature photosynthetic cells (summarized in Thomson and Whatley, 1980). The accumulated material, which stains positively for protein, is concentrated in a membrane-bound body which can be amorphous or crystalline. In mature chloroplasts this body has been shown to contain Fraction I protein (Sprey and Lambert, 1977). In differentiating cells diversion of plastids to proteinoplasts takes place at the time when grana are first being formed and is often accompanied by an unusual staining response (Platt-Aloia and Thomson, 1977).

3. *Amyloplasts and Other Storage Plastids (Fig. 2d)*

The capacity to store starch in the plastid stroma has been lost by all classes except the green algae and land plants. In these starch is not only accumulated on

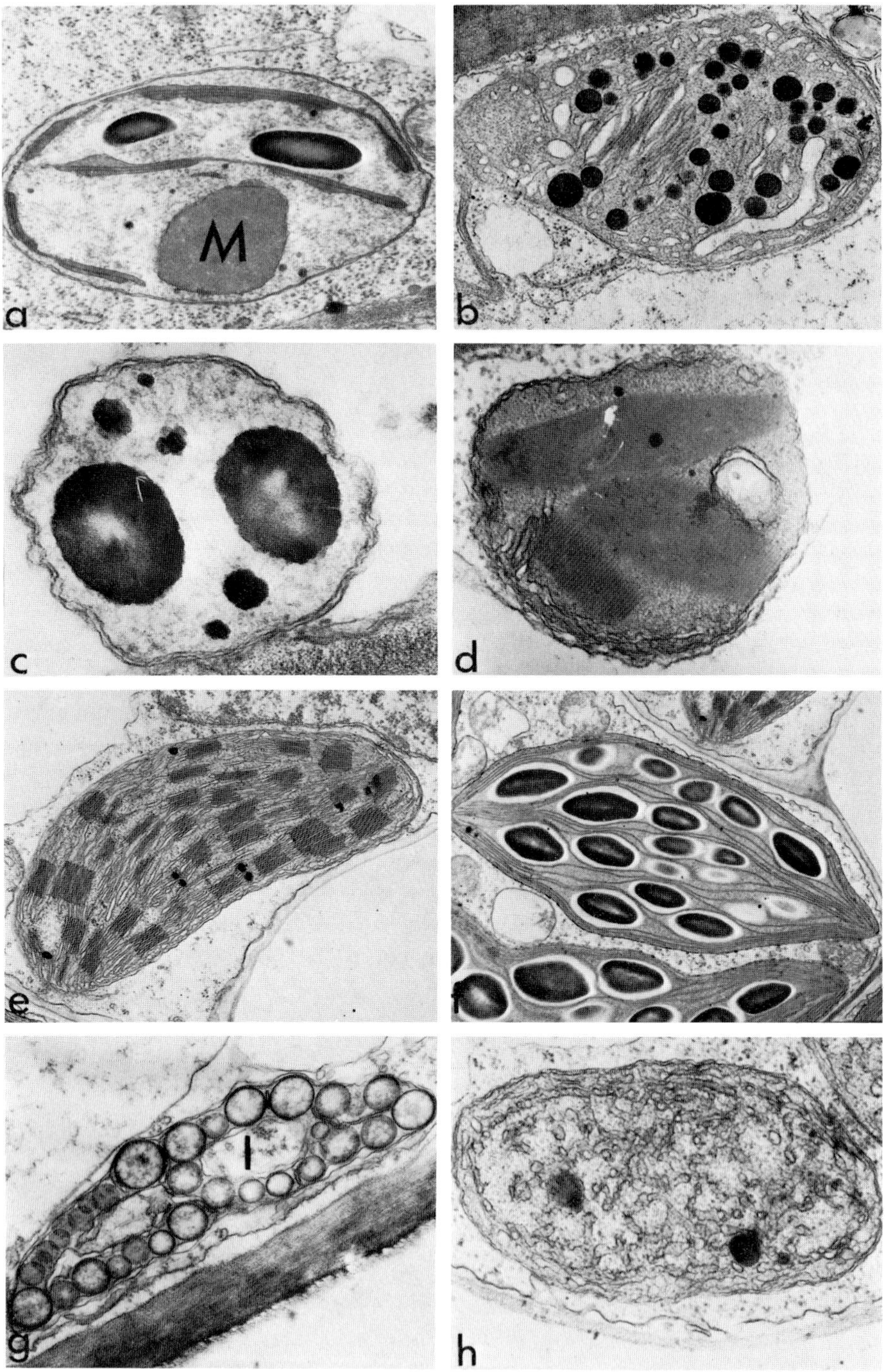
a
M
b
c
d
e
f
g
I
h

a transient daily basis inside mature chloroplasts in direct response to photosynthetic activity (Fig. 2a and b), but it can also accumulate on a more permanent basis in some specialized cells and as a regular process during the normal course of plastid development (Fig. 2d). In some coenocytic algae (members of the Caulerpales) chloroplasts and amyloplasts are, unusually, found together in the same cell (e.g., Dawes and Barilotti, 1969). In land plants amyloplasts are, for example, found in guard cells, in some storage organs, and in cells of the root cap where they act as statoliths, organs of gravity perception (Kirk and Tilney-Bassett, 1978). In such amyloplasts the thylakoid system is often poorly developed. During plastid development there is commonly at least one brief phase of starch accumulation and degradation immediately prior to the onset of the rapid extension of the thylakoid system (Senser *et al.*, 1975; Whatley, 1977). As starch accumulates the plastids are, for a short time, partly encircled by endoplasmic reticulum.

In at least one angiosperm, *Cecropia peltata*, polysaccharides are accumulated not as starch, but as phytoglycogen (Rickson, 1976). Specialized structures (Müllerian bodies) develop which are used by ants as a source of food. It is within these that the plastids accumulate rosettes of glycogen. Elsewhere in the plant, the plastids accumulate starch in the normal way.

The plastids of some angiosperms accumulate oil instead of starch. These elaioplasts have, for example, been found in photosynthetic cells of some cacti (Thomson and Platt, 1973). The plastids of some conifers, e.g., *Pinus pinea*, contain osmiophilic droplets which are thought to contain resin precursors which may later be exported to the cell cytoplasm (Wooding and Northcote, 1965). Osmiophilic droplets are also common in plastids of excretory glands and the tapetum of angiosperms and are believed to represent lipophilic materials, e.g., oils, resins, and "pollenkitt" which are later exported. These plastids are again commonly partially surrounded by endoplasmic reticulum (e.g., Dumas, 1974; Hesse, 1978).

FIG. 4. (a) Proteinoplast in a young leaf of *Impatiens* sp. The thylakoids show a reverse staining image, a reaction sometimes found at the stage of development when grana and the membrane-bound protein body (M) are first formed. ×25,000. (b) A plastid from the bundle sheath of a CAM plant, the gymnosperm, *Welwitschia*. This plastid has a form of peripheral reticulum. ×18,000. (c) Plastid from a sieve element of *Phaseolus*. It lacks thylakoids and ribosomes but contains sieve element starch. ×44,000. (d) Plastid from a sieve element of *Zea mays*. It contains proteinaceous crystalline deposits. ×43,000. (e) Mesophyll chloroplast from the C-4 plant *Zea mays* contains grana but lacks starch. ×12,000. (f) Bundle sheath chloroplast from *Zea* has single thylakoids and many starch granules but no grana. The peripheral reticulum is inconspicuous. ×8,000. (g) Chromoplast from the corona of *Narcissus* contains many large globules and is pleomorphic: note the cytoplasmic invagination (I). ×17,500. (h) Although it lacks carotenoids this plastid from the white petal of *Phaseolus* contains tubular structures similar to those in some true chromoplasts. This *Phaseolus* plastid developed from a mature chloroplast following dedifferentiation of the thylakoid system. ×35,000.

4. *Sieve Element Plastids (Fig. 4c and d)*

The plastids in the developing sieve elements of some brown algae may become pleomorphic and undergo some dedifferentiation; at maturity they contain few thylakoids, but, unlike the plastids in sieve elements of land plants, they appear to undergo no further modification (Schmitz and Srivastava, 1974, 1975). In the functionally similar sieve elements of land plants the plastid thylakoid system dedifferentiates, ribosomes appear to be lost, and normal starch is not produced (e.g., Perry and Evert, 1975; Warmbrodt and Evert, 1979; Thomson and Whatley, 1980). Sieve element plastid storage products include sieve element starch and proteinaceous material which can be deposited as crystals or in filamentous arrays, but this protein is never surrounded by a membrane as it is in proteinoplasts (Fig. 4d). In some angiosperm species the storage materials of sieve element plastids have been so characteristically modified that Behnke (1972, 1975) has been able to use these differences in relation to angiosperm systematics.

5. *Bundle Sheath Plastids of C-4 Plants (Fig. 4e and f)*

One distinctive innovation in plastid structure affects only those plastids which are present in the bundle sheath cells of some angiosperm species. During photosynthesis these plants produce organic acids with four carbon atoms as their first recognizable products of CO_2 fixation. These plants are commonly referred to as C-4 plants, in contrast to most other species which first produce compounds with three carbon atoms—C-3 plants. C-4 plants are of three types, depending on whether the primary product of photosynthesis is oxalacetate, malate, or aspartate. In C-4 plants the mesophyll cells have chloroplasts with a normal thylakoid system though they usually lack starch (Fig. 4e). The bundle sheath plastids initially develop in much the same way as mesophyll plastids, but, in some C-4 species, e.g., the grass maize (*Zea mays*), as the cells become fully differentiated, the grana essentially disappear, leaving the large bundle sheath plastids with only single thylakoids (Fig. 4f). These are the only mature ''green'' chloroplasts in which single continuous thylakoids are found. In other forms of C-4 plant (e.g., the dicotyledon, *Portulaca oleracea*) the grana are retained (Laetsch, 1974).

All C-4 bundle sheath chloroplasts accumulate starch and have at their periphery, and developed to varying extents, a system of anastomosing membranous tubules which may link the thylakoid system to the plastid envelope. This membranous system is called the peripheral reticulum.

The families in which C-4 metabolism is found are phylogenetically diverse and include both monocotyledons and dicotyledons; species which carry out either C-3 or C-4 metabolism can be found within the same genus, e.g., *Atriplex patula* (C-3) and *A. rosea* (C-4). During development or in response to external factors some species can change from one mode of operation to the other.

6. *CAM Plants (Fig. 4b)*

Another variation of the normal mode of photosynthesis is exhibited by plants which carry out crassulacean acid metabolism (CAM plants). CAM plants are able to store CO_2 temporarily as organic acid and release it later for use in photosynthesis. This capacity is of advantage in areas where the rate of transpiration is high; guard cells are kept closed during the day when potential water loss is great, but open to take in CO_2 during the night. The stored CO_2 is released when it becomes light and photosynthesis can begin. Most CAM plants are angiosperms but this form of metabolism has also been found in the primitive vascular plant genus *Isoetes* (Keeley, 1981) and in the gymnosperm *Welwitschia* (Dittrich, 1974).

The chloroplasts of CAM plants are not particularly distinctive in structure though they often have a peripheral reticulum of sorts (Fig. 4b). However, a similar poorly developed peripheral reticulum can sometimes also be found in species which show neither CAM nor C-4 activity, e.g., in plastids in epidermal cells of some species.

7. *Chromoplasts (Fig. 4g and h)*

Clusters or layers of carotenoid globules are found in the eyespot-containing plastids of many motile algal cells. In the less complex species of algae the eyespot plastid retains the normal complement of thylakoids. In more complex algae the eyespot is usually found only in the male gamete, and then often in a plastid which has undergone dedifferentiation. In such cells (e.g., the sperm of *Fucus*) the carotenoid globules may be the only structural feature of note in the plastid (Bouck, 1970). In some morphologically complex algae such as *Chara* no eyespot is present at any stage of the life cycle.

Chromoplasts can also contain carotenoid globules. Among algae they are found only in some advanced "green" genera such as *Chara* but they are common in land plants. Chromoplasts differ from eyespots in that they are not specifically associated with motile cells and the accumulation of carotenoids is not restricted to a single plastid but is characteristic of all plastids in the cell.

The chromoplasts of *Chara* are primitive. They retain their thylakoid system and the carotenoids are concentrated in globules often arranged in layers. In *Chara* chromoplasts are found in a layer of sterile cells (shield cells) which surround the antheridium (Pickett-Heaps, 1975). In bryophytes structurally similar chromoplasts are also found in a sterile jacket surrounding the antheridium (e.g., Duckett, 1975). "Primitive" chromoplasts with a thylakoid system and globular carotenoids are also found in the winter leaves of some ferns and gymnosperms, e.g., *Selaginella, Cryptomeria,* and *Ephedra*. Carotenoids again accumulate as globules in the leaf chloroplasts of many deciduous species of angiosperms and some gymnosperms but the plastids in these senescing leaves are not usually considered to be true chromoplasts.

The chromoplasts of angiosperms usually appear to be structurally more advanced. They commonly lack chlorophyll and are varied in the composition and structural form of their carotenoids, some of which are not found in other types of plants, e.g., lycopene and capsanthin. Sitte (1977) has distinguished several forms of angiosperm chromoplast on the basis of the structural organization of their carotenoids. The carotenoids can be assembled in globules (as in algae or lower land plants), in tubular, fibrillar, or crystalline form, or in association with membranes. Sitte has correlated these differences in structural organization with the ratio of semipolar to nonpolar components. A few membranes are commonly present in most angiosperm chromoplasts but these membranes are not photosynthetic thylakoids and they have a three-dimensional arrangement which is very different from that of the normal thylakoid system.

Chromoplasts in angiosperms can develop from several other types of plastid—from eoplasts, amyloplasts, and, following dedifferentiation, from mature chloroplasts (Thomson and Whatley, 1980). Sometimes chromoplasts are also able to redifferentiate into chloroplasts, e.g, in ripe oranges (*Citrus sinensis*) subjected to low temperatures and in the spadices of several members of the Araceae (Thomson, 1966; Thomson *et al.,* 1967., Grönegress, 1974; Whatley, 1978). Chromoplasts are most commonly found in organs associated with reproduction, e.g., in bracts, sepals, and petals of flowers and in fruits, but they are also found in some roots, e.g., carrot and sweet potato. Chromoplasts of different carotenoid content and structure can develop in different organs of the same plant, e.g., petals may have crystalline chromoplasts and fruits the globular form. However the evolution of chromoplasts in flowers and fruits seems to be linked with the evolution of animals, the bright reds, oranges, and yellows of the chromoplasts acting as attractants particularly to insects, which disperse pollen and seed while gathering food.

E. Summary

Eukaryotic algae are immensely diverse in the extent of variation in their body structure, in their methods of reproduction (both sexual and asexual), and in the complexity of their life cycles. They range from simple unicells and colonies through filamentous forms to complex, multicellular parenchymatous species with specialized organs. A single alga may have several different types of body structure which develop at different stages during a complicated life cycle. Indeed it is not uncommon for one "genus" of alga to be recognized later as being only a stage during the life cycle of another. Among multicellular eukaryotic algae, therefore, the potential for plastid variation between different types of cell but within the same organism would appear to be very great. Yet this is not so. For the vast majority of eukaryotic algae the basic chloroplast is the form in which the plastid is present in most cells throughout most of the life cycle.

The principal modification which has taken place in eukaryotic algae has been the evolution of the proplastid, an event which may have begun as a direct response to adverse environmental conditions and later have been brought under internal control and closely linked to the evolution of sexual reproduction. The extension of a proplastid mode, especially forward through the life cycle in the embryo and in meristems, was only possible in those algae with a large enough body of photosynthetic tissue to support a population of nonphotosynthetic cells. Among algae, extension of a proplastid mode seems to have proceeded least in the browns and farthest in the green, the only class in which further major modifications in plastid structure have taken place.

The evolution of algae which lived in water, into plants which were able to colonize the adverse terrestrial environment was accompanied by further considerable modification in body structure. It might again be thought that the stimuli so provided would result in the rapid emergence of some different types of plastid in the newly evolving specialized organs, but, once more, this is not so. Some of the plastid variations associated with land plants are also found in algae (e.g., the dedifferentiated plastids in the rhizoids and the chromoplasts in the antheridial jacket of *Chara*). Many of the remainder appear only in angiosperms, without any obvious preliminary "attempts" in more primitive classes of land plant. Indeed the obviously polyphyletic C-4 syndrome, for example, is not even found in primitive angiosperms. Only the specialized sieve element plastids and the structurally rather undistinguished plastids associated with CAM metabolism are found in lower plants as well as in angiosperms. Of the distinctive plastids in angiosperms only some sieve element plastids appear to fall into recognizable patterns related to angiosperm systematics. In spite of the long evolutionary history of land plants, and the many apparent stimuli for change, most variations in plastid structure have been quantitative only; the number of real *innovations* which have taken place have been remarkably few.

IV. Control of Symbiont and Organelle Populations

As more is learned about the interrelationships between organisms, it becomes clear that a continuum exists in the states of integration between associated pairs of species which extends between the two extremes of host and prey on the one hand and cell and chloroplast-like organelle on the other (Smith, 1979). The sequence as given here in Fig. 5 is not meant to imply (1) that intermediate stages necessarily follow in the order given or last for a precise length of time; (2) that all symbiotic partnerships will one day complete the progression; or (3) that the intermediate stages represent a notably less efficient partnership than that of cell and organelle. Along this continuum it is difficult to find a single feature which clearly distinguishes between any of the adjacent states of partnership (e.g., between blue-green algal endosymbionts and cyanelles or between cyanelles and red algal chloroplasts) (Fig. 5). Many of the methods used to control these

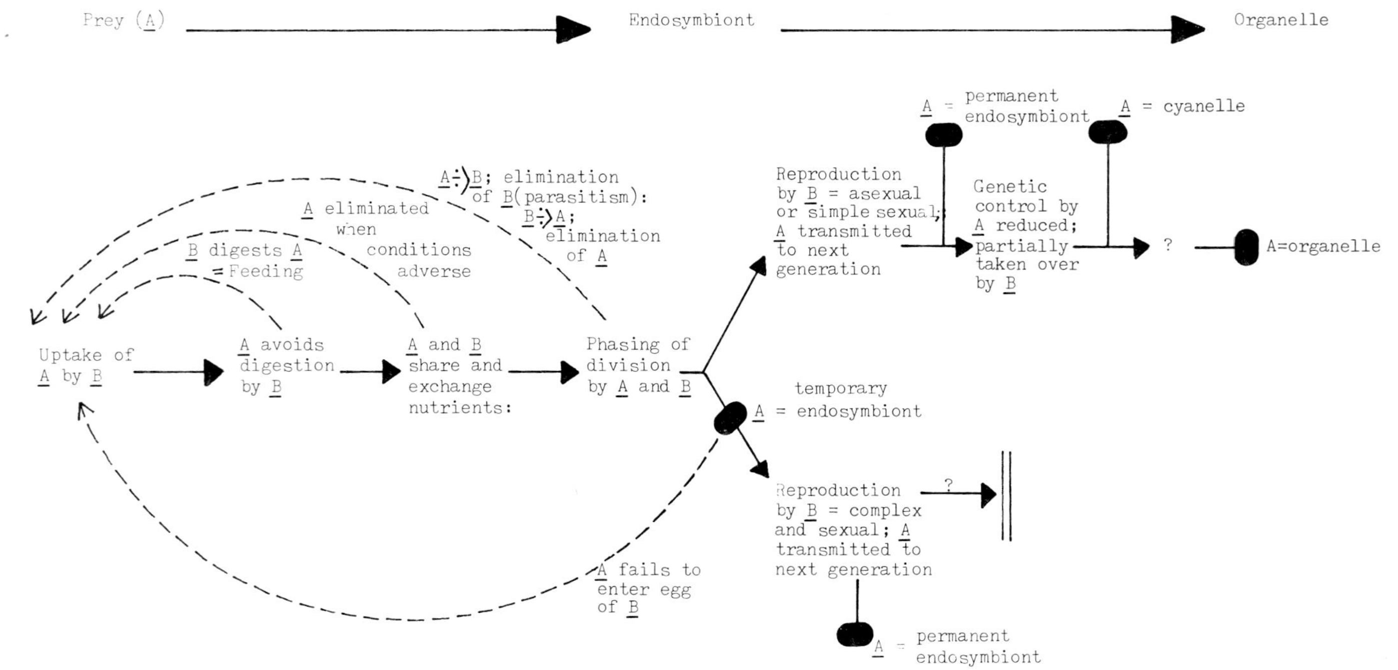

FIG. 5. Successive steps involved in the association between pairs of species which may have led from an initial relationship between prey (A) and its host (B) to a state of complete integration between a cell (B) and its symbiotically derived organelles (A).

partnerships are common to all. For example, the mechanisms used to control the size of a chloroplast population within a cell are essentially the same as those used to control endosymbiont populations and these same mechanisms may be used to deal with prey.

The population of endosymbionts in a cell is usually in a state of dynamic equilibrium. Maintenance of a stable relationship between symbionts and host is accomplished in three ways: (1) digestion or ejection of excess symbionts; (2) uptake of new symbionts; and (3) control of the rate of division of both partners by nutritional or other means (Muscatine and Pool, 1979). Several of these same methods are also used to control the plastid population of a cell and come into play most prominently during sexual reproduction and thereby influence patterns of plastid inheritance. The precise control mechanism employed varies from species to species and from one type of cell to another within the same organism.

A. Digestion

When a cell takes up prey into a phagocytic vacuole, digestion commonly takes place directly inside that vacuole, often at a particular locality within the cell. Endosymbionts lying inside the vacuole in which they were ingested tend to become concentrated in those parts of the cell where digestion does not normally take place, e.g., the cell periphery in *Paramecium* or the subepidermal region in *Convoluta*. Direct digestion of healthy endosymbionts is rare, as is degradation of functional chloroplasts. However, paternal plastids in the fertilized egg of some primitive green algae like *Ulva mutabilis* (Braten, 1973, Whatley, 1982) are known to degenerate *in situ*.

The other form of digestion—an indirect form—is autophagy. Prey which is able to enter the cytoplasm of a cell escapes immediate digestion but may subsequently be encompassed by and then digested inside an autophagic vacuole (Bannister, 1979). Trench (1980) has suggested that the isolated chloroplast symbionts of the saccoglossan mollusc, *Elysia,* may escape from their phagocytic vacuoles into the host cytoplasm but may subsequently be digested in autophagic vacuoles. Plastids, too, may sometimes be digested following autophagy, but only in a few specialized types of cell, and the process seldom involves the specific selection of plastids only. For example, the maternal plastids in the egg cells of some conifers (which have an essentially paternal pattern of plastid inheritance) are apparently removed in this way (e.g., Chesnoy and Thomas, 1971; Gianordoli, 1973).

B. Ejection

Ejection of healthy symbionts from host cells has been reported for sea anemones and *Hydra* (Thorington and Margulis, 1980; Trench, 1979). Plastids, too, may be ejected from cells but this is unusual and again nonspecific. It is more common for plastids to be eliminated by being "left behind" or debarred from entry into daughter cells. The only examples of true ejection of which I am aware take place during sexual reproduction in some lower land plants—bryophytes and ferns. During spermatogenesis in bryophytes the single plastid migrates to the base of the cell. As the mature sperm moves toward the egg it discards a vesicle derived from the basal cytoplasm and containing the plastid (most recently reviewed in Whatley, 1982). In ferns the process is even less selective; some mitochondria as well as several plastids are included in the discarded vesicle.

C. Uptake

In many partnerships the endosymbiont is lost during sexual reproduction and new symbionts must be taken up by the new generation of hosts. Other endosymbiotic partnerships are of shorter duration and uptake must occur repeatedly as with the chloroplast symbionts in saccoglossan molluscs. For plastids (and cyanelles) the partnership is permanent and uptake of new organelles is unnecessary. In any case most plant cells are enclosed within secondary walls and are incapable of phagocytosis. "Alien" plastids are, however, sometimes taken up into "host" cells during sexual reproduction.

Plastid inheritance is most commonly uniparental (usually maternal where this can be determined); paternal plastids usually fail to enter the egg. Nevertheless paternal plastids are occasionally successful in invading the egg cell, though only when plastid inheritance is biparental or essentially paternal do the "alien" plastids survive there. The method of uptake of paternal plastids is again nonspecific for the entire contents of the male gamete are ingested. "Uptake" seems most commonly to be by conventional plasmogamy and therefore is unlike normal endocytosis. However in a few angiosperms (e.g., in the genus *Plumbago*) the male gametes (including any plastids which are present) may be injected directly into the egg (Cass and Karas, 1974), a form of entry which resembles the method of "direct invasion" described by Bannister (1979) for microsporeans. In some gymnosperms (cycads and conifers), too, fertilization may possibly follow direct penetration by the pollen tube, but the components which enter the egg include not only the functional male gamete and its plastids but also other pollen tube contents (e.g., Chamberlain, 1919; Chesnoy and Thomas, 1971; Whatley, 1982). However, the way in which fertilization takes place in these gymnosperms requires further investigation.

There is yet one other occasion when a form of endosymbiosis takes place in plants and one cell is totally engulfed by another. The uninucleate pollen grain of an angiosperm divides to form a large vegetative cell and a small generative cell (which includes plastids when inheritance is biparental). The vegetative cell subsequently engulfs the generative cell which later divides to form the two male gametes. (A somewhat similar process also takes place in gymnosperms.) The engulfed generative cell or, later, the two gametes are separated from the "host" vegetative cell by two membranes—their own plasma membrane and that derived by invagination from the "host" (Whatley, 1982). The generative cell and the two gametes may thus be regarded as eukaryotic endosymbionts which have been taken up by the vegetative cell in the same way that chlorellae have been taken up as eukaryotic symbionts of *Hydra*. However, the angiosperm gametic symbionts belong to the same species as the "host," indeed to the same individual.

D. Division

The most effective method of population control for both symbionts and plastids is through the division process. To achieve a balance within the cell not only the rates but also the times of division of both the host cell and the symbionts or plastids must be closely integrated. An effective means of distributing symbionts or plastids to daughter cells is also required, and for this their position within the cell is often important. Finally the two integrated systems of division must be adjusted to suit prevailing environmental conditions—e.g., the nutrient supply, light intensity etc. The cell assumes overall control of all these processes, both for itself and on behalf of the plastids or symbionts.

It has often been observed that the numbers of symbionts maintained by a particular species of host tend to be constant. McAuley (1981) working with five strains of *Hydra* cultured under the same conditions found that in some strains the algal symbionts were numerous but small; in others they were fewer but larger. Further analysis showed that in all five strains the algae occupied about 13% of the total host volume (as measured by the amount of protein). Although equivalent measurements for plastids are hard to come by, and precise information is not available, it is nevertheless clear that the main photosynthetic cells of a plant each contain numbers of chloroplasts characteristic of the species. When the chloroplasts are small, their numbers are large and when the chloroplasts are larger, their numbers are smaller. The very small dedifferentiated plastids of gametes, embryonic cells, and meristems are usually few in number, but they may still occupy a proportion of the total cytoplasm of these small cells which is similar to that occupied by chloroplasts in the large but vacuolated photosynthetic cells. Attempts have been made to correlate the plastid complement with the cell surface area, the cell volume, and the total amount of nuclear DNA (e.g., Butterfass, 1980; Honda *et al.*, 1971; Whatley, 1980). Though no wholly suitable means of measurement has yet been applied, the impression remains that there is a direct relationship between the volume of cytoplasm and the volume of the plastids.

When there is only a single symbiont or plastid in each cell, its rate and time of division must be closely synchronized with that of host cell division. Such synchrony is customary in many of the less complex algae. The dinoflagellate, *Peridinium balticum,* has a single chrysophyte algal symbiont and the nuclei of the two organisms divide together (Tomas *et al.*, 1973). In species which have several symbionts or plastids in each cell precisely phased division is less essential. In the multinucleate giant amoeba, *Pelomyxa palustris,* for example, division of one group of endosymbiotic bacteria appears to reach a maximum just prior to cell division (Chapman-Andresen, 1971). In *Hydra,* too, division of the symbiotic algae and the host is closely linked (Pardy, 1974; McAuley, 1981; Cook, this volume). In land plants the usually numerous plastids of a leaf cell do not all divide together. However, the overall period during which plastid division takes place more or less coincides with the period of cell division.

Meristematic leaf cells of the lower vascular plant, *Isoetes lacustris,* contain only one plastid and so a well-regulated system for plastid division and distribution is important. Plastids usually divide during prophase of nuclear division and then migrate to opposite poles of the cell (Whatley, 1974). Fully differentiated leaf cells of *Isoetes* may contain up to eight chloroplasts, i.e., plastid division continues for two or three cycles after cell division has ceased. In the primary leaf primordia in the angiosperm, *Phaseolus vulgaris,* division of the comparatively few plastids in each cell (about 10) more or less keeps pace with cell division. However, as the leaf expands following germination the rate of plastid division exceeds that of cell division and division of the developing chloroplasts exceeds that of the cells by about two cycles (Whatley, 1980). This excess may perhaps be linked to increased nuclear ploidy during leaf maturation. In *Cyanophora paradoxa* cyanelle division may also outstrip cell division (Trench *et al.*, 1978).

E. Position

The continuity of the symbiont–host or the plastid–host partnerships and the proportion of symbionts or plastids contributed to each of the daughter cells can be influenced by the position of the symbionts or plastids within the mother cell. Continuity and equal distribution are both promoted when symbionts or plastids closely encircle the mother cell nucleus. This arrangement of the bacterial symbionts is found in the amoeba, *Pelomyxa palustris,* for example (Whatley, 1976) and there is a similar arrangement of chloroplasts in many algae and lower land plants. The egg cells of many angiosperms contain comparatively few plastids and these often encircle the nucleus (e.g., Jensen, 1967); the numerous chloroplasts of more fully differentiated cells do not.

Sometimes plastids are unequally distributed to two daughter cells which later undergo divergent differentiation. In the mother cell the plastids tend to migrate through the cytoplasm and become concentrated toward one pole; this pattern of behavior is usually followed by asymmetrical cell division, e.g., during the formation of hairs in the green alga, *Bulbochaete* (Pickett-Heaps, 1974), and during pollen development in the angiosperm, *Phaseolus vulgaris* (Whatley, 1982). Before the mother cell divides the plastids migrate to a portion of the cytoplasm which will not be incorporated in the future smaller daughter cell, i.e., in the hair or in the pollen generative cell. The megaspore mother cell of the gymnosperm, *Ginkgo biloba,* undergoes two successive divisions to form a linear tetrad of spores; the plastids meanwhile have migrated in such a way that they become concentrated in the basal (i.e., the only functional) spore (Stewart and Gifford, 1967). By contrast, during microsporogenesis, the plastids are divided equally among the daughter cells (Wolniak, 1976).

Occasionally the distribution of symbionts or plastids to daughter cells seems to be assisted by "tying" them to the nuclear membrane. Thus, prior to cytokinesis in *Pelomyxa,* one group of endosymbiotic bacteria first takes up a position close to the nuclei and, later, their vacuolar membranes become linked to the nuclear envelope by membranous tubules which are, in effect, endoplasmic reticulum (Whatley, 1976). Each daughter amoeba is thus ensured its own complement of bacteria. In several classes of chromophyte algae the complex chloroplasts, too, are linked to the nuclear envelope by endoplasmic reticulum; in those species in which the nuclear envelope persists during mitosis retention of the chloroplasts must again be promoted in the progeny (reviewed by Gibbs, 1981a). Again the parallelism between plastids and endosymbionts is very striking.

V. Symbiosis and the Environment

The establishment of new endosymbiotic partnerships and the evolution of new types of organelle appears to be closely linked to the problem of diminishing food resources, a problem (assuming the existence of a primeval "soup") for which the organisms themselves were largely responsible. As one food supply became exhausted, new resources were exploited: as organisms in turn became modified to cope with new forms of metabolism or to extend the area which they were able to exploit, so the environment was changed even more. (Fig. 6). The establishment of a symbiotic relationship between two organisms which had undergone differential evolution could have led in a single evolutionary step to an expansion in the sources of food or range of habitat exploited. However, as organisms became more complex and the environment underwent further modification, both the forms of symbiotic partnership and the functions for which a host cell "required" a symbiont would be altered. For all those endosymbionts which

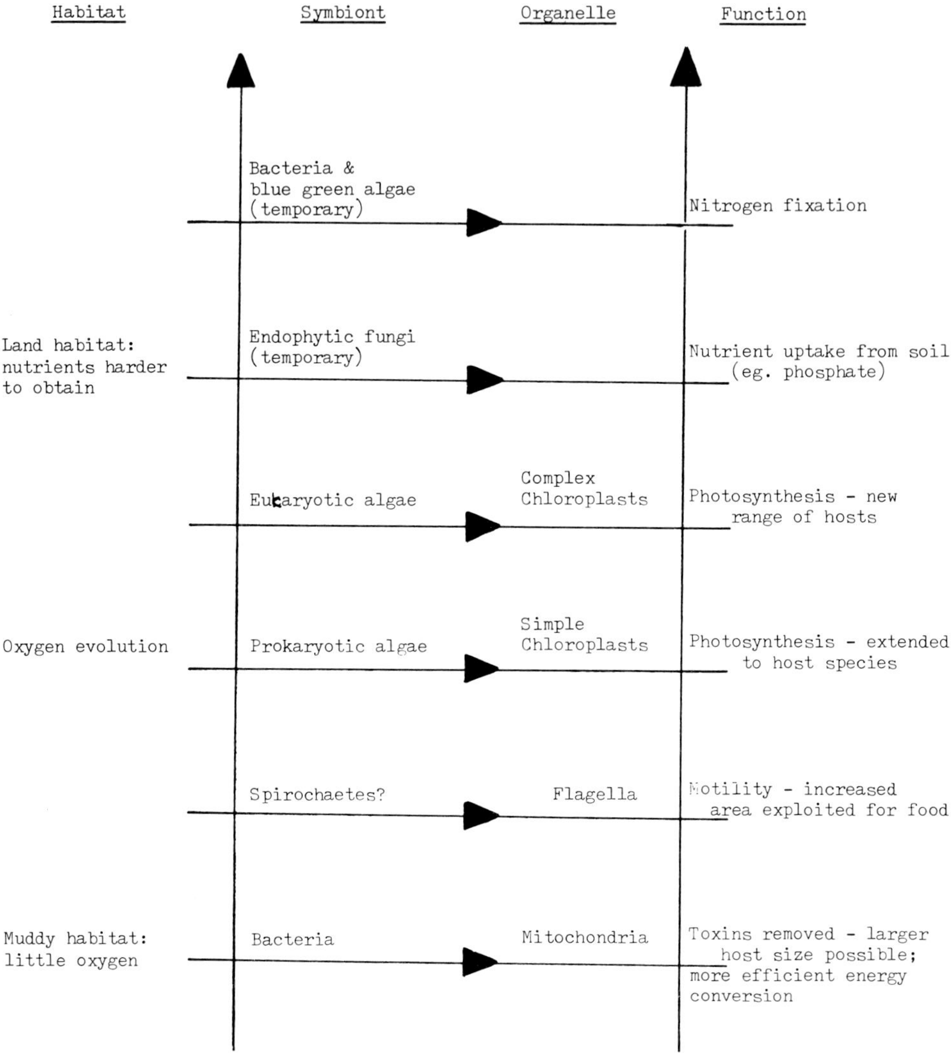

FIG. 6. The "requirements" of host cells have changed as the environment has changed and this is reflected in the functions in which symbionts have specialized.

have been proposed as precursors of organelles, the initial hosts were unicells. The subsequent evolution of these hosts into the complex plants and animals of today took place in conjunction with organelle evolution, but only after the organelles had already been integrated into the host life cycles. Many present day endosymbionts are associated with multicellular hosts which have already evolved a highly complex body structure and a complicated reproductive system. It seems unlikely that symbionts in such large and complex species will ever evolve into organelles like plastids, which are to be found in virtually all the cells within one plant and are regularly transmitted to the next generation during reproduction. Only those symbionts which are nowadays associated with unicellular hosts would appear to have any real potential for evolving in the future into true organelles.

A. Use of Resources

The evolution by host cells of a capacity for phagocytosis not only provided directly a new source of food but also led indirectly to more efficient exploitation of other food sources. During the course of evolution the functions of symbionts have changed in response to changes in their hosts and in the environment (Fig. 6). If one takes the most extreme view of the endosymbiotic theory, this progression would include (1) the removal of toxic waste allowing an increase in host size, and, later, more efficient energy conversion (bacteria → mitochondria), (2) provision of a wider area from which to obtain food (spirochaetes → flagella), and (3) provision of a new and continuing food source, using photosynthesis (algae → chloroplasts). Subsequent associations between symbionts and hosts have not led to the development of new forms of organelle but have continued to bring about new ways of making use of the changing environment. Indeed colonization of the adverse terrestrial environment by plants may only have been possible as a result of their symbiotic cooperation with fungi (Pyrozinski and Mulloch, 1975). For example, lichens are the first colonizers of bare rock habitats and primitive vascular plants like *Psilotum* and *Lycopodium* do not normally grow unless they are associated with endophytic fungi. Today fungi in the form of mycorrhiza promote the uptake of phosphorus and other nutrients by the roots of many if not all vascular plants.

The pattern of progressive modification of hosts and their symbionts is reflected in the changing forms and activities of photosynthetic prokaryotes and plastids. Early photosynthetic bacteria may have lived in an anaerobic, muddy habitat in which most of the penetrating light would have been of higher wavelengths. The evolution in prokaryotic algae of a new system of photosynthesis using different pigments not only released oxygen to the environment but may also have allowed additional wavelengths of light to be used in a now presumably less murky habitat. Uptake by various hosts of a range of different algal symbionts extended the range of species carrying out photosynthesis. Between them, the various photosynthetic pigments of the marine algae make use of virtually the entire spectrum of light which penetrates water. By contrast the light used directly in photosynthesis by land plants is restricted to only a small part of the visible spectrum; additional wavelengths are absorbed by carotenoids which can act as antenna pigments and transfer energy for indirect use in photosynthesis. Even so only part of the potential spectrum is used, though some of the remainder is made use of for other purposes. Does this reflect the comparatively recent colonization of the land? It certainly emphasizes the fact that the land was colonized only by species derived from green algae, and, in these land plants, no new photosynthetic pigments have evolved. However, land plant evolution generally has been accompanied by a decline in the proportion of tissue given over to photosynthesis. For example a large part of the biomass of a grass may be made up of nonphotosynthetic roots, and a tree, particularly one in a tropical rain forest, may have the entire burden of photosynthesis falling on a comparatively small number of leaves. The switch in emphasis from photosynthetically active chloroplasts to non-green plastids may represent a saving in energy which might otherwise have been used in the unnecessary formation of a thylakoid system and may result from a change in function in response to the evolution of land-oriented organs (e.g., roots) or to specific environmental factors.

B. The Evolutionary Future of Photosynthetic Symbionts

Ancient symbioses have led directly to the evolution of comparatively few different types of completely integrated, photosynthetically functional chloroplasts, in spite of the apparent opportunities offered to photosynthetic symbionts by a multiplicity of simple unicellular phagocytic hosts and strong environmental pressures. Today photosynthesis is widespread and commonplace and most potential hosts are multicellular and complex. The best host prospects for producing new forms of chloroplast from symbionts would, as in the past, appear to be (1) unicells (like *Cyanophora paradoxa*) which already contain cyanelles; (2) nonphotosynthetic unicells like some non-green euglenoids and dinoflagellates and many protozoa which are capable of phagocytosis; and (3) phagocytic unicells the primary plastids of which (the eyespots) are photosynthetically ineffective (like *Peridinium balticum* and its chrysophyte algal symbionts). However, as suggested in Section IV, the progression from prey to organelle seldom seems to be successfully completed. Nevertheless some present day complex host species have evolved a form of symbiotic partnership with photosynthetic symbionts which may point the way toward a new evolutionary trend in keeping with the modern land environment.

All or almost all symbiotic associations with complex host species are temporary and many are extracellular. For these the greatest barrier to the permanent retention of a symbiont population seems to be sexual reproduction by the host (Fig. 5), the symbiont being unable to gain access to the egg cell. However, the released symbionts are sometimes retained at sites immediately outside the host but readily available for reinfection, e.g., on the outer surface of the egg of *Hydra* or on specialized hairs of the aquatic fern, *Azolla* (Thorington and Margulis, 1980; Sprent, 1979). Inside the host the symbionts are commonly concentrated in compartments close to the zone of infection. In plants these compartments are frequently, but not always, extracellular and are, in effect, culture chambers for the symbionts. Formation of the chambers often results from meristematic activity of the host induced by symbiont entry and the host cells surrounding the compartments sometimes develop as transfer cells (e.g., in the bryophytes, *Blasia pusilla* and *Anthoceros* spp.) (Duckett and Prasad, 1974; Duckett *et al.*, 1975, 1977). The duration of the subsequent symbiosis may be brief (as between nitrogen-fixing bacteria and some legumes) or prolonged (as between nitrogen-fixing actinomycetes and some woody perennials).

In the land environment a major plant requirement is for nitrogen and nowadays several types of plant symbioses are directed toward supplementing the comparatively short supply available from the soil. A number of different classes of prokaryote are able to fix atmospheric nitrogen and some of these are found in symbiotic association with plants. The most common of these nitrogen-fixing symbionts become concentrated in nodules of plant roots, e.g., bacteria, *Rhizobium* spp., almost always in legumes (Verma and Long, this volume) and actinomycetes, *Frankia* spp., in woody perennials like *Alnus glutinosa, Casuarina cunninghamii,* and *Myrica gale*. However, filamentous blue-green algae, usually *Nostoc* or *Anabaena* spp., also form nitrogen-fixing symbiotic partnerships, e.g., in lichens and with a few species of phylogenetically diverse land plants.

Some lichens are an association between a fungus (usually an ascomycete) and a blue-green alga. In these, the algae are active in photosynthesis and over 70% of the fixed carbon is transferred to the surrounding fungus (Smith, 1980): nitrogen fixation appears to be limited as only about 4% of the cells are heterocysts. However, other lichens include a eukaryotic green alga as well as the other two partners. In these, photosynthetic activity is carried out mainly by the eukaryotic algae; the blue-green algae show reduced photosynthetic activity but nitrogen fixation is increased, the number of heterocysts rising to 20–30% of cells (Sprent, 1979).

Land plants other than lichens in which nitrogen-fixing blue-green algae have been reported as symbionts are shown in Table I. All these host land plants are able to carry out photosynthesis for themselves. As in the lichens with green as well as blue-green algal symbionts, the blue-green algae apparently make use of photosynthetic products provided by their partner (which may be released into the culture cavities as mucilage) and switch their own metabolism from a primarily photosynthetic to a nitrogen-fixing mode (Stewart and Rodgers, 1977).

Perhaps the future evolution of photosynthetic symbionts lies not in the evolution of completely integrated organelles as we know them today, or even in specialization in photosynthesis. Rather it may be directed toward the growth of temporary blue-green algal symbionts in highly specialized plant host compart-

TABLE I
BLUE-GREEN ALGAE AND LAND PLANTS[a]

Blue-green algal symbionts	Host genera	Site within host
	Mosses	
Nostoc	*Sphagnum*	
	Liverworts	
	Aneura	
	Blasia (pusilla)	
	Cavicularia	
Nostoc	*Diplolaena*	Cavities in thallus
	Pellia	
	Riccia	
	Hornworts	
Nostoc	*Anthoceros*	Cavity in thallus
	Ferns	
Anabaena (+ *Nostoc*)	*Azolla*	Cavity in leaf
	Gymnosperms (cyads)	
	Cycas	
	Encephalartos	Cavities in cortex of coralloid roots
Anabaena + *Nostoc*	*Macrozamia*[b]	
	Zamia	
	Angiosperms	
Nostoc	*Gunnera*[b]	Cavities in roots, stems, and callus outgrowths of leaf bases
Nostoc + other genera	*Lemna (trisulca)*	Cavities in thallus
Nostoc	*Trifolium (alexandrinum)*	Root nodules

[a]Summarized from Caiola (1975), Nathanielsz and Staff (1975a,b), and Silvester and McNamara (1976).
[b]Intracellular symbionts reported.

ments in which their main function vis-à-vis the host is nitrogen fixation. For example, the photosynthetically active plant host might form in its roots, stems, or leaves mucilage-filled culture chambers bordered by transfer cells. During sexual reproduction by the host the algae might be kept outside but close to these chambers, held there at modified host retention sites such as hairs. Reinfection might be by way of special pores (e.g., as in *Azolla*). If there were only a brief period during which the symbionts remained free-living and outside the host the alga might permanently lose its capacity for photosynthesis (cf. the plastids in many "free-living" male gametes today), and instead restrict its operation entirely to nitrogen fixation, the attribute most "needed" by the host in the present-day terrestrial environment. Indeed such a partnership does not differ greatly from one or another of those which already exist among land plants today. Unfortunately the number of plant host species with nitrogen-fixing algal symbionts is small, and these relationships are often highly specific. The experimental manipulation of such partnerships for agricultural purposes, though highly desirable, is, as with other current attempts to extend the range of species in which nitrogen can be fixed, likely to be extraordinarily difficult!

VI. Concluding Remarks

Information is becoming increasingly available about the ways in which endosymbionts interact with their host cells. Less consideration has been given to subsequent evolutionary changes which may have taken place within symbionts during prolonged periods of permanent association with their host cells. If it is true that chloroplasts evolved from photosynthetic symbionts, then the variations in structure which one finds among plastids today should reflect these long-term evolutionary modifications, It is with this particular aspect of endosymbiotic relationships that the present article is concerned.

Most people now accept that chloroplasts evolved from algae (both prokaryotic and eukaryotic) which were once free-living but were later taken up as permanent endosymbionts by a variety of different unicellular eukaryotic host species which already contained mitochondira. On the basis of their photosynthetic pigments and their thylakoid systems, the modern descendants of the presumed predecessors of chloroplasts and the chloroplasts themselves fall into three major groups: (1) blue-green algae and the chloroplasts of red algae (chlorophyll a and phycobilins arranged on single thylakoids); (2) prochlorons and the chloroplasts of green algae, land plants, and euglenoids (chlorophylls a and b located in irregular thylakoid stacks); and (3) chloroplasts of chromophyte algae (chlorophylls a and c located in broad triple thylakoid bands). For the chloroplasts in the first two groups major differences in pigmentation and in the photosynthetic apparatus appear to have become established prior to symbiotic uptake, i.e., in prokaryotic ancestors capable of producing all their own metabolites. For the chloroplasts in the third group, fundamental modification of the pigment system and photosynthetic apparatus may have taken place *either* prior to uptake in a prokaryote of which no descendant is known today *or* soon after uptake in the evolving chloroplast of a eukaryotic (perhaps a red) alga. Additional widespread and therefore presumably early evolutionary modifications seem to have taken place after the chloroplasts were already established in each of the three main groups, e.g., the girdle lamella in plastids of some chromophyte algae. However, widespread but postuptake structural modifications like this are few. Within each of the three main algal groups it is the similarities in chloroplast structure rather than the differences which are most conspicuous and should be emphasized. This conservatism of chloroplast structure presumably reflects (1) the retention of their photosynthetic apparatus, (2) the loss of control of other metabolic functions during their early stages of transformation into organelles, and (3) the early assumption by the "host" cell of the overall control of all these functions.

On the basis of morphological similarities which they observed between blue-green algae and the proplastids of *Euglena* and of the angiosperm, *Phaseolus lunatus*, Klein *et al*. (1972) and Schiff (1980) have suggested that if plastids arose from endosymbionts, these symbionts resembled proplastids rather than mature chloroplasts. However, most prokaryotic algae and the chloroplasts of most unicellular eukaryotic algae retain their thylakoid systems at all times; the thylakoid systems of blue-green algae and of prochlorons closely resemble those of their presumed relatives, the mature chloroplasts of the simpler red and green algae; the establishment and proliferation of a proplastid mode is only apparent in more complex algae or in some more recently derived plastids such as those of *Euglena*, which seem to be the end result of two successive acts of endosymbiosis. For these reasons, I believe (1) that chloroplasts evolved initially from symbiotic prokaryotic algae which contained a more or less permanent, fully developed thylakoid system and (2) that only later did proplastids evolve.

More recent evolutionary modifications of plastids, too, seem to have continued the trend toward the loss rather than major modification of the photosynthetic apparatus. Though the evolution of proplastids may first have taken place in unicellular eukaryotes, the more widespread later proliferation of a proplastid mode could only have taken place after the already photosynthetic hosts had evolved into complex multicellular organisms able to support some photosynthetically inactive cells. The starting point for the evolution of proplastids may have been an innate capacity of the former prokaryotic symbionts to modulate their photosynthetic apparatus in response to changes in environmental conditions. In chloroplasts of unicells this capacity for modulation of the thylakoid system was perhaps, at first, regularly expressed only during dormancy of the zygote. Important early steps in evolution of the proplastid would then have been (1) the assumption of control by the nucleus of the processes of dedifferentiation and redifferentiation of the plastid thylakoid system, and (2) the expression of this control in response to changes in internal cellular rather than external environmental conditions. During subsequent evolution loss or absence of photosynthetic activity of the plastids seems to have been extended to stages of the host life cycle which both immediately preceded (first in the male and later in the female gamete) and, more conspicuously, succeeded fertilization (in the embryo and in meristems). The parallel evolution of proplastids in meristems of red, brown, and green algae is noteworthy.

We do not know why the evolution of nonphotosynthetic plastids has been promoted. Perhaps it represents a saving of energy which otherwise would have been used in thylakoid development or perhaps thylakoid turnover was unable to keep pace with rapid cell and plastid growth and replication in meristems, for example. In any event it seems clear that among multicellular algae, but more particularly during the evolution of land plants, there has been retained for photosynthetic purposes both a smaller proportion of the plant body and (with the evolution in land plants of the deciduous habit and other seasonal forms of dormancy) apparently also a smaller proportion of the plant life cycle. It is perhaps surprising that in spite of an apparent trend among the more complex algae and land plants toward an increase in the proportion of cells which are permanently nonphotosynthetic, cells which completely lack plastids are uncommon.

It might have been expected that, at first, the evolution of complex multicellular algae and, later, the colonization of the very different land environment would have brought about the evolution of many new types of plastid. This does not seem to have taken place. Almost all of the many more recent variations in plastid structure and function are "quantitative" and seem to be essentially superficial. For the most part, variations in plastid structure result from the production or accumulation of larger or smaller amounts of normal plastid components or their forerunners, e.g., a less extensive thylakoid system in proplastids or an increased accumulation of starch in amyloplasts. Few truly innovative changes have taken place. It is somewhat startling to find that most of those innovations which have appeared in plastids of land plants, for example, are to be found only in angiosperms, and, as with the C-4 syndrome, they seem to be polyphyletic in origin and are even absent from the most primitive angiosperms.

The retention by plastids of responsibility for photosynthesis (albeit under overall nuclear control) is presumably reflected in the conservatism of the functionally active photosynthetic apparatus. A similar conservatism seems to be shown by the mechanisms used by an alga or plant to transfer plastids from one generation to the next, to select those which will be transferred during sexual reproduction (plastid inheritance), and then to maintain a characteristically balanced population of plastids within each different type of cell. These mechanisms closely parallel those used by one host or another to control its population of endosymbionts. Endocytosis is employed not only by hosts when they ingest symbionts but also by higher plants, for example, when the pollen vegetative cells engulf their sister generative cells. Just as symbionts may be ejected or digested by their hosts, so, for example, paternal plastids of some algae and lower land plants are ejected from the sperm and the maternal plastids of some conifers undergo autophagy within the egg. For both symbionts and plastids the times and rates of division are closely phased with nuclear division, particularly when only one or a few are present in each cell. In addition, characteristic numbers of symbionts and plastids are maintained by particular types of cell or by cells occupying particular sites. Equal distribution to daughter cells of both symbionts and plastids is often promoted by their concentration around the mother cell nucleus prior to division and, even, occasionally, by their being tied to permanently closed nuclear envelopes by membrane links.

There exist today many host species which contain either prokaryotic or eukaryotic algal endosymbionts. By judicious arrangement of the prokaryotic algal symbionts, for example, one can build up a sequence to suggest a possible pattern of progressive integration with unicellular hosts which might resemble that followed during the evolution of simple chloroplasts (Kies, 1980; Whatley and Whatley, 1981). It might be thought that the numerous algal symbionts which now exist will all follow such a progression and eventually evolve into new types of chloroplast: a few might even become more complex than the present day chloroplasts of chromophyte algae, having developed following more than two successive acts of endosymbiosis. However, the many examples of photosynthetic symbionts known at the present time, when compared with the few different basic types of chloroplast which now exist, suggest that the integration sequence is seldom completed and that the ''status'' of organelle is seldom achieved. In the marine or freshwater environment today, phagocytic unicells are potentially still ''available'' as future hosts of photosynthetic symbionts. However, most niches for photosynthetic organisms may well already be occupied and so the presumably strong selective ''advantages'' of the distant past probably no longer exist. Thus even the future establishment of new forms of symbiotic partnerships which specialize in photosynthesis seems unlikely.

Many of today's host species are already multicellular and complex. In these hosts, the sites of symbiotic infection tend to be far removed from the reproductive organs; direct transmission of symbionts to the next host generation is seldom, if ever, accomplished. The future evolution of new chloroplasts or other symbiotically derived organelles within such hosts would seem to be even less probable than in unicellular hosts. However, a distinctive new form of symbiotic partnership now seems to have evolved; the symbionts are unicellular or filamentous and include blue-green algae and several different types of bacteria: the phylogenetically diverse hosts range from fish to flowering plants. These symbiotic pairings show a considerable degree of integration between the partners although direct transmission of the symbiont to the next host generation by way of the egg cell cytoplasm has not been achieved. One organ of the host develops in such a way that a special ''culture chamber'' is made available for the symbionts. The host may even provide external sites close to the reproductive organs to which released symbionts become attached prior to their uptake by the next host generation. Perhaps such partnerships are the starting point for a new evolutionary trend for both photosynthetic and other types of symbiont within modern complex hosts.

Among land plants only a few unrelated species produce ''culture chambers'' for their symbionts. These symbionts include bacteria, actinomycetes, and blue-green algae. In spite of their very different forms of metabolic activity outside the host, once inside, almost all of these particular symbionts begin to specialize in the fixation of atmospheric nitrogen, a useful attribute in the modern land environment. The blue-green algae when established in the "culture chamber" of an independently photosynthetic host reduce their own photosynthetic activity (probably obtaining additional carbohydrate requirements from their hosts) and concentrate instead on nitrogen fixation. Thus for these present day blue-green algal symbionts a capacity to provide their hosts with photosynthate has become a less pressing "requirement" than it was for their supposed predecessors when they became established as red algal chloroplasts.

This apparent selection by the host of an algal symbiont which is able to specialize in a function other than photosynthesis is paralleled among land plants in particular by a declining emphasis on photosynthetic activity in all cells. Proplastids and other forms of nonphotosynthetic plastid are now found in many different types of cell. The proportion of the plant body and of the plant growth cycle given over to photosynthesis seems to have declined progressively during the period of land plant evolution, and other needs, like fixation of atmospheric nitrogen, appear to have become more important. It may thus be concluded that on the land, as in water, new forms of chloroplast are unlikely to evolve. The future evolution of today's photosynthetic (blue-green algal) symbionts would seem to be directed toward their packaging within localized and specialized host "culture chambers," where they carry out nitrogen fixation; their possible role as photosynthetic organelles seems already to have been preempted.

References

Aghajanian, J. G., and Hommersand, M. H. (1980). *J. Phycol.* **16,** 15.

Altmann, R. (1890). "Die Elementarorganismen und ihre Beziehung zu den Zellen." Veit, Leipzig.

Badenhuizen, N. P., and Salema, R. (1967). *Rev. Biol. (Lisbon)* **6,** 139.

Bannister, L. H. (1979). *Proc. R. Soc. London Ser. B* **204,** 141.

Behnke, H. D. (1972). *Bot. Rev.* **38,** 155.

Behnke, H. D. (1975). *Protoplasma* **83,** 91.

Bisalputra, T., Shields, C. M., and Markham, J. W. (1971). *J. Microsc.* **10,** 83.

Bold, H. C., and Wynne, M. J. (1978). "Introduction to the Algae." Prentice-Hall, New York.

Borowitzka, M. A. (1978). *Protoplasma* **95,** 217.

Borowtizka, M. A., and Larkum, A. W. D. (1974). *Protoplasma* **81,** 131.

Bouck, G. B. (1962). *J. Cell Biol.* **12,** 553.

Bouck, G. B. (1970). *Ann. N.Y.Acad. Sci.* **175,** 673.

Braten, T. (1973). *J. Cell Sci.* **12,** 385.

Burr, F. A., and West, J. A. (1970). *Phycologia* **9,** 17.

Butterfass, Th. (1980). *In* "Chloroplasts" (J. Reinert, ed.), pp. 29–44. Springer-Verlag, Berlin and New York.

Cacares, E. J., and Robinson, D. G. (1981). *J. Phycol.* **17,** 173.

Caiola, M. G. (1975). *Phycologia* **14,** 25.

Cass, D. D., and Karas, I. (1974). *Protoplasma* **81,** 49.

Cavalier-Smith, T. (1975). *Nature (London)* **256,** 463.

Cavalier-Smith, T. (1976). *Protoplasma* **87,** 297.

Cavalier-Smith, T. (1980). *In* "Endocytobiology" (W. Schwemmler and H.E.A. Schenk, eds.), pp. 893–916. De Gruyter, Berlin.
Cavalier-Smith, T. (1981). *Symp. Sco. Gen. Microbiol.* **32,** 33.
Chamberlain, C. J. (1919). "The Living Cycads." Univ. of Chicago Press, Chicago, Illinois.
Chapman-Andresen, C. (1971). *Annu. Rev. Microbiol.* **24,** 27.
Chesnoy, L., and Thomas, M.-J. (1971). Phytomorphology **21,** 50.
Christensen, T. (1964). *In* "Algae and Man" (D. F. Jackson, ed.), pp. 59–64. Plenum, New York.
Coombs, J., and Greenwood, A. D. (1976). *In* "The Intact Chloroplast" (J. Barber, ed.), pp. 1–50. Elsevier, Amsterdam.
Craig, I. W., and Gibor, A. (1971). *J. Cell Biol.* **49,** 950.
Dawes, C. J., and Barilotti, D. C. (1969). *Am. J. Bot.* **56,** 8.
Dawes, C. J., and Rhamstine, E. L. (1967). *J. Phycol.* **3,** 117.
Dittrich, P. (1974). *Int. Congr. Photosynth. Res., 3rd, Israel.*
Dodge, J. D. (1973). "The Fine Structure of Algal Cells." Academic Press, New York.
Dodge, J. D. (1979). *Biochem. Physiol. Protozoa.* **1,** 7.
Dodge, J. D. (1980). *In* "Endocytobiology" (W. Schwemmler and H.E.A. Schenk, eds.), pp. 33–50. De Gruyter, Berlin.
Dodge, J. D., and Crawford, R. M. (1969). *J. Cell Sci.* **5,** 479.
Duckett, J. G. (1975). *Cytobiologie* **10,** 432.
Duckett, J. G., and Peel, M. C. (1978). *In* "The Systematics Association Special Volume No. 10—Modern Approaches to the Taxonomy of Red and Brown Algae" (D. E. G. Irvine and J. H. Price, eds.), pp. 157–204. Academic Press, New York.
Duckett, J. G., and Prasad, A. K. S. K. (1974). *J. Bryol.* **8,** 181.
Duckett, J. G., Toth, R., and Soni, S. L. (1975). *New Phytol.* **75,** 111.
Duckett, J. G., Prasad, A. K. S. K., Davies, D. A., and Walker, J. (1977). *New Phytol.* **79,** 349.
Dumas, C. (1974). *Botaniste* **56,** 81.
Evans, L. V. (1968). *New Phytol.* **67,** 173.
Evans, L. V., and Holligan, M. S. (1972). *New Phytol.* **71,** 1173.
Francis, D. (1967). *J. Exp. Biol.* **47,** 495.
Fredrick, J. F. (1980). Phytochemistry **19,** 2611.
Fredrick, J. F., ed. (1981). *Ann. N.Y. Acad. Sci.* **361.**
Gaffal, K. P. (1978). *Protoplasma* **94,** 175.
Gianordoli, M. (1973). *Caryologia* **25** (Suppl.), 135.
Gibbs, S. P. (1962). *J. Cell Biol.* **15,** 343.
Gibbs, S. P. (1970). *Ann. N.Y. Acad. Sci.* **175,** 454.
Gibbs, S. P. (1978). *Can. J. Bot.* **56,** 2883.
Gibbs, S. P. (1981a). *Int. Rev. Cytol.* **72,** 49.
Gibbs, S. P. (1981b). *Ann. N.Y. Acad. Sci.* **361,** 193.
Giles, K. L., and Sarafis, V. (1974). *Nature (London)* **248,** 512.
Gillot, M. A., and Gibbs, S. P. (1980). *J. Phycol.* **16,** 558.
Greenwood, A. D. (1974). *Int. Congr. Electron Microsc., 8th; Canberra* 566.
Grevet, C. (1978). *Protistologica* **13,** 127.
Gronegress, P. (1974). *J. Microsc.* **19,** 183.
Gunning, B. E. S., and Steer, M. W. (1975). "Ultrastructure and the Biology of Plant Cells." Arnold, London.
Hesse, M. (1978). *Plant Syst. Evol.* **129,** 13.
Hibberd, D. J. (1979). *Biosystems* **11,** 243.
Honda, S. I., Hongladarom-Honda, T., Kwanyven, P., and Wildman, S. G. (1971). *Planta* **97,** 1.
Jeffrey, S. W. (1976). *J. Phycol.* **12,** 349.
Jeffrey, S. W., and Vesk, M. (1976). *J. Phycol.* **12,** 450.

Jensen, W. A. (1967). *Planta* **79,** 346.
John, P., and Whatley, F. R. (1975). *Nature (London)* **254,** 495.
Kageyama, A., Yokohama, Y., Shimura, S., and Ikawa, T. (1977). *Plant Cell Physiol.* **18,** 477.
Keeley, J. E. (1981). *Int. Bot. Congr., 13th, Sydney* 85.
Kies, L. (1980). *In* "Endocytobiology" (W. Schwemmler and H. E. A. Schenk, eds.), pp. 7–19. De Gruyter, Berlin.
Kirk, J. T. O., and Tilney-Bassett, R. A. E. (1978). "The Plastids." Elsevier, Amsterdam.
Kivic, P. A., and Vesk, M. (1974). *Can. J. Bot.* **52,** 695.
Klein, S., Schiff, J. A., and Holowinsky, A. W. (1972). *Dev. Biol.* **28,** 253.
Laetsch, W. M. (1974). *Annu. Rev. Plant Physiol.* **25,** 27.
Lang, N. J., and Whitton, B. A. (1973). *In* "The Biology of Blue-Green Algae" (N. G. Carr and B. A. Whitton, eds.), pp. 66–79. Blackwell, Oxford.
Lewin, R. A., and Withers, N. W. (1975). *Nature (London)* **256,** 735.
Liaaen-Jensen, S. (1977). *In* "Marine Natural Products Chemistry" (D. J. Faulkner and W. H. Fenical, eds.), pp. 239–259. Plenum, New York.
Liaaen-Jensen, S. (1979). *Pure Appl. Chem.* **51,** 661.
McAuley, P. J. (1981). *Experientia* **37,** 346.
McLean, R. (1967). *Physiol. Plant.* **20,** 41.
Mahler, H. R. (1980). *In* "Endocytobiology" (W. Schwemmler and H. E. A. Schenk, eds.), pp. 869–892. De Gruyter, Berlin.
Margulis, L. (1981). "Symbiosis in Cell Evolution." Freeman, San Francisco, California.
Margulis, L., Chase, D., and To, L. P. (1979). *Proc. R. Soc. London Ser. B* **204,** 189.
Margulis, L., To, L. P., and Chase, D. (1981). *Ann. N.Y. Acad. Sci.* **361,** 356.
Mencke, W., and Fricke, B. (1962). *Port. Acta Biol.* **6,** 243.
Mereschkowsky, C. (1905). *Biol. Zentralbe.* **25,** 593; 689.
Moestrup, Ø. (1970). *J. Mar. Biol. Assoc. U. K.* **50,** 513.
Müller, M. (1975). *Annu. Rev. Microbiol.* **29,** 467.
Muscatine, L., and Pool, R. R. (1979). *Proc. R. Soc. London Ser. B* **204,** 131.
Myles, D. G. (1978). *J. Cell Sci.* **30,** 265.
Nathanielsz, C. P., and Staff, I. A. (1975a). *Am. J. Bot.* **62,** 232.
Nathanielsz, C. P., and Staff, I. A. (1975b). *Ann. Bot.* **39,** 363.
Newcomb, E. H. (1967). *J. Cell Biol.* **33,** 143.
Newcomb, E. H., and Pugh, T. D. (1975). *Nature* (London) **253,** 533.
Ohad, I., Siekevitz, P., and Palade, G. E. (1967). *J. Cell Biol.* **35,** 553.
Pardy, R. L. (1974). *Biol. Bull.* **147,** 105.
Peel, M. C., and Duckett, J. G. (1975). *Bot. J. Linn. Soc.* **7,** Suppl. 1, 1.
Pellegrini, L. (1980). *J. Cell Sci.* **41,** 209.
Perry, J. W., and Evert, R. F. (1975). *Am. J. Bot.* **62,** 1038.
Pickett-Heaps, J. D. (1974). *J. Phycol.* **10,** 148.
Pickett-Heaps, J. D. (1975). "Green Algae." Sinnauer, Sunderland, Massachusetts.
Pringsheim, E. G. (1941). *Biol. Rev.* **16,** 191.
Pringsheim, E. G. (1963). "Farblose Algen." Fischer, Stuttgart.
Platt-Aloia, K. A., and Thomson, W. W. (1977). *New Phytol.* **78,** 599.
Pyrozinski, K. A., and Mulloch, D. W. (1975). *Biosystems* **6,** 153.
Raff, R. A., and Mahler, H. R. (1972). *Science* **177,** 575.
Raff, R. A., and Mahler, H. R. (1975). Symp. *Soc. Exp. Biol.* **29,** 41.
Ragan, M. A., and Chapman, D. J. (1978). "A Biochemical Phylogeny of the Protists." Academic Press, New York.
Raven, P. H. (1970). *Science* **169,** 641.
Richmond, M. H., and Smith, D. C., eds. (1979). *Proc. R. Soc. London Ser. B.* **204.**

Rickson, F. R. (1976). *Am. J. Bot.* **63,** 1272.
Ris, H., and Plaut, W. (1962). *J. Cell Biol.* **13,** 383.
Salema, R. (1971). *An. Fac. Ciênc(Porto)* **54,** 1.
Schiff, J. A. (1980). *In* "The Biochemistry of Plants" (P. K. Stumpf and E. E. Conn, eds.), pp. 209–272. Academic Press, New York.
Schiff, J. A., and Epstein, H. T. (1965). *In* "Reproduction: Molecular, Subcellular and Cellular" (M. Locke, ed.), pp. 131–189. Academic Press, New York.
Schimper, A. F. W. (1883). *Bot. Ztg.* **41,** 105; 121; 137; 153.
Schmitz, K., and Srivastava, L. M. (1974). *Cytobiology* **10,** 66.
Schmitz, K., and Srivastava, L. M. (1975). *Can. J. Bot.* **53,** 861.
Schnepf, E. (1969). *Öst. Bot. Z.* **116,** 65.
Schnepf, E. (1980). *In* "Chloroplasts" (J. Reinert, ed.), pp. 1–28. Springer-Verlag, Berlin and New York.
Schonbeck, M. W. and Norton, T. A. (1979). *Bot. Mar.* **23,** 233.
Schwemmler, W., and Schenk, H. E. A., eds. (1980). "Endocytobiology." De Gruyter, Berlin.
Seewaldt, E., and Stackebrandt, E. (1982). *Nature (London)* **295,** 618.
Senser, M., Schötz, F., and Beck, E. (1975). *Planta* **126,** 1.
Sepsenwohl, S. (1973). *Exp. Cell Res.* **76,** 395.
Sievers, A. (1965). *Z. Pflanzenphysiol.* **53,** 193.
Silvester, W. B., and McNamara, P. J. (1976). *New Phytol.* **77,** 135.
Sitte, P. (1977). *Biol. Unserer Zeit* **7,** 65.
Smith, D. C. (1979). *Proc. R. Soc. London Ser. B* **204,** 115.
Smith, D. C. (1980). *In* "Cellular Interactions in Symbiosis and Parasitism" (C. B. Cook, P. W. Pappas, and E. D. Rudolph, eds.), pp. 197–227. Ohio State Univ. Press, Columbus.
Sonneborn, T. M. (1959). *Adv. Virus Res.* **6,** 229.
Sprent, J. I. (1979). "The Biology of Nitrogen-fixing Organisms." McGraw-Hill, New York.
Sprey, B., and Lambert, C. (1977). *Z. Pflanzenphysiol.* **83,** 227.
Stewart, K. D., and Gifford, E. M., Jr. (1967). *Am. J. Bot.* **54,** 375.
Stewart, W. D. P., and Rodgers, G. A. (1977). *New Phytol.* **78.** 459.
Taylor, F. J. R. (1979). *Proc. R. Soc. London Ser. B* **204,** 267.
Taylor, F. J. R. (1980). *Biosystems* **13,** 65.
Thomson, W. W. (1966). *Bot. Gaz.* **127,** 133.
Thomson, W. W., and Platt, K. (1973). *New Phytol.* **72,** 791.
Thomson, W. W., and Whatley, J. M. (1980). *Annu. Rev. Plant Physiol.* **31,** 375.
Thomson, W. W., Lewis, L. N., and Coggins, C. W. (1967). *Cytologia* **32,** 117.
Thorington, G., and Margulis, L. (1980). *In* "Endocytobiology" (W. Schwemmler and H. E. A. Schenk, eds.), pp. 175–222. De Gruyter, Berlin.
Tomas, R. N., and Cox, E. R. (1973). *J. Phycol.* **9,** 304.
Tomas, R. N., Cox, E. R., and Steidinger, K. A. (1973). *J. Phycol.* **9,** 91.
Trainor, F. (1970). *Ann. N.Y. Acad. Sci.* **175,** 749.
Trench, R. K. (1979). *Annu. Rev. Plant Physiol.* **30,** 485.
Trench, R. K. (1980). *In* "Endocytobiology" (W. Schwemmler and H. E. A. Schenk, eds.), pp. 703–727. De Gruyter, Berlin.
Trench, R. K., Boyle, J. E. and Smith, D. C. (1973). *Proc. R. Soc. London Ser. B* **184,** 51.
Trench, R. K., Pool, R. R., Logan, M., and Engelland, A. (1978). *Proc. R. Soc. London Ser. B* **202,** 423.
Uzzell, T., and Spolsky, C. (1974). *Am. Sci.* **62,** 334.
Uzzell, T., and Spolsky, C. (1981). *Ann. N.Y. Acad. Sci.* **361,** 481.
Walne, P. L., and Arnott, H. J. (1967). *Planta* **77,** 325.
Warmbrodt, R. D., and Evert, R. F. (1979). *Am. J. Bot.* **66,** 412.

Weitzman, P. D. J. (1981). *Ann. N.Y. Acad. Sci.* **361,** 505.
Wetherbee, R. (1980). *J. Ultrastruct. Res.* **70,** 259.
Whatley, J. M. (1974). *New Phytol.* **73,** 139.
Whatley, J. M. (1976). *New Phytol.* **76,** 111.
Whatley, J. M. (1977). *New Phytol.* **78,** 407.
Whatley, J. M. (1978). *New Phytol.* **80,** 494.
Whatley, J. M. (1979). *New Phytol.* **82,** 1.
Whatley, J. M. (1980). *New Phytol.* **86,** 1.
Whatley, J. M. (1982). *Biol. Rev.* (in press).
Whatley, J. M., and Gunning, B. E. S. (1981). *New Phytol.* **89,** 129.
Whatley, J. M. and Whatley, F. R. (1981). *New Phytol.* **87,** 233.
Whatley, J. M., John, P., and Whatley, F. R. (1979). *Proc. R. Soc. London Ser. B* **204,** 165.
Wheeler, A. E., and Page, J. Z. (1974). *J. Phycol.* **10,** 336.
Withers, N. W., and Haxo, F. T. (1978). *Plant Physiol.* **62,** 36.
Withers, N. W., Cox, E. R., Tomas, R., and Haxo, F. T. (1977). *J. Phycol.* **13,** 454.
Withers, N. W., Alberte, R. S., Lewin, R. A., Thornber, J. P., Britton, G., and Goodwin, T. W. (1978). *Proc. Natl. Acad. Sci. U.S.A.* **75,** 2301.
Wolf, F. R., and Cox, E. R. (1981). *J. Phycol.* **17,** 395.
Wolniak, S. (1976). *Am. J. Bot.* **63,** 251.
Wooding, F. B. P. and Northcote, D. H. (1965). *Z. Naturforsch.* **C31,** 456.
Yokohama, Y., and Kageyama, A. (1977). *Bull. Jpn. Soc. Phycol.* **25,** 168.

NOTE ADDED IN PROOF. It has come to my attention (Norstog, personal communication) that membrane-bound bodies similar to those found in some angiosperm plastids (Section III,D,2) have been observed in leaf plastids of a gymnosperm, the cycad, *Encephalartos ferox*.

Index

O

P

R

S

T

U

V

X

Z